Natural Biomarkers for Cellular Metabolism

Series in Cellular and Clinical Imaging

Series Editor
Ammasi Periasamy

PUBLISHED

Coherent Raman Scattering Microscopy
edited by Ji-Xin Cheng and Xiaoliang Sunney Xie

Imaging in Cellular and Tissue Engineering
edited by Hanry Yu and Nur Aida Abdul Rahim

Second Harmonic Generation Imaging
edited by Francesco S. Pavone and Paul J. Campagnola

The Fluorescent Protein Revolution
edited by Richard N. Day and Michael W. Davidson

Natural Biomarkers for Cellular Metabolism:
Biology, Techniques, and Applications
edited by Vladimir V. Gukasyan and Ahmed A. Heikal

FORTHCOMING

Optical Probes in Biology
edited by Jin Zhang and Carsten Schultz

SERIES IN CELLULAR AND CLINICAL IMAGING
AMMASI PERIASAMY, SERIES EDITOR

Natural Biomarkers for Cellular Metabolism

Biology, Techniques, and Applications

Edited by
Vladimir V. Ghukasyan
Ahmed A. Heikal

CRC Press
Taylor & Francis Group
Boca Raton London New York

CRC Press is an imprint of the
Taylor & Francis Group, an **informa** business

CRC Press
Taylor & Francis Group
6000 Broken Sound Parkway NW, Suite 300
Boca Raton, FL 33487-2742

First issued in paperback 2017

ISBN-13: 978-1-4665-0998-6 (hbk)
ISBN-13: 978-1-138-19878-4 (pbk)

Library of Congress Cataloging-in-Publication Data

Natural biomarkers for cellular metabolism : biology, techniques, and applications / edited by Vladimir V. Ghukasyan and Ahmed A. Heikal.
p. ; cm. -- (Series in cellular and clinical imaging)
Includes bibliographical references and index.
ISBN 978-1-4665-0998-6 (alk. paper)
I. Ghukasyan, Vladimir V., editor. II. Heikal, Ahmed A., editor. III. Series: Series in cellular and clinical imaging.
[DNLM: 1. Biological Markers. 2. Cell Physiological Processes. 3. Cells--metabolism. 4. Optical Imaging. QW 541]

QH583
571.6072'1--dc23 2014022167

Visit the Taylor & Francis Web site at
http://www.taylorandfrancis.com

and the CRC Press Web site at
http://www.crcpress.com

To my family: my mother Hratsin, my late grandmother Nellie, my soulmate—my wife Jessie, and my incredible children—my son Jivan and my daughter Adriana. You are my love, my drive, my cause, and my inspiration.

Vladimir V. Ghukasyan

To Watt Wetmore Webb—a mentor, a preeminent scholar, and a man with utmost integrity.

Ahmed A. Heikal

Contents

Series Preface

A picture is worth a thousand words.

This proverb says everything. Imaging began in 1021 with the use of a pinhole lens in a camera in Iraq; later, in 1550, the pinhole was replaced by a biconvex lens developed in Italy. This mechanical imaging technology migrated to chemical-based photography in 1826 with the first successful sunlight-picture made in France. Today, digital technology counts the number of light photons falling directly on a chip to produce an image at the focal plane; this image may then be manipulated in countless ways using additional algorithms and software. The process of taking pictures ("imaging") now includes a multitude of options—it may be either invasive or noninvasive, and the target and details may include monitoring signals in two, three, or four dimensions.

Microscopes are an essential tool in imaging used to observe and describe protozoa, bacteria, spermatozoa, and any kind of cell, tissue, or whole organism. Pioneered by Antoni van Leeuwenhoek in the 1670s and later commercialized by Carl Zeiss in 1846 in Jena, Germany, microscopes have enabled scientists to better grasp the often misunderstood relationship between microscopic and macroscopic behavior by allowing for the study of the development, organization, and function of unicellular and higher organisms as well as structures and mechanisms at the microscopic level. Furthermore, the imaging function preserves temporal and spatial relationships that are frequently lost in traditional biochemical techniques and gives two- or three-dimensional resolution that other laboratory methods cannot. For example, the inherent specificity and sensitivity of fluorescence; the high temporal, spatial, and three-dimensional resolution that is possible; and the enhancement of contrast resulting from the detection of an absolute rather than relative signal (i.e., unlabeled features do not emit) are several advantages of fluorescence techniques. Additionally, the plethora of well-described spectroscopic techniques providing different types of information and the commercial availability of fluorescent probes such as visible fluorescent proteins (many of which exhibit an environment- or analytic-sensitive response) increase the range of possible applications, such as the development of biosensors for basic and clinical research. Recent advancements in optics, light sources, digital imaging systems, data acquisition methods, and image enhancement, analysis, and display methods have further broadened the applications in which fluorescence microscopy can be applied successfully.

Another development has been the establishment of multiphoton microscopy as a three-dimensional imaging method of choice for studying biomedical specimens from single cells to whole animals with submicron resolution. Multiphoton microscopy methods utilize naturally available endogenous fluorophores—including NADH, TRP, FAD, and so on—whose autofluorescent properties provide a label-free approach. Researchers may then image various functions and organelles at molecular levels using two-photon and fluorescence lifetime imaging microscopy to distinguish normal from cancerous conditions. Other widely used nonlabeled imaging methods are coherent anti-Stokes Raman scattering spectroscopy and stimulated Raman scattering microscopy, which allow imaging of molecular function using the molecular vibrations in cells, tissues, and whole organisms. These techniques have been widely used in gene therapy, single molecule imaging, tissue engineering, and stem cell research. Another non-labeled method is harmonic generation (SHG and THG), which is also widely used in clinical imaging, tissue engineering, and stem cell research. There are many more advanced technologies developed for cellular and clinical imaging, including multiphoton tomography, thermal imaging in animals, ion imaging (calcium, pH) in cells, etc.

The goal of this series is to highlight these seminal advances and the wide range of approaches currently used in cellular and clinical imaging. Its purpose is to promote education and new research across a broad spectrum of disciplines. The series emphasizes practical aspects, with each book focusing on a particular theme that may cross various imaging modalities. Each book covers basic to advanced imaging methods as well as detailed discussions dealing with interpretations of these studies. The series also provides cohesive, complete state-of-the-art, cross-modality overviews of the most important and timely areas within cellular and clinical imaging.

Since my graduate student days, I have been involved and interested in multimodal imaging techniques applied to cellular and clinical imaging. I have pioneered and developed many imaging modalities throughout my research career. Series manager Luna Han recognized my genuine enthusiasm and interest to develop a series on cellular and clinical imaging. This project would not have been possible without her support. I am sure that all the book editors, chapter authors, and I have benefited greatly from her continuous input and guidance to make this series a success.

Equally important, I personally thank the book editors and chapter authors. This has been an incredible experience, working with colleagues who demonstrate such a high level of interest in educational projects, even though they are all fully occupied with their own academic activities. Their work and intellectual contributions based on their deep knowledge of the subject matter will be appreciated by everyone who reads this book series.

Ammasi Periasamy, PhD
Series Editor
Professor and Center Director
W.M. Keck Center for Cellular Imaging
University of Virginia
Charlottesville, Virginia

Foreword: Britton Chance—Appreciation

Later in life, Briton Chance wrote on his pioneering discovery of reduced nicotinamide adenine dinucleotide (NADH) and flavin adenine dinucleotide (FAD) fluorescence in mitochondria of both in vivo and ex vivo: "This was perhaps the most important discovery of my career because, for the first time, we could obtain optical signals from living mitochondrial tissues. A series of papers, exploring this discovery in the liver, kidney, adrenal glands, and brain, opened a new field of metabolic research." This book on intrinsic biomarkers for cellular metabolism would be incomplete without an appreciation of the late Britton Chance (1913–2010), a preeminent scientist, innovator, mentor, educator, humanist, visionary, and sailing enthusiast.

To work with this giant scientist can be a challenging, yet rewarding, experience. Working on the first fiber optics–based fluorometer/reflectometer, Mayevsky remembers the daily meetings in Chance's office late in the evening on the fifth floor of the Richard Building, where they would review the results of the day. Chance would usually dictate the results interpretation into a tape recorder, and a printed summary of the discussion would be ready the next day. While he was driven to develop a device for NADH monitoring from the brains of awake, active rats, this was typical of his work ethic and enthusiasm for science. He expected no less from his students. While working on the development of the fiber optics–based fluorometer, students would do whatever changes were needed in optimizing and testing the design on a daily basis.

Perhaps his passion was to help patients by translating his scientific bench discoveries into clinical usage as he extended his research toward monitoring mitochondrial metabolic state in patients using NADH fluorometry or phosphorus nuclear magnetic resonance (NMR). His most rewarding research experience was when he and his coworkers helped an Indian woman with mitochondrial genetic disease that limited her ability to walk. Using ^{31}P-NMR, they identified a missing link in her mitochondrial metabolism, which was treated with vitamins (Eleff et al., 1984). The woman started to walk and continued to live well to old age. His handheld breast cancer detector is another example of his vision of translational research, which is a low-cost, effective strategy for helping people.

Believing in the peaceful use of science and technology, his team at the MIT radiation lab worked on radar research in order to fight against fascism during World War II. It was his strong conscience that prevented him and presumably his team of 300 members from participating in the atomic bomb project. Later, Chance would always advocate against the use of atomic bombs.

Britton Chance was also an educator with a passion for the development of scientific research communities all over the world, and many hundreds or even thousands of his students and collaborators are now leaders in different research fields. He was the consummate mentor and advocate for his students, regardless of their gender, ethnicity, nationality, religion, or social status. In 1996, at the age of 83, Chance devoted himself to a summer program for high school minority students at the University of

Pennsylvania. Over the next ten years, he taught them the basics of science research, despite running a multitude of projects at the time.

Chance has tirelessly promoted scientific cooperation and academic exchanges between the East and West. He believed in the future of science for the betterment of human life as he tirelessly helped multiple institutions and research labs in the Far East. For example, he helped to establish the Britton Chance Center for Biomedical Photonics (BC-CBMP) at the Huazhong University of Science and Technology in Wuhan, Hubei Province, China. He maintained solid collaborations with BC-CBMP in various scientific research programs, personnel training, management, and international cooperations. Thanks to his commitments, biomedical photonics research in China is developing at an amazing pace and catching up with the most advanced research known globally.

F.1 A Short Biography of a Giant

Born to an aristocratic family in Wilkes-Barre, Pennsylvania, on July 24, 1913, Britton Chance developed an early passion for both science and sailing. At the young age of 15, he built an optoelectric device for steering ships that was later patented and tested for large cargo ships, such as the *Texas Sun* and *New Zealand Star.* He earned a bachelor of science (1935), a master of science (1936), and a PhD in physical chemistry (1940) at the University of Pennsylvania, where he developed a syringe-driven rapid flow system for chemical mixing and reaction kinetics studies. In 1937, he joined Glenn Millikan's group at the University of Cambridge to develop a micro stop-flow device for studying enzyme-substrate kinetics, where he eventually earned a PhD in physiology (1942). In 1941, Chance accepted an assistant professorship at the University of Pennsylvania. He stayed at UPenn for the rest of his professional career.

During World War II (1941–1945) he worked at the Radiation Lab at the Massachusetts Institute of Technology, Boston, to develop radar systems, which were used against the V-1 flying bombs at Anzio and Normandy. From 1946 to 1948, he worked at the Nobel Institute, Stockholm, Sweden, and the Moleno Institute at Cambridge as a Guggenheim Fellow. During his spare time he continued sailing and earned a gold medal at the Olympics (Sailing, Men's 5.5 Meter Class) in 1952 at Finland. He also won several world championships for sailing in the 1950s and 1960s.

In 1966, he was the foreign fellow in the Churchill College at the University of Cambridge. After he officially retired as the Eldridge Reeves Johnson emeritus professor in 1983, his active research continued fruitfully at the University of Pennsylvania for nearly 30 years. Britton Chance remained active in research until the last few days of his life, riding a bicycle to work six days a week even in his 90s.

F.2 Honors and Awards

The rare distinction of Britton Chance in his scientific research earned him the President's Certificate of Merit (1950), U.S. National Medal of Science (1974), and memberships in the academy of sciences in Argentina, Germany, Italy, Sweden, the United Kingdom, and the United States. He also earned a doctor of science at the University of Cambridge in 1952 and more than 10 honorary PhDs and MDs throughout his life. Among over 50 honors and awards he received are the Morlock Award from IEEE in 1961, the Heineken Medal of the Netherlands Academy of Science and Letters in 1970, the Canada Gairdner Award in 1972, the Gold Medal for Distinguished Service to Medicine from the American College of Physicians in 1987, the Gold Medal from the Society of Magnetic Resonance in Medicine in 1988, the Benjamin Franklin Medal for Distinguished Achievement in the Sciences in 1990, the Christopher Columbus Discovery Award in Biomedical Research of the National Institutes of Health in 1992, the International Society for Optical Engineering (SPIE) Lifetime Achievement Award in 2005, the Gold Medal from the American Roentgen Ray Society, and a Distinguished Achievement Award from the American Aging Association in 2006. In recognition to his global effort to promote scientific innovation and collaboration, the State Administration of Foreign Experts Affairs of China granted Britton Chance the Friendship Award. He was also awarded the Chime Bell Award in September 2008 by the government of Hubei Province. In 2009, he was recognized with the International Science and Technology Cooperation Award, China's highest national honor given to foreign scientists.

F.3 A Glimpse of the Scientific Discoveries and Contributions to the Field of Bioenergetics

Considering the limited space here, it is not our intention to provide a detailed account of all the scientific contributions and discoveries (Figure F.1) of Britton Chance, which may run into a whole book. Rather, we celebrate his life with a glimpse at the landmark discoveries by Britton Chance, who represented the essence of an interdisciplinary researcher over eight decades in physics, engineering, chemistry, biology, and medicine with extraordinary achievements and high impact.

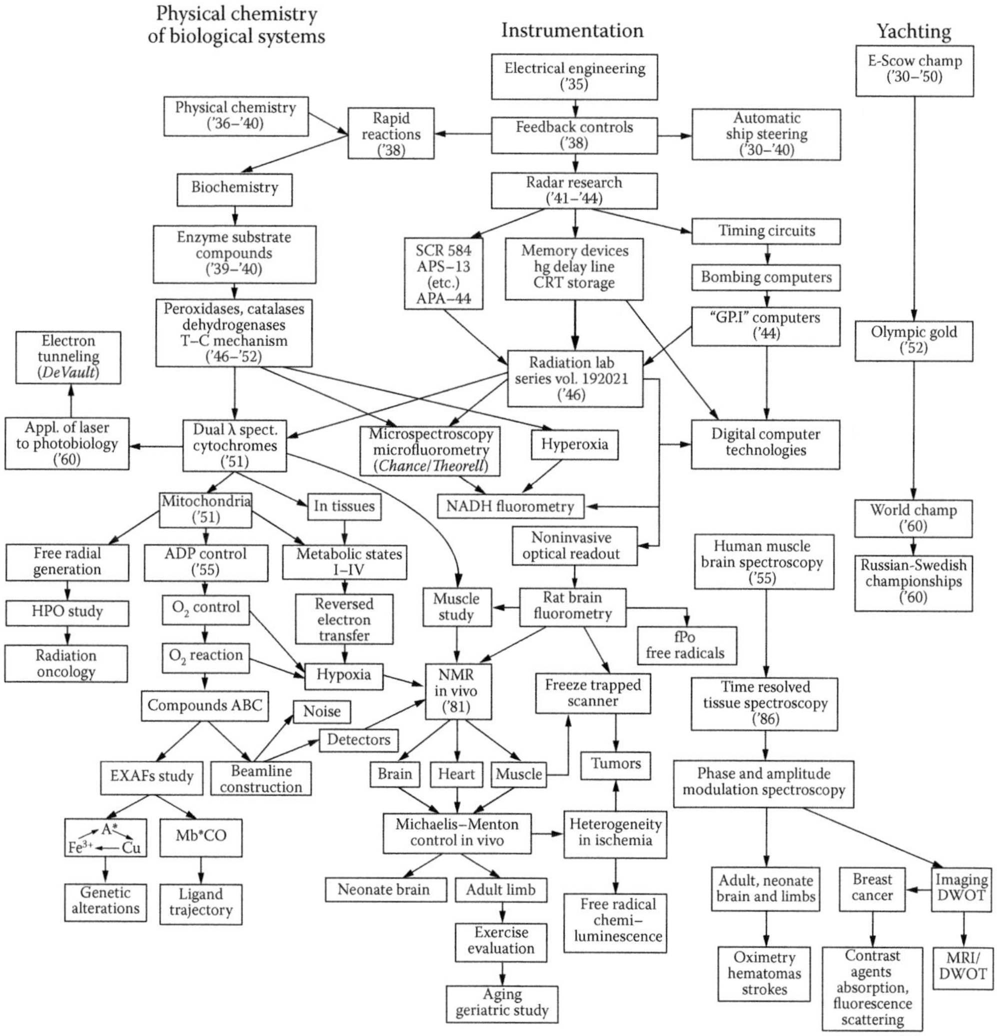

FIGURE F.1 The chart of life achievements drawn by Britton Chance. (Reproduced from the Johnson Research Foundation, Department of Biochemistry and Biophysics, University of Pennsylvania, Philadelphia, PA.)

Early in his career in the 1940s, Chance demonstrated the first experimental proof of the existence of Michaelis–Menten enzyme-substrate complex (Chance et al., 1940; Chance, 1943). In the 1950s and 1960s, Chance and his coworkers made a series of key discoveries in the electron transport chain and bioenergetics of mitochondria using novel optical techniques including NADH and flavoprotein fluorometers. In the 1960s, Chance and his coworkers identified the first electron-tunneling phenomenon in biological systems (DeVault and Chance, 1966); later he discovered that hydrogen peroxide was generated by mitochondria respiration as a by-product (Loschen et al., 1971). Both discoveries have later flourished into tremendous research activity on photosynthesis and reactive oxygen species, a key factor for many conditions including cancer, diabetes, and aging. During the 1950s and 1960s, he discovered calcium uptake by mitochondria and demonstrated that calcium ions act as stoichiometric uncouplers. From the late 1970s to 1980s, he also pioneered the development of in vivo NMR spectroscopy for studying tissue phosphor energetics and related diseases in intact organ tissues and human subjects (Garlick et al., 1977; Chance et al., 1981; Eleff et al., 1984). Since the late 1980s, Chance and his coworkers also developed near-infrared spectroscopy and imaging methods for the measurement of blood volume, oxygenation, and flow; muscle functions; brain activities; and cancer detections and diagnosis. These methods have been widely applied in both laboratory and clinic studies. During the last decade of his life he worked on novel molecular beacons for cancer detection and diagnosis (Zheng et al., 2002), prediction of tumor aggressiveness by redox scanning (Li et al., 2009; Xu et al., 2010), and development of an oral optical metabolometer (Ao et al., 2005; Chance et al., 2010) for human nutritional status.

He was a leader and one of the founders of biomedical photonics as well as bioenergetics with numerous inventions that include the micro stop-flow device, dual-beam spectrometer, NADH fluorometer and redox scanner, photon diffusion tomography, phased array, and handheld NIR breast cancer detector. Table F.1 is a summary of the early historical milestones for NADH monitoring with instrumental contributions from Chance and his coworkers. Reflecting on his interest in metabolic markers such as NADH, flavoproteins, and cytochromes since the 1950s, Chance wrote (Chance et al. 1973):

> The accumulation of evidence since the pioneer work of Otto Warburg an "Atmungsferment" (Warburg, 1949) and David Keilin on cytochromes (Keilin, 1966) as the keystones of cellular oxygen utilization led us to the study of the redox states of electron carriers in isolated mitochondria as a function of oxygen concentration and to develop techniques to measure the states of anoxia and normoxia in living tissues.

TABLE F.1 Milestones in the Early Period of NADH Monitoring

Year	Discovery	Author(s)
1904	Involvement of adenine containing nucleotide in fermentation by yeast	Harden and Young
1935	Description of full structure of hydrogen transferring coenzyme in erythrocyte	Warburg et al.
1936	Naming the two cofactors DPN (NAD in the notation of this book) and TPN (NADP in the notation of this book). The reduced form NAD(P)H has an optical absorption at 340 nm	Warburg and Christian
1948	First observation of NADH fluorescence	Warburg
1951	Shift in the absorption spectrum of DPNH with alcohol dehydrogenase	Theorell and Bonnichsen
1951	Development of a rapid and sensitive spectrophotometer	Chance and Legallias
1951	Development of the double-beam spectrophotometer	Chance
1952	Monitoring of pyridine nucleotide enzymes	Chance
1957	First detailed study on NADH in cells using fluorescence spectrophotometry	Duysens and Amesz
1958	Measurement of NADH fluorescence in isolated mitochondria and identification of five redox states	Chance and Baltscheffsky
1959	Measurements of muscle NADH fluorescence in vitro	Chance and Jobsis
1962	In vivo monitoring of NAD(P)H fluorescence from the brain and kidney	Chance et al.

The breadth, depth, and impact of Britton Chance's research make him stand tall among the scientific giants of the world. Our incomplete search and compilation of his publications resulted in over 1500 manuscripts. As of July 2013, his h-index was 130 according to the ISI Science Citation Index. Six of his papers have been cited more than a thousand times.

F.4 Closing Thoughts

Many people who had the privilege to work with Britton Chance regarded him as a role model, great mentor, and dear friend. Since 2010, when he passed away, there have been a number of memorial articles published in newspapers and journals to commemorate his life and legacy. Numerous conferences and special journal issues/sections were organized to honor him. Two Chance memorial symposiums and a workshop were organized in 2011 and 2013 at the University of Pennsylvania, attracting hundreds of researchers and clinicians from all over the world. By reflecting on his life and achievements, we hope the scientific community will carry on his legacy and spirit of scientific curiosity, strong motivation for innovation, love for peace, and unyielding effort to serve the community and society.

References

Ao, T., L. Zhou, and B. Chance. 2005. Miniature fluorescent metabolometer. In *Conference on Lasers and Electro-Optics*, Baltimore, MD.

Chance, B. 1943. The kinetics of the enzyme-substrate compound of peroxidase. *Journal of Biological Chemistry* 151(2):553–577.

Chance, B., J. G. Brainerd, F. A. Cajori, and G. A. Millikan. 1940. The kinetics of the enzyme-substrate compound of peroxidase and their relation to the Michaelis theory. *Science* 92: 455.

Chance, B., S. Eleff, J. S. Leigh Jr., D. Sokolow, and A. Sapega. 1981. Mitochondrial regulation of phosphocreatine/inorganic phosphate ratios in exercising human muscle: A gated 31P NMR study. *Proceedings of the National Academy of Sciences of the United States of America* 78(11 II): 6714–6718.

Chance, B., S. Nioka, A. Duo, and J.-R. Horng. 2010. A novel time-shared fluorometer gives the mitochondrial redox state as the ratio of two components of the respiratory chain of the animal and human buccal cavity with quantitative measures of the redox energy state. *Journal of Innovative Optical Health Sciences* 3(4): 235–245.

Chance, B., N. Oshino, T. Sugano, and A. Mayevsky. 1973. Basic principles of tissue oxygen determination from mitochondrial signals. In *Advances in Experimental Medicine and Biology*, vol. 37A, pp. 277–292. New York: Springer.

DeVault, D. and B. Chance. 1966. Studies of photosynthesis using a pulsed laser. I. Temperature dependence of cytochrome oxidation rate in chromatium. Evidence for tunneling. *Biophysical Journal* 6(6): 825–847.

Eleff, S., N. G. Kennaway, N. R. Buist et al. 1984. 31P NMR study of improvement in oxidative phosphorylation by vitamins K3 and C in a patient with a defect in electron transport at complex III in skeletal muscle. *Proceedings of the National Academy of Sciences of the United States of America* 81(11): 3529–3533.

Garlick, P. B., G. K. Radda, P. J. Seeley, and B. Chance. 1977. Phosphorus NMR studies on perfused heart. *Biochemical and Biophysical Research Communications* 74(3): 1256–1262.

Li, L. Z., R. Zhou, H. N. Xu et al. 2009. Quantitative magnetic resonance and optical imaging biomarkers of melanoma metastatic potential. *Proceedings of the National Academy of Sciences of the United States of America* 106(16): 6608–6613.

Loschen, G., L. Flohé, and B. Chance. 1971. Respiratory chain linked H_2O_2 production in pigeon heart mitochondria. *FEBS Letters* 18(2): 261–264.

Xu, H. N., S. Nioka, J. D. Glickson, B. Chance, and L. Z. Li. 2010. Quantitative mitochondrial redox imaging of breast cancer metastatic potential. *Journal of Biomedical Optics* 15(3): 036010.
Zheng, G., H. Li, M. Zhang et al. 2002. Low-density lipoprotein reconstituted by pyropheophorbide cholesteryl oleate as target-specific photosensitizer. *Bioconjugate Chemistry* 13(3):392–396.

Avraham Mayevsky
Bar-Ilan University

Lin Z. Li
University of Pennsylvania

Preface

Recent advances in noninvasive imaging techniques and fluorescent labeling methods have revolutionized our knowledge of cellular processes, signaling and metabolic pathways, underlying mechanisms for health problems, and the identification of new therapeutic targets for drug discoveries. Endogenous biomolecules such as amino acid tryptophan, coenzymes (e.g., NADH, FAD), and structural proteins (e.g., collagen, elastin) are intrinsically fluorescent and therefore have potential as natural biomarkers for cellular processes as well as clinical diagnosis. In contrast with fluorescent dyes, these native biomarkers minimize toxicity, interference with biomolecular machinery, and nonspecific binding in living cells.

This book is divided into four sections and is designed to capture the current state of knowledge and challenges associated with using cellular/tissue autofluorescence in quantifying metabolic activities under physiological and pathological conditions toward clinical diagnosis. Section I highlights the fundamentals of cellular energy metabolism as well as natural biomarkers within the context of their biological functions. Section II outlines the theoretical and technical background of quantitative, noninvasive, autofluorescence microscopy and spectroscopy methods, which include experimental design, data acquisition, calibration, pitfalls, and remedies. Finally, Sections III and IV highlight selected applications in biochemistry, cell biology, and medicine. The book is also suitable for advanced undergraduate and graduate students, biophysics instructors, as well as researchers at the postdoctoral level.

We are grateful to the authors who contributed to this book as leaders in the field. We are also thankful to the senior editor, Dr. Ammasi Periasamy, for inviting us to edit this book as a part of the Series in Cellular and Clinical Imaging. This book would not have been possible without the guidance and support of Luna Han, senior publishing editor at Taylor & Francis Group, LLC.

Editors

Vladimir V. Ghukasyan is a research assistant professor at the Department of Cell Biology and Physiology and a director of the Confocal and Multiphoton Imaging Facility of the Neuroscience Center, University of North Carolina at Chapel Hill.

He graduated from the Yerevan State University and earned a PhD in 2002 in biology from the Institute of Biotechnology, Yereva, Armenia. Prior to joining UNC, he completed postdoctoral training at the Institute of Biophotonics, Taipei, Taiwan.

Ahmed A. Heikal is a professor in the Department of Chemistry and Biochemistry, Swenson College of Science and Engineering, University of Minnesota Duluth.

His research activities include biomolecular dynamics, intermolecular interactions, energy metabolism, biomembranes, and cell signaling using a range of fluorescence microscopy and spectroscopy methods. His interest in biophysics started when he joined the research group of Watt W. Webb at Cornell University (1997–2003).

He earned a PhD in applied physics from the California Institute of Technology in the field of ultrafast laser spectroscopy and molecular dynamics under the supervision of Nobel Laureate Ahmed H. Zewail, Division of Chemistry and Chemical Engineering.

Contributors

Efrat Barbiro-Michaely
The Mina and Everard Goodman Faculty of Life Sciences
and
The Leslie and Susan Gonda Multidisciplinary Brain Research Center
Bar-Ilan University
Ramat-Gan, Israel

Alzbeta Marcek Chorvatova
Department of Biophotonics
International Laser Centre
Bratislava, Slovakia

Irene Georgakoudi
Department of Biomedical Engineering
Tufts University
Medford, Massachusetts

Richard Hallworth
Department of Biomedical Sciences
Creighton University
Omaha, Nebraska

Ilmo E. Hassinen
Faculty of Biochemistry and Molecular Medicine
University of Oulu
Oulu, Finland

Ahmed A. Heikal
Department of Chemistry and Biochemistry
Swenson College of Science and Engineering
University of Minnesota Duluth
Duluth, Minnesota

Kuravi Hewawasam
Department of Physics
University of Minnesota Duluth
Duluth, Minnesota

Vinod Jyothikumar
W.M. Keck Center for Cellular Imaging
University of Virginia
Charlottesville, Virginia

Alan K. Lam
Institute of Biomaterials and Biomedical Engineering
University of Toronto
and
Toronto General Research Institute
University Health Network
Toronto, Ontario, Canada

Lin Z. Li
Department of Radiology
Molecular Imaging Laboratory
and
Britton Chance Laboratory of Redox Imaging
Department of Biochemistry and Biophysics
Johnson Research Foundation
and
Perelman School of Medicine
University of Pennsylvania
Philadelphia, Pennsylvania

Avraham Mayevsky
The Mina and Everard Goodman Faculty of Life Sciences
and
The Leslie and Susan Gonda Multidisciplinary Brain Research Center
Bar-Ilan University
Ramat-Gan, Israel

Takakazu Nakabayashi
Graduate School of Pharmaceutical Sciences
Tohoku University
Sendai, Japan

Michael G. Nichols
Department of Physics
Creighton University
Omaha, Nebraska

Nobuhiro Ohta
Research Institute for Electronic Science
Hokkaido University
Sapporo, Japan

Ammasi Periasamy
W.M. Keck Center for Cellular Imaging
and
Department of Biology
and
Department of Biomedical Engineering
University of Virginia
Charlottesville, Virginia

Kyle P. Quinn
Department of Biomedical Engineering
Tufts University
Medford, Massachusetts

Narasimhan Rajaram
Department of Biomedical Engineering
Duke University
Durham, North Carolina

Nirmala Ramanujam
Department of Biomedical Engineering
Duke University
Durham, North Carolina

V. Krishnan Ramanujan
Metabolic Photonics Laboratory
Department of Surgery
and
Department of Biomedical Sciences
Biomedical Imaging Research Institute
Cedars-Sinai Medical Center
Los Angeles, California

Jonathan V. Rocheleau
Institute of Biomaterials and Biomedical Engineering
and
Department of Physiology
and
Department of Medicine
University of Toronto
and
Toronto General Research Institute
University Health Network
Toronto, Ontario, Canada

Jan Rupp
Institute of Medical Microbiology and Hygiene
University of Lübeck
and
Medical Clinic III/University Hospital Schleswig-Holstein
Lübeck, Germany

Dietrich Schweitzer
Department of Ophthalmology, Experimental Ophthalmology
University of Jena
Jena, Germany

Heather Jensen Smith
Department of Biomedical Sciences
Creighton University
Omaha, Nebraska

Nannan Sun
Department of Radiology
and
Britton Chance Laboratory of Redox Imaging
Department of Biochemistry and Molecular Biophysics
Perelman School of Medicine
Johnson Research Foundation
University of Pennsylvania
Philadelphia, Pennsylvania

and

Britton Chance Center for Biomedical Photonics
Wuhan National Laboratory for Optoelectronics
and
Key Laboratory of Biomedical Photonics of Ministry of Education
Department of Biomedical Engineering
Huazhong University of Science and Technology
Wuhan, Hubei, People's Republic of China

Yuansheng Sun
W.M. Keck Center for Cellular Imaging
University of Virginia
Charlottesville, Virginia

Márta Szaszák
Institute of Medical Microbiology and Hygiene
University of Lübeck
Lübeck, Germany

Harshad D. Vishwasrao
Department of Neuroscience
Columbia University
New York, New York

Hsing-Wen Wang
Fischell Department of Bioengineering
University of Maryland
College Park, Maryland

Kristina Ward
Department of Physics
Creighton University
Omaha, Nebraska

Yi Yang
East China University of Science and Technology
Xuhui, Shanghai, People's Republic of China

Qianru Yu
Lung Biology Imaging Core
Dartmouth College
Hannover, New Hampshire

Lyandysha V. Zholudeva
Department of Physics
Creighton University
Omaha, Nebraska

Abbreviations

A2E	*N*-Retinylidene-*N*-retinylethanolamine
AcAc	Acetoacetate
ACN	Aconitase
ACSF	Artificial cerebrospinal fluid
ADC	Analog-to-digital converter
ADP	Adenosine dinucleotide phosphate
AF	Autofluorescence
AGE	Advanced glycation end product
AIDS	Acquired immune deficiency syndrome
AIF	Apoptosis-inducible factor
ALA 5	Aminolevulinic acid
AMD	Age-related macular degeneration
AMP	Adenosine monophosphate
AMPK	AMP-activated protein kinase
ANT	Adenine nucleotide translocator
AOM	Acousto-optic modulator
AREDS	Advanced age-related macular degeneration
ASAT	Aspartate aminotransferases
ATG proteins	Autophagy-related proteins
ATP	Adenosine triphosphate
ATPS	Adenosine triphosphate synthase
BChl	Bacteriochlorophyll
β-OHB	β-Hydroxybutyrate
cAMP	Cyclic adenosine monophosphate
CARS	Coherent anti-Stokes Raman spectroscopy
CBF	Cerebral blood flow
CBP	CREB-binding protein
CCD	Charge-coupled device
CF	Corrected fluorescence
Chl	Chlorophyll
CMV	Cytomegalovirus
CPT	Carnitine palmitoyltransferase
cpYFP	Circularly permuted yellow fluorescent protein
CS	Citrate synthase
CTBP	Carboxyl-terminal-binding protein 1
CVD	Cardiovascular disease
CVS	Cardiovascular system

CW	Continuous wave
DAMP	Damage-associated molecular pattern
DC	Direct current
DIC	Differential interference contrast
DMEM	Dulbecco's modified Eagle's medium
dMRI	Diffusion magnetic resonance imaging
DMSO	Dimethyl sulfoxide
DTF	Detrended fluctuation analysis
ECM	Extracellular matrix
EMCCD	Electron-multiplier charge-coupled device
EOM	Electro-optic modulator
ER	Endoplasmic reticulum
ESCs	Embryonic stem cells
ETC	Electron transport chain
ETF	Electron transfer flavoprotein
ETF-QO	Electron transfer protein–coenzyme Q oxidoreductase
FABP	Fatty acid–binding protein
FABPc	Fatty acid–binding protein (cytosolic)
FABPpm	Plasma membrane fatty acid–binding protein
FAD	Flavin adenine dinucleotide
FAS	Fatty acid synthase
FATP	Fatty acid transport protein
FCCP	Carbonyl cyanide *p*-trifluoromethoxy-phenylhydrazone
FDA	Food and Drug Administration
FDG	2-Fluoro-deoxy-glucose
FFA	Free fatty acid
FFT	Fast Fourier transform
FH	Fumarate hydratase
FIFO	First in, first out
FLIM	Fluorescence lifetime imaging microscopy
FMN	Flavin mononucleotide
FOXO	Forkhead box O protein
FRET	Förster resonance energy transfer
GFP	Green fluorescent protein
GLS	Glutaminase
GLUD	Glutamate dehydrogenase
GPDH	Glyceraldehyde-3-phosphate dehydrogenase
GSIS	Glucose-stimulated insulin secretion
GTP	Guanosine triphosphate
hASCs	Human adipose-derived mesenchymal stem cells
Hb	Hemoglobin
HCMV	Human cytomegalovirus
HCV	Hepatitis C virus
HEPES	4-(2-Hydroxyethyl)-1-piperazineethanesulfonic acid
HIC1	Hypermethylated in cancer
HIF-1	Hypoxia-inducible factor-1
HIV	Human immunodeficiency virus
HMG	CoA-1, 3-hydroxy-3-methylglutaryl-CoA
hMSCs	Human bone marrow–derived mesenchymal stem cells
hMVECs	Human microvascular endothelial cells

HSV	Herpes simplex virus
IBA	Isobutyramide
ICCD	Intensified charge-coupled device
ICD	Isocitrate dehydrogenase
IET	Intramolecular electron transfer
IFN-γ	Interferon gamma
IHC	Immunohistochemistry
IMS	Inter membrane space
IR	Infrared
IRF	Instrument response function
IRP	Iron regulatory protein
KSHV	Kaposi's sarcoma herpesvirus
LADH	Liver alcohol dehydrogenase
LDH	Lactate dehydrogenase
LED	Light-emitting diode
LHC	Light-harvesting complex
LipDH	Lipoamide dehydrogenase
MAP	Mean arterial pressure
Mb	Myoglobin
MCP-PMT	Microchannel plate photomultiplier tube
MCU	Mitochondrial Ca^{2+} uniporter
MDR1	Multidrug transporter protein
ME	Malic enzyme
MELAS	Mitochondrial encephalomyopathy, lactic acidosis, and stroke-like episode syndrome
MFE	Multifunctional enzyme
MIM	Mitochondrial inner membrane
mMDH	Mitochondrial malate dehydrogenase
MNNG	*N*-Methyl-*N*′-nitro-*N*-nitrosoguanidine
MOC	Malate-oxoglutarate carrier
MOM	Mitochondrial outer membrane
MPC	Mitochondrial pyruvate carrier
MPE	Multiphoton excitation
MRI	Magnetic resonance imaging
mtAKAP	Mitochondrial A-kinase anchor protein
mtDNA	Mitochondrial DNA
MTERF1	Mitochondrial transcription termination factor
mTOR	Mammalian target of rapamycin
NADH	Nicotinamide adenine dinucleotide reduced
NADPH	Nicotinamide adenine dinucleotide phosphate reduced
NDUFA10	NADH dehydrogenase [ubiquinone] 1 alpha subcomplex subunit 10
ND:YAG	Neodymium-doped yttrium aluminum garnet
NE	Norepinephrine
NIR	Near infrared
NLR	NOD-like receptor
NMR	Nuclear magnetic resonance
OCT	Optical coherence tomography
OGD	2-oxoglutarate dehydrogenase
OHBDH	NAD^+-linked β-hydroxybutyrate dehydrogenase
OHC	Outer hair cell
OPO	Optical parametric oscillator

OS	Oxygen saturation
OSCP	Olygomycin sensitivity conferral protein
PAMP	Pathogen-associated molecular pattern
PBS	Phosphate-buffered saline
PCP	Pentachlorophenol
PDC	Pyruvate dehydrogenase complex
PDK	Pyruvate dehydrogenase kinase
PDMS	Polydimethylsiloxane
PDP	PDH phosphatase
PDT	Photodynamic therapy
PEI	Polyethylenimine
PET	Positron emission tomography
PFK1	6-Phosphofructo-1-kinase
PGC1A	Peroxisome proliferator-activated receptor-γ coactivator-1α
PHD	Prolyl hydroxylase
PI3K	Phosphatidylinositol-3-kinase
PI4-P	Phosphatidylinositol-4-phosphate
PKA	Protein kinase
PMT	Photomultiplier tube
POLG	Polymerase-γ
POLRMT	Mitochondrial RNA polymerase
PPAR	Peroxisome proliferator-activated receptor
PPi	Pyrophosphate
PPP	Pentose phosphate pathway
PSD	Power spectral density
PSI	Photosystem I
PSII	Photosystem II
PTP	Permeability transition pore
qPCR	Quantitative polymerase chain reaction
RBL	Rat basophilic leukemia
RC	Reaction center
RLD	Rapid lifetime determination
ROC	Receiver-operating characteristics
ROI	Region of interest
ROS	Reactive oxygen species
RPE	Retinal pigment epithelium
SAM	Sorting and assembly machinery
SCS	Succinyl-CoA synthetase
SDH	Succinate dehydrogenase
SHG	Second-harmonic generation
shRNA	Short hairpin RNA
SIRT1	Protein deacetylase, sirtuin 1
SNR	Signal-to-noise ratio
SOD	Superoxide dismutases
SPCM	Single photon counting module
SPR	Surface plasmon resonance
SREBP	Sterol regulatory element-binding protein
STS	Staurosporine
TBF	Total blood flow
TCA	Tricarboxylic acid

TCSPC	Time-correlated single-photon counting
TDP	Thiamine diphosphate
TFAM	Mitochondrial transcription factor A
TFBM	Mitochondrial transcription factor B
TIM	Translocase of the inner membrane
TIRF	Total internal reflection fluorescence
TLR	Toll-like receptor
TMRM	Tetramethylrhodamine methyl ester
TNF	Tumor necrosis factor
TOM	Translocase of the outer membrane
TPEF	Two-photon-excited fluorescence
TRAF6	Tumor necrosis factor receptor-associated factor 6
Trp	Tryptophan
Trx	Thioredoxin
TSFR	Time-sharing fluorometer/reflectometer
UCP	Uncoupling protein
UV	Ultraviolet
VDAC	Voltage-dependent anion carrier
VHL	von Hippel–Lindau (tumor suppressor)

I

Biochemical, Biological, and Biophysical Background

1

Mitochondria and Energy Metabolism: Networks, Mechanisms, and Control

Ilmo E. Hassinen
University of Oulu

1.1 Introduction

Life creates and maintains order, which defies the general tendency of entropy increase and therefore demands energy supplies from the surrounding. In living organisms, endergonic processes are running at the expense of a universal cellular energy currency, which must be continuously regenerated. The enzymatic machinery for this provision can be defined as energy metabolism. In aerobic cells, this is accomplished by means of oxidation–reduction reactions, which are linked to conservation and conversion of the combustion energy to be saved in "energy-rich" molecules, such as adenosine triphosphate (ATP). This molecule functions as a universal energy currency whose breakdown can be coupled to the endergonic reactions of biosynthesis, transport, and motion. The majority (~95%) of the biological energy demand in aerobic cells is supplied by the machinery of oxidative phosphorylation residing in mitochondria. Cells, tissues, and organs with continuously high energy consumption are particularly dependent on mitochondria and also vulnerable from their disorders. This chapter highlights the different aspects of mitochondrial role in energy metabolism needed for cellular function and survival and outlines the optical methods of energy metabolism network studies.

1.2 Energy Supply and Conversion Grid

1.2.1 Mitochondria

1.2.1.1 Overview

Mitochondria are cytoplasmic organelles classically visualized as ellipsoid vesicles with a diameter of 0.5–1 μm, as found after isolation from tissue homogenates. However, mitochondria are subjected to regulated fusion and fission, and as a result, they can also attain a tubular or reticular form (Ogata and Yamasaki 1997). For example, the reticular mitochondrial configuration is prominent in neural cells, where axonal transport of mitochondria occurs along microtubules (Cai et al. 2011). Several fusion/fission proteins have been identified, although their mechanisms of action still remain to be established (for a review, see Palmer et al. 2011). Evidently, the ellipsoid vesicular structure in mitochondrial suspensions is caused by rupture and sealing (Picard et al. 2011).

Mitochondria are composed of an outer and an inner membrane (Figure 1.1). The inner membrane encloses the matrix space containing the central enzyme systems of intermediary energy metabolism, including the tricarboxylic acid (TCA) cycle (see Section 1.3.5) and the fatty acid β-oxidation system (Section 1.3.3), in which the terminal reactions of breakdown and oxidation of fuel molecules to carbon dioxide occur with concomitant provision of reducing equivalents to the mitochondrial electron transfer chain (ETC, see Section 1.2.2). A series of enzyme complexes in the mitochondrial inner membrane (MIM) reduce oxygen to water and convert the combustion energy to ATP, which has a large negative free energy change of hydrolysis capable of driving otherwise endergonic reactions in coupled reaction networks. The mechanism for converting combustion energy is based on redox-driven pumping of protons from the matrix space with a formation of transmembrane electrochemical potential across the MIM. This potential is transformed to "chemical" energy of ATP by means of F_1F_o-ATP synthase (described in Section 1.2.4), a nanomechanical engine, which, by means of rotatory motion brought about by proton backflow, produces protein conformation changes capable of energizing the ATP synthesis from adenosine dinucleotide phosphate (ADP) and inorganic phosphate (P_i). To prevent undue draining of energy as heat, the electrochemical potential should not be allowed to relax due to uncontrolled permeability of the membrane to charged molecules. Even under this condition, substrates and some key metabolites must be able to enter or leave the mitochondria,

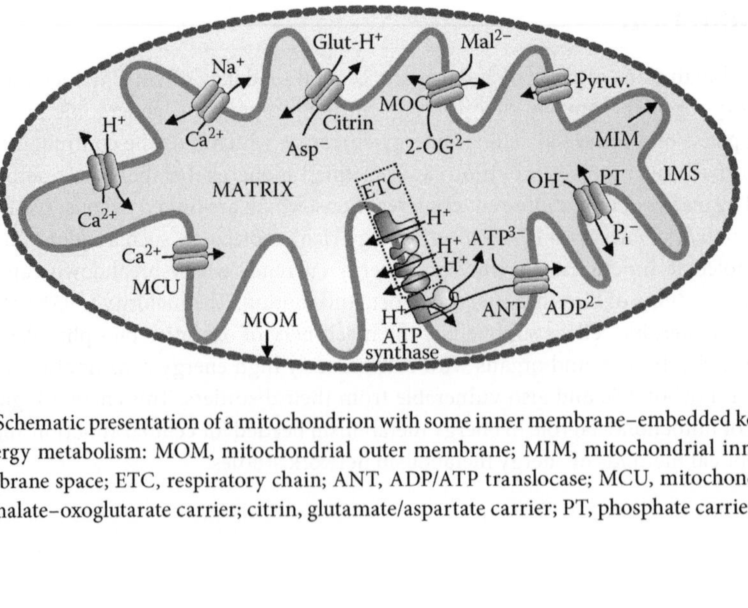

FIGURE 1.1 Schematic presentation of a mitochondrion with some inner membrane–embedded key components involved in energy metabolism: MOM, mitochondrial outer membrane; MIM, mitochondrial inner membrane; IMS, intermembrane space; ETC, respiratory chain; ANT, ADP/ATP translocase; MCU, mitochondrial Ca^{2+} uniporter; MOC, malate–oxoglutarate carrier; citrin, glutamate/aspartate carrier; PT, phosphate carrier.

which are dependent on cytosolic metabolism. This kind of relationships imposed by compartmentation has led to the buildup of an intricate metabolic grid of energy metabolism encompassing the cytosol and the mitochondria.

1.2.1.2 Biogenesis

The biosynthesis of mitochondrial components is encoded in two genomes—nuclear chromosomes and the mitochondrial DNA (mtDNA) housed in the mitochondrial matrix. The circular mtDNA and the mitochondrial ribosomes are descendants of bacterial ancestors and therefore have retained some prokaryotic properties, such as their sensitivity profile to certain antibiotics.

The 16.5 kb mtDNA houses 13 genes for the hydrophobic subunits of the mitochondrial respiratory complexes, 2 genes for mitochondrial ribosomal RNA, and a set of 22 tRNAs (in contrast to the 31 tRNAs of the nucleus-guided system) (Figure 1.2). The rest of the mitochondria are encoded in the nucleus. The peptides encoded in mtDNA include seven subunits of respiratory Complex I (NADH–ubiquinone oxidoreductase), one subunit of Complex III (ubiquinol–cytochrome *c*–oxidoreductase, also called

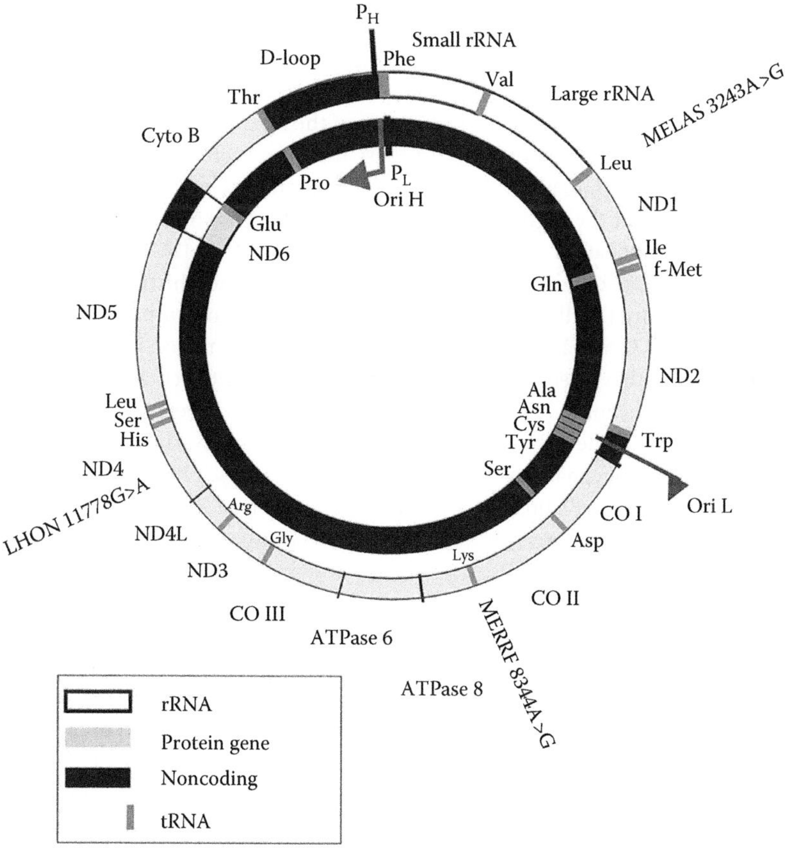

FIGURE 1.2 Gene map of mtDNA. The three-letter amino acid symbols refer to the corresponding tRNAs. Ori H and Ori L mark the replication origins and P_H and P_L the translation promoters of the heavy and light chains, respectively. P_H and P_L show the transcription origins of the heavy and light chains, respectively. Positions of three pathogenic point mutations are shown: MELAS, Mitochondrial encephalomyopathy, lactic acidosis, and stroke-like episodes; MERRF, Myoclonic epilepsy with ragged red fibers; LHON, Leber hereditary optic neuropathy.

cytochrome bc_1 complex), three subunits of Complex IV (cytochrome *c* oxidase, or cytochrome *aa*3), and two subunits of Complex V (F_1F_o-ATPase, or F_1F_o-ATP synthase). Because mitochondria have only 22 different tRNAs, the mitochondrial genetic code and the wobble rules for translation are more relaxed than the "universal code" of the nuclei or prokaryotes. It is because of that reason that mitochondrial tRNAs can read a higher number of synonymous codes—a necessity that follows from the limited tRNA assortment (Watanabe and Yokobori 2011).

Both gene distribution and replication of mammalian mtDNA strands are highly asymmetric (for references, see Xia 2012). In higher eukaryotes, the space needed by the genes in mtDNA is minimized by the lack of introns. Even the translation termination codons are incomplete to be only completed with the synthesis of the poly-A tail, and tRNA genes are used as separation marks in the so-called tRNA punctuation model (Ojala et al. 1981).

For synthesis and maintenance of mtDNA, the mitochondria contain DNA polymerase γ (POLG), a heteromer of one catalytic and two accessory subunits. For transcription, the mitochondria possess the RNA polymerase (POLRMT), transcription factors A (TFAM) and transcription factors B (TFBM), and the transcription termination factor MTERF1. All enzymes needed for mtDNA replication, transcription, and RNA translation are encoded in the nucleus.

The mtDNA is not protected by histones. However, it is not completely naked and has been recently shown to be compacted into structures called nucleoids (Chen and Butow 2005). The nucleoids contain 2–8 mtDNA molecules that are cross-linked by a proteinaceous core of TFAM, POLG, POLRMT, and ancillary proteins and covered by some peripheral proteins. It is considered that mtDNA replication and transcription are initiated simultaneously. Because of the lack of histones, the mutation frequency of mtDNA is 10-fold higher than that of nuclear genes. Thus, the evolution of mtDNA haplotypes and their polymorphs is fast enough to allow demographic research as genetic distances in the framework of population migrations, but the high mutation rate is also the cause of mitochondrial diseases. Ever since the first precise identification of a pathogenic mtDNA mutation 11778G>A, which changes the 340th amino acid in the ND4 subunit of Complex I from an arginine to a histidine and causes the Leber hereditary optic neuropathy (Wallace et al. 1988), the number of known pathogenic mtDNA mutations has grown rapidly. At least 267 point mutations in the coding region and 283 mutations in the mitochondrial RNA genes have been reported to be pathogenic (Mitoweb 2013). The mtDNA-linked diseases are considered rare, but only for a few of them prevalence has been determined. For example, the prevalence of only one of the $tRNA^{Leu(UUR)}$ gene mutations, namely, the A3243G mutation, which leads to the mitochondrial encephalomyopathy, lactic acidosis, and stroke-like episode (MELAS) syndrome, has been found to be 16.3/100,000 in an adult population of 245,201 individuals in northern Finland (Majamaa et al. 1998).

1.2.1.3 Transmembrane Carriers

The mitochondrial outer membrane (MOM) is permeable to small molecular weight compounds due to the presence of mitochondrial porin, a pore-forming protein with a 19-strand β-barrel structure, also called voltage-dependent anion channel (VDAC). The translocase of the outer mitochondrial membrane (TOM complex) and the sorting and assembly machinery (SAM complex) of the outer membrane are needed for mitochondrial protein import. The peptides passing through the TOM complex are destined to the inner membrane or pass to the matrix through the translocase of the inner membrane, the TIM23 complex, which recognizes the mitochondrial signal sequence and employs the inner membrane potential for energizing the transfer.

To allow linking of cytosolic and mitochondrial metabolism without uncontrolled dissipation of the membrane potential and proton gradient, the inner membrane is equipped with an assortment of carriers, used by fuel substances, precursors, and products of oxidative phosphorylation. These carriers belong to a large family of six transmembrane helix proteins with similar secondary structure and can be divided into subclasses according to substrate specificity and symmetry-related amino acid triplets

(Palmieri et al. 2011). The electrochemical potential can be retained by coupled transport of a counter-ion or exchange of ions. When carrier operation is not electroneutral, the membrane potential or the proton gradient can also be utilized to build transmembrane substrate gradients of selected molecules. As a result, the electrophoretic action of adenine nucleotide translocator (ANT) builds a cytosolic ATP/ADP concentration ratio, which is much higher than the mitochondrial one (Kauppinen et al. 1980), although the mitochondrion is the ATP producer.

A nonspecific channel of the inner membrane, permeability transition pore (PTP), allows permeation of molecules with a mass up to 1.5 kDa. Opening of this pore, which is normally closed or flickering, is a major event in the triggering of apoptosis. It is regulated by several factors, such as the apoptotic regulators Bax and the Bcl-2 protein family. The opening of PTP is triggered by calcium influx to the matrix, while NAD(P)H oxidation, production of reactive oxygen species (ROS), and loss of adenine nucleotides seem to be important. Cyclosporin A is able to protect mitochondria from PTP opening (for a review, see Halestrap 2009).

Early work suggested that the PTP is formed upon association of the inner membrane ANT with the outer membrane VDAC, matrix cyclophilin-D (peptidyl-prolyl *cis–trans* isomerase), and intermembrane space (IMS) hexokinase. However, recent work with gene deletion or silencing has shattered the conventional picture of PTP composition. It has been found that PTP operates even in *Ant1* and *Ant2* knockout mitochondria, albeit a higher Ca^{2+} concentration is needed for opening (Kokoszka et al. 2004). The knockout was not complete because there are other ANT isomorphs. PTP opening has been observed also in *Vdac1/3* knockouts, treated with *Vdac2* small interfering RNA, which indicates that those VDAC isomorphs are not indispensable either (Baines et al. 2007). Inactivation of the cyclophilin-D-coding gene increases the calcium concentration needed for PTP opening, but does not totally prevent it (Basso et al. 2005). Thus, the molecular composition of the PTP is still unknown.

Calcium is an important metabolic regulator in the mitochondria (see Section 1.4.2) and has been also shown to be crucial in the initiation of apoptosis, the "programmed" cell death. Mitochondrial and cytosolic free calcium concentrations are linked by means of mitochondrial calcium translocators. The early observation of mitochondrial Ca^{2+} uptake (De Luca and Engstrom 1961) led to the concept of an electrogenic calcium uniporter, which still holds. The concept postulates that Ca^{2+} enters mitochondria through a Ca^{2+} uniporter driven by the negative inside membrane potential, which may lead to Ca^{2+} concentration gradients of the order of 10^5–10^6 if allowed to reach equilibrium. This is intolerable, and to protect from damaging calcium sequestration, the MIM is equipped with Ca^{2+}/Na^+ and Ca^{2+}/H^+ exchange carriers (Figure 1.1). Both of them appear to be electrogenic, because the sodium gradient alone would be insufficient for an effective calcium extrusion by electroneutral $Ca^{2+}/2Na^+$ exchange. More energy must be available for the outward-directed calcium pumping by coupling the exchange to the proton motive force (Jung et al. 1995), and according to the present consensus, the calcium–sodium exchange operates in the electrogenic mode with stoichiometries of $Ca^{2+}/(3–4)Na^+$. The identity of the mitochondrial calcium–sodium exchanger was recently revealed as the NCLX protein, previously considered to be related to plasma membrane (Palty et al. 2010).

The calcium uniporter protein, now called mitochondrial Ca^{2+} uniporter (MCU) (previously known as NP_001028431), is a 40 kDa protein containing two transmembrane helices and a linker region with acidic residues. One of its putative components, a calcium-binding motif-containing peptide MICU1, was first identified by using tools of molecular genetics (Perocchi et al. 2010). The gene and its homologs are conserved in eukaryotes, some protozoa, and bacteria (Bick et al. 2012). Gene silencing with the shRNA and siRNA technology, and the use of the mitochondrially targeted calcium probe aequorin, confirmed the MCU role in calcium influx to mitochondria (Baughman et al. 2011). For a recent review, see Drago et al. (2011).

Investigation of calcium research has recently been greatly benefited from calcium indicators based on targeted recombinant luminescent proteins, which allow compartment-specific imaging, as exemplified by the earlier-described work on translocators.

1.2.2 Respiratory Chain

Traditionally, the mitochondrial respiratory chain is defined as the group of enzyme complexes involved in electron transfer to oxygen with simultaneous conservation of energy by means of redox-driven proton pumps and the utilization of the proton and electric gradient in ATP synthesis (Figure 1.3). The components are Complex I, Complex III, Complex IV, and Complex V.

Closely related enzymes feeding into the respiratory chain are the ubiquinone-reducing, flavin adenine dinucleotide (FAD)-containing enzymes (i.e., succinate–ubiquinone oxidoreductase [Complex II], electron transfer flavoprotein [ETF]–ubiquinone oxidoreductase, and glycerol-3-phosphate–ubiquinone oxidoreductase), but they are not capable of proton pumping.

The proton pumping function of the respiratory chain redox enzymes generates a proton gradient (ΔpH) and a membrane potential ($\Delta\psi$) across the inner membrane. Together, the proton and membrane potential gradients represent an electrochemical potential difference of protons ($\Delta\tilde{\mu}_{H^+}$)—a free energy charge (kJ/mol) driving the ATP synthesis from ADP and P_i by the F_1F_o-ATPase:

$$\Delta\tilde{\mu}_{H^+} = F \cdot \Delta\psi + 2.3RT \cdot \Delta pH \quad (1.1)$$

where

F is the Faraday constant
R is the gas constant
T is the temperature in Kelvin

The terms on the right side of the equation stand for the electric and chemical components of the electrochemical potential, respectively. In mitochondria respiring under basal conditions, ΔpH is about

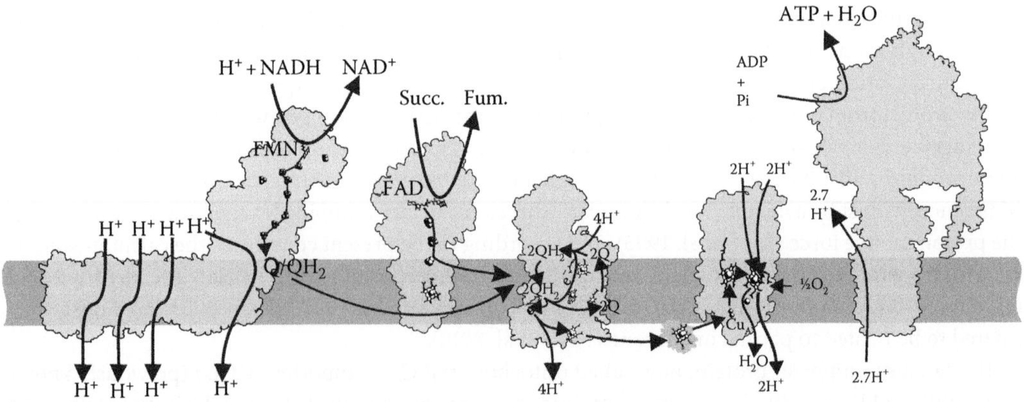

FIGURE 1.3 Outline of the mitochondrial oxidative phosphorylation system. The shapes and relative positions of the redox centers are drawn on the basis of the protein 3-D coordinate files: 4HEA.ENT for *T. thermophilus* Complex I (Baradaran et al. 2013), 3SFE.ENT for *Sus scrofa* Complex II (Zhou et al. 2011), 1NTZ.ENT for *Bos taurus* Complex III (Gao et al. 2003), 1OCC.PDB for *Bos taurus* cytochrome *c* oxidase (Tsukihara et al. 1996), and 4B2Q.ENT for *Saccharomyces cerevisiae* F_1F_o-ATP synthase (Davies et al. 2012). Fe–S clusters and metal atoms are depicted as black balls, hemes, flavins, and quinones in structural sticks. The proton pumping stoichiometry is calculated for two electrons transferred. The chemistry shown is only indicative. For Complex I, the locations of four proton channels proposed by Baradaran et al. (2013) are approximated. See Figure 1.5 for the principle of Q-cycle in Complex III.

0.5 *pH* units (alkaline inside) and $\Delta\psi = 150–180$ mV (negative inside). The proton electrochemical gradient is occasionally expressed in units of electric potential (volts) and called proton motive force (Δp), where

$$\Delta p = \Delta\psi + 2.3RT \cdot \frac{\Delta pH}{F} \tag{1.2}$$

1.2.2.1 Complex I

Complex I (NADH–ubiquinone oxidoreductase) is one of the largest multisubunit enzymes in the inner membrane of mitochondria. The mammalian enzyme is a heteromer of 44 dissimilar subunits with a total molecular mass of 980 kDa (Balsa et al. 2012). Its bacterial counterpart is relatively small (550 kDa) and has only 14 subunits (or 13 due to gene fusion), which are considered to be the functional core of the enzyme. Both mammalian and bacterial enzymes are L shaped with a hydrophilic matrix-directed arm and a hydrophobic arm embedded in the MIM. All seven of the hydrophobic subunits are encoded in mtDNA. The enzyme contains noncovalently bound flavin mononucleotide (FMN), or riboflavin-5′-phosphate, in the NADH-binding active site of the hydrophilic domain and an ubiquinone-binding site in the membrane domain. NADH donates reducing equivalents to FMN, from which a sequence of nine iron–sulfur centers conveys electrons to the ubiquinone-reducing site located at the hinge region between the membranous and hydrophilic arms.

Complex I functions as a redox-driven proton pump. An $H^+/2e^-$ stoichiometry of four has been generally accepted. However, this has been recently challenged on the basis of thermodynamic data and a stoichiometry of three has been proposed (Wikström and Hummer 2012). Although gathering structural information of this enzyme has been rapid in recent years, the mechanism of its proton pumping remains speculative. The most recent models are based on parallel pumps in the membrane arm with conformation-mediated energy transfer from the hydrophilic arm assisted by a longitudinal amphipathic helix on the matrix side of hydrophobic arm (Efremov and Sazanov 2011). This has been challenged by Wikström and Hummer (2012), who suggested that the pump function is achieved by concerted action of the ND2, ND4, and ND5 subunits driven by charges in the ubiquinone-binding domains. However, the complete 3-D structure of *Thermus thermophilus* Complex I on the basis of x-ray crystallography (Baradaran et al. 2013) reveals that there is only one ubiquinone-binding pocket in the enzyme and therefore rules out the proton pumping model proposed by Wikström and Hummer. Based on amino acid sequences and 3-D structures in the membrane arm, a train of buried charged residues (Kervinen et al. 2004) and proline hinges in the middle of transmembrane helices of the antiporter type membrane subunits (Figure 1.3) have been visualized as structures allowing the propagation of conformation changes and proton pumping in the enzyme (Baradaran et al. 2013).

Because 7 of the 13 peptide genes in mtDNA encode Complex I subunits (Figure 1.2), majority of the mtDNA mutations affect Complex I. Indeed, of the 260 known pathogenic peptide gene point mutations in mtDNA, 139 affect Complex I (Mitoweb 2013).

Some Complex I subunits encoded in the nucleus allow covalent modification of the enzyme by phosphorylation/dephosphorylation. Both cyclic adenosine monophosphate (cAMP) and cAMP-dependent protein kinase (PKA) exist in the mitochondrial matrix (reviewed in Feliciello et al. 2005). The MIM is impervious to cAMP (Müller and Bandlow 1987), and it has been suggested that the cytosolic cAMP signal reaches the mitochondrion by mediation of PKAII (a subtype of PKA having type RII regulatory subunits). This kinase is bound to the outer face of the MIM with mitochondrial A-kinase anchor proteins (mtAKAPs) by means of a "transductosome" (Feliciello et al. 2005). The 42 kDa (NDUFA10) subunit can be phosphorylated at certain tyrosine residues that remain phosphorylated even during isolation of the complex (Schilling et al. 2005). The 18 kDa (AQDQ) subunit in the iron–sulfur protein fraction of Complex I is phosphorylated by a mitochondrial PKA (Papa et al. 1996) and dephosphorylated by a mitochondrial serine phosphatase, which is inhibited by calcium. This is in contrast to the mitochondrial pyruvate dehydrogenase (PDH) phosphatase, which is activated by calcium (Signorile et al. 2002). The 18 kDa subunit of Complex I is coded by the nuclear *NDUFS4* gene, and its mutation results in Complex I deficiency when homozygous. A 5 bp duplication in the *NDUF4* gene destroys the phosphorylation site and abolishes the cAMP-dependent activation of Complex I in fibroblast cultures. Another mutation in the *NDUF4* gene leads to translation termination after the mitochondrial targeting sequence and results in a defective assembly of Complex I (Papa et al. 2002).

Because Complex I contains the flavin chromophore FMN, which has redox-sensitive light absorption bands in the UV-VIS region, its redox changes could, in principle, be probed by spectrophotometry. However, the iron–sulfur clusters have a broad absorption peak in the same region. The reduced vs. oxidized spectrum of isolated Complex I has an absorbance minimum at 460 nm, but both FMN and Fe–S clusters contribute to the structureless "iron–flavoprotein trough" in the reduced-form spectrum, where the Fe–S clusters with their multitude swamp the flavin contribution (Gutman et al. 1970). Although free FMN is fluorescent, it is heavily quenched in Complex I and therefore its fluorescence-based monitoring is not easy to achieve in mitochondrial suspensions, cells, or organs (Ragan and Garland 1969). Redox behavior of the Fe–S clusters is better probed by means of electron paramagnetic resonance (Sinegina et al. 2005).

1.2.2.2 Complex III

The mammalian Complex III (ubiquinol–cytochrome *c*–oxidoreductase, or cytochrome bc_1 complex) is a heteromeric enzyme complex of 11–12 dissimilar subunits that relay electrons from ubiquinone to cytochrome *c*. The bacterial counterpart consists of only three subunits, which contain all of the redox-active components. Central to the function of Complex III is cytochrome *b*, a membrane protein with eight transmembrane helices, two heme B groups, and two ubiquinone-binding domains (outer oxidizing, Q_o, and inner reducing, Q_i) in the vicinity of the hemes. The low-potential heme (B_L, −30 mV) is located near the outer face of the membrane and the high-potential heme (B_H, +90 mV) near the inner face of the membrane (Figure 1.4). In addition, the complex contains an iron–sulfur cluster (Rieske Fe_2–S_2 center) and cytochrome c_1. Cytochrome *b* is encoded in mtDNA and the rest of Complex III in the nuclear genome.

Complex III operates as a redox-driven proton pump with $4H^+/2e^-$ ratio. It is based on the so-called Q-cycle (Mitchell 1975), where the ubiquinol molecule is oxidized at Q_o in a bifurcated path. In this case, one electron is donated to the B_L heme to be further conveyed to B_H and another to the Rieske Fe_2–S_2 center to be transferred to cytochrome c_1 and further to cytochrome *c* on the outer facet of the MIM. Thus, cytochrome *b* with its two hemes functions as a vectorial transmembrane electron wire for the oxidation of ubiquinol at the Q_o site along with simultaneous release of a proton and reduction of ubiquinone and its semiquinone (sequentially in two cycles) in the Q_i pocket with simultaneous uptake of protons. The Rieske iron–sulfur protein is linked to the complex by a hinge domain, which allows a pendulum movement of the Fe_2–S_2-containing domain to relay electrons from the Q_o site to cytochrome c_1. This cyclic process implies translocation of one ubiquinol molecule from Q_i to Q_o, one molecule of ubiquinone and one electron from Q_o to Q_i. The experimental data gathered so far and the known 3-D structure of the bc_1 complex resolved at atomic level and 3 Å resolution (Gao et al. 2003; Iwata et al. 1998) conform to the original Q-cycle formulated by Peter Mitchell (1975) as a component of his chemiosmotic principle of oxidative phosphorylation (Figure 1.5).

Complex III crystallizes as a dimer, which is considered as the functional unit of the enzyme. There is experimental evidence of redox interactions between the joined monomers, which allow modifications of the Q-cycle (Covian and Trumpower 2008).

Cytochrome *b* has a characteristic spectrum in the UV-VIS range. Under favorable conditions, the absorption maximum of the reduced form of the low-potential heme B_L at 566 nm can be resolved from the 562 nm maximum of the high-potential heme B_H in mitochondrial suspensions. In routine mitochondrial metabolic work, the changes of absorbance difference between 563 and 575 nm are monitored to observe oxidation–reduction changes of cytochrome *b* (Vanneste 1966).

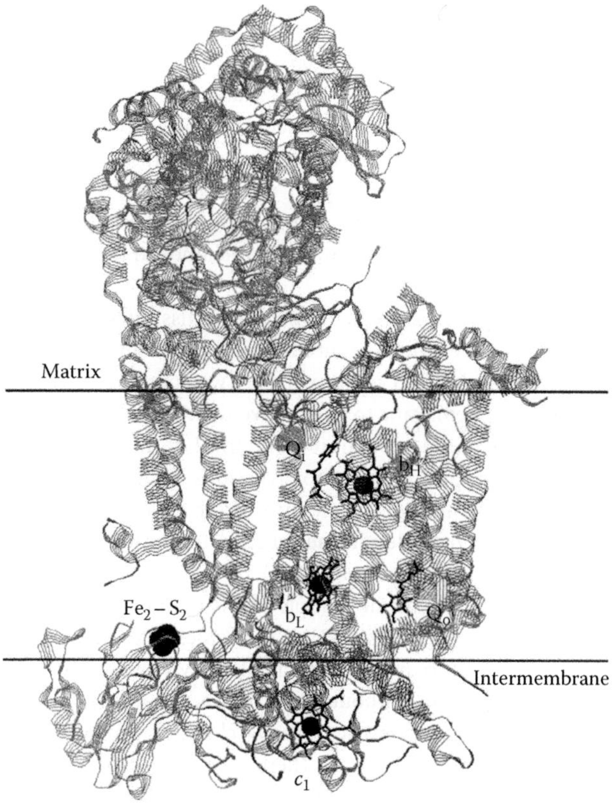

FIGURE 1.4 Locations of the redox-active groups in Complex III (ubiquinol–cytochrome *c* oxidoreductase). The figure is constructed on the basis of coordinates (1NTZ.ENT) from crystalline bovine Complex III containing bound ubiquinone (Gao et al. 2003). b_H and b_L are the high-potential and low-potential hemes of cytochrome *b*.

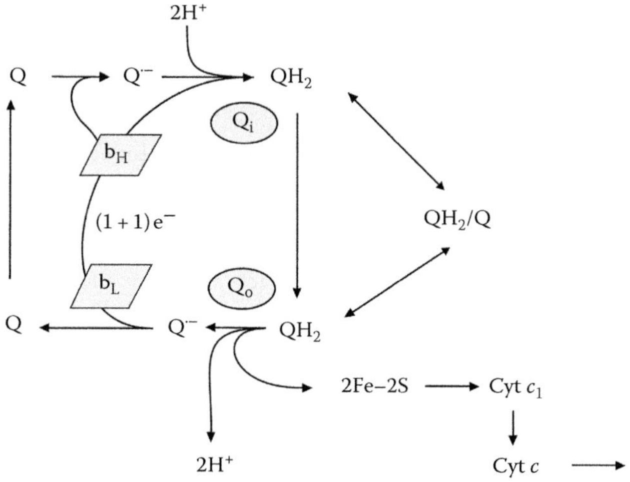

FIGURE 1.5 Simplified scheme of the Q-cycle in mitochondrial Complex III. b_H and b_L are the high-potential and low-potential hemes of cytochrome *b*. Q_i an Q_o are the inner (reducing) and outer (oxidizing) ubiquinone-binding sites, respectively. 2Fe–2S is the Rieske binuclear iron–sulfur center. Note that because of sequential involvement of one-electron and two-electron carriers the net reaction cannot be accurately described in one picture.

1.2.2.3 Complex IV

Complex IV (cytochrome *c* oxidase; cytochrome aa_3) catalyzes the reduction of oxygen to water by the reduced cytochrome *c* (ferrocytochrome *c*). This cosubstrate diffuses laterally from Complex III and attaches to the outer face of the MIM so that electron transfer to the oxygen-reducing site is vectorial across the membrane. Because the direction of electron transfer is inwards and the capture of protons for forming water from the reduced oxygen occurs near the inward facet of the membrane, the enzyme produces an electrochemical proton gradient. This gradient is further stepped up by redox-driven transmembrane pumping of protons from the mitochondrial matrix with the stoichiometry of $2H^+/2e^-$ (Wikström and Krab 1979).

The mammalian enzyme is a 220 kDa size heteromer of 13 dissimilar subunits, of which three most hydrophobic ones (I, II, and III) are encoded in mtDNA. Subunits I and II retain all of the redox-active centers of the enzyme: the copper-A (Cu_A) center with its two copper atoms resides in subunit II near the cytosolic surface of the inner membrane to be reactive with cytochrome *c*, whereas heme *a* and the binuclear heme a_3-Cu_B center are found in subunit I. Subunit I of the eukaryotic Complex IV contains bound Mg^{2+}, which does not participate in the redox reactions. Subunit III is not necessary for proton pumping but contributes to the stability of the enzyme. From x-ray crystallographic data, the spatial structure of bovine Complex IV has been determined at 2.8 Å resolution and found to be dimeric, which is also a prerequisite for function (Tsukihara et al. 1996).

The number of redox-active components in Complex IV is in line with the requirement of four electrons for reducing one oxygen molecule to two water molecules in the binuclear heme a_3-Cu_B cluster, although the enzyme is supplied with reducing equivalents only sequentially by a one-electron carrier, cytochrome *c*. Complex IV manages this without releasing reactive, partially reduced oxygen species.

Two input proton paths have been documented: the D-channel, which contains a conserved aspartate, and the K-channel with a conserved lysine. There is also an exit channel for the pumped protons (Sharpe and Ferguson-Miller 2008). In addition to amino acid residues capable to protonation/deprotonation, bound water contributes to the formation of "proton wires" by a mechanism proposed by de Grotthuss (2006). The binuclear aa_3/Cu_B center resides in a hydrophobic region of the enzyme, but there must be an exit pathway toward the IMS for the water produced through oxygen reduction. Experiments on the effect of $^{17}O_2$ and ^{17}O-enriched water on the Mn^{2+} EPR spectrum of manganese-substituted Mg^{2+}-binding site in Complex IV suggest that bound Mn^{2+}/Mg^{2+} is accessible to the produced water by using a specific channel rather than random diffusion (Schmidt et al. 2003) and that the water exit channel ribbons near the Mg^{2+}-binding site. The mechanism of proton pumping remains hypothetical (Figure 1.6).

The function of the additional 10 subunits of the mammalian enzyme is not known with certainty, but they may contribute to regulation and stability of the complex, since the purified bacterial enzyme is more labile than the eukaryotic one. Two of these additional subunits are involved in allosteric regulation of the enzyme. The matrix domain of subunit IV contains a binding site for ATP, which acts as an allosteric inhibitor of the enzyme (Arnold and Kadenbach 1997), whereas another site is accessible from the IMS (Napiwotzki and Kadenbach 1998). ATP acts also as an allosteric inhibitor of the heart isomorph of subunit VIa resulting in the decrease of the proton pumping efficiency, whereas ADP is an allosteric activator of the enzyme (Anthony et al. 1993).

Although the MIM is impervious to cAMP, covalent modification of Complex IV by means of PKA has been observed. It has been shown by modeling and mutagenesis work that Complex IV subunit IV-1 becomes phosphorylated by PKA, resulting in an alleviation of the ATP inhibition of the enzyme (Acin-Perez et al. 2011).

Both hemes *a* and a_3 of Complex IV represent structurally heme A (here, the capital letter refers to the heme structure and lower case to hemoprotein name) and should have roughly same optical properties in the near-UV-VIS region, but distinct characteristics in their local environment within the assembled enzyme. Both have an α band at around 605 nm and a γ band at 444 nm, but their contributions to

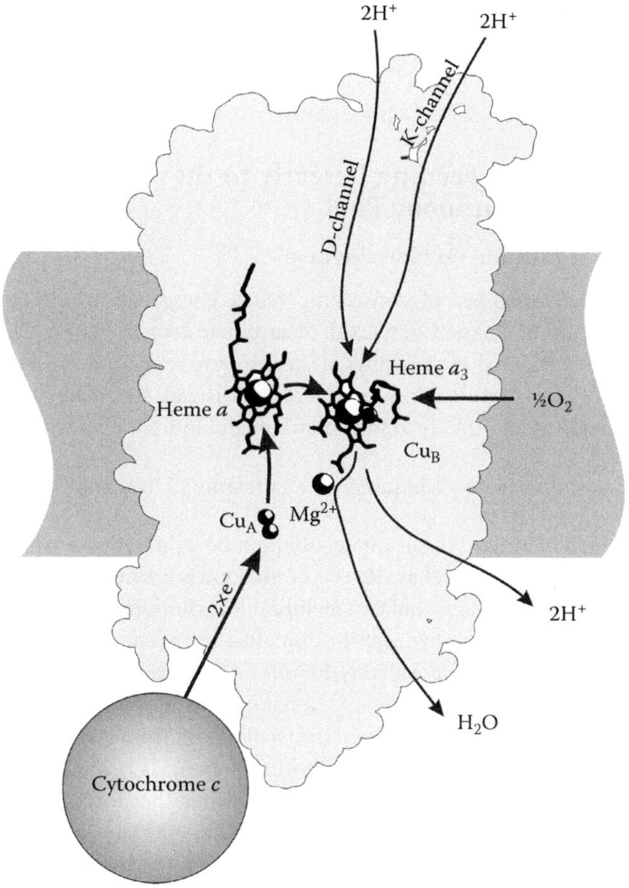

FIGURE 1.6 Schematics of the function of cytochrome *c* oxidase. The picture is an outline projection and the relative locations of the redox-active groups on the basis of spatial coordinates given in 2ASN.PDB (Suga et al. 2011). Only subunits 1 and 2 are included. The redox-active components and Mg^{2+} are superimposed.

individual bands differ. The contribution of heme *a* to the α band is 80%, while both hemes contribute equally to the γ band. The metabolic redox changes of Complex IV are conventionally monitored as an absorbance difference at 605 vs. 630 nm.

The 605 nm absorbance peak of cytochrome *a* has been successfully used for monitoring the redox state of cytochrome oxidase in isolated perfused organs, such as liver and heart (Hassinen et al. 1971; Tamura et al. 1976). The high myoglobin content in the myocardium may impose a potential source of interference, but experiments with myoglobin knockout hearts (Liimatta et al. 2004) have largely corroborated the data obtained with the conventional dual-wavelength spectrometry technique (Hassinen et al. 1981; Tamura et al. 1976).

There is also a weak absorbance band at 830 nm due to oxidized Cu_A, which decreases upon reduction. Cu_B is not visible for optical readout. Attempts have been made for 30 years to develop a method, which would allow reliable measurement of the Cu_A redox state in vivo by means of IR spectroscopy to obtain an estimate of cell oxygenation or hypoxia. Although IR light has the advantage of deep penetrations into tissues because of low scattering, a major obstacle is the high concentration of hemoglobin and its spectral changes in the IR region upon oxygenation/deoxygenation. Several algorithms for data analysis are currently available for isolating the hemoglobin background signal but the methodology remains unreliable for clinical use (Piantadosi et al. 1997; Springett et al. 2000). The copper centers of Complex IV are also visible in electron paramagnetic resonance spectrometry, but the method necessitates cryostat work at low temperatures and is applicable only for discrete samples (Pezeshk et al. 2001).

1.2.3 Metabolic Enzymes Feeding Directly to the Mitochondrial Ubiquinone Pool

1.2.3.1 Succinate–Ubiquinone Oxidoreductase

Although not having the capability of conserving redox energy as an electrochemical potential, succinate–ubiquinone oxidoreductase (Complex II, or succinate dehydrogenase [SDH]), a key enzyme of the TCA cycle, has been considered as a member of the respiratory chain for historical reasons, because it is an intrinsic membrane enzyme localized in the MIM. Instead, succinate–ubiquinone oxidoreductase conveys reducing equivalent directly to the ubiquinone/ubiquinol pool with the net reaction.

$$\text{Succinate} + \text{Ubiquinone} \leftrightarrow \text{Fumarate} + \text{Ubiquinol}$$

The structure of mammalian and avian succinate–ubiquinone oxidoreductase has been determined by means of x-ray crystallography (Zhou et al. 2011). The enzyme is a heteromer of four proteins: a FAD-binding protein, an iron–sulfur protein, and two membrane anchor proteins. The complex has an asymmetric, q-shaped form with a globular hydrophilic domain and a membrane domain (Figure 1.3).

The globular domain contains one covalently bound FAD molecule in subunit A and three iron–sulfur clusters in subunit B: a dinuclear (2Fe–2S), a trinuclear (3Fe–4S), and a tetranuclear (4Fe–4S). The iron–sulfur clusters form a 40 Å long electron transport chain from FAD to ubiquinone and heme B in the membrane domain. The heterodimeric anchoring domain is composed of a 140-residue large subunit (CybL) and a 103-residue small subunit (CybS). Together, they form a six-helix transmembrane bundle linked to a single central heme B_{560} and are tied together with other helices. The anchor domain contains two ubiquinone-binding sites. A proximal one is located between the 3Fe–4S cluster and heme B, but the distance from the iron–sulfur cluster to the bound ubiquinone is shorter than to the heme. Moreover, the redox potential of the heme is lower than that of the 3Fe–4S cluster, which would mean uphill electron transport to the heme, whose role in electron transfer remains uncertain. Another ubiquinone-binding site is located close to the intermembrane facet of the membrane. In contrast to the respiratory Complexes I, III, and IV, none of SDH subunits is encoded in mtDNA.

The fluorescence efficiency of FAD in the succinate–ubiquinone oxidoreductase is so low that it can be neglected when measuring mitochondrial flavin fluorescence in cells or mitochondrial suspensions (Kunz and Kunz 1985).

1.2.3.2 Electron Transfer Flavoprotein–Ubiquinone Oxidoreductase

Electrons from the fatty acid β-oxidation pathway are conducted to the respiratory chain via two routes. The first route for fatty acid oxidation is catalyzed by the FAD-containing fatty acyl-CoA dehydrogenase enzyme, which becomes reduced. The second route employs NADH as a carrier of reducing equivalents (as described in Section 1.3.3). The reduced form of fatty acyl-CoA dehydrogenase is reoxidized by the ETF, which in turn is oxidized by ETF–ubiquinone oxidoreductase (ETF dehydrogenase). In fact, the ETF/ETF–ubiquinone system transfers electrons from nine acyl-CoA dehydrogenases participating in fatty acid and amino acid oxidation (Watmough and Frerman 2010).

The ETF–ubiquinone oxidoreductase is a single-subunit membrane metalloflavoprotein containing one FAD and one 4Fe–4S cluster. The structure of the enzyme from pig liver has been solved by x-ray crystallography at 2.1 Å resolution (Zhou et al. 2011). The enzyme has three functional (not necessarily structural) domains: the FAD binding, the ubiquinone binding, and the iron–sulfur cluster domains.

ETF–ubiquinone oxidoreductase lacks an intrinsic membrane anchor but is attached to the matrix side on the inner mitochondrial membrane by means of a hydrophobic plain surface, which includes an entrance to the ubiquinone-binding site.

The electron transfer trajectory in the ETF–ubiquinone oxidoreductase has been probed by mutagenesis and characterized by the distances of the redox centers from each other and from the protein surface. It has become apparent that the primary electron acceptor is the iron–sulfur center. The flavin and the iron–sulfur cluster form an isopotential group at +38 mV, indicating rapid electron equilibration. However, the 19 Å distance between the Fe–S center and ubiquinone is a little too long for biological electron transfer, which may suggest a role for FAD as a mediator between these two redox centers.

ETF fluorescence changes can be resolved from total flavin fluorescence in mitochondrial suspensions under certain conditions (Kunz 1986).

1.2.3.3 Glycerol-3-Phosphate–Ubiquinone Oxidoreductase

There are two different glycerol-3-phosphate dehydrogenases in cells. The cytosolic NAD^+-linked glycerol phosphate dehydrogenase is a heterodimer of two subunits that is important in glycerolipid synthesis. Mitochondria contain an ubiquinone-reducing glycerol-3-phosphate dehydrogenase with FAD as the prosthetic group. Together with the cytosolic NAD^+-linked glycerol-3-phosphate dehydrogenase, it completes the glycerophosphate cycle for reoxidation of cytosolic NADH.

This enzyme is bound to the outer surface of the inner membrane so that the glycerol-3-phosphate-binding site faces the IMS and is accessible to cytosolic glycerol phosphate. The 3-D structure of this membrane enzyme has been determined only in some bacteria by means of x-ray crystallography (Yeh et al. 2008). Their structure may be used as a template for visualization of the mammalian enzyme. In contrast to the ubiquinone-linked oxidoreductases mentioned earlier, the presence of an iron–sulfur cluster is unknown.

The K_m of the enzyme for glycerol-3-phosphate is substantially lowered by calcium, which has a K_m of 0.1 mM. Expression of the glycerol-3-phosphate dehydrogenase is highly variable and depends on the species and tissue type. The expression of mitochondrial glycerol-3-phosphate dehydrogenase in the liver has been shown to be stimulated by the thyroid hormone (Lee and Lardy 1965). High activities of this enzyme are found in the insect flight muscle (Crabtree and Newsholme 1972). The activity of the mitochondrial enzyme is also constitutively high in the β-cells of the pancreatic islets, which show high rates of glucose oxidation (Sekine et al. 1994). In contrast, the mitochondrial enzyme activity is low in the heart muscle, where the main mechanism of aerobic oxidation of cytosolic NADH is the malate-aspartate cycle (Williamson et al. 1973).

1.2.4 ATP Synthesis

F_1F_o-ATPase (Complex V, or F_1F_o-ATP synthase) is the mitochondrial version of proton ATPases, "nanomechanical engines," which employ rotatory motors to pump protons across a membrane at the expense of ATP hydrolysis. Conversely, Complex V is capable of ATP synthesis when energized by an electrochemical proton gradient. The detailed spatial structure of bovine and yeast F_1F_o-ATPase has been determined by crystallography and electron cryotomography (Abrahams et al. 1994; Davies et al. 2012).

The enzyme complex consists of three domains: a membrane-embedded rotor (F_o), a hydrophilic catalytic domain (F_1) in the matrix space, and a joining stalk. The F_1 part consists of three α-subunits and three β-subunits and one of each of γ-, δ-, and ε-subunits. Pairs of α- and β-subunits form three catalytic sites capable of ATP hydrolysis or synthesis. Subunits γ, δ, and ε are fixed to the F_o part and therefore rotate with it and form an axle extending to the center of the F_1 part. Subunits oligomycin sensitivity conferral protein (OSCP), F_6, b, and d constitute a lateral stalk, which immobilizes the F_1 part and prevents it from following the rotation of the central stalk. There is also a mitochondrial F_1F_o-ATPase inhibitor IF_1, which binds to the central stalk upon the loss of mitochondrial proton motive force and prevents its rotation in wasteful proton pumping as an ATPase (for references, see Garcia-Trejo and Morales-Rios 2008).

In the F_0 domain, there is a "rotor" composed of 8–15 c-subunits depending on the species (Dmitriev et al. 1999). The subunit count in the F_0 part of the bovine enzyme is $abc_8defg(A6L)F_6$. The rotation of the rotor is mechanically coupled to proton flux, which is channeled to the c-subunit oligomer by the a-subunit. The stepped, proton-driven rotation of the rotor subunit is linked to an ordered protonation/deprotonation of an exposed glutamic acid residue in the c-subunit. The number of protons needed for a full revolution of the rotor equals to the number of c-subunits in it. The rotor drives the rotation of the asymmetric γ-subunit, which produces conformation changes in the three active sites of the F_1 domain. Those provide the energy for ATP synthesis from ADP and P_i. One turn of the F_0 rotor results in the synthesis of three ATP molecules, whereas the ATP/H^+ stoichiometry depends on the number of c-subunits in the rotor. The eight-c-subunit F_0 rotor in mammalian cells gives an H^+/ATP ratio of 8/3 = 2.7.

Recently, it has been shown that the mitochondrial F_1F_0-ATPase is able to attain a dimeric form, which may be advantageous in preventing its stator rotation and in modeling of the cristae (Garcia-Trejo and Morales-Rios 2008).

1.2.5 Oxidative Phosphorylation as an Entity

As described earlier, the respiratory enzyme complexes and F_1F_0-ATP synthase together carry out the process of conserving and converting the combustion energy of fuel nutrients to a high-energy phosphate bond in ATP. The overall efficiency of the process can be calculated from the $H^+/2e^-$ ratio of the respiratory complexes utilized by a specific substrate, the ATP/H^+ ratio of the F_1F_0-ATPase, and the proton transferred concomitantly with the import of P_i to the matrix space. For NADH oxidation involving Complexes I + III + IV, the ATP/$2e^-$ ratio is (3 + 4 + 2)/(2.7 + 1) = 2.4. According to the current consensus concerning the value of 4 $H^+/2e^-$ for Complex I, the ATP/$2e^-$ ratio is (4 + 4 + 2)/(2.7 + 1) = 2.7. Because Complex II is not a proton pump, oxidation of succinate, glycerol-3-phosphate, or ETF yields only (4 + 2)/(2.7 + 1) = 1.6 ATP/$2e^-$ (through Complexes III + IV). Regulation of cell respiration as an entity is discussed in Section 1.4.

1.2.6 Reactive Oxygen Species

Partially reduced oxygen species, such as superoxide anion ($O_2^{\cdot-}$), hydrogen peroxide (H_2O_2), hydroxyl radical ($OH^{\cdot}$), peroxyl radical ($ROO^{\cdot}$), and nitrogen oxide radical ($NO^{\cdot}$), are reactive. These ROS can be produced under certain conditions in mitochondria, peroxisomes, and the plasma membrane. As reactive species, they are short lived and rapidly converted to less reactive products (Brand 2010).

In mitochondria, the semiquinones of FMN and ubiquinone in Complex I and the ubisemiquinone-binding Q_o pocket in Complex III are the main O_2-producing sites (Kussmaul and Hirst 2006; Ohnishi et al. 2005). There is evidence showing that Complex III is the biggest source of $O_2^{\cdot-}$ (Lanciano et al. 2013), but even Complex II is able to generate superoxide, particularly during reversed electron flow to Complex I (Muller et al. 2008). It is remarkable that Complex IV, the main physiological oxygen-reducing constituent, is capable of the four-electron reduction of O_2 to $2H_2O$ without any release of partially reduced oxygen species. The mitochondria-produced superoxide is converted to hydrogen peroxide by means of mitochondrial (Mn^{2+}-containing) or cytosolic (Zn^{2+}-, Cu^{2+}-containing) superoxide dismutases (SODs).

ROS formation has been implicated in metabolic regulation, initiation of apoptotic and necrotic cell death, and ischemic/anoxic tissue damage or protection (Circu and Aw 2010). The multitude of detection methods for superoxide has brought up a paradoxical consensus of ROS formation during complete ischemia, but some of the paradoxes seem to result from erratic probe behavior (for references, see Näpänkangas et al. 2012).

1.3 Intermediary Energy Metabolism

1.3.1 Carbohydrate Oxidation

1.3.1.1 Glycolysis

Energy conversion from glucose starts with the cytosolic process of glycolysis, which provides a net yield of two molecules of ATP per one molecule of glucose according to the following net reaction:

$$\text{Glucose} + 2\text{NAD}^+ + 2\text{ADP} + 2\text{P}_i \rightarrow 2\text{ATP} + 2\text{H}_2\text{O} + 2\text{NADH} + 2\text{pyruvate} + 2\text{H}^+$$

Catabolism of glucose is initiated by phosphorylation performed by ubiquitous hexokinase or hepatic glucokinase with the formation of glucose-6-phosphate and consumption of one ATP molecule. After isomerization by phosphohexose isomerase to fructose-6-phosphate, one more ATP molecule is consumed in the 6-phosphofructo-1-kinase (phosphofructokinase-1 [PFK1]) reaction to produce fructose-1,6-bisphosphate. The PFK1 has three different isomorphs in mammals and various oligomeric states with dimer being the smallest active form. It is considered the rate-limiting enzyme of glycolysis with a feedback control from mitochondria. The regulation is enhanced by hormonal control mediated through 6-phosphofructo-2-kinase/phosphatase (PFK2/PFP), a bifunctional enzyme, which adjusts the concentration of fructose-2,6-bisphosphate, the most potent activator of PFK1. The PFK1 also has several adenine nucleotide binding sites where AMP and ADP act as activators, while ATP and ADP serve as inhibitors. At least the heart enzyme can be phosphorylated by cAMP-dependent PKA, which makes the enzyme less sensitive to inhibition by ATP and citrate with fructose-2,6-bisphosphate-dependent activities (Narabayashi et al. 1985).

PFK2 can be phosphorylated by 5′-AMP-activated protein kinase (AMPK) to its active form, so that the glycolysis rate becomes linked to the cellular energy state. AMP is a sensitive reporter of the ATP/ADP ratio because of the equilibrium poise of the adenylate kinase reaction ATP + AMP ↔ 2 ADP, where $[\text{AMP}] = [\text{ADP}]^2/([\text{ATP}]\cdot K)$ so that the AMP concentration is proportional to the square of ADP concentration. The ADP, whose concentration is much higher than that of AMP, is bound to AMPK. However, due to the amplifier action of adenylate kinase, the fractional changes in [AMP] would be larger than that of [ADP]. For a review on AMPK, see Carling et al. (2012).

It is remarkable that the brain neurons have very low activities of PFK2/PFP, which limits their capacity to generate ATP glycolytically during the lack of oxygen and predisposes them to apoptosis in anoxia. However, astrocytes in brain produce the lactate during anoxia, and it has been recently suggested that lactate is a major oxidative substrate in neurons, in which glucose is mainly metabolized through the pentose phosphate pathway (PPP) to supply NADPH for the protection of these postmitotic cells from oxidative damage (see also Section 1.3.2). Still, the neurons are critically dependent on glucose (for references, see Bolaños et al. 2009).

The 1,6-bisphosphate produced in the PFK1 reaction is converted to two glyceraldehyde 3-phosphate (also known as triose phosphates) molecules. Through oxidative phosphorylation, catalyzed by the glyceraldehyde-3-phosphate dehydrogenase, glyceraldehyde 3-phosphate is converted to 1,3-bisphosphoglycerate by using NAD^+ as the electron acceptor. This 1,3-bisphosphoglycerate is an energy-rich compound, which can be used to generate ATP in two subsequent kinase reactions that yield pyruvate as a product. Thus, four ATP molecules are generated after two molecules of ATP are used for glucose phosphorylations, so that the ATP gain in glycolysis is two per glucose.

For continual running of glycolysis, however, NADH must be reoxidized. Under anaerobic conditions, such reoxidation of NADH occurs by using pyruvate as the oxidant in the lactate dehydrogenase (LDH)-catalyzed reaction resulting in lactate accumulation. Pyruvate reduction to lactate is a metabolic dead end, so that lactate elimination must proceed through dehydrogenation to pyruvate for further metabolism, which necessitates other means for NADH reoxidation. Under aerobic conditions, lactate accumulation is avoided by directing the reducing equivalents of NADH to mitochondrial oxidative phosphorylation. The MIM is not permeable to NADH or NAD^+, so that the hydrogens must be conveyed by means of NAD^+-linked substrate cycles. In most tissues, this is provided by the malate–aspartate "shuttle" employing cytosolic and mitochondrial malate dehydrogenases (MDHs) and aspartate aminotransferases (ASAT), respectively, the mitochondrial glutamate/aspartate exchange carrier (Palmieri et al. 2001), and the 2-oxoglutarate/malate carrier (Runswick et al. 1990). Transaminations are necessary in the cycle because MDHs are employed for accepting and donating reducing equivalents. Namely, oxaloacetate penetrates the MIM poorly. The glutamate/aspartate exchange carrier is rather peculiar. It has calcium-binding isomorphs (citrin and aralar), which have a calcium-binding activation site in their N-terminal domain. As a result, it reaches the IMS and thus is able to mediate the signal of extramitochondrial calcium to the mitochondrion. The calcium-binding isomorphs are expressed particularly in excitable cells (Palmieri et al. 2001), where the malate/aspartate shuttle becomes Ca^{2+} activated (Contreras and Satrústegui 2009). The cycle appears symmetric, but in fact, the transfer of reduced equivalents is uphill a thermodynamic gradient, because the mitochondrial free NADH/NAD^+ pool is much more reduced than the cytosolic one. The cycle itself is therefore endergonic, but because the carrier swaps anionic aspartate for protonated glutamate, the glutamate/aspartate carrier is electrogenic and driven by both ΔpH and $\Delta\psi$. In fact, the malate–aspartate shuttle is in near equilibrium with the proton motive force. In an isolated perfused rat heart, the redox potentials of the cytosolic and mitochondrial NADH/NAD^+ couples are −226 and −354 mV, respectively. The difference (128 mV) is not far from the 155 mV proton motive force of the MIM (Hassinen 1986).

In some tissues, reoxidation of cytosolic NADH is arranged with the glycerol-3-phoshate shuttle, which feeds into the mitochondrial ubiquinone pool as described earlier in connection with the glycerol-3-phosphate–ubiquinone oxidoreductase (Section 1.2.3.3). However, even in the pancreatic β-cells, which have high mitochondrial glycerophosphate oxidase activity, the malate–aspartate shuttle is more important than the glycerophosphate shuttle (Eto et al. 1999). Glycolysis is intricately regulated by feedback from the mitochondrial energy state, mainly at the PFK step.

1.3.1.2 Redox Probing of Glycolysis

Cellular oxidation–reduction states can be defined as the redox states of cytosolic and mitochondrial NADH/NAD^+ pair. In principle, NADH can be employed as an endogenous fluorescent probe for cellular redox state. There is, however, a problem imposed by the inequality of the cytosolic and mitochondrial NADH/NAD^+ ratios. The cytosolic free NADH/NAD^+ ratio calculated from the [lactate]/[pyruvate] ratio is low, and particularly in the heart muscle, cytosolic NADH fluorescence is swamped by mitochondrial NADH fluorescence (Chapman 1972). Indeed, during titrations with rising [pyruvate]/[lactate] ratios, the NADH fluorescence of isolated perfused rat hearts and livers shifts to opposite direction: an increase is observed in the heart muscle and a decrease in the liver, which indicates that majority of the myocardial NADH fluorescence originates from mitochondria (Nuutinen 1984). For the estimation of compartment-specific cytosolic NADH/NAD^+ ratios, metabolites of compartment-confined near-equilibrium enzymes, such as LDH, must be used as indicators (Bücher et al. 1972). The fluorescence signal can be calibrated against the metabolite redox ratio even in circumstances where the fully oxidized and fully reduced states cannot be achieved, as shown by Bücher et al. (1972).

LDH is established as a cytosolic enzyme with only a few exceptions (e.g., the mitochondrial LDH of spermatozoa). However, the paradigm of a strictly cytosolic LDH may need reevaluation because of recent claims of mitochondrial LDH in some tissues (Pizzuto et al. 2012).

1.3.1.3 Pyruvate Oxidation

Under aerobic conditions, pyruvate produced by glycolysis is oxidized in the mitochondria. It permeates the MIM by means of free diffusion, a monocarboxylate translocator, or a specific pyruvate carrier (Hildyard et al. 2005). It was recently shown by means of MIM proteomics and gene silencing that a specific mitochondrial pyruvate carrier (MPC) is necessary and is composed of two 15 kDa proteins Mpc1 and Mpc2, which form a 150 kDa functional complex in mammals. Moreover, the study of three families with lactate acidosis identified the carriers of point mutation in a conserved Mpc1 amino acid residue (Bricker et al. 2012). The MPC is evidently able to translocate also some other 2-oxo-carboxylic acids (Paradies and Papa 1975).

In plasma membrane, a family of eight monocarboxylate carrier isoforms with distinctive tissue and substrate specificities exists, which allow for the exchange of lactate and pyruvate molecules between organs and tissues. Pyruvate is oxidized to acetyl-coA in the PDH reaction:

$$\text{Pyruvate} + \text{CoA-SH} + \text{NAD}^+ \rightarrow \text{acetyl-CoA} + \text{NADH} + \text{H}^+ + \text{CO}_2$$

The decarboxylating dehydrogenation makes the reaction exergonic and produces a high-energy thioester in addition to NADH. PDH (or PDC) is a 9.5 MDa multienzyme complex consisting of a PDH (E1), dihydrolipoyl transacetylase (E2), and dihydrolipoyl dehydrogenase (E3), which has FAD as the prosthetic group. Lipoic acid is bound to a lysyl residue of the E2 component, whereas E1 uses thiamin diphosphate (TDP) as a cofactor. In the decarboxylation step, pyruvate is transformed to a hydroxyethyl derivative of TDP that reacts with the covalently bound and oxidized lipoamide in the E2 unit producing an acetyl-lipoamide moiety, which then reacts with CoA-SH to form a free acetyl-CoA. E3 catalyzes the oxidation of dihydrolipoamide in the presence of NAD^+. The core moiety of PDC is composed of 60 E1 units and 12 E3 binding proteins (E3BPs) (Jordan and Cronan 2002; Vijayakrishnan et al. 2010).

Irreversible enzyme reactions are often metabolic pacemakers and must be subject to efficient regulation. In the case of PDH, both product concentrations and covalent enzyme modification are used for this purpose. PDH exists in an inactive, phosphorylated form (PDH_b) and an active, dephosphorylated form (PDH_a). Four different PDH kinases (PDKs) with tissue-specific expression, targeted to three different phosphorylation sites in E1, are capable of converting PDH_a to its inactive form PDH_b, whereas two isomorphs of PDH phosphatase (PDP) convert PDH_b to the active form. The short-term regulation is based on the inhibition of PDK by pyruvate and activation by acetyl-CoA and NADH. The $NADH/NAD^+$ and acetyl-CoA/CoA ratios regulate the acetylation and reduction of lipoate in E2, which increases the PDK activity (Figure 1.7).

The PDPs are dependent on Mg^{2+} and, because intracellular ATP is complexed with Mg^{2+}, the free concentration of the latter is dependent on ATP and increases upon ATP decrease. When bound to its regulatory subunit, the K_m of the catalytic subunit of PDP1 is 3.5 mM for Mg^{2+}. This means that the PDP activity becomes dependent on the ATP/ADP ratio so that the PDH activity increases upon reduction of the mitochondrial energy state. Ca^{2+} regulates the K_m of PDP1 for Mg^{2+} (Midgley et al. 1987), whereas PDP2 is not Ca^{2+} sensitive. In excitable cells, such as muscle, the action potentials produce sharp free Ca^{2+} spikes. The MIM encompasses an efficient calcium transport system consisting of an electrogenic calcium uniporter and a proton/calcium exchange carrier. Both are energized by the proton motive force. However, they work in opposite directions so that an excessive rise of the matrix calcium concentration by the uniporter action is prevented by calcium export with the proton/calcium exchanger. The net ion fluxes become balanced in such a way that the mitochondrial calcium concentration follows its average in the cytosol. This means that the mitochondrial energy-supplying carbohydrate oxidation is regulated by the same messenger as the energy-consuming processes in the cytosol.

The irreversible PDH reaction is situated at the crossroads of carbohydrate and fatty acid metabolism. As a result, the corresponding multitude of the regulatory links contributes to the mutual regulation of carbohydrate oxidation, lipogenesis, and fatty acid oxidation.

1.3.1.4 Biomarker Aspects of Pyruvate Oxidation Machinery

The dihydrolipoyl dehydrogenase enzyme (E3) contains FAD as its fluorescent prosthetic group that has a midpotential of −305 mV at pH 7.4 (Hassinen and Chance 1968). It is noteworthy that it is in rapid equilibrium with the matrix free $NADH/NAD^+$ pool having an $E_{m(7.4)}$ of −320 mV. This means that the $FADH_2/FAD$ ratio closely follows the $NADH/NAD^+$ ratio, so that the prominent flavin fluorescence can be employed as a redox indicator of the mitochondrial matrix. This close reciprocity gives also

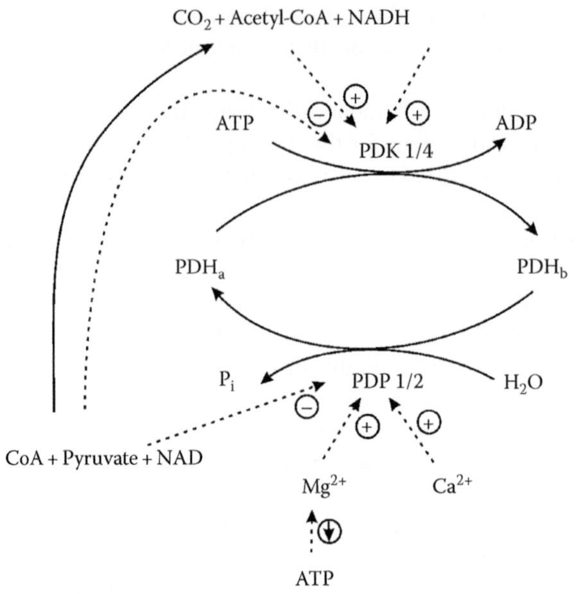

FIGURE 1.7 Regulation of PDH. PDH_a, dephosphorylated pyruvate dehydrogenase (active); PDH_b, phosphorylated pyruvate dehydrogenase (inactive); PDK1/4, pyruvate dehydrogenase kinases 1, 2, 3, and 4; PDP1/2, pyruvate dehydrogenase phosphatases 1 and 2. The dashed arrows show regulatory interactions.

the advantage of using the flavin/NADH fluorescence ratio as compensated oxidation–reduction readout (Huang et al. 2002). However, in intact tissues, the fluorescence quenching by internal filtering due to other chromophores at wavelengths used for flavin is not necessarily the same as at those used for NADH. For this reason, "normalized fluorescence ratios," such as NADH/(flavin + NADH) or flavin/(flavin + NADH), have been employed (Li et al. 2007; Xu et al. 2011).

1.3.2 Pentose Phosphate Pathway

In addition to glycolysis, glucose-6-phosphate can take an alternative, markedly different oxidative route, namely, the PPP, which provides ribose precursors for nucleotide synthesis and reduces power in the form of NADPH for reductive syntheses (e.g., lipogenesis). It also participates in the regulation of cellular growth and protection against oxidative damage. The first, committed step of PPP is catalyzed by glucose-6-phosphate dehydrogenase according to this reaction:

$$\text{Glucose-6-phosphate} + NADP^+ \rightarrow \text{6-Phosphogluconolactone} + NADPH + H^+$$

The product of this reaction, 6-phosphogluconolactone, is hydrolyzed by means of the 6-phosphogluconolactonase to 6-phosphogluconate, which reacts in the decarboxylating, NADP-linked 6-phosphogluconate dehydrogenase reaction:

$$\text{6-Phosphogluconate} + NADP^+ \rightarrow \text{Ribulose-5-phosphate} + CO_2 + NADPH + H^+$$

In addition to this oxidative branch, PPP can be visualized to have a nonoxidative branch, which metabolizes pentose, tetrose, and triose phosphate intermediates back to glucose-6-phosphate in order to enable complete oxidation of glucose-6-phosphate in PPP (for description, see Kruger and von Schaewen 2003).

Glucose-6-phosphate dehydrogenase is the rate-limiting and regulatory enzyme of PPP. The regulation occurs mainly at transcriptional level in parallel with other lipogenic enzymes. Acetyl-CoA carboxylase is the only one of lipogenic enzymes to be under control by covalent modifications.

An additional role for PPP in the antioxidative defense is by conserving thiol groups and reduced glutathione by means of glutathione reductase for converting hydrogen peroxide to water in the glutathione peroxidase reaction. This has also a bearing on the integrity of mitochondria. It is evident that defects in the nonoxidative part of PPP affect the flux through the two NADPH-yielding steps in the oxidative branch, because transaldolase deficiency results in functional defects of mitochondria, such as a reduction of mitochondrial mass and membrane potential (Perl et al. 2006). This is the obvious reason for male infertility and liver damage in transaldolase deficiency. As described earlier, the loss of NADPH or decrease in NADPH/$NADP^+$, oxidative stress, and calcium overload predispose to the mitochondrial permeability transition leading to disintegration of mitochondria and initiation of apoptosis (for references, see Perl et al. 2011).

1.3.3 Fatty Acid Oxidation

1.3.3.1 Compartmentation and Uptake into Cells and Mitochondria

The oxidation of fatty acids has dual compartmentation: major part of it proceeds in mitochondria, whereas prior processing in peroxisomes is necessary in certain cases because of limitations imposed by the substrate specificities of the mitochondrial system. The main oxidative path is the mitochondrial β-oxidation that proceeds with the release of acetyl-CoA (or propionyl-CoA in the last oxidation of odd carbon number fatty acid). The peroxisomal α-, β-, and ω-oxidation systems initiate the mitochondrial system when the fatty acid backbone is not acceptable without prior modification. In α- and ω-oxidations, one carbon from the carboxylic or noncarboxylic end is removed, respectively.

Compartmentation by membrane-permeability barriers necessitates a transmembrane carrier system (Figure 1.8). As a fuel, fatty acids usually reach the cells through a humoral route, and in blood plasma, they are transported as nonesterified "free" fatty acids (FFAs) after lipolysis in adipose tissue. In blood plasma, long-chain FFAs are bound to albumin, and it has been suggested that a plasma membrane albumin receptor operates to enhance the release of the FFA from albumin for transmembrane passage (Hütter et al. 1984). To this effect, a fatty acid translocase (FA/CD36) and a fatty acid–binding protein in the plasma membrane (FABPpm) operate in concert to facilitate the release of FFA from its albumin complex. The fatty acid transport proteins (FATPs 1–6) are integral membrane proteins with an acyl-CoA synthase and a fatty acid uptake carrier activities located in separate moieties of the protein (for a review, see Gimeno 2007). There is also a cytosolic fatty acid–binding protein (FABPc).

To avoid intracellular accumulation of fatty acids in their potentially toxic free form, they are rapidly converted to their CoA esters. They contain a reactive thioester high-energy bond, which is formed by acyl-CoA synthase in a reaction whose irreversibility is ensured by ATP hydrolysis to AMP and pyrophosphate (PP_i). This occurs mainly in the endoplasmic reticulum or the MOM. The breakage of ATP to AMP and PP_i is equivalent to a loss of two molecules to ADP and P_i.

1.3.3.2 Mitochondrial β-Oxidation

Long-chain fatty acyl-CoA cannot penetrate the MIM. Therefore, the acyl-CoAs are converted to carnitine esters by means of carnitine palmitoyltransferase (CPT) in a reversible exchange reaction. The CPTs have dual location: in the MOM (CPT1) and on the inner face of the inner membrane (CPT2) to regenerate acyl-CoA in the matrix space. Thus, the fatty acid penetrates the membrane as a carnitine ester through the carnitine/acylcarnitine translocase located within the inner membrane.

First reaction of the β-oxidation is catalyzed by the acyl-CoA dehydrogenase. The enzyme contains FAD as the prosthetic group donating a pair of electrons to the ETF and oxidizes the acyl-CoA moiety

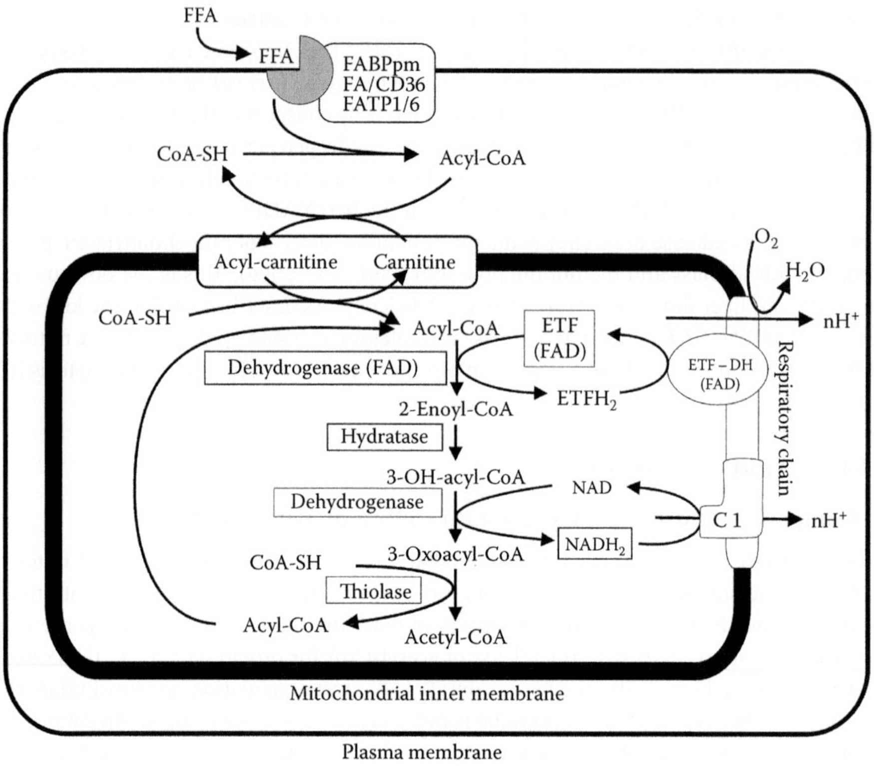

FIGURE 1.8 Mitochondrial fatty acid oxidation and substrate entry into the cell. FBPpm, plasma membrane fatty acid–binding protein; FA/CD36, fatty acid translocase; FATP1/6, fatty acid transport proteins 1–6.

to Δ2-*trans*-enoyl-CoA. ETF, in turn, is a substrate of ETF–ubiquinone oxidoreductase, so that the net reaction is

$$\text{Acyl-CoA} + \text{Ubiquinone} \rightarrow \Delta 2\text{-}\textit{trans}\text{-enoyl-CoA} + \text{Ubiquinol}$$

The three reactions that follow are catalyzed by a trifunctional protein, which has the following sequence of three enzymatic activities: (1) the double bond is removed in the Δ2-*trans*-enoyl-CoA hydratase reaction producing 3-hydroxyacyl-CoA; (2) the 3-hydroxyacyl-CoA dehydrogenase produces 3-ketoacyl-CoA in an NAD^+-linked reaction; (3) the acyl chain is shortened by a C_2 fragment in the thiolase reaction with CoA so that acetyl-CoA is released (for references, see Houten and Wanders 2010).

Thus, one β-oxidation cycle reduces one ubiquinone molecule to ubiquinol, which can be reoxidized by the Complex III + IV chain, and one NAD^+ molecule to NADH, which is reoxidized by the Complex I + III + IV sequence. The average ATP/$2e^-$ ratio for a β-oxidation step is then 2.57, but for a complete fatty acid molecule, the ATP/O ratio depends on its chain length because of the loss of two energy-rich phosphates in the acyl-CoA synthase step.

The regulation of the β-oxidation machinery is not fully understood because the spiral-shaped sequence of its acetyl-CoA production usually runs to completion without accumulation of partially oxidized intermediates. Regulation is possible in the uptake by cells, import to mitochondria, and in the β-oxidation proper. It has been shown that the semiquinone of ETF is a feedback inhibitor of fatty acyl-CoA dehydrogenase (Beckmann and Frerman 1985). NADH inhibits 3-hydroxyacyl-CoA dehydrogenase and acetyl-CoA inhibits thiolase (Schulz 1991). This behavior suggests that β-oxidation is under feedback control by the respiratory chain. However, the substrate preference of mitochondrial oxidations in heart muscle, for example, is greatly influenced by the availability and uptake of FFAs.

To prevent futile cycling, the rates of fatty acid oxidation and synthesis pathways are reciprocally regulated at the level of fatty acid transport to mitochondria. Malonyl-CoA, a product of the first reaction of fatty acid synthesis (see Section 1.3.4 and Figure 1.11), is an allosteric inhibitor of CPT1, which is considered to be important in the regulation of mitochondrial long-chain fatty acid oxidation (McGarry and Foster 1979).

1.3.3.3 Ketogenesis and Ketone Body Oxidation

Hepatic ketogenesis is an overflow process that operates under the conditions of excessive FFA import and leads to mitochondrial fatty acid oxidation rates exceeding the capacity of the TCA cycle to handle acetyl-CoA and the electron transfer rate in the respiratory chain, which is regulated by the ATP consumption. Instead, acetyl-CoA is used for acetoacetyl-CoA synthesis by reversal of the thiolase reaction. Acetoacetyl-CoA is condensed with acetyl-CoA by means of the mitochondrial isoenzyme of 1,3-hydroxy-3-methylglutaryl-CoA (HMG-CoA) synthase to yield 3-hydroxy-3-methyglutaryl-CoA, which is converted by HMG-CoA lyase to acetyl-CoA and free acetoacetate (AcAc). The latter is not metabolized further in the liver except being reduced to β-hydroxybutyrate (β-OHB) with the NAD^+-linked β-hydroxybutyrate dehydrogenase (β-OHBDH). The hepatic mitochondrial redox state poises the reaction toward β-OHB formation. Because of this metabolic dead end and the high hepatic OHBDH activity, the β-OHB/AcAc ratio can be used as an indicator of the redox state of mitochondrial matrix $NADH/NAD^+$ ratio in the liver. In addition to the regulation of fatty acid oxidation at the CPT1 level, HMG-CoA synthase is also a regulator of ketogenesis activated and inactivated by desuccinylation and succinylation, respectively. The ketogenic hormone glucagon lowers the mitochondrial concentration of succinyl-CoA and thus decreases the extent of HMG-CoA synthase succinylation (Quant et al. 1990). Transcription of the gene of the mitochondrial HMG-CoA synthase is enhanced in starvation, a condition that leads to ketosis (for references, see Hegardt 1999).

β-OHB and AcAc traverse membranes through the monocarboxylate transporters 1 and 2 to be metabolized in extrahepatic tissues. AcAc is converted back to AcAc-CoA in the succinyl-CoA–3-ketoacid CoA-transferase-catalyzed exchange with succinyl-CoA and splits to two acetyl-CoAs in the thiolase reaction to be able to enter the TCA cycle. Ketone bodies are of utmost importance in the process of adaptation to fasting, because the brain, which is not able to oxidize long-chain fatty acids, can substitute them for glucose, provided that their concentration is high enough.

1.3.3.4 Peroxisomal Fatty Acid Oxidation

The peroxisomes are vesicular organelles with a single membrane and contain a selection of dehydrogenases and oxidases, which react with molecular oxygen to produce H_2O_2. They also contain the H_2O_2-eliminating enzyme catalase. Fatty acids with excessive chain length must be shortened by means of the peroxisomal fatty acid oxidation system to be able to enter mitochondrial β-oxidation. In addition, long-chain dicarboxylic fatty acids are oxidized in peroxisomes (Figure 1.9).

The fatty acid CoA esters do not require prior conversion to carnitine esters for penetration of the peroxisomal membrane, which is also permeable to NAD^+ and NADH. The peroxisomal acyl-CoA dehydrogenase is a flavoenzyme, which reacts with molecular oxygen producing hydrogen peroxide and Δ2-enoyl-CoA. The following two reactions are catalyzed by one and the same multifunctional enzyme (MFE):

$$\Delta 2\text{-Enoyl-CoA} + H_2O \xrightarrow{\Delta 2\text{-}trans\text{-enoyl-CoA hydratase}} \text{3-OH-acyl-CoA}$$

$$\text{3-OH-acyl-CoA} + \text{NAD} \xrightarrow{\text{3-hydroxyacyl-CoA dehydrogenase}} \text{3-Ketoacyl-CoA} + \text{NADH}$$

Mammalian peroxisomes contain two isomorphs (MFE-1 and MFE-2) of the enzyme, which catalyze the hydratase with mirror-image stereochemistry and have different evolutionary origins. MFE-1 is

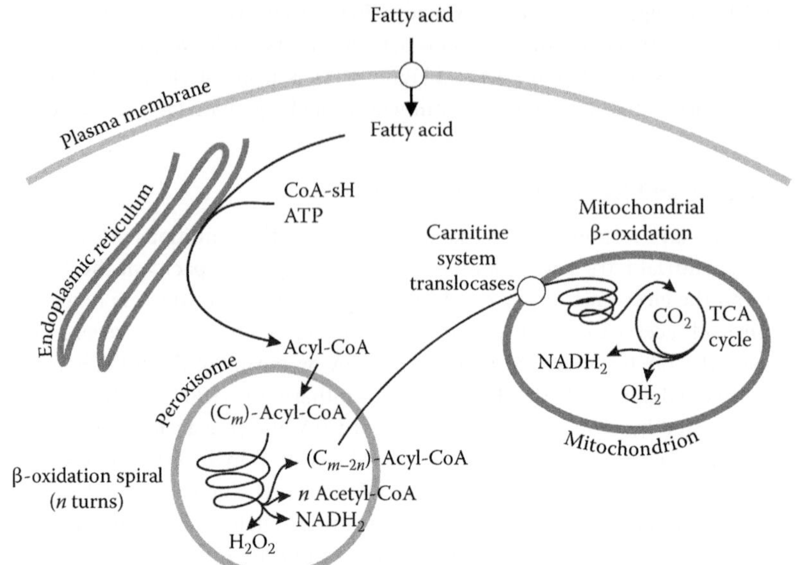

FIGURE 1.9 Collaboration of peroxisomes and mitochondria in metabolism of very long-chain fatty acids. The fatty acids are activated to acyl-CoA, which penetrates the peroxisomal membrane freely, as do also NAD^+ and NADH. After shortening in the peroxisomal β-oxidation, the acyl-CoA enters the mitochondrion to be fully oxidized in the β-oxidation and TCA cycle.

the most abundant one. Otherwise, the peroxisomal β-oxidation is reminiscent of its mitochondrial counterpart.

Branched fatty acid backbones cannot be processed by mitochondrial β-oxidation. Phytanic acid, formed from the phytol side chain of chlorophyll, has a methyl group in the position 3 that must be removed before further degradation. This is accomplished by the peroxisomal α-oxidation that relies on a 2-oxoglutarate- and Fe^{2+}-dependent dioxygenase, phytanoyl-CoA hydroxylase, which converts phytanoyl-CoA into 2-hydroxyphytanoyl-CoA. As a result, CO_2 and succinate are produced with the consumption of one molecule oxygen. Subsequently, a formyl moiety is removed so that the chain is shortened by one carbon atom. The residue is oxidized to pristanic acid, which proceeds in peroxisomal or mitochondrial β-oxidation.

H_2O_2 production constitutes oxidative stress, and for protection, peroxisomes contain catalase, a hydrogen peroxide–eliminating enzyme. Mammalian catalase is an enzyme with Fe(III)-containing protoporphyrin IX (i.e., heme B) as the prosthetic group and catalyzes the dismutation of hydrogen peroxide to water and oxygen in the following catalytic reaction:

$$2H_2O_2 \rightarrow 2H_2O + O_2$$

When the electron donor for H_2O_2 reduction is other than hydrogen peroxide, the reaction can be regarded as peroxidatic:

$$H_2O_2 + 2AH \rightarrow 2H_2O + 2A$$

For catalase, several aliphatic alcohols can perform as peroxidatic substrates to be oxidized to corresponding aldehydes.

Catalase is present in erythrocytes, for which they lend protection in their environment of high oxygen concentration (Gaetani et al. 1996); it was also found in rat liver mitochondria (Salvi et al. 2007).

Catalase was the first enzyme, for which the formation of primary and secondary enzyme-substrate complexes (Compounds I and II) was demonstrated by Britton Chance (1948) using sensitive dual-wavelength, rapid mixing technology. The catalase Compound I has been shown to be an oxoiron(IV) porphyrin π-cation radical species (for references, see Nicholls 2012).

1.3.3.5 Optical Probing of Fatty Acid Oxidation

FAD in isolated ETF is fluorescent. It has an $E_{7.4}$ of −56 mV (Voltti and Hassinen 1978) and fatty acid–induced flavin fluorescence quenching is observable in mitochondrial suspension, cells, and isolated organs. Because β-oxidation is a heavy producer of NADH, even dihydrolipoyl dehydrogenase (E3) is reduced because of its near equilibrium with the NADH/NAD^+ couple (see Section 1.3.1.3). The emission maximum of E3 is redshifted by 27 nm in comparison with ETF (Kunz 1986); on this basis, ETF fluorescence has been used to trace cellular fatty acid oxidation (Lam et al. 2012). However, the fluorescence quenching of ETF plus fatty acyl-CoA dehydrogenase upon reduction, obtained by a fatty acylcarnitine ester in liver mitochondria, is only 5% of that of dihydrolipoyl dehydrogenase upon addition of an NAD^+-linked substrate, when expressed as Δ-fluorescence/Δ-absorbance ratio (Garland et al. 1967). Since the flavin fluorescence emission bands are broad and lack characteristic detail, selective ETF fluorometry in tissues is probably prone to errors. In the isolated perfused heart, flavoprotein fluorescence changes very precisely following the free NADH/NAD^+ ratio (Nuutinen et al. 1981), which suggest that dihydrolipoyl dehydrogenase is the main fluorescent species in that organ. In tissues where the activity of the β-OHBDB is high, the β-OHB/AcAc ratio can be used as an indicator of the mitochondrial free NADH/NAD^+ ratio.

The concentration of catalase Compound I can be easily monitored using spectroscopy even in intact tissues (Oshino et al. 1975). For example, catalase spectroscopy can be used to assess the contribution of peroxisomes to the oxidation of different fatty acids in the liver (Hiltunen et al. 1986; Kärki et al. 1987) or total H_2O_2 production by means of titration with methanol (Hassinen and Kähönen 1974).

1.3.4 Fatty Acid Synthesis

The anabolic route of fatty acid synthesis is considered here in connection with energy-yielding metabolism. This is simply because its enzyme regulation and a metabolite are also involved in fatty acid catabolism. Fatty acid synthesis is a cytosolic process that utilizes acetyl-CoA as a precursor and reducing power in the form of NADPH. Because the simultaneous running of reciprocal pathways will result in an energy-wasting futile cycle, the routes need efficient reciprocal regulation, preferably in an early, committed step of the pathway.

De novo synthesis of fatty acid is initiated from acetyl-CoA that originates from carbohydrate catabolism through PDH in mitochondria. Acetyl-CoA does not easily penetrate the MIM. Therefore, a shuttle-based mechanism compatible with the carrier repertoire of the MIM must be employed. The acetyl backbone can be exported from the matrix space to cytosol after the citrate synthase (CS) step by means of the tricarboxylate translocator as citrate and split into acetyl-CoA and oxaloacetate by means of the cytosolic ATP citrate lyase. Oxaloacetate is reduced by the NAD^+-linked MDH before being further metabolized to pyruvate and CO_2 by the NADP-linked MDH (malic enzyme [ME]) in order to provide NADPH for the fatty acid synthesis. Transport as a carnitine ester is also possible, but the citrate lyase is perhaps more important, because its expression follows that of fatty acid synthase (FAS). The hexose monophosphate pathway provides 60% of the reducing power, while the remaining 40% is provided by the NADPH-linked ME in accord with the role of citrate in acetyl-CoA export.

Acetyl-CoA in the cytosol is converted to malonyl-CoA according to the acetyl-CoA carboxylase reaction:

$$\text{ATP} + \text{acetyl-CoA} + \text{CO}_2 \rightarrow \text{Malonyl-CoA} + \text{ADP} + \text{P}_\text{i}$$

The malonyl-CoA moiety will become covalently bound to FAS, a homodimer of two multifunction proteins with seven different enzyme activities. The enzyme is primed by acetyl-CoA, which will be sequentially lengthened by two carbon units in a condensation reaction with malonyl-CoA with liberation of CO_2 and two subsequent reduction steps per cycle.

The FAS is under transcriptional control and can undergo 20-fold concentration changes (for references, see Jensen-Urstad and Semenkovich 2012). Acetyl-CoA carboxylase is a covalently interconvertible enzyme, which can be phosphorylated to an inactive form by the AMP-activated PKA and activated by a protein phosphatase. Malonyl-CoA is an inhibitor of the mitochondrial palmitoyl carnitinetransferase-1 (CPT1), which is rate limiting in mitochondrial fatty acid oxidation (McGarry and Foster 1979). Thus, malonyl-CoA has been considered as a major regulator—particularly of ketogenesis. All this means that regulations of both fatty acid synthesis and their oxidation are geared to the cellular energy state but in opposite directions.

A mitochondrial pathway for medium-chain length fatty acid synthesis also exists. Mitochondria possess a type II FAS, which is composed of separate monofunctional enzymes that produce mainly octanoate for lipoic acid synthesis, but it is also involved in mitochondrial RNA processing (for a review, see Hiltunen et al. 2010).

1.3.5 Tricarboxylic Acid Cycle

The proximal provider of reducing equivalents to the respiratory chain is the TCA cycle in the mitochondrial matrix (also known as citric acid cycle and Krebs cycle). It is the melting pot of the catabolites of carbohydrates, lipids, and amino acids. Net products of the cycle are carbon dioxide and reducing equivalent for the respiratory chain (Figure 1.10).

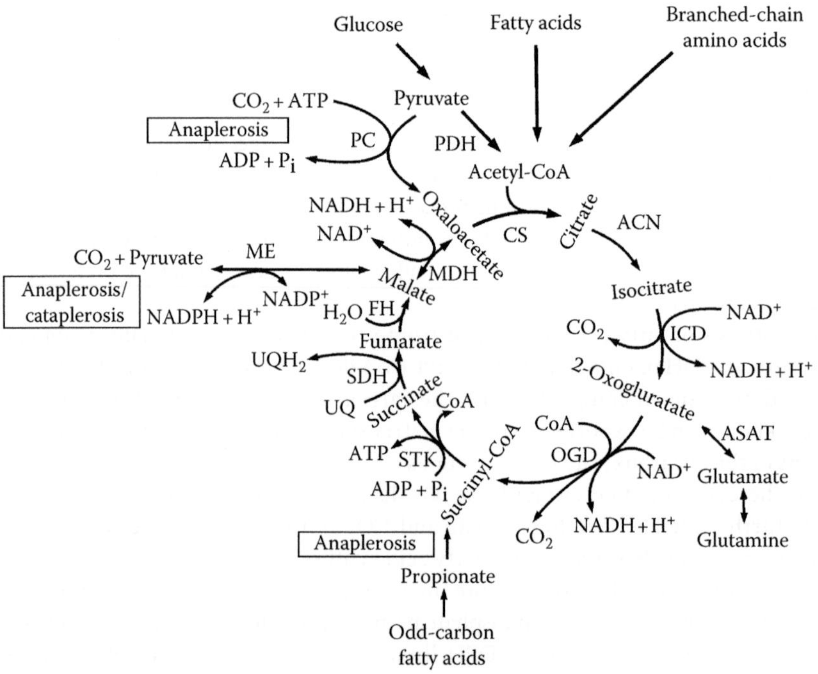

FIGURE 1.10 TCA cycle with connecting pathways related to maintenance of cycle intermediates. CS, citrate synthase; ACN, aconitase; ICD, isocitrate dehydrogenase; OGD, 2-oxoglutarate dehydrogenase; STK, succinate thiokinase; SDH, succinate dehydrogenase; FH, fumarate hydratase; MDH, malate dehydrogenase, PDH, pyruvate dehydrogenase; ASAT, aspartate aminotransferase; ME, malic enzyme; PC, pyruvate carboxylase.

The principal substrate consumed by the TCA cycle is acetyl-CoA. Although a few metabolic routes join the cycle by forming one of its intermediates, this would only increase the cycle pool size in case that the intermediates do not exit the cycle as a metabolite, which can be converted to acetyl-CoA for proper entering. On the other hand, cycle intermediates can be utilized in other pathways, in which case, their pool would be drained if not regulated in the balanced processes of anaplerosis and cataplerosis (for references, see Ala-Rämi et al. 2005).

The following input reaction of CS employs oxaloacetate and acetyl-CoA as precursors:

$$\text{Oxaloacetate} + \text{Acetyl-CoA} \rightarrow \text{Citrate} + \text{CoA}$$

The K_m of oxaloacetate, the product of the MDH reaction, is within the range of its mitochondrial concentration (Siess et al. 1984). This means that under certain conditions, oxaloacetate concentration becomes a limiting factor. An extreme situation is the oxidation of ethanol to acetate that takes place in the liver and produces high amounts of NADH. In this case, the equilibrium state of MDH reaction shifts heavily toward reduction so that oxaloacetate concentration becomes limiting. Under these conditions, the respiratory chain is fully fuelled by NADH, although the TCA cycle comes to halt as evidenced by unaltered oxygen consumption and stop of CO_2 formation (Forsander 1967).

Citrate is converted to isocitrate by means of aconitase (ACN) (aconitate hydratase), a metalloenzyme that contains a tetranuclear iron–sulfur cluster (4Fe–4S). In this reaction, citrate is first dehydrated to *cis*-aconitate and then hydrated to isocitrate:

$$\text{Citrate} \leftrightarrow \textit{cis}\text{-Aconitate} \leftrightarrow \text{Isocitrate}$$

ACN has also a cytosolic isoform that has an important function as an iron regulatory protein (IRP1), in which formation of an iron–sulfur cluster has a central role (for references, see Anderson et al. 2012). Citrate not only acts as a "catalyst" in the TCA cycle but is also a carbon conduit from mitochondrial acetyl-CoA to cytosolic fatty acid synthetase system. Although the exported citrate in cytosol is mainly metabolized by means of citrate lyase, part of it is converted to isocitrate to be used as the substrate of cytosolic NADP-linked isocitrate dehydrogenase (ICD), so that cytosolic ACN activity is needed.

The NAD^+-linked mitochondrial ICD is a central enzyme in the TCA cycle because of its regulatory properties that fine-tune the cycle activity to match ATP consumption and oxidative phosphorylation:

$$\text{Isocitrate} + \text{NAD}^+ \rightarrow \text{2-Oxoglutarate} + \text{NADH} + \text{H}^+$$

The ICD activity is enhanced by ADP and inhibited by ATP, NADH, and NADPH (Gabriel et al. 1986). The enzyme must also bind ADP because it binds Mg-isocitrate. ICD is also activated by calcium, which lowers its K_m for substrate in the presence of ADP; the calcium effect vanishes at high substrate concentration. The ADP/ATP ratio is a more effective regulator than calcium. The regulation is complex, because the substrate also competes with the adenine nucleotides for Mg^{2+} and Ca^{2+}. Activation by Mg^{2+} suggests that mitochondrial Mg^{2+} concentration is also an important regulator (see the role of free Mg^{2+} in the regulation of PDH).

2-Oxoglutarate undergoes oxidative decarboxylation catalyzed by 2-oxoglutarate dehydrogenase (OGD) in a reaction yielding NADH and a high-energy thioester compound succinyl-CoA:

$$\text{2-Oxoglutarate} + \text{CoA-SH} + \text{NAD}^+ \rightarrow \text{Succinyl-CoA} + \text{CO}_2 + \text{NADH} + \text{H}^+$$

The reaction is analogous to the PDH reaction (described in Section 1.3.1.3) and has a similar multienzyme assembly: OGD (E1), dihydrolipoyl succinyl transferase (E2), and dihydrolipoyl dehydrogenase (E3). The conglomerate has a core of 24 E2 units, around which the E1 and E3 components are attached (McLain et al. 2011). Although the enzyme complex is analogous to the PDH, no evidence yet exists for covalent modification of the OGD (Perham 1991).

The enzyme is under efficient control—the substrates and effectors show cooperative interactions as expected from the oligomeric nature of the enzyme. All three component enzymes display product inhibition. The lipoyl/dihydrolipoyl moiety of the enzyme provides a connection to the cellular S–S/SH balance. An increase in the NADH/NAD$^+$ and succinyl-CoA/CoA-SH ratios inhibits the enzyme, whereas ADP and calcium are an activator by increasing the affinity of the enzyme to 2-oxoglutarate.

Succinyl-CoA is converted to free succinate by means of two isoforms of succinyl-CoA synthetase—SCS-A and SCS-G (Johnson et al. 1998):

$$\text{Succinyl-CoA} + \text{ADP} + \text{P}_\text{i} \xrightarrow{\text{SCS-A}} \text{Succinate} + \text{CoA-SH} + \text{ADP}$$

$$\text{Succinyl-CoA} + \text{GDP} + \text{P}_\text{i} \xrightarrow{\text{SCS-G}} \text{Succinate} + \text{CoA-SH} + \text{GTP}$$

The enzyme is composed of two dissimilar subunits (α and β), of which the α-subunit is shared in both isoforms. The β-subunit has two isoforms, which determine the nucleotide specificity of the enzyme, and the expression ratio of the ATP- and guanosine triphosphate (GTP)-specific isoforms depends on the organ. The SCS-A β-subunit is encoded by the *SUCLA2* gene, whose mutations have been shown to result in mitochondrial disease characterized by methylmalonic aciduria, multiple mtDNA deletions, and neurological symptoms. This was suggested to be related to a close link between SCS-A and nucleoside diphosphate kinase, which is necessary for dNTP replenishment during mtDNA synthesis. Thus, part of the combustion energy of 2-oxoglutarate, which was conserved in the thioester bond of succinyl-CoA, is in the SCS-A reaction converted to ATP.

The next TCA cycle reaction produces fumarate in the succinate–ubiquinone oxidoreductase, which donates electrons to the mitochondrial respiratory chain.

Fumarate is converted to malate in the fumarate hydratase (FH) reaction. Defects caused by homozygous mutations in succinate–ubiquinone oxidoreductase subunits or FH lead to neurodegeneration. Heterozygous defects, however, may predispose to a cancer onset and lead to leiomyomas, renal cell cancer, and neurological deficit. The cancerous trend has been suggested to be caused by a "pseudohypoxia" and further stabilization of the hypoxia-induced transcription factors HIF-1α and HIF-2α, which regulate antiapoptotic and proliferative genes (for references, see Eng et al. 2003).

Malate is converted to oxaloacetate in the NAD$^+$-linked MDH reaction. As pointed out earlier in connection with the CS enzyme, the MDH reaction is in near equilibrium with the mitochondrial NADH/NAD$^+$ pool. As a result, oxaloacetate concentration and the rate of the CS reaction become dependent on the NADH/NAD$^+$ ratio.

Malate can also be converted to pyruvate by the decarboxylating NAD(P)$^+$-linked ME, which is probably connected to adjustment of the pool size in the TCA cycle by replenishment or elimination of its intermediates (Sundqvist et al. 1987, 1989). In human liver, however, 90% of the ME is extramitochondrial (Zelewski and Swierczynski 1991), in accord with its role in production of reducing power for fatty acid synthesis.

The dihydrolipoyl dehydrogenase component of OGD is an NAD$^+$-linked, FAD-containing enzyme and displays flavin fluorescence applicable to compartment-specific monitoring of the redox state of the NADH/NAD$^+$ couple in mitochondrial matrix.

1.3.6 Amino Acids

The catabolism of amino acids typically takes place through amphibolic intermediates, which can be used for energy-yielding purposes or in the anabolic processes of gluconeogenesis and lipogenesis. Only some aspects relevant to energy metabolism are discussed here.

The backbones of aspartate and glutamate can be converted by means of transamination to oxaloacetate or 2-oxoglutarate but only at the expense of forming another amino acid, which must be catabolized further. Because 2-oxoglutarate and oxaloacetate are intermediates in the TCA cycle, glutamate- or aspartate-linked transaminations affect the TCA cycle pool size.

Because of its high concentration and preferential location in mitochondria, glutamate is a potential marker for studying the substrate preference of oxidative energy metabolism by means of ^{13}C isotopomer analysis of the labeling pattern of 2-oxoglutarate in intact tissues (Malloy et al. 1988). This is because of the poise of the glutamate/aspartate translocator and the high activity of the mitochondrial glutamate/2-oxoglutarate-linked transaminases. The method is based on the fact that the labeling of individual carbons of glutamate can be deduced from its ^{13}C nuclear magnetic resonance (NMR) spectrum without any purification or destructive chemistry. Because of the transaminase equilibrium, the glutamate labeling pattern reflects that of the TCA cycle intermediate 2-oxoglutarate. During the metabolism of carbohydrates and fatty acids that are specifically ^{13}C labeled to a single carbon atom, the fractional contributions of the pathways feeding into the TCA cycle can be calculated (Ala-Rämi et al. 2005).

After an initial transamination to oxoacids, branched-chain amino acids are oxidized by the mitochondrial branched-chain 2-oxoacid dehydrogenase, which is a heterooligomeric complex consisting of a decarboxylase unit (E1), dihydrolipoyl transacetylase unit (E2), and dihydrolipoyl dehydrogenase (E3) unit, which has FAD as the prosthetic group. Thus, the product acetyl-CoA enters the TCA cycle. Similar to PDC, the branched-chain 2-oxoacid dehydrogenase is regulated by means of phosphorylation/dephosphorylation. It is worth noting that mutation in the branched-chain ketoacid dehydrogenase is the underlying cause for the maple syrup urine disease (for references, see Brunetti-Pierri et al. 2011).

The catabolism of methionine, isoleucine, and valine produces propionyl-CoA, which is transformed to methylmalonyl-CoA by propionyl-CoA carboxylase in a biotin-dependent reaction. This is converted to succinyl-CoA and expands the pool of the TCA cycle intermediates (Figure 1.10).

The branched-chain 2-oxoacid dehydrogenase is closely reminiscent of and mechanistically similar to the 2-oxoacid dehydrogenases that deliver electrons to NAD^+ by mediation of the FAD-containing dihydrolipoyl dehydrogenase. The latter, together with NADH, forms a pair of reciprocal fluorescence emission during changes in redox state and is useful in compartment-specific monitoring of mitochondrial activities in tissues. Glutamate can be employed in ^{13}C tracer isotopomer studies as a probe for label incorporation into specific carbon atoms of 2-oxoglutarate (Malloy et al. 1988). The analysis can be performed by direct ^{13}C NMR spectrometry of tissue extracts without their prior purification.

1.3.7 Oxidative Fuel Selection

Selection of fuels for energy conversion is an important aspect of nutrition, particularly because the interconversions between main substrate and store categories can be unidirectional and some organs are critically dependent on a certain substrate. To cope with these constraints, there are organ-specific rules for oxidative fuel consumption. The most appreciated is the competition between carbohydrate and fatty acid oxidation.

In 1963, Philip Randle, Peter Garland, Nick Hales, and Eric Newsholme proposed the "glucose–fatty acid cycle" to emphasize the mutual regulation of the oxidation of these substrate classes (reviewed in Randle 1998). In fact, the interaction is not cyclic but rather very tight and involves both metabolite and hormonal control. Mutual regulation mainly saves glucose for tissues that are not responsible to insulin and not capable of fatty acid oxidation. This cycle reigns particularly between adipose tissue, muscle, and liver. In the muscle, the fatty acids and/or ketone bodies derived from them suppress glucose and pyruvate oxidation.

The main regulatory mechanisms have been described in Sections 1.3.1.3 and 1.3.3. If present, the fatty acids dominate the biological oxidation reactions because of their excessive reducing power and acetyl-CoA production rates. The glucose–fatty acid cycle has also a bearing in adaptation to fasting, because the brain is not capable of oxidizing fatty acids. Here, the hepatic ketogenesis during high fatty acid oxidation rates comes to rescue, since the brain is able to use the ketone bodies as a fuel that can

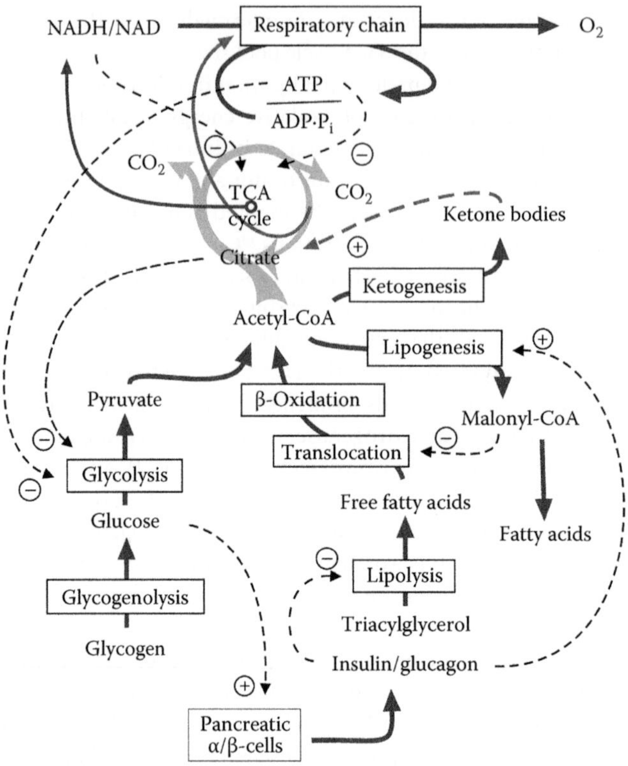

FIGURE 1.11 Interactions of the main pathways of energy-yielding metabolism. Dashed arrows indicate regulatory effects.

be oxidized at high enough concentration. In vivo, the mutual regulation between fatty acids and carbohydrates becomes reciprocal because of the involvement of insulin (Figure 1.11). Glucose stimulates the secretion of insulin, which decreases lipolysis with concomitant decrease in fatty acid oxidation and increase in lipogenesis, which provide a switch in fuel preference toward carbohydrates.

1.4 Flux Regulation of Terminal and Intermediary Energy Metabolism

Because the duty of energy-yielding metabolism is ATP production, the size of ATP reserves acts as its major regulator. The short-term storage of ATP energy is best expressed as the [ATP]/[ADP] ratio or [ATP]/([ADP]·[P_i]) ratio, because the free energy available is dependent on how much the ATP hydrolysis deviates from its equilibrium state.

Mitochondrial respiration is characterized by the “respiratory control,” which is numerically expressed as the respiratory control ratio—the ratio of oxygen consumption rates in the presence (or absence) of P_i and ADP (Lardy and Wellman 1952). At the whole organ level, this respiratory control comprises an immense regulatory network. In general, regulation is effectively focused on irreversible reactions, but a multitude of regulatory mechanisms are needed to prevent unnecessary accumulation of metabolic intermediates. In general, the regulation of cell respiration can be subdivided into (1) control over availability of ADP and P_i; (2) control by substrate supply, that is, provision of reducing equivalents in the form of NADH or ubiquinol; (3) substrate level control over dehydrogenases in the mitochondrial matrix; (4) control of cytochrome *c* oxidase through its reduction grade determined by some equilibrium reactions in respiratory Complexes I and III; (5) allosteric or covalent regulation of cytochrome *c* oxidase or Complex I; and (6) control through gene regulation of the respiratory enzymes or mitochondrial biogenesis. An important burden of substrate level regulation is to prevent excessive reduction of the low-potential end of the respiratory chain. This in return prevents leakage of reducing equivalents, which leads to the formation of partially reduced oxygen species and ROS.

1.4.1 NADH Limitation

The redox state of the NADH/NAD^+ couple determines the Gibbs free energy across the respiratory chain, that is, the driving force of oxygen reduction. There is at least one irreversible reaction in the respiratory chain and the cytochrome *c* oxidase reaction. Linear irreversible thermodynamics predicts that the reaction rate is proportional to the driving force (Westerhoff and van Dam 1987). The effect of the redox state of the NADH/NAD^+ couple may be generated in an indirect way. For example, there is experimental evidence that suggests a near equilibrium over the stretch between Complex I and Complex III and the [ATP]/([ADP]·[P_i]) system. This means that during unchanged [ATP]/([ADP]·[P_i]) ratio, reduction of NAD^+ leads to an increase in concentration of ferrocytochrome *c*, the cosubstrate of cytochrome *c* oxidase (Wilson et al. 1974), which would increase the oxygen consumption rate. Due to the near equilibria in the mitochondrial chain, the oxygen consumption rate would be indefinite without control of the mitochondrial free NADH/NAD^+ ratio. Optical readout data from isolated, perfused heart are in accord with the near-equilibrium hypothesis (Hassinen and Hiltunen 1975).

1.4.2 Calcium as a Regulator

As described earlier, calcium is involved in several processes related to cellular respiration. For example, it is involved in the covalent modification of PDC through activation of the PDPs by regulating the K_m-value of PDP1 for Mg^{2+}. As a result, PDH is converted to its active, dephosphorylated form upon the increase of free Ca^{2+} concentration. The ICD and OGD enzymes are also activated by calcium. In spite of the close similarity of the reaction mechanism and subunit structure of for the latter enzyme with PDH, the Ca^{2+} effect is not mediated by covalent interconversions (see Sections 1.3.1.2 and 1.3.5). As a second messenger, calcium is an initiator of several ATP-consuming processes such as muscular contraction and exocytosis. Therefore, calcium activation of NADH generation processes in mitochondria provides a feedforward system of regulating ATP production.

1.5 Mitochondrial Energy State and Regulation of Cell Respiration

The mitochondrion-based mechanism for regulating cell respiration is strictly focused on the respiratory chain and its oxygen-consuming terminal reaction, cytochrome *c* oxidase. The nomenclature used for describing the phenomenon of respiratory control is defined in terms of ATP demands as well as the availability of ADP as a phosphate acceptor. The mitochondrial availability of ADP is influenced by the ANT, and because this translocator is electrogenic, the poise of the ADP/ATP ratio becomes linked to the mitochondrial membrane potential. The phosphate translocator is a P_i^-/OH^- exchanger and becomes coupled to the ΔpH (Palmieri et al. 1993).

ANT is an abundant protein, but experiments with its inhibitor atractyloside allow probing of its regulatory role. The controlling power of a reaction in a given metabolic path can be evaluated by using the principle defined by Kacser, where the fractions of total control are given to each of the enzymes in the sequence (Kacser and Burns 1995). Analysis and modeling of skeletal muscle mitochondria yield an estimated flux control coefficient of 0.14 for ANT, when defined as the ratio of fractional flux change to fractional enzyme concentration change, where the sum of coefficients over the whole pathway is unity (Kacser and Burns 1995).

It is paradoxical that during phenylephrine-induced work transitions, the concentrations of phosphocreatine, ATP, and P_i in the myocardium remain almost constant (Katz et al. 1989). This may suggest that the respiratory chain is extremely sensitive to these potential regulators. Indeed, it has been shown that the rate of oxidative phosphorylation does not follow simple Michaelis–Menten kinetics but is at least of second order with respect to ADP (Jeneson et al. 1996). So it is likely that an amplification mechanism exists. However, there are other mechanisms, such as the calcium stimulation of the TCA cycle, that contribute to the relative stability of the energy state in the muscle. In modeling muscle metabolism, a concept of "each-step activation" has also been proposed as an explanation of the great sensitivity of metabolism to the energy demand. The model postulates a common signal molecule influencing all steps of a pathway (Liguzinski and Korzeniewski 2006).

Indeed, the control of the respiratory chain is at least partially distributed, even without postulating a common signal molecule. Complex I is subject to cAMP-dependent activation (Papa et al. 1996), and Complex IV is under allosteric inhibitory control by ATP (Arnold and Kadenbach 1997), which in turn can be alleviated by phosphorylation by a PKA-mediated mechanism (Acin-Perez et al. 2011). The earlier-described energy-linked regulation of NADH production in dehydrogenations at the substrate level, superimposed to the link of the irreversible cytochrome *c* oxidase reaction to the ATP/(ADP·P_i) ratio, indicates that the phenomenon of respiratory control is efficiently layered at cellular level.

Oxygen consumption can also be regulated independently of the energy requirements to adjust heat production. For this purpose, mitochondria have a provision to regulate the coupling between redox-driven proton pumping and ATP synthesis by short-circuiting the transmembrane proton cycle in the phenomenon of uncoupling with protonophores. Physiologically, this is accomplished by means of the uncoupling protein (UCP1), the expression of which is prominent in brown adipose tissue and stimulated by cold exposure. In the MIM, UCP1 represents a potential proton conductance channel, which is kept closed in its resting state by purine nucleotides (ATP), but the presence of FFAs alleviates this purine nucleotide interaction opening the channel (for a review, see Nedergaard and Cannon 2010). UCP homologs are expressed also in other tissues where they may have other functions (for references, see Ledesma et al. 2002).

1.6 Concluding Remarks

Oxidative energy metabolism, which takes place predominantly in mitochondria, is amenable to spectroscopic monitoring because of the favorable properties of its redox carriers. Many of them (cytochromes and other hemoproteins, flavoproteins, and nicotinamide nucleotides) contain chromophores that absorb in the visible, UV, or IR range. Some of the carriers (flavoproteins and nicotinamide nucleotides) are fluorescent with discernible, sufficiently unique excitation and emission bands allowing their use as natural biomarkers for monitoring the oxidation–reduction state in intact cells or organs. They may be confined to a specific cell compartment enabling their use as compartment-specific optical redox probes. Particularly useful are those natural probes that can be monitored using long wavelengths in transmission spectrophotometry for deep-tissue penetration (e.g., cytochrome *c* oxidase). For those biomarkers, which absorb in the UV region (e.g., NADPH), two-photon excitation in near IR proved to be useful for deep-tissue imaging while avoiding UV-based damages (discussed in details in Chapter 4). These new spectroscopy and microscopy methods show promise for nondestructive in vivo monitoring of cellular redox state and metabolism (Vielhaber et al. 1999; Yu and Heikal 2009). In tissues that contain a mixture of chromophores at high concentrations, internal filter effects on fluorescence detection are an outstanding challenge. However, new fluorescence lifetime methodologies (Vishwasrao et al. 2005) could provide an alternative approach for eliminating such challenge and are highlighted in Chapters 4 and 5.

References

Abrahams, J. P., A. G. Leslie, R. Lutter, and J. E. Walker. 1994. Structure at 2.8 Å resolution of F_1-ATPase from bovine heart mitochondria. *Nature* 370(6491): 621–628.

Acin-Perez, R., D. L. Gatti, Y. Bai, and G. Manfredi. 2011. Protein phosphorylation and prevention of cytochrome oxidase inhibition by ATP: Coupled mechanisms of energy metabolism regulation. *Cell Metabolism* 13(6): 712–719.

Ala-Rämi, A., M. Ylihautala, P. Ingman, and I. E. Hassinen. 2005. Influence of calcium-induced workload transitions and fatty acid supply on myocardial substrate selection. *Metabolism: Clinical and Experimental* 54(3): 410–420.

Anderson, C. P., M. Shen, R. S. Eisenstein, and E. A. Leibold. 2012. Mammalian iron metabolism and its control by iron regulatory proteins. *Biochimica et Biophysica Acta* 1823(9): 1468–1483.

Anthony, G., A. Reimann, and B. Kadenbach. 1993. Tissue-specific regulation of bovine heart cytochrome-c oxidase activity by ADP via interaction with subunit VIa. *Proceedings of the National Academy of Sciences of the United States of America* 90(5): 1652–1656.

Arnold, S. and B. Kadenbach. 1997. Cell respiration is controlled by ATP, an allosteric inhibitor of cytochrome-c oxidase. *European Journal of Biochemistry* 249(1): 350–354.

Baines, C. P., R. A. Kaiser, T. Sheiko, W. J. Craigen, and J. D. Molkentin. 2007. Voltage-dependent anion channels are dispensable for mitochondrial-dependent cell death. *Nature Cell Biology* 9(5): 550–555.

Balsa, E., R. Marco, E. Perales-Clemente, R. Szklarczyk, E. Calvo, M. O. Landázuri, and J. A. Enríquez. 2012. NDUFA4 is a subunit of complex IV of the mammalian electron transport chain. *Cell Metabolism* 16(3): 378–386.

Baradaran, R., J. M. Berrisford, G. S. Minhas, and L. A. Sazanov. 2013. Crystal structure of the entire respiratory complex I. *Nature* 494(7438): 443–448.

Basso, E., L. Fante, J. Fowlkes, V. Petronilli, M. A. Forte, and P. Bernardi. 2005. Properties of the permeability transition pore in mitochondria devoid of Cyclophilin D. *The Journal of Biological Chemistry* 280(19): 18558–18561.

Baughman, J. M., F. Perocchi, H. S. Girgis et al. 2011. Integrative genomics identifies MCU as an essential component of the mitochondrial calcium uniporter. *Nature* 476(7360): 341–345.

Beckmann, J. D. and F. E. Frerman. 1985. Reaction of electron-transfer flavoprotein with electron-transfer flavoprotein-ubiquinone oxidoreductases. *Biochemistry* 24(15): 3922–3925.

Bick, A. G., S. E. Calvo, and V. K. Mootha. 2012. Evolutionary diversity of the mitochondrial calcium uniporter. *Science* 336(6083): 886.

Bolaños, J. P., M. A. Moro, I. Lizasoain, and A. Almeida. 2009. Mitochondria and reactive oxygen and nitrogen species in neurological disorders and stroke: Therapeutic implications. *Advanced Drug Delivery Reviews* 61(14): 1299–1315.

Brand, M. D. 2010. The sites and topology of mitochondrial superoxide production. *Experimental Gerontology* 45(7): 466–472.

Bricker, D. K., E. B. Taylor, J. C. Schell et al. 2012. A mitochondrial pyruvate carrier required for pyruvate uptake in yeast, *Drosophila*, and humans. *Science* 337(6090): 96–100.

Brunetti-Pierri, N., B. Lanpher, A. Erez et al. 2011. Phenylbutyrate therapy for maple syrup urine disease. *Human Molecular Genetics* 20(4): 631–640.

Bücher, T., B. Brauser, A. Conze, F. Klein, O. Langguth, and H. Sies. 1972. State of oxidation-reduction and state of binding in the cytosolic NADH-system as disclosed by equilibration with extracellular lactate-pyruvate in hemoglobin-free perfused rat liver. *European Journal of Biochemistry* 27(2): 301–317.

Cai, Q., M. L. Davis, and Z.-H. Sheng. 2011. Regulation of axonal mitochondrial transport and its impact on synaptic transmission. *Neuroscience Research* 70(1): 9–15.

Carling, D., C. Thornton, A. Woods, and M. J. Sanders. 2012. AMP-activated protein kinase: New regulation, new roles? *The Biochemical Journal* 445(1): 11–27.

Chance, B. 1948. The enzyme-substrate compounds of catalase and peroxides. *Nature* 161(4102): 914–917.

Chapman, J. B. 1972. Fluorometric studies of oxidative metabolism in isolated papillary muscle of the rabbit. *Journal of General Physiology* 59: 135–157.

Chen, X. J. and R. A. Butow. 2005. The organization and inheritance of the mitochondrial genome. *Nature Reviews Genetics* 6(11): 815–825.

Circu, M. L. and T. Y. Aw. 2010. Reactive oxygen species, cellular redox systems, and apoptosis. *Free Radical Biology & Medicine* 48(6): 749–762.

Contreras, L. and J. Satrústegui. 2009. Calcium signaling in brain mitochondria: Interplay of malate aspartate NADH shuttle and calcium uniporter/mitochondrial dehydrogenase pathways. *The Journal of Biological Chemistry* 284(11): 7091–7099.

Covian, R. and B. L. Trumpower. 2008. Regulatory interactions in the dimeric cytochrome bc_1 complex: The advantages of being a twin. *Biochimica et Biophysica Acta* 1777(9): 1079–1091.

Crabtree, B. and E. A. Newsholme. 1972. The activities of phosphorylase, hexokinase, phosphofructokinase, lactate dehydrogenase and the glycerol 3-phosphate dehydrogenases in muscles from vertebrates and invertebrates. *The Biochemical Journal* 126(1): 49–58.

Davies, K. M., C. Anselmi, I. Wittig, J. D. Faraldo-Gómez, and W. Kühlbrandt. 2012. Structure of the yeast F_1F_o-ATP synthase dimer and its role in shaping the mitochondrial cristae. *Proceedings of the National Academy of Sciences of the United States of America* 109(34): 13602–13607.

de Grotthuss, C. J. T. 2006. Memoir on the decomposition of water and of the bodies that it holds in solution by means of galvanic electricity. *Biochimica et Biophysica Acta* 1757(8): 871–875.

De Luca, H. F. and G. W. Engstrom. 1961. Calcium uptake by kidney mitochondria. *Proceedings of the National Academy of Sciences of the United States of America* 47(11): 1744–1750.

Dmitriev, O. Y., P. C. Jones, and R. H. Fillingame. 1999. Structure of the subunit c oligomer in the F_1F_o ATP synthase: Model derived from solution structure of the monomer and cross-linking in the native enzyme. *Proceedings of the National Academy of Sciences of the United States of America* 96(14): 7785–7790.

Drago, I., P. Pizzo, and T. Pozzan. 2011. After half a century mitochondrial calcium in- and efflux machineries reveal themselves. *The EMBO Journal* 30(20): 4119–4125.

Efremov, R. G. and L. A. Sazanov. 2011. Respiratory complex I: 'Steam engine' of the cell? *Current Opinion in Structural Biology* 21(4): 532–540.

Eng, C., M. Kiuru, M. J. Fernandez, and L. A. Aaltonen. 2003. A role for mitochondrial enzymes in inherited neoplasia and beyond. *Nature Reviews Cancer* 3(3): 193–202.

Eto, K., S. Suga, M. Wakui et al. 1999. NADH shuttle system regulates K_{ATP} channel-dependent pathway and steps distal to cytosolic Ca^{2+} concentration elevation in glucose-induced insulin secretion. *The Journal of Biological Chemistry* 274(36): 25386–25392.

Feliciello, A., M. E. Gottesman, and E. V. Avvedimento. 2005. cAMP-PKA signaling to the mitochondria: Protein scaffolds, mRNA and phosphatases. *Cellular Signalling* 17(3): 279–287.

Forsander, O. A. 1967. Influence of some aliphatic alcohols on the metabolism of rat liver slices. *The Biochemical Journal* 105(1): 93–97.

Gabriel, J. L., P. R. Zervos, and G. W. Plaut. 1986. Activity of purified NAD-specific isocitrate dehydrogenase at modulator and substrate concentrations approximating conditions in mitochondria. *Metabolism: Clinical and Experimental* 35(7): 661–667.

Gaetani, G. F., A. M. Ferraris, M. Rolfo, R. Mangerini, S. Arena, and H. N. Kirkman. 1996. Predominant role of catalase in the disposal of hydrogen peroxide within human erythrocytes. *Blood* 87(4): 1595–1599.

Gao, X., X. Wen, L. Esser et al. 2003. Structural basis for the quinone reduction in the bc_1 complex: A comparative analysis of crystal structures of mitochondrial cytochrome bc_1 with bound substrate and inhibitors at the Q_i site. *Biochemistry* 42(30): 9067–9080.

Garcia-Trejo, J. J. and E. Morales-Rios. 2008. Regulation of the F_1F_0-ATP synthase rotary nanomotor in its monomeric-bacterial and dimeric-mitochondrial forms. *Journal of Biological Physics* 34(1–2): 197–212.

Garland, P. B., B. Chance, L. Ernster, C.-P. Lee, and D. Wong. 1967. Flavoproteins of mitochondrial fatty acid oxidation. *Proceedings of the National Academy of Sciences of the United States of America* 58(4): 1696–1702.

Gimeno, R. E. 2007. Fatty acid transport proteins. *Current Opinion in Lipidology* 18(3): 271–276.

Gutman, M., T. P. Singer, H. Beinert, and J. E. Casida. 1970. Reaction sites of rotenone, piericidin A, and amytal in relation to the nonheme iron components of NADH dehydrogenase. *Proceedings of the National Academy of Sciences of the United States of America* 65(3): 763–770.

Halestrap, A. P. 2009. What is the mitochondrial permeability transition pore? *Journal of Molecular and Cellular Cardiology* 46(6): 821–831.

Hassinen, I. and B. Chance. 1968. Oxidation-reduction properties of the mitochondrial flavoprotein chain. *Biochemical and Biophysical Research Communications* 31(6): 895–900.

Hassinen, I. and M. T. Kähönen. 1974. Hydrogen peroxide formation and catalase regulation in rats treated with ethyl-alpha-p-chlorophenoxy-isobutyrate (Clofibrate). In: *Alcohol and Aldehyde Metabolizing Systems*, R. G. Thurman, T. Yonetani, J. R. Williamson, and B. Chance (eds.), pp. 199–206. New York: Academic Press.

Hassinen, I. E. 1986. Mitochondrial respiratory control in the myocardium. *Biochimica et Biophysica Acta* 853(2): 135–151.

Hassinen, I. E. and K. Hiltunen. 1975. Respiratory control in isolated perfused rat heart. Role of the equilibrium relations between the mitochondrial electron carriers and the adenylate system. *Biochimica et Biophysica Acta* 408(3): 319–330.

Hassinen, I. E., J. K. Hiltunen, and T. E. Takala. 1981. Reflectance spectrophotometric monitoring of the isolated perfused heart as a method of measuring the oxidation-reduction state of cytochromes and oxygenation of myoglobin. *Cardiovascular Research* 15(2): 86–91.

Hassinen, I. E., R. H. Ylikahri, and M. T. Kähönen. 1971. Regulation of cellular respiration by thyroid hormone. Spectroscopic evidence of mitochondrial control in intact rat liver. *Archives of Biochemistry and Biophysics* 147(1): 255–261.

Hegardt, F. G. 1999. Mitochondrial 3-hydroxy-3-methylglutaryl-CoA synthase: A control enzyme in ketogenesis. *The Biochemical Journal* 338(3): 569–582.

Hildyard, J. C. W., C. Ämmälä, I. D. Dukes, S. A. Thomson, and A. P. Halestrap. 2005. Identification and characterisation of a new class of highly specific and potent inhibitors of the mitochondrial pyruvate carrier. *Biochimica et Biophysica Acta* 1707(2): 221–230.

Hiltunen, J. K., K. J. Autio, M. S. Schonauer, V. A. Kursu, C. L. Dieckmann, and A. J. Kastaniotis. 2010. Mitochondrial fatty acid synthesis and respiration. *Biochimica et Biophysica Acta* 1797(6): 1195–1202.

Hiltunen, J. K., T. Kärki, I. E. Hassinen, and H. Osmundsen. 1986. Beta-oxidation of polyunsaturated fatty acids by rat liver peroxisomes. A role for 2,4-dienoyl-coenzyme A reductase in peroxisomal beta-oxidation. *The Journal of Biological Chemistry* 261(35): 16484–16493.

Houten, S. M. and R. J. Wanders. 2010. A general introduction to the biochemistry of mitochondrial fatty acid beta-oxidation. *Journal of Inherited Metabolic Disease* 33(5): 469–477.

Huang, S., A. A. Heikal, and W. W. Webb. 2002. Two-photon fluorescence spectroscopy and microscopy of NAD(P)H and flavoprotein. *Biophysical Journal* 82(5): 2811–2825.

Hütter, J. F., H. M. Piper, and P. G. Spieckermann. 1984. Myocardial fatty acid oxidation: Evidence for an albumin-receptor-mediated membrane transfer of fatty acids. *Basic Research in Cardiology* 79(3): 274–282.

Iwata, S., J. W. Lee, K. Okada et al. 1998. Complete structure of the 11-subunit bovine mitochondrial cytochrome bc1 complex. *Science* 281(5373): 64–71.

Jeneson, J. A., R. W. Wiseman, H. V. Westerhoff, and M. J. Kushmerick. 1996. The signal transduction function for oxidative phosphorylation is at least second order in ADP. *The Journal of Biological Chemistry* 271(45): 27995–27998.

Jensen-Urstad, A. P. and C. F. Semenkovich. 2012. Fatty acid synthase and liver triglyceride metabolism: Housekeeper or messenger? *Biochimica et Biophysica Acta* 1821(5): 747–753.

Johnson, J. D., J. G. Mehus, K. Tews, B. I. Milavetz, and D. O. Lambeth. 1998. Genetic evidence for the expression of ATP- and GTP-specific succinyl-CoA synthetases in multicellular eucaryotes. *The Journal of Biological Chemistry* 273(42): 27580–27586.

Jordan, S. W. and J. E. Cronan Jr. 2002. Chromosomal amplification of the *Escherichia coli* lipB region confers high-level resistance to selenolipoic acid. *Journal of Bacteriology* 184(19): 5495–5501.

Jung, D. W., K. Baysal, and G. P. Brierley. 1995. The sodium-calcium antiport of heart mitochondria is not electroneutral. *The Journal of Biological Chemistry* 270(2): 672–678.

Kacser, H. and J. A. Burns. 1995. The control of flux. *Biochemical Society Transactions* 23(2): 341–366.

Kärki, T., E. Hakkola, I. E. Hassinen, and J. K. Hiltunen. 1987. β-Oxidation of polyunsaturated fatty acids in peroxisomes. Subcellular distribution of Δ^3,Δ^2-enoyl-CoA isomerase activity in rat liver. *FEBS Letters* 215(2): 228–232.

Katz, L. A., J. A. Swain, M. A. Portman, and R. S. Balaban. 1989. Relation between phosphate metabolites and oxygen consumption of heart in vivo. *The American Journal of Physiology* 256(1): H265–H274.

Kauppinen, R. A., J. K. Hiltunen, and I. E. Hassinen. 1980. Subcellular distribution of phosphagens in isolated perfused rat heart. *FEBS Letters* 112(2): 273–276.

Kervinen, M., J. Pätsi, M. Finel, and I. E. Hassinen. 2004. A pair of membrane-embedded acidic residues in the NuoK subunit of *Escherichia coli* NDH-1, a counterpart of the ND4L subunit of the mitochondrial complex I, are required for high ubiquinone reductase activity. *Biochemistry* 43(3): 773–781.

Kokoszka, J. E., K. G. Waymire, S. E. Levy et al. 2004. The ADP/ATP translocator is not essential for the mitochondrial permeability transition pore. *Nature* 427(6973): 461–465.

Kruger, N. J. and A. von Schaewen. 2003. The oxidative pentose phosphate pathway: Structure and organization. *Current Opinion in Plant Biology* 6(3): 236–246.

Kunz, W. S. 1986. Spectral properties of fluorescent flavoproteins of isolated rat liver mitochondria. *FEBS Letters* 195(1): 92–96.

Kunz, W. S. and W. Kunz. 1985. Contribution of different enzymes to flavoprotein fluorescence of isolated rat liver mitochondria. *Biochimica et Biophysica Acta* 841(3): 237–246.

Kussmaul, L. and J. Hirst. 2006. The mechanism of superoxide production by NADH:ubiquinone oxidoreductase (complex I) from bovine heart mitochondria. *Proceedings of the National Academy of Sciences of the United States of America* 103(20): 7607–7612.

Lam, A. K., P. N. Silva, S. M. Altamentova, and J. V. Rocheleau. 2012. Quantitative imaging of electron transfer flavoprotein autofluorescence reveals the dynamics of lipid partitioning in living pancreatic islets. *Integrative Biology* 4(8): 838–846.

Lanciano, P., B. Khalfaoui-Hassani, N. Selamoglu, A. Ghelli, M. Rugolo, and F. Daldal. 2013. Molecular mechanisms of superoxide production by Complex III: A bacterial versus human mitochondrial comparative case study. *Biochimica et Biophysica Acta* 1827(11–12): 1332–1339.

Lardy, H. A. and H. Wellman. 1952. Oxidative phosphorylations: Role of inorganic phosphate and acceptor systems in control of metabolic rates. *Journal of Biological Chemistry* 195(1): 215–224.

Ledesma, A., M. G. de Lacoba, and E. Rial. 2002. The mitochondrial uncoupling proteins. *Genome Biology* 3(12): 3015.

Lee, Y.-P. and H. A. Lardy. 1965. Influence of thyroid hormones on L-α-glycerophosphate dehydrogenases and other dehydrogenases in various organs of the rat. *The Journal of Biological Chemistry* 240(3): 1427–1436.

Li, L. Z., R. Zhou, T. Zhong et al. 2007. Predicting melanoma metastatic potential by optical and magnetic resonance imaging. In: *Oxygen Transport to Tissue XXXVIII*, D. J. Maguire, D. F. Bruley, and D. K. Harrison (eds.), Vol. 599 of Advances in Experimental Medicine and Biology, I. R. Cohen, A. Lajtha, J. D. Lambris, and R. Paoletti (eds.), pp. 67–78. New York: Springer.

Liguzinski, P. and B. Korzeniewski. 2006. Metabolic control over the oxygen consumption flux in intact skeletal muscle: In silico studies. *American Journal of Physiology. Cell Physiology* 291(6): C1213–C1224.

Liimatta, E. V., A. Gödecke, J. Schrader, and I. E. Hassinen. 2004. Regulation of cellular respiration in myoglobin-deficient mouse heart. *Molecular and Cellular Biochemistry* 256–257(1–2): 201–208.

Majamaa, K., J. S. Moilanen, S. Uimonen et al. 1998. Epidemiology of A3243G, the mutation for mitochondrial encephalomyopathy, lactic acidosis, and strokelike episodes: Prevalence of the mutation in an adult population. *American Journal of Human Genetics* 63(2): 447–454.

Malloy, C. R., A. D. Sherry, and F. M. H. Jeffrey. 1988. Evaluation of carbon flux and substrate selection through alternate pathways involving the citric acid cycle of the heart by ^{13}C NMR spectroscopy. *The Journal of Biological Chemistry* 263(15): 6964–6971.

McGarry, J. D. and D. W. Foster. 1979. In support of the roles of malonyl-CoA and carnitine acyltransferase I in the regulation of hepatic fatty acid oxidation and ketogenesis. *The Journal of Biological Chemistry* 254(17): 8163–8168.

McLain, A. L., P. A. Szweda, and L. I. Szweda. 2011. α-Ketoglutarate dehydrogenase: A mitochondrial redox sensor. *Free Radical Research* 45(1): 29–36.

Midgley, P. J. W., G. A. Rutter, A. P. Thomas, and R. M. Denton. 1987. Effects of Ca^{2+} and Mg^{2+} on the activity of pyruvate dehydrogenase phosphate phosphatase within toluene-permeabilized mitochondria. *The Biochemical Journal* 241(2): 371–377.

Mitchell, P. 1975. The protonmotive Q cycle: A general formulation. *FEBS Letters* 59(2): 137–139.

Mitoweb. 2013. Mitoweb: A human mitochondrial genome database. http://www.mitomap.org/MITOMAP. Accessed February 27, 2013.

Muller, F. L., Y. Liu, M. A. Abdul-Ghani et al. 2008. High rates of superoxide production in skeletal muscle mitochondria respiring on both complex I and complex II linked substrates. *The Biochemical Journal* 409(2): 491–499.

Müller, G. and W. Bandlow. 1987. cAMP-dependent protein kinase activity in yeast mitochondria. *Zeitschrift für Naturforschung C (A Journal of Biosciences)* 42: 1291–1302.

Näpänkangas, J. P., E. V. Liimatta, P. Joensuu, U. Bergmann, K. Ylitalo, and I. E. Hassinen. 2012. Superoxide production during ischemia-reperfusion in the perfused rat heart: A comparison of two methods of measurement. *Journal of Molecular and Cellular Cardiology* 53(6): 906–915.

Napiwotzki, J. B. and B. Kadenbach. 1998. Extramitochondrial ATP/ADP-ratios regulate cytochrome c oxidase activity via binding to the cytosolic domain of subunit IV. *Biological Chemistry* 379(3): 335–339.

Narabayashi, H., J. W. R. Lawson, and K. Uyeda. 1985. Regulation of phosphofructokinase in perfused rat heart. Requirement for fructose 2,6-bisphosphate and a covalent modification. *The Journal of Biological Chemistry* 260(17): 9750–9758.

Nedergaard, J. and B. Cannon. 2010. The changed metabolic world with human brown adipose tissue: Therapeutic visions. *Cell Metabolism* 11(4): 268–272.

Nicholls, P. 2012. Classical catalase: Ancient and modern. *Archives of Biochemistry and Biophysics* 525(2): 95–101.

Nuutinen, E. M. 1984. Subcellular origin of the surface fluorescence of reduced nicotinamide nucleotides in the isolated perfused rat heart. *Basic Research in Cardiology* 79(1): 49–58.

Nuutinen, E. M., J. K. Hiltunen, and I. E. Hassinen. 1981. The glutamate dehydrogenase system and the redox state of mitochondrial free nicotinamide adenine dinucleotide in myocardium. *FEBS Letters* 128(2): 356–360.

Ogata, T. and Y. Yamasaki. 1997. Ultra-high-resolution scanning electron microscopy of mitochondria and sarcoplasmic reticulum arrangement in human red, white, and intermediate muscle fibers. *The Anatomical Record* 248(2): 214–223.

Ohnishi, S. T., T. Ohnishi, S. Muranaka et al. 2005. A possible site of superoxide generation in the complex I segment of rat heart mitochondria. *Journal of Bioenergetics and Biomembranes* 37(1): 1–15.

Ojala, D., J. Montoya, and G. Attardi. 1981. tRNA punctuation model of RNA processing in human mitochondria. *Nature* 290(5806): 470–474.

Oshino, N., D. Jamieson, T. Sugano, and B. Chance. 1975. Optical measurement of the catalase-hydrogen peroxide intermediate (Compound I) in the liver of anaesthetized rats and its implication to hydrogen peroxide production in situ. *The Biochemical Journal* 146(1): 67–77.

Palmer, C. S., L. D. Osellame, D. Stojanovski, and M. T. Ryan. 2011. The regulation of mitochondrial morphology: Intricate mechanisms and dynamic machinery. *Cellular Signalling* 23(10): 1534–1545.

Palmieri, F., F. Bisaccia, L. Capobianco et al. 1993. Transmembrane topology, genes, and biogenesis of the mitochondrial phosphate and oxoglutarate carriers. *Journal of Bioenergetics and Biomembranes* 25(5): 493–501.

Palmieri, F., C. L. Pierri, A. De Grassi, A. Nunes-Nesi, and A. R. Fernie. 2011. Evolution, structure and function of mitochondrial carriers: A review with new insights. *The Plant Journal* 66(1): 161–181.

Palmieri, L., B. Pardo, F. M. Lasorsa et al. 2001. Citrin and aralar1 are Ca^{2+}-stimulated aspartate/glutamate transporters in mitochondria. *The EMBO Journal* 20(18): 5060–5069.

Palty, R., W. F. Silverman, M. Hershfinkel et al. 2010. NCLX is an essential component of mitochondrial Na^+/Ca^{2+} exchange. *Proceedings of the National Academy of Sciences of the United States of America* 107(1): 436–441.

Papa, S., A. M. Sardanelli, T. Cocco, F. Speranza, S. C. Scacco, and Z. Technikova-Dobrova. 1996. The nuclear-encoded 18 kDa (IP) AQDQ subunit of bovine heart complex I is phosphorylated by the mitochondrial cAMP-dependent protein kinase. *FEBS Letters* 379(3): 299–301.

Papa, S., A. M. Sardanelli, S. Scacco et al. 2002. The NADH: Ubiquinone oxidoreductase (complex I) of the mammalian respiratory chain and the cAMP cascade. *Journal of Bioenergetics and Biomembranes* 34(1): 1–10.

Paradies, G. and S. Papa. 1975. The transport of monocarboxylic oxoacids in rat liver mitochondria. *FEBS Letters* 52(1): 149–152.

Perham, R. N. 1991. Domains, motifs, and linkers in 2-oxo acid dehydrogenase multienzyme complexes: A paradigm in the design of a multifunctional protein. *Biochemistry* 30(35): 8501–8512.

Perl, A., R. Hanczko, T. Telarico, Z. Oaks, and S. Landas. 2011. Oxidative stress, inflammation and carcinogenesis are controlled through the pentose phosphate pathway by transaldolase. *Trends in Molecular Medicine* 17(7): 395–403.

Perl, A., Y. Qian, K. R. Chohan et al. 2006. Transaldolase is essential for maintenance of the mitochondrial transmembrane potential and fertility of spermatozoa. *Proceedings of the National Academy of Sciences of the United States of America* 103(40): 14813–14818.

Perocchi, F., V. M. Gohil, H. S. Girgis et al. 2010. MICU1 encodes a mitochondrial EF hand protein required for Ca^{2+} uptake. *Nature* 467(7313): 291–296.

Pezeshk, A., J. Torres, M. T. Wilson, and M. C. R. Symons. 2001. The EPR spectrum for Cu_B in cytochrome *c* oxidase. *Journal of Inorganic Biochemistry* 83(2): 115–119.

Piantadosi, C. A., M. Hall, and B. J. Comfort. 1997. Algorithms for in vivo near-infrared spectroscopy. *Analytical Biochemistry* 253(2): 277–279.

Picard, M., T. Taivassalo, D. Ritchie et al. 2011. Mitochondrial structure and function are disrupted by standard isolation methods. *PloS One* 6(3): e18317.

Pizzuto, R., G. Paventi, C. Porcile, D. Sarnataro, A. Daniele, and S. Passarella. 2012. l-Lactate metabolism in HEP G2 cell mitochondria due to the l-lactate dehydrogenase determines the occurrence of the lactate/pyruvate shuttle and the appearance of oxaloacetate, malate and citrate outside mitochondria. *Biochimica et Biophysica Acta* 1817(9): 1679–1690.

Quant, P. A., P. K. Tubbs, and D. Brand. 1990. Glucagon activates mitochondrial 3-hydroxy-3-methylglutaryl-CoA synthase in vivo by decreasing the extent of succinylation of the enzyme. *European Journal of Biochemistry* 187(1): 169–174.

Ragan, C. I. and P. B. Garland. 1969. The intra-mitochondrial localization of flavoproteins previously assigned to the respiratory chain. *European Journal of Biochemistry* 10(3): 399–410.

Randle, P. J. 1998. Regulatory interactions between lipids and carbohydrates: The glucose fatty acid cycle after 35 years. *Diabetes/Metabolism Reviews* 14(4): 263–283.

Runswick, M. J., J. E. Walker, F. Bisaccia, V. Iacobazzi, and F. Palmieri. 1990. Sequence of the bovine 2-oxoglutarate/malate carrier protein: Structural relationship to other mitochondrial transport proteins. *Biochemistry* 29(50): 11033–11040.

Salvi, M., V. Battaglia, A. M. Brunati et al. 2007. Catalase takes part in rat liver mitochondria oxidative stress defense. *The Journal of Biological Chemistry* 282(33): 24407–24415.

Schilling, B., R. Aggeler, B. Schulenberg et al. 2005. Mass spectrometric identification of a novel phosphorylation site in subunit NDUFA10 of bovine mitochondrial complex I. *FEBS Letters* 579(11): 2485–2490.

Schmidt, B., J. McCracken, and S. Ferguson-Miller. 2003. A discrete water exit pathway in the membrane protein cytochrome c oxidase. *Proceedings of the National Academy of Sciences of the United States of America* 100(26): 15539–15542.

Schulz, H. 1991. Beta oxidation of fatty acids. *Biochimica et Biophysica Acta* 1081(2): 109–120.

Sekine, N., V. Cirulli, R. Regazzi et al. 1994. Low lactate dehydrogenase and high mitochondrial glycerol phosphate dehydrogenase in pancreatic beta-cells. Potential role in nutrient sensing. *The Journal of Biological Chemistry* 269(7): 4895–4902.

Sharpe, M. A. and S. Ferguson-Miller. 2008. A chemically explicit model for the mechanism of proton pumping in heme-copper oxidases. *Journal of Bioenergetics and Biomembranes* 40(5): 541–549.

Siess, E. A., R. I. Kientsch-Engel, and O. H. Wieland. 1984. Concentration of free oxaloacetate in the mitochondrial compartment of isolated liver cells. *The Biochemical Journal* 218: 171–176.

Signorile, A., A. M. Sardanelli, R. Nuzzi, and S. Papa. 2002. Serine (threonine) phosphatase(s) acting on cAMP-dependent phosphoproteins in mammalian mitochondria. *FEBS Letters* 512(1): 91–94.

Sinegina, L., M. Wikström, M. I. Verkhovsky, and M. L. Verkhovskaya. 2005. Activation of isolated NADH:ubiquinone reductase I (complex I) from *Escherichia coli* by detergent and phospholipids. Recovery of ubiquinone reductase activity and changes in EPR signals of iron-sulfur clusters. *Biochemistry* 44(23): 8500–8506.

Springett, R., J. Newman, M. Cope, and D. T. Delpy. 2000. Oxygen dependency and precision of cytochrome oxidase signal from full spectral NIRS of the piglet brain. *American Journal of Physiology, Heart and Circulatory Physiology* 279: H2202–H2209.

Suga, M., N. Yano, K. Muramoto et al. 2011. Distinguishing between Cl^- and O_2^{2-} as the bridging element between Fe^{3+} and Cu^{2+} in resting-oxidized cytochrome c oxidase. *Acta Crystallographica Section D: Biological Crystallography* 67(8): 742–744.

Sundqvist, K. E., J. Heikkilä, I. E. Hassinen, and J. K. Hiltunen. 1987. Role of $NADP^+$-linked malic enzymes as regulators of the pool size of tricarboxylic acid-cycle intermediates in the perfused rat heart. *The Biochemical Journal* 243: 853–857.

Sundqvist, K. E., J. K. Hiltunen, and I. E. Hassinen. 1989. Pyruvate carboxylation in the rat heart. Role of biotin-dependent enzymes. *The Biochemical Journal* 257: 913–916.

Tamura, M., N. Oshino, and B. Chance. 1976. A new spectroscopic approach to cardiac energy metabolism. *Recent Advances in Studies on Cardiac Structure and Metabolism* 11: 307–312.

Tsukihara, T., H. Aoyama, E. Yamashita et al. 1996. The whole structure of the 13-subunit oxidized cytochrome c oxidase at 2.8 Å. *Science* 272(5265): 1136–1144.

Vanneste, W. H. 1966. Molecular proportion of the fixed cytochrome components of the respiratory chain of Keilin-Hartree particles and beef heart mitochondria. *Biochimica et Biophysica Acta* 113(1): 175–178.

Vielhaber, S., K. Winkler, E. Kirches et al. 1999. Visualization of defective mitochondrial function in skeletal muscle fibers of patients with sporadic amyotrophic lateral sclerosis. *Journal of the Neurological Sciences* 169(1): 133–139.

Vijayakrishnan, S., S. M. Kelly, R. J. Gilbert et al. 2010. Solution structure and characterisation of the human pyruvate dehydrogenase complex core assembly. *Journal of Molecular Biology* 399(1): 71–93.

Vishwasrao, H. D., A. A. Heikal, K. A. Kasischke, and W. W. Webb. 2005. Conformational dependence of intracellular NADH on metabolic state revealed by associated fluorescence anisotropy. *Journal of Biological Chemistry* 280(26): 25119–25126.

Voltti, H. and I. E. Hassinen. 1978. Oxidation-reduction midpoint potentials of mitochondrial flavoproteins and their intramitochondrial localization. *Journal of Bioenergetics and Biomembranes* 10(1–2): 45–58.

Wallace, D. C., G. Singh, M. T. Lott et al. 1988. Mitochondrial DNA mutation associated with Leber's hereditary optic neuropathy. *Science* 242(4884): 1427–1430.

Watanabe, K. and S. Yokobori. 2011. tRNA modification and genetic code variations in animal mitochondria. *Journal of Nucleic Acids* 2011: 623095.

Watmough, N. J. and F. E. Frerman. 2010. The electron transfer flavoprotein: Ubiquinone oxidoreductases. *Biochimica et Biophysica Acta* 1797(12): 1910–1916.

Westerhoff, H. V. and K. van Dam. 1987. *Thermodynamics and Control of Biological Free Energy Transduction.* Amsterdam, the Netherlands: Elsevier.

Wikström, M. and G. Hummer. 2012. Stoichiometry of proton translocation by respiratory complex I and its mechanistic implications. *Proceedings of the National Academy of Sciences USA* 109(12): 4431–4436.

Wikström, M. and K. Krab. 1979. Proton-pumping cytochrome c oxidase. *Biochimica et Biophysica Acta* 549(2): 177–122.

Williamson, J. R., B. Safer, K. F. LaNoue, C. M. Smith, and E. Walajtys. 1973. Mitochondrial-cytosolic interactions in cardiac tissue: Role of the malate-aspartate cycle in the removal of glycolytic NADH from the cytosol. *Symposia of the Society for Experimental Biology* 27: 241–281.

Wilson, D. F., M. Stubbs, R. L. Veech, M. Erecinska, and H. A. Krebs. 1974. Equilibrium relations between the oxidation-reduction reactions and the adenosine triphosphate synthesis in suspensions of isolated liver cells. *The Biochemical Journal* 140: 57–64.

Xia, X. 2012. DNA replication and strand asymmetry in prokaryotic and mitochondrial genomes. *Current Genomics* 13(1): 16–27.

Xu, H. N., S. Nioka, B. Chance, and L. Z. Li. 2011. Heterogeneity of mitochondrial redox state in premalignant pancreas in a PTEN null transgenic mouse model. In: *Oxygen Transport to Tissue XXXII*, J. C. LaManna, M. A. Puchowicz, K. Xu, D. K. Harrison, and D. F. Bruley (eds.), Vol. 701 of Advances in Experimental Medicine and Biology, I. R. Cohen, A. Lajtha, J. D. Lambris, and R. Paoletti (eds.), pp. 207–213. New York: Springer.

Yeh, J. I., U. Chinte, and S. Du. 2008. Structure of glycerol-3-phosphate dehydrogenase, an essential monotopic membrane enzyme involved in respiration and metabolism. *Proceedings of the National Academy of Sciences of the United States of America* 105(9): 3280–3285.

Yu, Q. and A. A. Heikal. 2009. Two-photon autofluorescence dynamics imaging reveals sensitivity of intracellular NADH concentration and conformation to cell physiology at the single-cell level. *Journal of Photochemistry and Photobiology B: Biology* 95(1): 46–57.

Zelewski, M. and J. Swierczynski. 1991. Malic enzyme in human liver. Intracellular distribution, purification and properties of cytosolic isozyme. *European Journal of Biochemistry* 201(2): 339–345.

Zhou, Q., Y. Zhai, J. Lou, M. Liu, X. Pang, and F. Sun. 2011. Thiabendazole inhibits ubiquinone reduction activity of mitochondrial respiratory complex II via a water molecule mediated binding feature. *Protein & Cell* 2(7): 531–542.

2

Intracellular Autofluorescent Species: Structure, Spectroscopy, and Photophysics

Nobuhiro Ohta
Hokkaido University

Takakazu Nakabayashi
Tohoku University

2.1 Introduction

Fluorescence microscopy along with engineered fluorescent probes has become a very powerful and indispensable method for investigating the biology and pathology of cells and tissues (Lichtman and Conchello 2005). A variety of fluorescent dyes and fluorescent proteins have been developed for imaging various biological processes at the cellular and subcellular level. However, introduction of exogenous dyes and even least invasive fluorescent proteins into cells and tissues has the risk of interfering with the cell biology and the activities of target biomolecules (e.g., proteins). In addition, these dyes are likely to be toxic and bind nonspecifically; as a result, quantitative studies of cellular and molecular biophysics are undermined (Alford et al. 2009; Billinton and Knight 2001). Moreover, delivery of the fluorescent probes into the cell is time consuming and often difficult. The described limitations of exogenous labels highlight the importance and convenience of endogenous fluorophores that cause cells and tissues to be naturally fluorescent (autofluorescent).

Historically, the first account of autofluorescence was published in 1565 by a Spanish physician Nicolas Monardes (Harvey 1957; Valeur 2001). He reported blue opalescence from the infusion of *Lignum nephriticum*, Mexican kidneywood, which was later attributed to flavonoid oxidation. Since then, a large number of other endogenous fluorophores exhibiting fluorescence in the visible and UV range have been discovered.

At present, the most important and well-known autofluorescent chromophores in cells and tissues are the following (Berezin and Achilefu 2010; Richards-Kortum and Sevick-Muraca 1996):

- Coenzymes—reduced nicotinamide adenine dinucleotide (NADH) and flavin adenine dinucleotide (FAD)—major participants to cell energy metabolism. Following the seminal work by Britton Chance in the 1960s (Chance et al. 1962; Mayevsky and Chance 2007), these chromophores received much attention.
- Amino acids having an aromatic ring (tryptophan, tyrosine, and phenylalanine)—building blocks of proteins and enzymes.
- Structural proteins collagen, elastin, and keratin, responsible for rigidity and flexibility of tissues and organs.
- Porphyrins and their related compounds—side product in mammals as well as light-absorbing and energy conversion elements in plants and photosynthetic bacteria.
- Fluorescent pigments lipofuscin, a marker of age-related pathologies, and melanin.

Since these autofluorescence emitters are directly related to cell functions and metabolic activities, autofluorescence bears valuable potential as natural biomarker for biology, physiology, and pathology of cells and tissues and can be monitored without disturbing the native intracellular conditions. In addition, autofluorescence-based techniques have a potential for medical diagnostic tests without elaborate staining.

Different aspects of cellular autofluorescence (e.g., intensity, spectra, and fluorescence lifetime) can serve as observables for molecular identification, environmental probing, and monitoring of the binding state of endogenous fluorophores under given physiological state of the cell. Prior to the discussion of each of the tissue chromophores, we briefly explain these attributes and principles of fluorescence.

2.2 Basic Principles of Fluorescence

Following the absorption of a photon (photoexcitation), a molecule transitions to the higher singlet excited states (Figure 2.1). Because of the instability of the excited configuration, the molecule then dissipates the excess energy either radiatively or nonradiatively and returns to the singlet ground state (S_0) or undergoes intersystem crossing to the triplet state T, unless any chemical reaction occurs. In condensed phase, the first stage is typically a vibrational relaxation to the lowest excited singlet state (S_1). Fluorescence is the mechanism for energy conservation in which the excited molecule returns from the S_1 to the S_0 via the emission of a photon. Because of the partial dissipation of the absorbed energy during the initial vibrational relaxation, the energy of the emitted photon is usually smaller than the energy

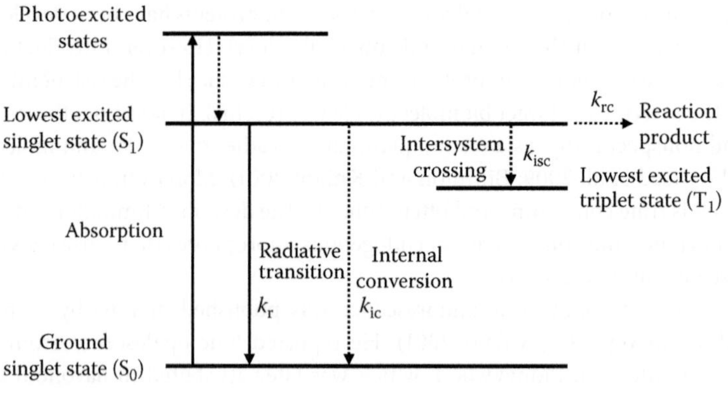

FIGURE 2.1 Energy levels of the S_0, S_1 and T_1 states and relaxation scheme from the S_1 state, following the relaxation from the photoexcited state to the S_1 state.

of the excitation photon; this shift of the emission spectrum toward the longer wavelengths compared to the absorption band is called the Stokes shift. Other nonradiative processes might compete with the excited-state depopulation via fluorescence.

Figure 2.1 shows a schematic energy diagram that describes photoexcitation and relaxation mechanisms of photon-driven cycling of molecules between different electronic states. In this energy diagram, there are two major competing pathways of the depopulation of the excited electronic S_1 state: radiative and nonradiative processes. The radiative decay represents the electronic transition from S_1 to S_0 with the emission of a photon at a rate constant of k_r, which depends solely on the chemical structure of a given fluorophore. The nonradiative process encompasses the transitions from the S_1 state to another electronic state without emitting a photon. The nonradiative rate constant (k_{nr}) collectively describes the overall processes that compete with fluorescence, such as charge transfer, isomerization, intersystem crossing, internal conversion, ionization, energy transfer, and dissociation, that is, k_{nr} is given by $k_{ic} + k_{isc} + k_{rc}$ in Figure 2.1.

In condensed phase, relaxation to the S_0 state through fluorescence photon emission is very efficient. The fluorescence intensity (I_f) is usually given by the following equation (Lakowicz 2006):

$$I_f = NLI_0\sigma(\lambda)\Phi_f, \tag{2.1}$$

where

N is the number of solute molecules per unit volume
L is the path length of the excitation light
I_0 is the excitation light intensity
$\sigma(\lambda)$ is the absorption cross section, which describes the ability of a molecule to absorb a photon of the excitation wavelength λ
Φ_f is the fluorescence quantum yield, defined as

$$\Phi_f = \frac{k_r}{k_r + k_{nr}} \tag{2.2}$$

where k_r and k_{nr} represent radiative and nonradiative decay rate constants, respectively. Depopulation of the excited electronic states does not happen instantaneously but rather follows an exponential law and depends on the residence time (i.e., fluorescence lifetime τ_f) of the molecule in the excited S_1 state. The τ_f inversely depends on the sum of k_r and k_{nr}:

$$\tau_f = \frac{1}{k_r + k_{nr}} \tag{2.3}$$

Suppose a group of molecules are excited with an infinitely sharp pulse. The population of the molecules in S_1 and correspondingly the fluorescence intensity will then exponentially decay as a function of time (t):

$$I_f(t) = I_0 \exp\left(-\frac{t}{\tau_f}\right) \tag{2.4}$$

where

I_0 is the initial intensity at the zero time of photoexcitation
τ_f is the average fluorescence lifetime

In this case, fluorescence exhibits a single-exponential decay with a lifetime of τ_f, which corresponds to the time in which the fluorescence intensity decays to 1/e (ca. 37%) of the initial intensity at $t=0$.

Time-resolved autofluorescence recorded in cells and tissues usually shows a multiexponential decay due to several chromophores with overlapping spectra, different states of chromophore species, and different environments. Then, fluorescence decay profile is given as follows:

$$I_f(t) = \sum_i C_i \exp\left(-\frac{t}{\tau_i}\right) \tag{2.5}$$

where

τ_i is the fluorescence lifetime of the ith component

C_i is the preexponential factor that represents the contribution of the ith component to the observed fluorescence decay curve

The average lifetime (τ_{av}) is then determined as follows:

$$\tau_{av} = \frac{\sum_i C_i \tau_i}{\sum_i C_i} \tag{2.6}$$

Fluorescence lifetime for specific experimental conditions and preexponential factors of selected chromophores responsible for cellular autofluorescence are shown in Table 2.1.

Among the nonradiative processes, the most probable in living systems are internal conversion, intersystem crossing, intramolecular charge transfer, and interactions with other molecules and environment in excited state, for example, electron transfer, energy transfer, and collisions. In particular, fluorescence resonance energy transfer (FRET)—transition of excitation energy from a photoexcited donor molecule to an acceptor molecule—is an important nonradiative process, currently utilized in microscopy for probing intermolecular interactions (Wallrabe and Periasamy 2005; see also Chapter 8 for a more detailed discussion of FRET and its applications).

Fluorescence intensity is the key observable in a range of fluorescence-based imaging methods, such as wide-field, confocal, and total internal reflection fluorescence (TIRF) microscopy, and provides valuable information on the cellular and intracellular environments. However, evaluation of fluorescence quantum yield or absolute fluorescence intensity is extremely difficult for such complicated samples as cells and tissues. Furthermore, it is hard to determine the concentration of fluorescent species that contribute to the observed fluorescence image. As recognized from Equation 2.1, the observed fluorescence intensity depends not only on the chemical structure of a given fluorophore and its concentration but also on its excitation/detection conditions. In contrast, the fluorescence lifetime is an inherent property of a fluorophore in a given environment; reliable quantitative analysis of the fluorescence lifetime can be carried out even when the magnitude of the observed change is small. Careful and systematic fluorescence lifetime measurements in well-designed environmental variations allow for the identification of the nonradiative processes associated with a given chromophore. For example, such environmental variables as ion concentrations, viscosity, amino acid polarity, pH, and the crowded nature of biological organelles/biomolecules are sensitive to the physiological state of a cell or tissue. As a result, fluorescence lifetime imaging microscopy (FLIM) is a powerful method for characterizing cells and tissues (see Chapters 4 and 5) and identification of individual chromophores in a mix even if their spectra are identical (Awasthi et al. 2012; Bastiaens and Squire 1999; Becker et al. 2007; Berezin and Achilefu 2010; Borst and Visser 2010; Chorvat and Chorvatova 2009; de Grauw and Gerritsen 2001; Ghukasyan and Kao 2009; Islam et al. 2013; Ito et al. 2009; Lakowicz and Berndt 1991; Levitt et al. 2009; Nakabayashi et al. 2008a,b,c, 2012; Ogikubo et al. 2011; Ohta et al. 2009, 2010; Wallrabe and Periasamy 2005; Wang et al. 2007).

TABLE 2.1 Absorption Peak, Fluorescence Peak, and Fluorescence Lifetime of Autofluorescence Species

Chromophore	Absorption Peak (nm)	Fluorescence Peak (nm)[a]	Lifetime (ns)[b]	Medium	References
NADH	260, 340	~465 (370)	0.14 (0.60), 0.43 (0.37), 1.1 (0.03)	Buffer solution	Ogikubo et al. (2011)
		~450 (370)	0.53 (0.73), 3.10 (0.25), 15.9 (0.02) [440 nm]	HeLa cells	Ogikubo et al. (2011)
		440–445 (355)	0.44 (0.63), 1.88 (0.30), 5.68 (0.07) [440 nm]	Intact cardiac mitochondria	Blinova et al. (2005)
ODH-NADH		~450 (336)	1.2 (0.46), 3.1 (0.54) [466 nm]	Buffer solution	Brochon et al. (1977)
FAD	375, 450	~530 (450)	0.007 (0.66), 0.22 (0.03), 2.09 (0.17), 3.97 (0.14)	Buffer solution	Islam et al. (2013)
		525–530 (450)	0.08 (0.60), 0.70 (0.25), 3.17 (0.13), 10.2 (0.02) [530 nm]	HeLa cells	Islam et al. (2013)
		504–522 (438)	0.18 (0.66), 0.75 (0.21), 2.10 (0.13) [531–549 nm]	Cardiomyocytes	Chorvat and Chorvatova (2006)
LipDH (FAD)		504 (438)	0.88 (0.35), 4.14 (0.65) [495–550 nm]	Buffer solution	Chorvat and Chorvatova (2006)
FMN	373, 445	~530 (450)	4.67	Buffer solution	Grajek et al. (2007)
Riboflavin	373, 445	~530 (450)	5.06	Buffer solution	Drössler et al. (2002)
Tryptophan	278	~355 (250)	0.43 (0.19), 3.3 (0.81)	Buffer solution	Robbins et al. (1980)
Tyrosine	274	~303 (250)	3.4	Buffer solution	Guzow et al. (2002)
Phenylalanine	257	280–285 (240)	7.5	Buffer solution	McGuinness et al. (2006)
Eumelanin		520–540 (335)	0.06 (0.54), 0.51 (0.22), 2.97 (0.16), 7.05 (0.08) [520 nm]	Buffer solution	Forest et al. (2000)
A2E	440–460	660–680 (434)	0.012 (0.954), 0.090 (0.043), 0.700 (0.001) [650 nm]	Methanol	Lamb et al. (2001)
Collagen		320–330 (270)		0.5 M acetic acid	Menter (2006)
		390–400 (325)			
		450–460 (370)			
		380–400 (337)	3.9 (0.65), 9.9 (0.35) [390 nm]	Dry form	Maarek et al. (2000)
Elastin		400–450 (337)	1.3 (0.63), 5.7 (0.37) [390 nm]	Dry form	Maarek et al. (2000)
Protoporphyrin IX			1.76 (0.62), 8.14 (0.38) [633 nm]	Normal oral mucosa	Chen et al. (2005)
Protoporphyrin IX	399, 500–630	629,697		Methanol	Marcelli et al. (2013)
Chlorophyll *a*	443, 671	677	6.3 [677 nm]	Pyridine	Niedzwiedzki and Blankenship (2010)
Chlorophyll *b*	473, 655	662	3.2 [662 nm]	Pyridine	Niedzwiedzki and Blankenship (2010)
Bacteriochlorophyll *a*	374, 613, 781	795	2.9 [795 nm]	Pyridine	Niedzwiedzki and Blankenship (2010)

[a] Excitation wavelength is shown in parentheses.

[b] The monitoring wavelength is shown in square brackets, and the value in parentheses is the preexponential factor of each lifetime component.

2.3 Spectroscopic Properties and Photophysics of Endogenous Fluorophores

Multitude of endogenous fluorophores with identical fluorescence properties makes autofluorescence observation and analysis a challenging task. Moreover, emission spectra, quantum yield, and fluorescence lifetimes of the autofluorescence emitters may be affected by their environment, conformational, and binding states. A good understanding of the fluorescence properties of each endogenous chromophore and their change upon observation condition is required to utilize autofluorescence for biomedical studies and medical diagnostics (Figure 2.2 and Table 2.1). In the following, we discuss in more details the photophysics and biological functions of selected endogenous fluorophores.

2.3.1 NADH

Reduced NADH is one of the major cofactors for hundreds of different dehydrogenases in the cell with the primary role of transporting electrons from one reaction to another. In particular, NADH is the principal electron donor for the electron transfer chain of oxidative phosphorylation in mitochondria: its oxidation, catalyzed by Complex I, initiates a series of oxidation–reduction (redox) reactions that eventually result in ATP generation (see Sections 1.2.2.1 and 1.3.3.1 for a detailed overview).

The molecular structure of NADH consists of two nucleotides—ribosylnicotinamide 5′-diphosphate and adenosine 5′-phosphate—linked by a pyrophosphate bridge (Figure 2.2). In aqueous solution, NADH has absorption bands at 259 and 337 nm (Figure 2.3). The 259 nm band is attributed to the absorption of adenine moiety and is also exhibited by the oxidized form of the cofactor (NAD^+). The electronic transition at 337 nm arises from the nicotinamide and is characteristic of the reduced state of the cofactor only (König et al. 1997). When excited at 337 nm, NADH exhibits fluorescence with the emission peak around 455 nm, while NAD^+ is not fluorescent when excited at this wavelength. Such spectral differences are utilized for imaging NAD^+/NADH redox ratio. The shape of the two-photon fluorescence spectrum of NADH was the same as that of the single-photon excited (Huang et al. 2002; Kierdaszuk et al. 1996). The majority of NADH autofluorescence is mainly observed in the mitochondria of living cells, where the concentration of NADH is higher than in other cellular areas.

Upon binding to an enzyme, the spectroscopic properties of NADH change: it exhibits a blueshifted fluorescence, increased average fluorescence lifetime, and up to fourfold higher quantum yield (Blinova et al. 2005; Brochon et al. 1977). As described previously and in Chapter 1, binding of NADH to Complex I of the electron transport chain (ETC) is the first event in a chain of reactions coupled to the synthesis of ATP. Therefore, monitoring of the population of the free NADH relative to the enzyme-bound NADH can provide insight into the metabolic state of the cell. Estimation of such ratio by analyzing the spectra of the two populations is rather challenging, given the small difference in the emission spectra of free and bound species. On the other hand, the fluorescence lifetime provides with a better contrast: the fluorescence of NADH in cells typically exhibits a biexponential decay with the short picosecond-lifetime component attributed to free NADH and long nanosecond-lifetime component considered to originate from the protein-bound NADH (Berezin and Achilefu 2010; Blinova et al. 2005; Chorvat and Chorvatova 2009; Ghukasyan and Kao 2009; Lakowicz et al. 1992; Niesner et al. 2004; Skala et al. 2007). This assignment comes from the fact that the fluorescence lifetimes of free and enzyme-bound NADH in aqueous solution were measured to be 0.3–0.4 ps and 2–3 ns, respectively (Brochon et al. 1977; Lakowicz et al. 1992). The relative concentration of the free and bound NADH species is then calculated from the preexponential factors of the short- and long-lifetime components, respectively, in Equation 2.5. The population ratio between free and protein-bound NADH calculated this way was shown to correlate with cellular metabolism (Ghukasyan and Kao 2009; Skala et al. 2007). The reason behind such difference in fluorescence lifetime is still unclear. It is suggested that interactions of nicotinamide moiety with solvent environment as an additional nonradiative relaxation pathway for free NADH lead to a decrease in the fluorescence lifetime

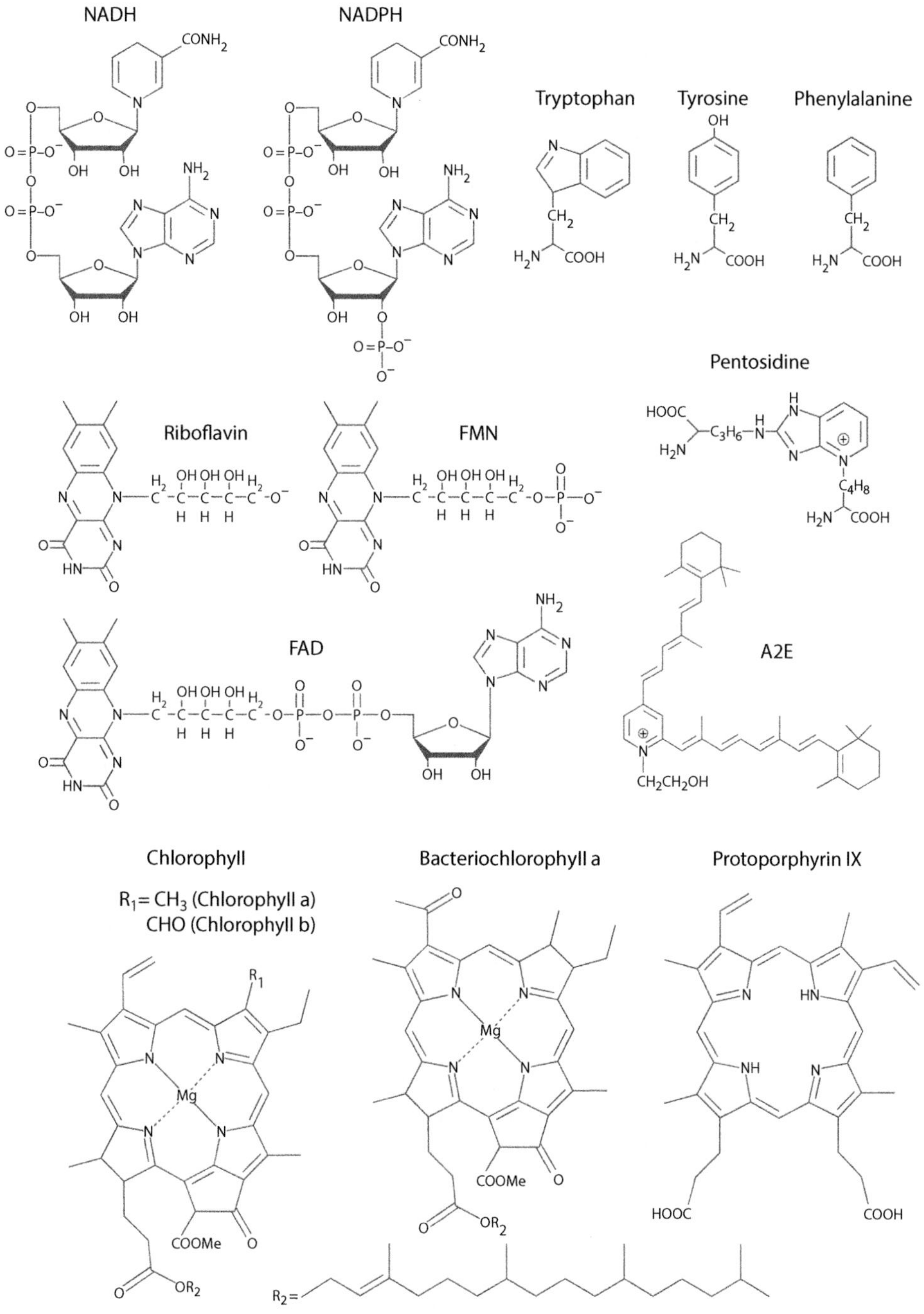

FIGURE 2.2 Chemical structures of autofluorescent chromophores.

(Gafni and Brand 1976). Quenching of nicotinamide by adenine moiety in folded conformation of free NADH was also proposed as the major mechanism behind the difference in fluorescence lifetime of the two species (Lakowicz 2006). It should be noted that the fluorescence of free NADH in aqueous solution has also been reported to show a biexponential decay (Couprie et al. 1994; Evans et al. 2005; Vishwasrao et al. 2005).

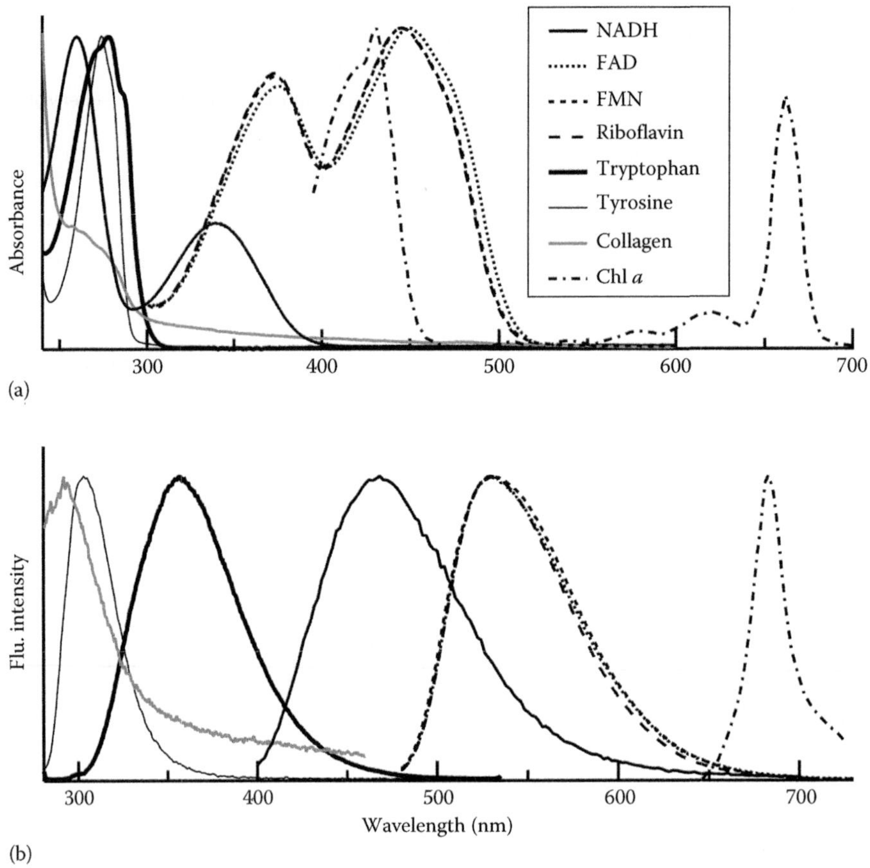

FIGURE 2.3 (a) Absorption and (b) fluorescence spectra of autofluorescent chromophores in neutral buffer solution. (The spectra of Chl *a* were reprinted with permission from Yamazaki, I., Tamai, N., Yamazaki, T., Murakami, A., Mimuro, M., and Fujita, Y., Sequential excitation energy transport in stacking multilayers: A comparative study between photosynthetic antenna and Langmuir-Blodgett multilayers, *J. Phys. Chem.*, 92(17), 5035–5044, 1998. Copyright 2013 American Chemical Society.)

As outlined in this book, a wide range of studies has utilized NADH autofluorescence for noninvasive investigation of cell physiology, pathology, and intracellular environment. For example, using FLIM of NADH, Bird et al. reported a reduction of the fluorescence lifetimes of free and protein-bound NADH with concurrent decrease of the protein-bound population as the cells confluence increases (Bird et al. 2005). Similar changes were also observed by potassium cyanide (KCN) treatment that inhibits electron transport in a cell. Since both the confluence and KCN treatment were expected to result in an increase in the ratio of $NADH/NAD^+$, it was suggested that the fluorescence lifetimes of both free and protein-bound NADH and their relative population were related to the change in the ratio of $NADH/NAD^+$ (Bird et al. 2005).

FLIM of NADH was also demonstrated as a perspective cancer diagnosis technique (reviewed in Chapter 13). Skala et al. have reported a reduction of the population and lifetime of protein-bound NADH in both low- and high-grade epithelial precancers as compared with normal epithelial tissues (Skala et al. 2007). It was also suggested that the relative free-to-bound population of NADH becomes larger in breast cancer cells than in normal breast cells (Yu and Heikal 2009). Wang et al. observed time-lapse changes of NADH fluorescence lifetime in cultured cells at apoptosis (Ghukasyan and Kao 2009; Wang et al. 2008; see an overview in Chapter 10).

The average autofluorescence lifetime of NADH in brain slices was found to decrease from normoxia to hypoxia, suggesting that the concentrations of free NADH, as well as protein-bound NADH with a relatively short lifetime, increased over that of NADH bound to other enzymes in a hypoxic state (Vishwasrao et al. 2005; see Chapter 5). The changes both in the population ratio and autofluorescence lifetime between free and protein-bound NADH in human cervical tumor cells were quantitatively evaluated at each cellular compartment with the addition of NaCN that is an inhibitor of the mitochondrial ETC and Cd that is expected to stimulate cell growth and DNA synthesis (Li et al. 2008). NADH autofluorescence has also been demonstrated as an important utility for the study of mitochondrial pathology in brain slices and mapping of neuronal activity (Shuttleworth 2010). These results, collectively, demonstrate that the autofluorescence lifetime of free and bound NADH is sensitive to cell physiology and pathology.

2.3.2 NADPH

Reduced nicotinamide adenine dinucleotide phosphate (NADPH) is a pyridine nucleotide existing in a variety of living systems. Although both NADH and NADPH are significant cofactors in living systems and participate as electron carriers in cell metabolism, the biochemical roles of these two nucleotides are different: NADH is primarily used for energy metabolism, whereas NADPH is largely used for reductive biosynthesis of fatty acids and steroids and is also vital to cellular defense systems (Yu and Heikal 2009; Pollak et al. 2007). The chemical structure of NADPH is almost the same as that of NADH: the difference in molecular structure between the two molecules is only a single phosphate group attached to one of the ribose units (Figure 2.2). Both the absorption and fluorescence spectra of NADPH are also identical to that of NADH; because of that reason, autofluorescence observed upon excitation around 350 nm is therefore often referred to as NAD(P)H fluorescence, since the two species cannot be distinguished from each other (Ghukasyan and Kao 2009; Huang et al. 2002; Visser and van Hoek 1981; Ziegenhorn et al. 1976). The similarity of the spectra is due to the fact that the nicotinamide moiety in these nucleotides is responsible both for the absorption band around 350 nm and for the fluorescence in the visible region (Figure 2.3).

Since the absorption and fluorescence properties are nearly same in NADH and NADPH, it is necessary to estimate the contribution of each to the observed autofluorescence. In some cases, the contribution of NADPH was regarded to be much smaller than that of NADH; autofluorescence of NADPH was considered to be insignificant (Blinova et al. 2005; Vishwasrao et al. 2005; Chorvat and Chorvatova 2009; Yu and Heikal 2009). In fact, the concentration of the fluorescent NADH was evaluated to be about five times larger than that of the fluorescent NADPH in mouse hippocampus, based on the measurements of liquid chromatography (Klaidman et al. 1995). The blue autofluorescence in heart mitochondria was assumed to dominantly originate from NADH, since the concentration of NADPH was much lower than that of NADH in this organelle (Blinova et al. 2005). Furthermore, the change in autofluorescence of NADPH resulting from metabolic perturbations in the heart was assumed to be negligibly small (Chance et al. 1965).

In photosynthetic organisms, NADPH plays a significant role in the ETC of photosynthesis (Shikanai 2007). Fluorescence of NADPH in leaves is generally too weak to be detected; however, blue fluorescence arising from NADPH could be detected in isolated chloroplasts (Latouche et al. 2000). The fluorescence of free NADPH exhibited a tri-exponential decay in buffer solution with the average fluorescence lifetime of 0.43 ns. On the other hand, the fluorescence decay of NADPH in reconstituted chloroplasts was complicated and fitted by a multiexponential decay, indicative of the existence of free and protein-bound species. The fluorescence lifetimes of 1.3 and 3.8 ns were obtained for protein-bound NADPH existing in reconstituted chloroplast (Latouche et al. 2000).

2.3.3 Flavins

Flavins is a collective term referring to a group of molecules that contain a tricyclic isoalloxazine ring (7,8-dimethylisoalloxazine). The basic precursor of all flavins in mammalian cells is riboflavin (vitamin B2), synthesized by bacteria, protists, fungi, plants, and some animals, while mammals obtain riboflavin from food. In cells, the precursor is phosphorylated at the 5′-hydroxyl group to yield the flavin mononucleotide (FMN). A following reaction, adenylylation, adds an adenosine monophosphate (AMP) moiety to FMN yielding the FAD (Figure 2.2).

Flavins are well-known cofactors in living systems and many oxidoreductases require flavins as their prosthetic group, forming complexes called flavoenzymes or flavoproteins. The majority of flavins in cells exist as part of flavoproteins: FAD is typically found in larger quantities than FMN; little is present as free riboflavin. When reduced, FMN and FAD form $FMNH_2$ and $FADH_2$, respectively. Unlike NADH and NADPH, which can only donate two electrons, flavins can also participate in single-electron transfers, forming FMN· and FAD·. The reduction state also determines the absorption and emission properties of flavins for which the isoalloxazine moiety is responsible. Fully oxidized flavins (FpOx) have absorption maxima at 370 and 440 nm and emit fluorescence with the peak at 520–530 nm, while semiquinones—partially oxidized free-radical forms of flavins—exhibit absorption maxima at 380, 480, 580, and 625 nm. Fully reduced flavins (Fp) are nonfluorescent and have one absorption maximum around 360 nm; another peak at 450 nm is added at partial reduction (one electron) (Nelson and Cox 2008). The similarity of flavin spectra (Figure 2.3) makes it challenging to distinguish FMN from FAD with absorption and emission spectroscopy methods. In contrast, fluorescence lifetime measurements provide with better distinguishing capacity. The fluorescence of riboflavin and FMN in water decays single exponentially with a lifetime of 5.1 ns (Drössler et al. 2002) and 4.7 ns (Grajek et al. 2007), respectively. On the other hand, the fluorescence of FAD exhibits a multiexponential decay in solution with two major lifetime components of 7 ps and 2.7 ns at 293 K (van den Berg et al. 2002). Such decay of the FAD fluorescence is due to the existence of two structural conformations in water: stacked (short-lifetime component, Figure 2.4a) and open (long-lifetime component, Figure 2.4b) (Kandoth et al. 2010; Nakabayashi et al. 2010; Sengupta et al. 2011; van den Berg et al. 2002; Weber 1950). In the stacked conformation of FAD, the adenine moiety is in close proximity to isoalloxazine and efficiently quenches fluorescence by intramolecular electron transfer (IET) (van den Berg et al. 2002). In the extended open conformation, the two aromatic rings are separated from each other and IET is less efficient. Medium polarity was found to affect the fluorescence lifetime of the open and especially stacked conformations: with decreasing dielectric constant, the fluorescence lifetime of the stacked conformation almost monotonically increased, suggesting that the picosecond component of FAD fluorescence lifetime can be used as a probe of environmental polarity. The picosecond- and nanosecond-lifetime components were also observed for FAD autofluorescence in cells, for example, in isolated cardiomyocytes (Chorvat and Chorvatova 2006) and HeLa cells (Islam et al. 2013).

The fluorescence lifetime pattern of flavins at binding to proteins is opposite to that of NADH: it is shorter in flavoproteins, which is due to efficient fluorescence quenching of photoexcited isoalloxazine by surrounding amino acids (Chosrowjan et al. 2010; Mataga et al. 2000). Most flavoproteins, therefore, exhibit weak fluorescence with short fluorescence lifetime in the range of femtoseconds to several picoseconds (Chosrowjan et al. 2010; Kao et al. 2008; Mataga et al. 2000). Only some of the flavoproteins that include FAD show autofluorescence with detectable intensity in cultured cells (Bastiaens et al. 1992; Chorvat and Chorvatova 2006; Hall and Kamin 1975; Kunz 1986; Kunz and Kunz 1985; Romashko et al. 1998). Among such flavoproteins is lipoamide dehydrogenase (LipDH), which has been shown to constitute as high as 50% of the total fluorescence from flavoproteins in some cells (Kunz and Kunz 1985). The LipDH fluorescence exhibits a multiexponential decay with subnanosecond and nanosecond fluorescence lifetimes in aqueous solution (Bastiaens et al. 1992; Chorvat and Chorvatova 2006).

The two-photon fluorescence spectra of FAD and LipDH were reported to be identical to their one-photon spectra (Huang et al. 2002). In contrast with the fluorescence spectra, one-photon absorption and two-photon excitation spectra of FAD and LipDH exhibited marked different shapes from each other (Huang et al. 2002). The two-photon excitation bands in the UV region (360–400 nm) and in the visible region (400–500 nm) shifted to a shorter wavelength from the bands observed in the one-photon absorption spectra (Figure 2.3). Furthermore, the intensity of the UV band relative to that of the visible

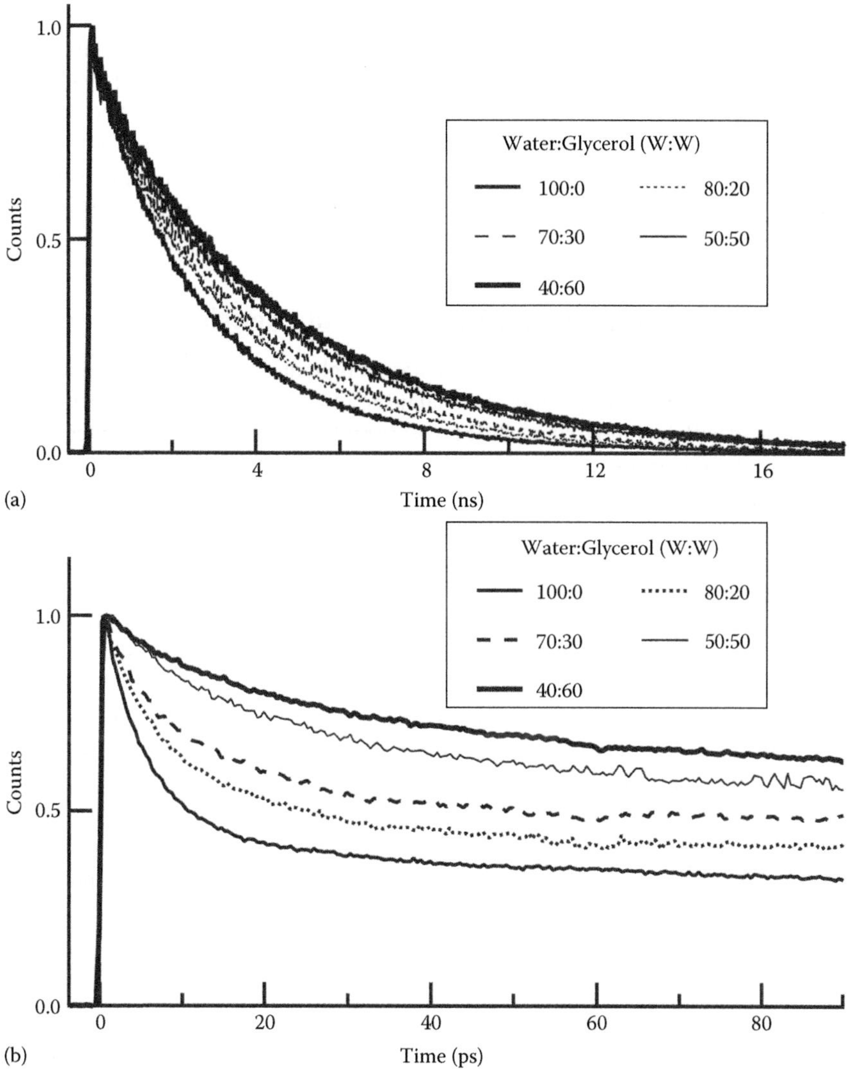

FIGURE 2.4 Fluorescence decay curves of FAD in a mixture of water and glycerol at different glycerol weight percent in the (a) picosecond–nanosecond region and in the (b) femtosecond–picosecond region. The glycerol weight% was 0, 20, 30, 50, and 60. (Reprinted with permission from Nakabayashi, T., Islam, Md.S., and Ohta, N., Fluorescence decay dynamics of flavin adenine dinucleotide in a mixture of alcohol and water in the femtosecond and nanosecond time range, *J. Phys. Chem. B*, 114(46), 15254–15260, 2010. Copyright 2013 American Chemical Society.)

band becomes much larger with two-photon excitation; the maximum two-photon excitation cross section (σ_{2P}) of the UV band was more than twice as large as that of the visible band. The σ_{2P} of LipDH at 720 nm was about 5 and 12 times larger than those of FAD and NADH (NADPH), respectively (Huang et al. 2002).

Similar to NADH, flavins fluorescence has been utilized in a wide range of studies. Complementary to NADH autofluorescence, the fluorescence lifetime of FAD was measured in both normal and precancer tissues (Skala et al. 2007). In those studies, an increase in the fluorescence lifetime of protein-bound FAD was reported in high-grade precancers along with concurrent reduction of the corresponding population fraction. The FAD autofluorescence lifetime was also used to characterize yeast strains (Bhatta and Goldys 2008). Flavins autofluorescence was also used as a reporter of cellular environment conditions: intracellular pH-dependence studies of FAD autofluorescence in HeLa cells showed that the average fluorescence lifetime became shorter with increasing intracellular pH (Islam et al. 2013). Although flavins were not as popular as NADH in the studies of brain due to lower signal and concerns about hemoglobin absorption (detailed in Chapter 6), their fluorescence have been shown to have utility in the studies of neuroenergetics and mapping patterns of neuronal activity (Reinert et al. 2007, 2011; Tohmi et al. 2009).

2.3.4 Amino Acids: Tryptophan, Tyrosine, and Phenylalanine

Proteins generally emit fluorescence under UV excitation light due to the existence of aromatic amino acid residues—tryptophan, tyrosine, and phenylalanine. As shown in Figure 2.2, tryptophan is an aromatic amino acid with an indole ring, while tyrosine and phenylalanine have a phenol and a phenyl ring, respectively. All the aromatic amino acids have an absorption band assigned to $\pi \rightarrow \pi^*$ transition in the UV region at around 260 nm. The absorption spectrum of tryptophan is slightly broader and redshifted to 280–300 nm due to the indole side chain (Lin and Sakmar 1996; Pierce and Boxer 1995). The fluorescence spectrum of tryptophan is sensitive to environmental polarity (redshift of the emission peak with increased polarity) of the surrounding medium. In aqueous solutions, the emission maximum of tryptophan near 360 nm is also significantly longer than those exhibited by tyrosine (~300 nm) and phenylalanine (~280 nm). The intrinsic emission of proteins mostly arises from tryptophan residues: tryptophan has a higher molar extinction coefficient of the $\pi \rightarrow \pi^*$ transition and a higher fluorescence quantum yield as compared with those of tyrosine and phenylalanine; fluorescence, detected from proteins at UV excitation, can thereby be mostly attributed to tryptophan. It should be noted that serotonin, a monoamine neurotransmitter derived from tryptophan, also has an indole ring and shows absorption and fluorescence spectra similar to those of tryptophan. The aromatic amino acids also have distinct fluorescence lifetimes: in buffer solution, tryptophan was measured to be 2.8 ns (Robbins et al. 1980); tyrosine, 3.3 ns (Guzow et al. 2002); and phenylalanine, 7.5 ns (McGuinness et al. 2006).

Both the fluorescence spectrum and the lifetime of tryptophan have been widely used to probe the protein environment surrounding the tryptophan residues (Beechem and Brand 1985; Callis and Burgess 1997; Tanaka and Mataga 1987; see also Chapter 8). In a polypeptide, tryptophan residues are generally surrounded by both apolar and polar residues that produce strong local electric field affecting the buried tryptophan. It was reported that redshift of tryptophan emission by 1 nm corresponds to the increase in the electric field of ~1.3 MV cm^{-1} (Callis and Burgess 1997). The steric hindrance around tryptophan residues can also be examined using time-resolved fluorescence anisotropy measurements (see Chapter 5) (Tanaka and Mataga 1987).

Intracellular autofluorescence of tryptophan can be used for investigating cell physiology and pathology such as skin cancers (Brancaleon et al. 2001; Diagaradjane et al. 2005) and for testing toxicity (Fritzsche et al. 2009).

Recent development of femtosecond pulse lasers and multiphoton imaging technologies enables imaging of autofluorescence of tryptophan in living systems without exposing live cells to the deleterious UV light (Balaji et al. 2004; Maiti et al. 1997). Li et al. measured FLIM of tryptophan in human cervical tumor cells by two-photon excitation using a supercontinuum generated from a photonic crystal fiber and detected nonuniformity of the fluorescence lifetime of tryptophan residues in a cell (Li et al. 2009). FLIM of serotonin was also measured with two-photon excitation using optical parametric oscillator (OPO) laser pulses (Botchway et al. 2008). By comparison, the application of tyrosine and phenylalanine autofluorescence to cell studies has been limited due to the low fluorescence quantum yield. Fluorescence from tyrosine is sensitive to protein structure because its fluorescence is influenced by nearby amino acid residues (Elofsson et al. 1991; Wu et al. 2013).

2.3.5 Collagen, Elastin, and Keratin

Collagen is one of the naturally occurring proteins and serves as a major component of connective tissues, skin, bones, and blood vessels. It is classified into 11 types according to the constituting amino acids and has a shape of a triple helix in its secondary structure (Miller and Gay 1987). Collagen of type I is the most abundant in the human body and contains a triple helix of approximately 300 nm in length and 1.5 nm in diameter (Douglas et al. 2008). Fibrillar collagens associate into the so-called collagen microfibril, which bundle together to make the structural collagen fiber.

Elastin is a fibrous protein of elastic connective tissues responsible for the abilities to stretch and recoil. It is found in the heart, skin, and blood vessels and is responsible for the support of their physical structure (Bailey 1978).

Keratin is a fibrous protein that is abundant in epidermis, hair, and nails. Autofluorescence of keratin is observed in these tissues with an emission spectrum that is similar to that of collagen (Wu et al. 2004).

Both collagen and elastin show detectable fluorescence, which has been used to characterize tissues. The collagen fluorescence is attributed to four fluorescent species (Menter 2006). Emission near 300 nm observed with the excitation wavelength of 270 nm was assigned to tyrosine residues constituting collagen. Emission around 360 nm is believed to arise from excimer-like species—product of intermolecular interactions of tyrosine residues. Broad fluorescence around 400 nm observed with the excitation wavelength of 325 nm was attributed to dityrosine formed via photochemical or enzymatic reaction between neighboring tyrosine residues. Finally, broad fluorescence at 450 nm, which was observed under 370 nm illumination, was found to originate from the advanced glycation end products (AGEs) in collagen—result of a series of glycation and oxidation reactions (Menter 2006). The accumulation of AGEs occurs during the aging process and the progression of a number of diseases. Pentosidine (Figure 2.2), one of the AGEs, has been proposed to be responsible for collagen emission in the visible region (Sady et al. 1995; Sell and Monnier 1989; Sell et al. 2010).

Fluorescence spectrum of elastin also shows a complicated behavior with the peak at 410–450 nm (Elson et al. 2007; Fang et al. 2004; Maarek et al. 2000; Thomas et al. 2010). Fluorescence lifetime, measured from collagen and elastin, depends on the sample preparation and experimental conditions. In particular, fluorescence lifetimes of both collagen and elastin were shown to vary with the emission wavelength (Elson et al. 2007; Fang et al. 2004; Maarek et al. 2000; Thomas et al. 2010), due to the number of fluorescent species involved. It is therefore difficult to give the standard values of the fluorescence lifetimes of collagen and elastin. For example, the fluorescence lifetimes of collagen at 390 and 450 nm are 3.26 and 2.66 ns, respectively (Elson et al. 2007).

FLIM of collagen and elastin has been used to characterize tissues (Fang et al. 2004; Maarek et al. 2000). For example, endoscopy-based FLIM was used for in situ identification of collagen and elastin areas in blood vessels (Elson et al. 2007; Thomas et al. 2010). The accumulation of AGEs inside the body increases during the progression of age-related diseases, such as diabetes, Alzheimer's disease, and renal failure. Thus autofluorescence of AGEs in collagen has been used for in situ diagnosis of such diseases. For example, autofluorescence of AGEs in skin collagen was used to investigate the progress of diabetes (Meerwaldt et al. 2004, 2005). In addition to autofluorescence, the second harmonic generation (SHG) of collagen has been reported in cells and tissues without exogenous probe and used for cancer diagnosis (Campagnola 2011).

2.3.6 Porphyrin Family

Some natural porphyrin derivatives can be recognized as autofluorescent species. For example, chlorophyll (Chl) and bacteriochlorophyll (BChl) are important autofluorescent chromophores in plant photosynthesis and in photosynthetic bacteria (Figure 2.2). In plant photosynthesis, for example, light is collected via the light-harvesting systems as a form of energy prior to being transferred to specialized reaction centers (RCs) of photosystems I and II (PSI, PSII). The pigment–protein interactions in the light-harvesting protein complexes of PSII vary greatly with the type of plant. The corresponding light-harvesting protein complex in higher plants and green algae contains Chl *a*, Chl *b*, and carotenoids, collectively termed light-harvesting Chl *a*/*b*–protein complex II (LHC II) (Krause and Weis 1991). Such light-harvesting Chl *a*/*b*–protein complex of PSI (LHC I) is tightly bound to RC, forming a highly organized PSI complex. On the other hand, the photosynthetic bacteria are characterized by the presence of BChl and a photosynthetic mechanism that is distinguishable from that in higher plants and algae by the presence of only one photosystem (Pullerits and Sundström 1996). Except for a few species, all photosynthetic bacteria contain BChl *a*.

The study of autofluorescence in plants and bacteria provides insights into the basic photosynthesis and the structure–function relationship in the photosynthetic apparatus. Following photoexcitation, the light energy reaches the RC of PSII where a charge separation takes place in the Chl *a* dimer. Autofluorescence measurements, therefore, provide valuable and noninvasive approach to elucidate the mechanism of both energy and electron transfer in PSII complex in photosynthetic organisms.

Spectral properties and fluorescence lifetime of well-known members of the porphyrin family are shown in Table 2.1. With the exception of carotenoids, all photosynthetic pigments, extracted from the photosynthetic membranes, exhibit high fluorescence quantum yield in polar organic solvents (Niedzwiedzki and Blankenship 2010; Steglich et al. 2003). Pigment bound to protein, however, have different spectroscopic properties. For example, fluorescence observed from LHC I and LHC II originates from Chl *a* only due to an efficient energy transfer from Chl *b* to Chl *a* (Visser et al. 1996). The fluorescence of intact cells and isolated chloroplasts of higher plants and green algae is mainly assigned to PSII emission (Krause and Weis 1991).

Protoporphyrin IX, formed from δ-aminolevulinic acid (5-ALA), is also one of the well-known autofluorescence species in mammalian cells with the emission peak around 630 nm and two-exponential fluorescence decay with estimated lifetimes of 1.76 and 8.14 ns (Chen et al. 2005). Protoporphyrin IX is produced in association with the cancerous tissue and accumulates in tumor cells in the presence of excess 5-ALA. As a result, the intensity and lifetime of the fluorescence of protoporphyrin IX can be used to identify and characterize tumor cells. Due to its high quantum yield of generating singlet oxygen, protoporphyrin IX has also been applied as a photosensitizer for photodynamic therapy (Chen et al. 2005). Finally, the heme in red blood cells of mammals is a porphyrin derivative with negligible autofluorescence that is efficiently quenched by the coordinated iron (Nagababu and Rifkind 1998). However, as described in Chapter 6, in vivo measurements of NADH and/or flavins require correction for the absorption of heme.

2.3.7 Other Autofluorescent Species

Melanin is a group of pigments produced by melanocytes and is composed of a complex of heterogeneous large molecules, the structure of which has still been unresolved. It is found in most organisms and is responsible for the color of human skin, hair, eyes, and organs. Melanin has a broad and structureless absorption spectrum in the UV region, protecting the body from exposure to UV rays of the sun. Furthermore, the photon energy absorbed by melanin is converted very rapidly into heat by internal conversion process, resulting in the prevention of the generation of radicals, singlet oxygen, or harmful chemical species (Berezin and Achilefu 2010).

Eumelanin is the most common form of melanin and is a black-to-brown pigment found in dark skin, black hair, and retina of the eye. It exhibits a weak fluorescence in the visible region (400–700 nm) in buffer solution with fluorescence lifetimes in the range of 60 ps to 7 ns (Forest et al. 2000).

Lipofuscin is a granule of pigments found in the skin, heart muscle, liver, nerve cells, and other vital organs. It accumulates with age as well as during the progression of age-related diseases (Berezin and Achilefu 2010). Lipofuscin granules contain a complex mixture of fluorophores. At least one component has been structurally characterized—a quaternary pyridinium salt A2E (Figure 2.2) with a broad emission spectrum in the visible region and a fluorescence lifetime of roughly 12 ps (Lamb et al. 2001). Contribution from other fluorophores, however, is present in lipofuscin fluorescence. Thus, in multiexponential fluorescence decays measured from retinal lipofuscin, along with the short-lifetime component, which still has the highest preexponential factor, subnanosecond and nanosecond components were identified (Cubeddu et al. 1990; Schweitzer et al. 2004, 2007). The correlation between the pathology of eyes and the autofluorescence lifetime of melanin or lipofuscin has been investigated. Chapter 16 describes the FLIM study of melanin and lipofuscin in the human eyeground (Schweitzer et al. 2004, 2007). It was also proposed that multiphoton FLIM has a potential to identify melanin and keratin in human hair (Ehlers et al. 2007).

2.4 Intracellular pH Dependence of Selected Autofluorescence Emitters

Intracellular pH is one of the most important factors in order to understand the physiological states of cells and tissues. For example, the intracellular pH influences the activities of proteins, homeostasis, energy metabolism, and enzymatic reactions in living systems (Paradiso et al. 1987; Yamashiro and Maxfield 1988). Significant change in intracellular pH has been observed with a variety of cellular functions such as ion transport and cell cycle. Mitochondrial matrix acidification was also suggested during synaptic activity (Azarias et al. 2011). Intracellular pH becomes acidic in cancer cells (Pouysségur et al. 2006) and imaging of intracellular pH may be used as a diagnostic tool to differentiate between cancer and normal tissues (Gallagher et al. 2008; Urano et al. 2009; Zhou et al. 2003). An important part in animal and plant development, cell death has been associated with compartmentalized pH changes. For example, mitochondrial matrix alkalinization was observed in an early event of staurosporine-induced apoptosis, but not in Fas-induced apoptosis (Matsuyama et al. 2000). The cytosolic pH has been also shown to decrease from 7.2 to 5.7 in U937 cells undergoing apoptosis in response to tumor necrosis factor-α (Nilsson et al. 2006).

As a result, much effort has been devoted to the development of imaging techniques to measure the spatial distribution of pH in living systems. Within this context, pH-sensitive fluorescent dyes have been developed and widely used to monitor intracellular pH (Llopis et al. 1998; Nedergaard et al. 1990; Whitaker et al. 1991). However, the staining of cells by exogenous fluorophores often induces an excess burden and perhaps toxicity to the living system. In contrast, the endogenous NADH or FAD can be used for probing the intracellular pH in living system as described in the following.

2.4.1 pH Dependence of NADH Autofluorescence

Figure 2.5a shows the pH dependence of the time-resolved fluorescence of NADH, excited at 370 and detected at 450 nm, in aqueous solution with the pH 5.0, 7.0, and 9.0 (Ogikubo et al. 2011). These fluorescence decays are triexponential with two decaying components in the picosecond range that were dominant at each pH with a minor (ca. 2%) component in the nanosecond range (ca. 1–2 ns). The magnitude of the pH-induced change in the average fluorescence lifetime (τ_{av}) is markedly smaller in aqueous solution (ca. 10%) than in HeLa cells (ca. 30%) as pH was increased from 5.0 to 9.0. The larger influence of pH change in cells was attributed to NADH–protein interactions.

The NADH autofluorescence decay was measured for a HeLa cell population with intracellular pH of 5.0, 7.0, and 9.0 (Figure 2.5b). The triple-exponential fitting parameters are summarized in Table 2.1 with a fast component of 500 ps as compared with the slow component of 3 ns. Additional component of more than 10 ns was also observed with a negligible preexponential factor (ca. 2%). Each of the fluorescence lifetime components became shorter as the intracellular pH increased with no significant change in the emission spectra (Ogikubo et al. 2011).

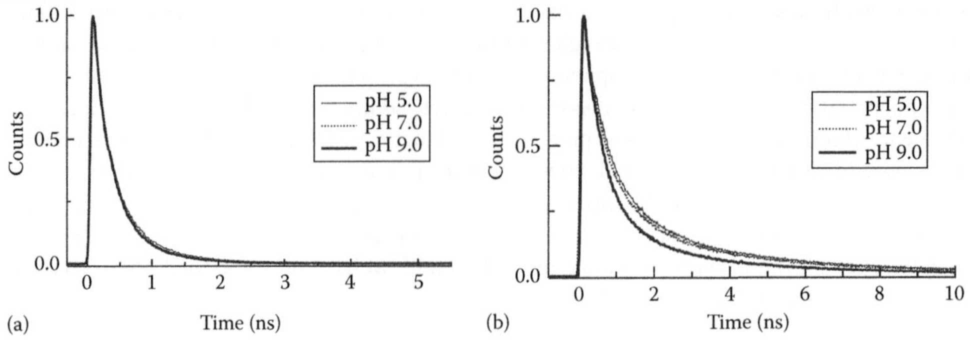

FIGURE 2.5 Representative fluorescence decay profiles of NADH in (a) buffer solution and in (b) HeLa cells at different pHs of 5.0, 7.0, and 9.0. Excitation wavelength was 370 nm. Detection wavelengths was (a) 450 and (b) 440 nm. (Reprinted with permission from Ogikubo, S., Nakabayashi, T., Adachi, T. et al., Intracellular pH sensing using autofluorescence lifetime microscopy, *J. Phys. Chem. B*, 115(34), 10385–10390, 2011. Copyright 2013 American Chemical Society.)

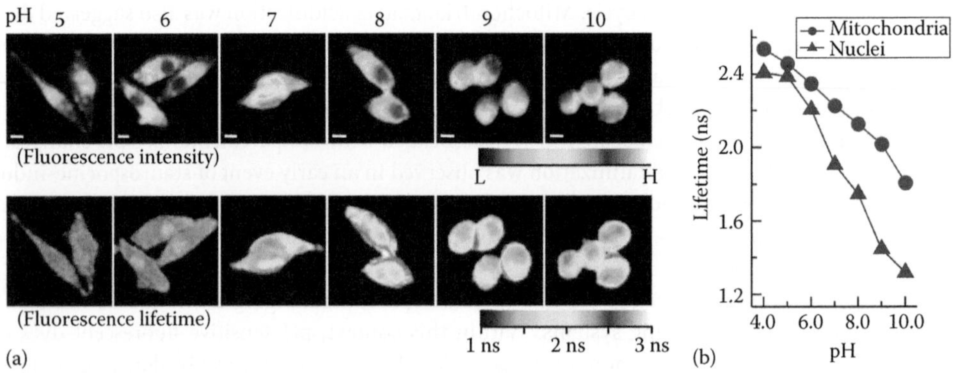

FIGURE 2.6 (a) Fluorescence intensity images (upper) and corresponding fluorescence lifetime images (lower) of NADH in HeLa cells at the intracellular pH indicated at the top. Scale bar is 10 μm. (b) Plots of the fluorescence lifetime as a function of intracellular pH using the values obtained at mitochondria (circles) and at nuclei (triangles) of the lifetime images. (Reprinted with permission from Ogikubo, S., Nakabayashi, T., Adachi, T. et al., Intracellular pH sensing using autofluorescence lifetime microscopy, *J. Phys. Chem. B*, 115(34), 10385–10390, 2011. Copyright 2013 American Chemical Society.)

Figure 2.6 shows the pH dependence of NADH fluorescence intensity and lifetime in HeLa cells (Ogikubo et al. 2011), which were incubated with the medium containing ionophore that equalizes intracellular and extracellular pH (Thomas et al. 1979).

2.4.2 pH Dependence of FAD Autofluorescence

Fluorescence decays of FAD in aqueous buffer at different pH levels are shown in Figure 2.7 (Islam et al. 2013). The multiexponential decay shape in buffer solution remains unchanged over the pH range of 5–9. Multiexponential fluorescence decays of FAD in HeLa cells, carried out with intracellular pH of 5.0, 7.0, and 9.0, are distinct from that in buffer solution (Islam et al. 2013). The fluorescence lifetime of intracellular FAD in HeLa cells becomes shorter with increasing intracellular pH. The picosecond-lifetime components (~80 and 800 ps) become dominant in HeLa cells.

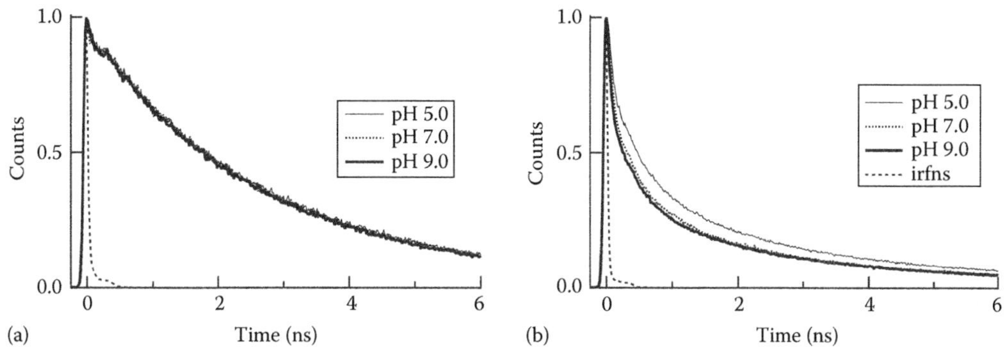

FIGURE 2.7 Representative fluorescence decay profiles of FAD in (a) buffer solution and in (b) HeLa cells at different pHs of 5.0, 7.0, and 9.0. The instrumental response function is shown in each panel by a dotted line. Excitation and detection wavelengths were 450 and 530 nm, respectively. (From Islam, Md.S. et al., *Int. J. Mol. Sci.*, 14(1), 1952, 2013. With permission.)

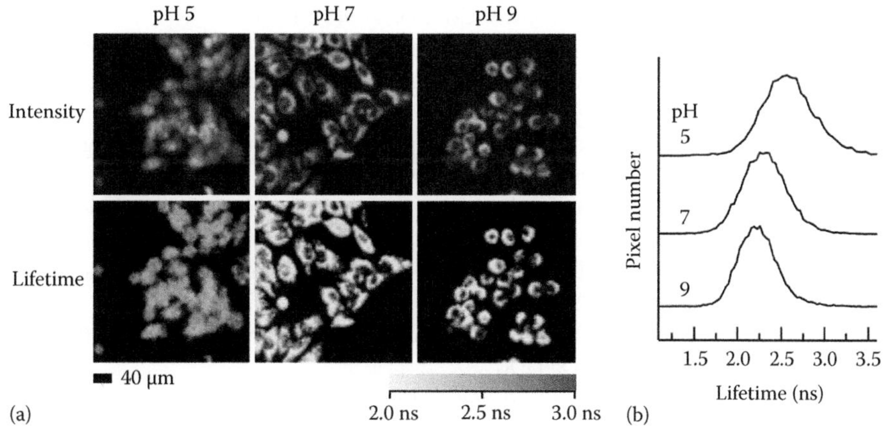

FIGURE 2.8 (a) Autofluorescence intensity images (upper) and corresponding lifetime images (lower) of HeLa cells at intracellular pH indicated at the top. Scale bar given in the bottom is 40 μm. (b) Intracellular pH dependence of the histogram of the fluorescence lifetime obtained from the whole area of the fluorescence lifetime image in (a). Excitation and monitoring wavelengths were 450 and 515–560 nm, respectively. (From Islam, Md.S. et al., *Int. J. Mol. Sci.*, 14(1), 1952, 2013. With permission.)

Autofluorescence lifetime images of FAD in HeLa cells at different intracellular pHs are shown in Figure 2.8 (Islam et al. 2013). The autofluorescence (515–560 nm) was excited at 450 nm and the pseudocolored fluorescence lifetime depends on intracellular pH (longer lifetime at low pH).

2.5 Summary

In this chapter, we surveyed the optical properties and photophysics of representative endogenous fluorophores in cells and tissues. The spectroscopy (absorption and fluorescence spectra) and fluorescence lifetime of endogenous fluorophores are sensitive to surrounding environment. As a result, a careful selection of both excitation and emission wavelengths would allow for selective monitoring of endogenous fluorophores both in vitro and in vivo. Noninvasive imaging of cells provides a unique opportunity to understand cellular condition, molecule–protein interaction, and dynamics and function of living systems with a potential for diagnostic application and therapy. Fluorescence lifetime imaging measurements of NADH and FAD are also applicable to determine the intracellular pH, which is strongly related with energy metabolism and enzymatic reactions.

References

Alford, R., H. M. Simpson, J. Duberman et al. 2009. Toxicity of organic fluorophores used in molecular imaging: Literature review. *Molecular Imaging* 8(6): 341–354.

Awasthi, K., T. Nakabayashi, and N. Ohta. 2012. Application of nanosecond pulsed electric fields into HeLa cells expressing enhanced green fluorescent protein and fluorescence lifetime microscopy. *The Journal of Physical Chemistry B* 116(36): 11159–11165.

Azarias, G., H. Perreten, S. Lengacher et al. 2011. Glutamate transport decreases mitochondrial pH and modulates oxidative metabolism in astrocytes. *The Journal of Neuroscience* 31(10): 3550–3559.

Bailey, A. J. 1978. Collagen and elastin fibres. *Journal of Clinical Pathology* 31: 49–58.

Balaji, J., R. Desai, and S. Maiti. 2004. Live cell ultraviolet microscopy: A comparison between two-and three-photon excitation. *Microscopy Research and Technique* 63(1): 67–71.

Bastiaens, P. I. H. and A. Squire. 1999. Fluorescence lifetime imaging microscopy: Spatial resolution of biochemical processes in the cell. *Trends in Cell Biology* 9(2): 48–52.

Bastiaens, P. I. H., A. Van Hoek, W. F. Wolkers, J. Claude Brochon, and A. J. W. G. Visser. 1992. Comparison of the dynamical structures of lipoamide dehydrogenase and glutathione reductase by time-resolved polarized flavin fluorescence. *Biochemistry* 31(31): 7050–7060.

Becker, W., A. Bergmann, and C. Biskup. 2007. Multispectral fluorescence lifetime imaging by TCSPC. *Microscopy Research and Technique* 70(5): 403–409.

Beechem, J. M. and L. Brand. 1985. Time-resolved fluorescence of proteins. *Annual Review of Biochemistry* 54(1): 43–71.

Berezin, M. Y. and S. Achilefu. 2010. Fluorescence lifetime measurements and biological imaging. *Chemical Reviews* 110(5): 2641–2684.

Bhatta, H. and E. M. Goldys. 2008. Characterization of yeast strains by fluorescence lifetime imaging microscopy. *FEMS Yeast Research* 8(1): 81–87.

Billinton, N. and A. W. Knight. 2001. Seeing the wood through the trees: A review of techniques for distinguishing green fluorescent protein from endogenous autofluorescence. *Analytical Biochemistry* 291(2): 175–197.

Blinova, K., S. Carroll, S. Bose et al. 2005. Distribution of mitochondrial NADH fluorescence lifetimes: Steady-state kinetics of matrix NADH interactions. *Biochemistry* 44(7): 2585–2594.

Bird, D. K., L. Yan, K. M. Vrotsos et al. 2005. Metabolic mapping of MCF10A human breast cells via multiphoton fluorescence lifetime imaging of the coenzyme NADH. *Cancer Research* 65(19): 8766–8773.

Borst, J. W. and A. J. W. G. Visser. 2010. Fluorescence lifetime imaging microscopy in life sciences. *Measurement Science and Technology* 21(10): 102002.

Botchway, S. W., A. W. Parker, R. H. Bisby, and A. G. Crisostomo. 2008. Real-time cellular uptake of serotonin using fluorescence lifetime imaging with two-photon excitation. *Microscopy Research and Technique* 71(4): 267–273.

Brancaleon, L., A. J. Durkin, J. H. Tu, G. Menaker, J. D. Fallon, and N. Kollias. 2001. In vivo fluorescence spectroscopy of nonmelanoma skin cancer. *Photochemistry and Photobiology* 73(2): 178–183.

Brochon, J. C., P. Wahl, M. O. Monneuse-Doublet, and A. Olomucki. 1977. Pulse fluorimetry study of octopine dehydrogenase-reduced nicotinamide adenine dinucleotide complexes. *Biochemistry* 16(21): 4594–4599.

Callis, P. R. and B. K. Burgess. 1997. Tryptophan fluorescence shifts in proteins from hybrid simulations: An electrostatic approach. *The Journal of Physical Chemistry B* 101(46): 9429–9432.

Campagnola, P. 2011. Second harmonic generation imaging microscopy: Applications to diseases diagnostics. *Analytical Chemistry* 83(9): 3224–3231.

Chance, B., P. Cohen, F. Jobsis, and B. Schoener. 1962. Intracellular oxidation-reduction states in vivo. *Science* 137(3529): 499–508.

Chance, B., J. R. Williamson, D. Jamieson, and B. Schoener. 1965. Properties and kinetics of reduced pyridine nucleotide fluorescence of the isolated and in vivo rat heart. *Biochemische Zeitschrift* 341: 357–377.

Chen, H.-M., C.-P. Chiang, C. You, T.-C. Hsiao, and C.-Y. Wang. 2005. Time-resolved autofluorescence spectroscopy for classifying normal and premalignant oral tissues. *Lasers in Surgery and Medicine* 37(1): 37–45.

Chorvat, Jr. D. and A. Chorvatova. 2006. Spectrally resolved time-correlated single photon counting: A novel approach for characterization of endogenous fluorescence in isolated cardiac myocytes. *European Biophysics Journal* 36(1): 73–83.

Chorvat, Jr. D. and A. Chorvatova. 2009. Multi-wavelength fluorescence lifetime spectroscopy: A new approach to the study of endogenous fluorescence in living cells and tissues. *Laser Physics Letters* 6(3): 175–193.

Chosrowjan, H., S. Taniguchi, N. Mataga et al. 2010. Effects of the disappearance of one charge on ultrafast fluorescence dynamics of the FMN binding protein. *The Journal of Physical Chemistry B* 114(18): 6175–6182.

Couprie, M. E., F. Mérola, P. Tauc et al. 1994. First use of the UV Super-ACO free-electron laser: Fluorescence decays and rotational dynamics of the NADH coenzyme. *Review of Scientific Instruments* 65(5): 1485–1495.

Cubeddu, R., F. Docchio, R. Ramponi, and M. Boulton. 1990. Time-resolved fluorescence spectroscopy of the retinal pigment epithelium: Age-related studies. *IEEE Journal of Quantum Electronics* 26(12): 2218–2225.

de Grauw, C. J. and H. C. Gerritsen. 2001. Multiple time-gate module for fluorescence lifetime imaging. *Applied Spectroscopy* 55(6): 670–678.

Diagaradjane, P., M. A. Yaseen, J. Yu, M. S. Wong, and B. Anvari. 2005. Autofluorescence characterization for the early diagnosis of neoplastic changes in DMBA/TPA-induced mouse skin carcinogenesis. *Lasers in Surgery and Medicine* 37(5): 382–395.

Douglas, T., S. Heinemann, U. Hempel et al. 2008. Characterization of collagen II fibrils containing biglycan and their effect as a coating on osteoblast adhesion and proliferation. *Journal of Materials Science: Materials in Medicine* 19(4): 1653–1660.

Drössler, P., W. Holzer, A. Penzkofer, and P. Hegemann. 2002. pH dependence of the absorption and emission behaviour of riboflavin in aqueous solution. *Chemical Physics* 282(3): 429–439.

Ehlers, A., I. Riemann, M. Stark, and K. König. 2007. Multiphoton fluorescence lifetime imaging of human hair. *Microscopy Research and Technique* 70(2): 154–161.

Elofsson, A., R. Rigler, L. Nilsson, J. Roslund, G. Krause, and A. Holmgren. 1991. Motion of aromatic side chains, picosecond fluorescence, and internal energy transfer in *Escherichia coli* thioredoxin studied by site-directed mutagenesis, time-resolved fluorescence spectroscopy, and molecular dynamics simulations. *Biochemistry* 30(40): 9648–9656.

Elson, D. S., J. A. Jo, and L. Marcu. 2007. Miniaturized side-viewing imaging probe for fluorescence lifetime imaging (FLIM): Validation with fluorescence dyes, tissue structural proteins and tissue specimens. *New Journal of Physics* 9(5): 127.

Evans, N. D., L. Gnudi, O. J. Rolinski, D. J. S. Birch, and J. C. Pickup. 2005. Glucose-dependent changes in NAD (P) H-related fluorescence lifetime of adipocytes and fibroblasts in vitro: Potential for non-invasive glucose sensing in diabetes mellitus. *Journal of Photochemistry and Photobiology B: Biology* 80(2): 122–129.

Fang, Q., T. Papaioannou, J. A. Jo, R. Vaitha, K. Shastry, and L. Marcu. 2004. Time-domain laser-induced fluorescence spectroscopy apparatus for clinical diagnostics. *Review of Scientific Instruments* 75(1): 151–162.

Forest, S. E., W. C. Lam, D. P. Millar, J. B. Nofsinger, and J. D. Simon. 2000. A model for the activated energy transfer within eumelanin aggregates. *The Journal of Physical Chemistry B* 104(4): 811–814.

Fritzsche, M., J. M. Fredriksson, M. Carlsson, and C.-F. Mandenius. 2009. A cell-based sensor system for toxicity testing using multiwavelength fluorescence spectroscopy. *Analytical Biochemistry* 387(2): 271–275.

Gafni, A. and L. Brand. 1976. Fluorescence decay studies of reduced nicotinamide adenine dinucleotide in solution and bound to liver alcohol dehydrogenase. *Biochemistry* 15(15): 3165–3171.

Gallagher, F. A., M. I. Kettunen, S. E. Day, D.-E. Hu, and J. H. Ardenkjær-Larsen. 2008. Magnetic resonance imaging of pH in vivo using hyperpolarized 13C-labelled bicarbonate. *Nature* 453(7197): 940–943.

Ghukasyan, V. V. and F.-J. Kao. 2009. Monitoring cellular metabolism with fluorescence lifetime of reduced nicotinamide adenine dinucleotide. *The Journal of Physical Chemistry C* 113(27): 11532–11540.

Grajek, H., I. Gryczynski, P. Bojarski, Z. Gryczynski, S. Bharill, and L. Kułak. 2007. Flavin mononucleotide fluorescence intensity decay in concentrated aqueous solutions. *Chemical Physics Letters* 439(1–3): 151–156.

Guzow, K., M. Szabelski, A. Rzeska, J. Karolczak, H. Sulowska, and W. Wiczk. 2002. Photophysical properties of tyrosine at low pH range. *Chemical Physics Letters* 362(5–6): 519–526.

Hall, C. L. and H. Kamin. 1975. The purification and some properties of electron transfer flavoprotein and general fatty acyl coenzyme A dehydrogenase from pig liver mitochondria. *The Journal of Biological Chemistry* 250(9): 3476–3486.

Harvey, E. N. 1957. *A History of Luminescence: From the Earliest Times Until 1900*. Philadelphia, PA: American Philosophical Society.

Huang, S., A. A. Heikal, and W. W. Webb. 2002. Two-photon fluorescence spectroscopy and microscopy of NAD (P) H and flavoprotein. *Biophysical Journal* 82(5): 2811–2825.

Islam, Md. S., M. Honma, T. Nakabayashi, M. Kinjo, and N. Ohta. 2013. pH Dependence of the fluorescence lifetime of FAD in solution and in cells. *International Journal of Molecular Sciences* 14(1): 1952–1963.

Ito, T., S. Oshita, T. Nakabayashi, F. Sun, M. Kinjo, and N. Ohta. 2009. Fluorescence lifetime images of green fluorescent protein in HeLa cells during TNF-α induced apoptosis. *Photochemical and Photobiological Sciences* 8(6): 763–767.

Kandoth, N., S. D. Choudhury, J. Mohanty, A. C. Bhasikuttan, and H. Pal. 2010. Inhibiting intramolecular electron transfer in flavin adenine dinucleotide by host–guest interaction: A fluorescence study. *The Journal of Physical Chemistry B* 114(8): 2617–2626.

Kao, Y.-T., C. Tan, S.-H. Song et al. 2008. Ultrafast dynamics and anionic active states of the flavin cofactor in cryptochrome and photolyase. *Journal of the American Chemical Society* 130(24): 7695–7701.

Kierdaszuk, B., H. Malak, I. Gryczynski, P. Callis, and J. R. Lakowicz. 1996. Fluorescence of reduced nicotinamides using one-and two-photon excitation. *Biophysical Chemistry* 62(1–3): 1–13.

Klaidman, L. K., A. C. Leung, and J. D. Adams, Jr. 1995. High-performance liquid chromatography analysis of oxidized and reduced pyridine dinucleotides in specific brain regions. *Analytical Biochemistry* 228(2): 312–317.

König, K., M. W. Berns, and B. J. Tromberg. 1997. Time-resolved and steady-state fluorescence measurements of β-nicotinamide adenine dinucleotide-alcohol dehydrogenase complex during UVA exposure. *Journal of Photochemistry and Photobiology B: Biology* 37(1–2): 91–95.

Krause, G. H. and E. Weis. 1991. Chlorophyll fluorescence and photosynthesis: The basics. *Annual Review of Plant Physiolocy and Plant Molecular Biology* 42(1): 313–349.

Kunz, W. S. 1986. Spectral properties of fluorescent flavoproteins of isolated rat liver mitochondria. *FEBS Letters* 195(1–2): 92–96.

Kunz, W. S. and W. Kunz. 1985. Contribution of different enzymes to flavoprotein fluorescence of isolated rat liver mitochondria. *Biochimica et Biophysica Acta (BBA)—General Subjects* 841(3): 237–246.

Lakowicz, J. R., ed. 2006. *Principles of Fluorescence Spectroscopy.* New York: Springer.

Lakowicz, J. R. and K. W. Berndt. 1991. Lifetime-selective fluorescence imaging using an rf phase-sensitive camera. *Review of Scientific Instruments* 62(7): 1727–1734.

Lakowicz, J. R., H. Szmacinski, K. Nowaczyk, and M. L. Johnson. 1992. Fluorescence lifetime imaging of free and protein-bound NADH. *Proceedings of the National Academy of Sciences of the United States of America* 89(4): 1271–1275.

Lamb, L. E., T. Ye, N. M. Haralampus-Grynaviski et al. 2001. Primary photophysical properties of A2E in solution. *The Journal of Physical Chemistry B* 105(46): 11507–11512.

Latouche, G., Z. G. Cerovic, F. Montagnini, and I. Moya. 2000. Light-induced changes of NADPH fluorescence in isolated chloroplasts: A spectral and fluorescence lifetime study. *Biochimica et Biophysica Acta (BBA)—Bioenergetics* 1460(2–3): 311–329.

Levitt, J. A., D. R. Matthews, S. M. Ameer-Beg, and K. Suhling. 2009. Fluorescence lifetime and polarization-resolved imaging in cell biology. *Current Opinion in Biotechnology* 20(1): 28–36.

Li, D., W. Zheng, and J. Y. Qu. 2008. Time-resolved spectroscopic imaging reveals the fundamentals of cellular NADH fluorescence. *Optics Letters* 33(20): 2365–2367.

Li, D., W. Zheng, and J. Y. Qu. 2009. Two-photon autofluorescence microscopy of multicolor excitation. *Optics Letters* 34(2): 202–204.

Lichtman, J. W. and J.-A. Conchello. 2005. Fluorescence microscopy. *Nature Methods* 2(12): 910–919.

Lin, S. W. and T. P. Sakmar. 1996. Specific tryptophan UV-absorbance changes are probes of the transition of rhodopsin to its active state. *Biochemistry* 35(34): 11149–11159.

Llopis, J., J. M. McCaffery, A. Miyawaki, M. G. Farquhar, and R. Y. Tsien. 1998. Measurement of cytosolic, mitochondrial, and Golgi pH in single living cells with green fluorescent proteins. *Proceedings of the National Academy of Sciences of the United States of America* 95(12): 6803–6808.

Maarek, J.-M. I., L. Marcu, W. J. Snyder, and W. S. Grundfest. 2000. Time-resolved fluorescence spectra of arterial fluorescent compounds: Reconstruction with the Laguerre expansion technique. *Photochemistry and Photobiology* 71(2): 178–187.

Maiti, S., J. B. Shear, R. M. Williams, W. R. Zipfel, and W. W. Webb. 1997. Measuring serotonin distribution in live cells with three-photon excitation. *Science* 275(5299): 530–532.

Marcelli, A., I. J. Badovinac, N. Orlic, P. R. Salvi, and C. Gellini. 2013. Excited-state absorption and ultrafast relaxation dynamics of protoporphyrin IX and hemin. *Photochemical and Photobiological Sciences* 12(2): 348–355.

Mataga, N., H. Chosrowjan, Y. Shibata, F. Tanaka, Y. Nishina, and K. Shiga. 2000. Dynamics and mechanisms of ultrafast fluorescence quenching reactions of flavin chromophores in protein nanospace. *The Journal of Physical Chemistry B* 104(45): 10667–10677.

Matsuyama, S., J. Llopis, Q. L. Deveraux, R. Y. Tsien, and J. C. Reed. 2000. Changes in intramitochondrial and cytosolic pH: Early events that modulate caspase activation during apoptosis. *Nature Cell Biology* 2(6): 318–325.

Mayevsky, A. and B. Chance. 2007. Oxidation–reduction states of NADH in vivo: From animals to clinical use. *Mitochondrion* 7(5): 330–339.

McGuinness, C. D., A. M. Macmillan, K. Sagoo, D. McLoskey, and D. J. S. Birch. 2006. Excitation of fluorescence decay using a 265 nm pulsed light-emitting diode: Evidence for aqueous phenylalanine rotamers. *Applied Physics Letters* 89(6): 063901.

Meerwaldt, R., R. Graaff, P. H. N. Oomen et al. 2004. Simple non-invasive assessment of advanced glycation endproduct accumulation. *Diabetologia* 47(7): 1324–1330.

Meerwaldt, R., J. W. L. Hartog, R. Graaff et al. 2005. Skin autofluorescence, a measure of cumulative metabolic stress and advanced glycation end products, predicts mortality in hemodialysis patients. *Journal of the American Society of Nephrology* 16(12): 3687–3693.

Menter, J. M. 2006. Temperature dependence of collagen fluorescence. *Photochemical and Photobiological Sciences* 5(4): 403–410.

Miller, E. J. and S. Gay. 1987. The collagens: An overview and update. In: *Structural and Contractile Proteins Part D: Extracellular Matrix*, L. W. Cunningham (ed.). Vol. 144 of Methods in Enzymology, J. N. Abelson, M. I. Simon, A. M. Pyle, and G. L. Verdine (eds.), pp. 3–41. San Diego, CA: Academic Press.

Nagababu, E. and J. M. Rifkind. 1998. Formation of fluorescent heme degradation products during the oxidation of hemoglobin by hydrogen peroxide. *Biochemical and Biophysical Research Communications* 247(3): 592–596.

Nakabayashi, T., Md. S. Islam, and N. Ohta. 2010. Fluorescence decay dynamics of flavin adenine dinucleotide in a mixture of alcohol and water in the femtosecond and nanosecond time range. *The Journal of Physical Chemistry B* 114(46): 15254–15260.

Nakabayashi, T., I. Nagao, M. Kinjo, Y. Aoki, M. Tanaka, and N. Ohta. 2008a. Stress-induced environmental changes in a single cell as revealed by fluorescence lifetime imaging. *Photochemical and Photobiological Sciences* 7(6): 671–674.

Nakabayashi, T., S. Oshita, R. Sumikawa, F. Sun, M. Kinjo, and N. Ohta. 2012. pH dependence of the fluorescence lifetime of enhanced yellow fluorescent protein in solution and cells. *Journal of Photochemistry and Photobiology A: Chemistry* 235: 65–71.

Nakabayashi, T., H.-P. Wang, M. Kinjo, and N. Ohta. 2008b. Application of fluorescence lifetime imaging of enhanced green fluorescent protein to intracellular pH measurements. *Photochemical and Photobiological Sciences* 7(6): 668–670.

Nakabayashi, T., H.-P. Wang, K. Tsujimoto, S. Miyauchi, N. Kamo, and N. Ohta. 2008c. Studies on effects of external electric fields on halobacteria with fluorescence intensity and fluorescence lifetime imaging microscopy. *Chemistry Letters* 37(5): 522–523.

Nedergaard, M., S. Desai, and W. Pulsinelli. 1990. Dicarboxy-dichlorofluorescein: A new fluorescent probe for measuring acidic intracellular pH. *Analytical Biochemistry* 187(1): 109–114.

Nelson, D. L. and M. M. Cox. 2008. *Lehninger Principles of Biochemistry*. New York: W. H. Freeman & Co.

Niedzwiedzki, D. M. and R. E. Blankenship. 2010. Singlet and triplet excited state properties of natural chlorophylls and bacteriochlorophylls. *Photosynthesis Research* 106(3): 227–238.

Niesner, R., B. Peker, P. Schlüsche, and K.-H. Gericke. 2004. Noniterative biexponential fluorescence lifetime imaging in the investigation of cellular metabolism by means of NAD(P)H autofluorescence. *ChemPhysChem* 5(8): 1141–1149.

Nilsson, C., U. Johansson, A.-C. Johansson, K. Kågedal, and K. Öllinger. 2006. Cytosolic acidification and lysosomal alkalinization during TNF-α induced apoptosis in U937 cells. *Apoptosis* 11(7): 1149–1159.

Ogikubo, S., T. Nakabayashi, T. Adachi et al. 2011. Intracellular pH sensing using autofluorescence lifetime microscopy. *The Journal of Physical Chemistry B* 115(34): 10385–10390.

Ohta, N., T. Nakabayashi, I. Nagao, M. Kinjo, Y. Aoki, and M. Tanaka. 2009. Fluorescence lifetime imaging study of a single cell: Stress-induced environmental change and electric field effects on fluorescence. *Proceedings of SPIE* 7190: 71900R1–71900R11.

Ohta, N., T. Nakabayashi, S. Oshita, and M. Kinjo. 2010. Fluorescence lifetime imaging spectroscopy in living cells with particular regards to pH dependence and electric field effect. *Proceedings of SPIE* 7576: 75760G1–75760G13.

Paradiso, A. M., R. Y. Tsien, and T. E. Machen. 1987. Digital image processing of intracellular pH in gastric oxyntic and chief cells. *Nature* 325(6103): 447–450.

Pierce, D. W. and S. G. Boxer. 1995. Stark effect spectroscopy of tryptophan. *Biophysical Journal* 68(4): 1583–1591.

Pollak, N., C. Dölle, and M. Ziegler. 2007. The power to reduce: Pyridine nucleotides-small molecules with a multitude of functions. *Biochemical Journal* 402(2): 205–218.

Pouysségur, J., F. Dayan, and N. M. Mazure. 2006. Hypoxia signalling in cancer and approaches to enforce tumour regression. *Nature* 441(7092): 437–443.

Pullerits, T. and V. Sundström. 1996. Photosynthetic light-harvesting pigment-protein complexes: Toward understanding how and why. *Accounts of Chemical Research* 29(8): 381–389.

Reinert, K. C., W. Gao, G. Chen, and T. J. Ebner. 2007. Flavoprotein autofluorescence imaging in the cerebellar cortex in vivo. *Journal of Neuroscience Research* 85(15): 3221–3232.

Reinert, K. C., W. Gao, G. Chen, X. Wang, Y.-P. Peng, and T. J. Ebner. 2011. Cellular and metabolic origins of flavoprotein autofluorescence in the cerebellar cortex in vivo. *The Cerebellum* 10(3): 585–599.

Richards-Kortum, R. and E. Sevick-Muraca. 1996. Quantitative optical spectroscopy for tissue diagnosis. *Annual Review of Physical Chemistry* 47(1): 555–606.

Robbins, R. J., G. R. Fleming, G. S. Beddard, G. W. Robinson, P. J. Thistlethwaite, and G. J. Woolfe. 1980. Photophysics of aqueous tryptophan: pH and temperature effects. *Journal of the American Chemical Society* 102(20): 6271–6279.

Romashko, D. N., E. Marban, and B. O' Rourke. 1998. Subcellular metabolic transients and mitochondrial redox waves in heart cells. *Proceedings of the National Academy of Sciences of the United States of America* 95(4): 1618–1623.

Sady, C., S. Khosrof, and R. Nagaraj. 1995. Advanced Maillard reaction and crosslinking of corneal collagen in diabetes. *Biochemical and Biophysical Research Communications* 214(3): 793–797.

Schweitzer, D., M. Hammer, F. Schweitzer et al. 2004. In vivo measurement of time-resolved autofluorescence at the human fundus. *Journal of Biomedical Optics* 9(6): 1214–1222.

Schweitzer, D., S. Schenke, M. Hammer et al. 2007. Towards metabolic mapping of the human retina. *Microscopy Research and Technique* 70(5): 410–419.

Sell, D. R. and V. M. Monnier. 1989. Structure elucidation of a senescence cross-link from human extracellular matrix. Implication of pentoses in the aging process. *The Journal of Biological Chemistry* 264(36): 21597–21602.

Sell, D. R., I. Nemet, and V. M. Monnier. 2010. Partial characterization of the molecular nature of collagen-linked fluorescence: Role of diabetes and end-stage renal disease. *Archives of Biochemistry and Biophysics* 493(2): 192–206.

Sengupta, A., R. V. Khade, and P. Hazra. 2011. pH dependent dynamic behavior of flavin mononucleotide (FMN) and flavin adenine dinucleotide (FAD) in femtosecond to nanosecond time scale. *Journal of Photochemistry and Photobiology A: Chemistry* 221(1): 105–112.

Shikanai, T. 2007. Cyclic electron transport around photosystem I: Genetic approaches. *Annual Review of Plant Biology* 58: 199–217.

Shuttleworth, C. W. 2010. Use of NAD(P)H and flavoprotein autofluorescence transients to probe neuron and astrocyte responses to synaptic activation. *Neurochemistry International* 56(3): 379–386.

Skala, M. C., K. M. Riching, A. Gendron-Fitzpatrick et al. 2007. In vivo multiphoton microscopy of NADH and FAD redox states, fluorescence lifetimes, and cellular morphology in precancerous epithelia. *Proceedings of the National Academy of Sciences of the United States of America* 104(49): 19494–19499.

Steglich, C., C. W. Mullineaux, K. Teuchner, W. R. Hess, and H. Lokstein. 2003. Photophysical properties of *Prochlorococcus marinus* SS120 divinyl chlorophylls and phycoerythrin in vitro and in vivo. *FEBS Letters* 553(1–2): 79–84.

Tanaka, F. and N. Mataga. 1987. Fluorescence quenching dynamics of tryptophan in proteins. Effect of internal rotation under potential barrier. *Biophysical Journal* 51(3): 487–495.

Thomas, J. A., R. N. Buchsbaum, A. Zimniak, and E. Racker. 1979. Intracellular pH measurements in Ehrlich ascites tumor cells utilizing spectroscopic probes generated in situ. *Biochemistry* 18(11): 2210–2218.

Thomas, P., P. Pande, F. Clubb, J. Adame, and J. A. Jo. 2010. Biochemical imaging of human atherosclerotic plaques with fluorescence lifetime angioscopy. *Photochemistry and Photobiology* 86(3): 727–731.

Tohmi, M., K. Takahashi, Y. Kubota, R. Hishida, and K. Shibuki. 2009. Transcranial flavoprotein fluorescence imaging of mouse cortical activity and plasticity. *Journal of Neurochemistry* 109(S1): 3–9.

Urano, Y., D. Asanuma, Y. Hama et al. 2009. Selective molecular imaging of viable cancer cells with pH-activatable fluorescence probes. *Nature Medicine* 15(1): 104–109.

Valeur, B. 2001. *Molecular Fluorescence: Principles and Applications*. Weinheim, Germany: Wiley-VCH.

van den Berg, P. A. W., K. A. Feenstra, A. E. Mark, H. J. C. Berendsen, and A. J. W. G. Visser. 2002. Dynamic conformations of flavin adenine dinucleotide: Simulated molecular dynamics of the flavin cofactor related to the time-resolved fluorescence characteristics. *The Journal of Physical Chemistry B* 106(34): 8858–8869.

Vishwasrao, H. D., A. A. Heikal, K. A. Kasischke, and W. W. Webb. 2005. Conformational dependence of intracellular NADH on metabolic state revealed by associated fluorescence anisotropy. *The Journal of Biological Chemistry* 280(26): 25119–25126.

Visser, A. J. W. G. and A. van Hoek. 1981. The fluorescence decay of reduced nicotinamides in aqueous solution after excitation with a UV-mode locked Ar ion laser. *Photochemistry and Photobiology* 33(1): 35–40.

Visser, H. M., F. J. Kleima, I. H. M. Van Stokkum, R. Van Grondelle, and H. Van Amerongen. 1996. Probing the many energy-transfer processes in the photosynthetic light-harvesting complex II at 77 K using energy-selective sub-picosecond transient absorption spectroscopy. *Chemical Physics* 210(1–2): 297–312.

Wallrabe, H. and A. Periasamy. 2005. Imaging protein molecules using FRET and FLIM microscopy. *Current Opinion in Biotechnology* 16(1): 19–27.

Wang, H.-P., T. Nakabayashi, K. Tsujimoto, S. Miyauchi, N. Kamo, and N. Ohta. 2007. Fluorescence lifetime image of a single halobacterium. *Chemical Physics Letters* 442(4–6): 441–444.

Wang, H.-W., V. Gukassyan, C.-T. Chen et al. 2008. Differentiation of apoptosis from necrosis by dynamic changes of reduced nicotinamide adenine dinucleotide fluorescence lifetime in live cells. *Journal of Biomedical Optics* 13(5): 054011.

Weber, G. 1950. Fluorescence of riboflavin and flavin-adenine dinucleotide. *Biochemical Journal* 47(1): 114–121.

Whitaker, J. E., R. P. Haugland, and F. G. Prendergast. 1991. Spectral and photophysical studies of benzo [*c*] xanthene dyes: Dual emission pH sensors. *Analytical Biochemistry* 194(2): 330–344.

Wu, K., W. Liu, and G. Li. 2013. The aggregation behavior of native collagen in dilute solution studied by intrinsic fluorescence and external probing. *Spectrochimica Acta Part A: Molecular Spectroscopy* 102: 186–193.

Wu, Y., P. Xi, J. Y. Qu, T.-H. Cheung, and M.-Y. Yu. 2004. Depth-resolved fluorescence spectroscopy reveals layered structure of tissue. *Optics Express* 12(14): 3218–3223.

Yamashiro, D. J. and F. R. Maxfield. 1988. Regulation of endocytic processes by pH. *Trends in Pharmacological Sciences* 9(6): 190–193.

Yamazaki, I., N. Tamai, T. Yamazaki, A. Murakami, M. Mimuro, and Y. Fujita. 1998. Sequential excitation energy transport in stacking multilayers: A comparative study between photosynthetic antenna and Langmuir-Blodgett multilayers. *The Journal of Physical Chemistry* 92(17): 5035–5044.

Yu, Q. and A. A. Heikal. 2009. Two-photon autofluorescence dynamics imaging reveals sensitivity of intracellular NADH concentration and conformation to cell physiology at the single-cell level. *Journal of Photochemistry and Photobiology B: Biology* 95(1): 46–57.

Zhou, J., J.-F. Payen, D. A. Wilson, R. J. Traystman, and P. C. M. van Zijl. 2003. Using the amide proton signals of intracellular proteins and peptides to detect pH effects in MRI. *Nature Medicine* 9(8): 1085–1090.

Ziegenhorn, J., M. Senn, and Th. Bücher. 1976. Molar absorptivities of b-NADH and b-NADPH. *Clinical Chemistry* 22(2): 151–160.

II

Autofluorescence Imaging Techniques: Fundamentals and Applications

3

One-Photon Autofluorescence Microscopy

Narasimhan Rajaram
Duke University

Nirmala Ramanujam
Duke University

3.1 Introduction

Autofluorescence from cells has been detected and utilized by scientists for well over 100 years for a variety of applications. The need for fluorescence as a noninvasive tool for cell studies was spurred by advances in optical microscopy, fueling the need for better contrasting agents to image samples. Initially, fluorescent dyes such as fluorescein, acridine orange, and neutral red provided a very high level of contrast due to their specificity in attaching to specific proteins in plant and animal cells. Cell and tissue autofluorescence was simply observed as a biological phenomenon that tended to interfere with the signal and reduce image contrast. It was not until the landmark studies of Britton Chance (Chance and Williams 1955a,b,c; Chance et al. 1962) that autofluorescence from cells was attributed to specific coenzymes, such as pyridine nucleotides, or flavoproteins in mitochondria. Further, it was argued that because the oxidized pyridine nucleotides (NAD^+) and the reduced form of flavins ($FADH_2$) were not fluorescent, it was possible to ascertain the reduction–oxidation (redox) state of cells. Since then, autofluorescence from cells in the form of NADH and FAD fluorescence has been used in many studies to investigate the metabolic activities of various organs. Specifically, some of the very first in vivo studies for examining intracellular redox states were performed using single-photon fluorescence microscopes (Chance et al. 1962). In this chapter, we discuss the important components that are essential in the design of fluorescence microscopy, calibration techniques, and image processing methods.

3.2 Working Principles of a Fluorescence Microscope

Optical microscopy has undergone vast advances in the last century. Microscopes are now equipped with onboard electronics and intuitive user-friendly software that allows a user to determine the

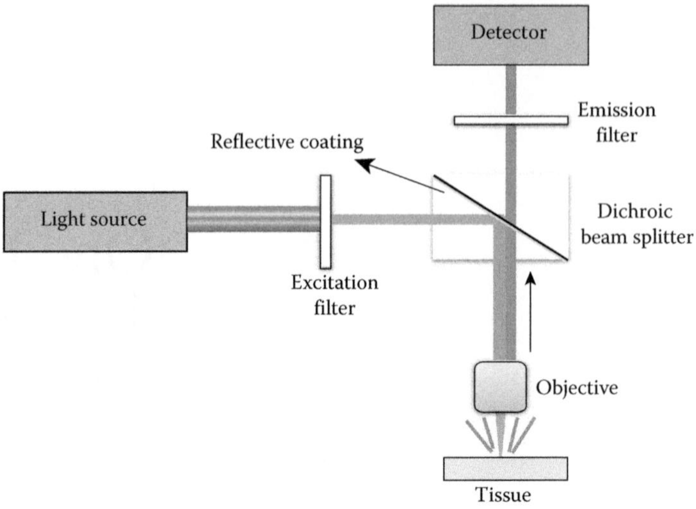

FIGURE 3.1 Schematic of an epifluorescence microscope.

acquisition over a specific area of the sample, perform time-lapse imaging, zoom into a region of interest, and acquire 3D image stacks by optically slicing the sample at regular intervals, just to mention a few features. Fluorescent microscopes have three major components (Figure 3.1): (1) the light source, (2) filter sets, and (3) detectors, such as charge-coupled devices (CCDs). In this chapter, we present a detailed overview of each component, the choices available to the user, and, based on the scientific question or hypothesis, how the selection of the right combination is crucial for recording good images from cell autofluorescence.

3.2.1 Major Designs

3.2.1.1 Wide-Field Microscopes

Two commercially available microscope designs are currently employed in fluorescence microscopy: inverted and upright epifluorescence with the microscope objective positioned below or above the sample, respectively. In both designs, the microscope objective is used both to focus the illumination light on the sample and to collect the emitted fluorescence photons. The same microscope objective is a key determinant of the projected image resolution. An alternative design is where the illumination and fluorescence collection are located on the opposite sides of the sample. This design is utilized in other microscopy modalities, such as bright-field microscopy and differential interference contrast (DIC) microscopy. It is recommended to use the least optical components between the sample and detector in order to avoid unnecessary loss of fluorescence photons.

Figure 3.1 shows a schematic representation of a typical fluorescence microscope without the optical details associated with the conditioning of the illumination light or epifluorescence photons. In epifluorescence geometry, excitation light (from arc lamps or lasers) is directed through a filter set (discussed in Section 3.2.4) onto the sample. The resulting fluorescence is collected by a detector, typically a CCD in wide-field microscopy, or a photomultiplier tube (PMT) in confocal microscopy. The output images are then recorded using specialized software for data storage and analysis.

Fluorescence contrast is an important issue that follows directly from the level of fluorescence intensity relative to the background signal. In a wide-field fluorescence microscope, secondary fluorescence from outside of the region of interest reduces the contrast and, correspondingly, the resolution of the features that are in focus. In cases where the background is nearly as bright as the signal obtained from the sample (i.e., very low contrast), a simple increase of the illumination light intensity will not solve the problem and might lead to photobleaching. Confocal fluorescence microscopy addresses these limitations and provides images at higher spatial resolution and with superior contrast as compared to wide-field fluorescence microscopy.

3.2.1.2 Laser-Scanning Confocal Fluorescence Microscopes

Confocal microscopy retains the same optical design as the fluorescence microscope shown in Figure 3.1, except for the inclusion of a variable aperture (pinhole) in front of the detector in order to reject out-of-focus light, and laser-scanning capability. The genius of the confocal pinhole is to reject the fluorescence photons that are generated from above and below the focal plane. The smaller the pinhole diameter, the better the spatial resolution and therefore the contrast. In confocal microscopy, the region of interest of the sample is excited by raster scanning of the excitation laser and the corresponding fluorescence signal is registered in a designated (x,y,z) voxel. The excitation laser scanning is achieved using two orthogonal galvanometer mirrors that are controlled using software provided by the manufacturers, where the excitation and fluorescence detection at each pixel are synchronized. Most commercially available confocal microscopes allow for variable pinhole diameter, providing researchers with the flexibility to optimize their images toward answering biological questions. Careful selection of the optimum pinhole diameter requires a balance between the signal level that can be detected (the larger the pinhole, the higher the signal level) and the image resolution (the smaller the pinhole, the higher the spatial resolution). It is also worth noting that the conditioning of illumination light in confocal microscopy is different from that in wide-field and total internal reflection fluorescence (TIRF) modalities.

3.2.2 Excitation Sources

A stable light source is essential for quantitative fluorescence microscopy in a wide range of applications. There are a number of illumination light sources that users can choose from, which in part depend on the financial resources available. We highlight some typical light sources used in optical microscopy in the following.

3.2.2.1 Arc Lamps

The most widely used light sources for NADH and FAD imaging are arc lamps, in particular mercury and xenon arc lamps. Mercury arc lamps have been used in fluorescence microscopy applications since the 1930s and have a unique spectral profile with several emission peaks located in the visible wavelength range. These lamps have a significantly higher output power at discrete 405, 436, 546, and 579 nm wavelengths compared to the background. Therefore, mercury lamps are ideal for fluorophores whose excitation wavelengths are located close to these spectral lines. However, mercury arc lamps have some disadvantages such as the need for specialized housing, regular maintenance and alignment of the internal optical components, and a relatively short lifetime. In addition, these lamps might not be ideal for imaging NADH and FAD autofluorescence due to the lamp's relatively weak output power around 350 and 450 nm, respectively. In contrast, xenon arc lamps provide a stable and nearly uniform emission profile (i.e., no prominent spectral lines or peaks) in the visible wavelength range (400–700 nm) as illustrated in Figure 3.2. Xenon lamps are therefore more suited to imaging NADH and FAD with proper choice of excitation filters.

3.2.2.2 Lasers

Lasers (an acronym for laser amplification by stimulated emission of radiation) provide yet another option as a powerful excitation light source for fluorescence applications, especially in confocal microscopy. Lasers possess several advantages suited for confocal microscopy: (1) high degree of

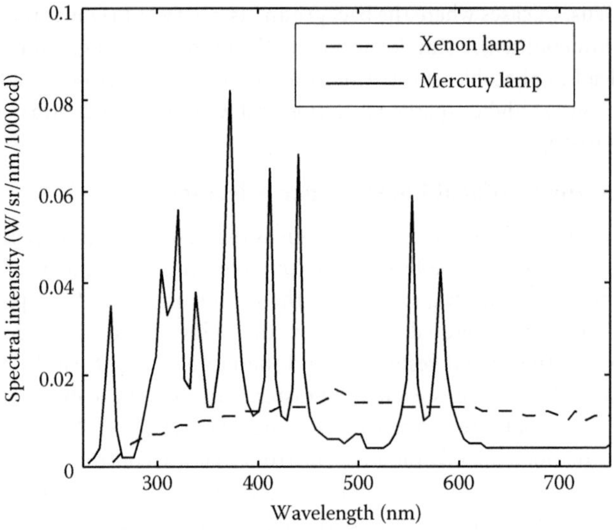

FIGURE 3.2 Spectral profiles of xenon and mercury arc lamps. (Courtesy of Zeiss Microscopy, Pleasanton, CA.)

monochromaticity (i.e., narrow spectral lines), (2) small divergence angle (i.e., collimated over a relatively long distance), (3) high brightness (or photon flux), and (4) high degree of spatial and temporal coherence. The invention of laser allowed the development of fast-scanning confocal microscopes that replaced slow mechanical scanning of the sample. Briefly, the laser consists of three important components: an active gain medium (a dye, a gas, a semiconductor, or a crystal); energy source for achieving the so-called population inversion, where the excited state population of the gain medium is larger than that of the ground state; and an optical resonator (or a laser cavity). The argon-ion and argon/krypton lasers are the most widely used in confocal microscopes due to the availability of several laser lines ranging from 351 to 514 nm as well as their reasonable price. The most commonly available laser lines are 458, 488, and 514 nm. Typically, the 458 nm laser line is best suited for imaging FAD. Exciting NADH fluorescence with the argon-ion laser requires a 351 nm laser line. However, laser lines in the UV region require specialized optics for efficient excitation and detection.

In addition to the established gas lasers, diode lasers can provide additional spectral lines at wavelengths as low as 405 nm. Recently, semiconductor diode lasers have begun to replace gas lasers due to their small footprint, ease of use, lower power consumption, and the ability to design smaller microscope packages. In any given application, the laser lines (or systems) can be chosen based on the fluorescent markers used for labeling target proteins or organelles in cells of tissues of interest.

3.2.3 Filter Sets

Knowledge of the absorption and fluorescence profiles of fluorophores of interest is the key for selecting excitation and emission filters, as well as the dichroic mirror for optimum imaging. A filter set for fluorescence microscopy applications consists of three parts: (1) an excitation filter for selecting the illumination wavelength, (2) a dichroic beam splitter that reflects a certain set of wavelengths while transmitting another range of wavelengths, and (3) an emission (barrier) filter for further restriction of the wavelengths that reach the detector. The dichroic beam splitter has a reflective coating on one side and a nonreflective coating on the other. The reflective coating always faces the excitation light source for efficient reflection (or steering) into the objective. The fluorescence is then passed through

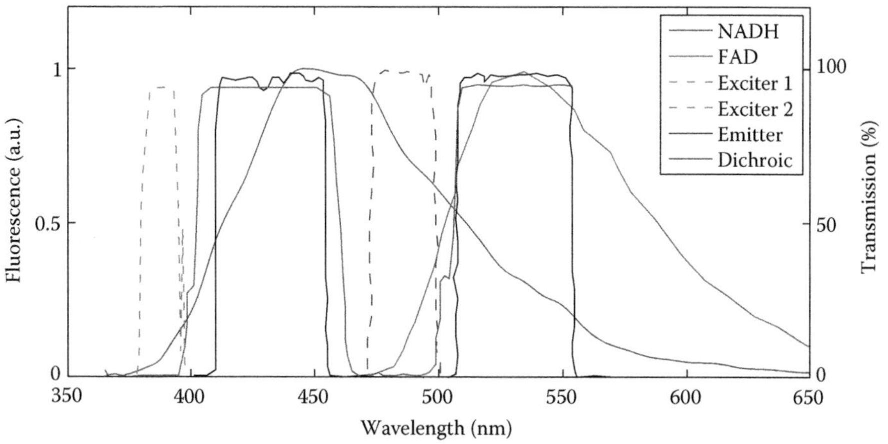

FIGURE 3.3 Spectral characteristics of the BrightLine Pinkel filter set and NADH and FAD fluorescence. Exciter 1 corresponds to the excitation pass band for the first fluorophore (NADH) and exciter 2 corresponds to the second fluorophore (FAD). The transmission band of the emission filters included in this set overlaps with the dichroic mirror. For the sake of clarity, absorption spectra of NADH and FAD are not shown.

both the dichroic mirror and the emission filter prior to detection. This combination of dichroic and emission filters represents a very efficient means for separating the high-intensity excitation light from the low-intensity fluorescence emission (i.e., minimum scattering signal and therefore higher image contrast).

Depending on the fluorophore of interest, it is important to select the right combination of excitation and emission filters to maximize fluorescence detection and image contrast while minimizing the background. Consider, for example, the endogenous FAD (excitation/emission max, ~450/525 nm). In this case, the recommended band-pass excitation filter can be centered at 450 nm with a bandwidth of 430–470 nm (also denoted as 450/40 nm). In addition, the recommended dichroic mirror should reflect the light of shorter wavelength range (430 and 470 nm) while transmitting fluorescence wavelength longer than 480 nm. Finally, the emission filter for FAD fluorescence should be centered around 525 nm with enough bandwidth to maximize the detection efficiency while minimizing spectral overlap with other fluorophores in case of multiple-channel microscopy.

Filter sets are available commercially from a number of manufacturers, such as Semrock and Chroma, which provide several options optimized for specific fluorophores and the microscope model. For imaging NADH and FAD autofluorescence, one may choose the BrightLine® Pinkel (Semrock, part number DA/Fl-2X-B-000). Figure 3.3 presents the spectral characteristics of this filter set and how it overlaps with the absorption and emission spectra of both NADH and FAD for efficient excitation and detection. This filter set consists of two single-band excitation filters with center wavelengths at 385 and 485 nm and bandwidths of 20 and 30 nm, respectively. The corresponding dichroic mirror transmits all wavelengths (1) between 410 and 460 nm and (2) longer than 510 nm. This allows simultaneous and sensitive detection of both NADH (at 450 nm) and FAD (at 525 nm) emissions. The transmission band of the emission filters included in this set overlaps with the dichroic mirror (Figure 3.3). Data for this figure are derived from the Semrock catalog documenting excitation and emission filter characteristics.

3.2.4 Detectors

The standard detector in modern microscopes for recording fluorescence images is a CCD, which converts the detected fluorescence photons at a given pixel (or a photoactive region) to a charge. A separate shift-register mechanism in each pixel shifts the charge accumulated into the next well or pixel and so on until the last well dumps the charges into a charge amplifier that converts the charges into voltage. An analog-to-digital conversion (ADC) unit digitizes the voltage and converts it into 0–(2^n – 1) scale (where n represents the bit count of the ADC circuitry). For example, a 12-bit camera will display the accumulated light/charge/voltage on a 0–4095 scale. It is important to note that there is a certain amount of readout noise associated with the amplifier step when the charges are being read out and converted to voltage. Most CCD manufacturers indicate the readout noise in a CCD camera as the number of electrons/pixel. The readout noise can sometimes be significant enough to overpower weak fluorescence signals and can be reduced by taking the average of multiple frames of the same field of view. Another option available with most CCDs is the ability to program bin pixels. For example, a 2D 1024 × 1024 pixel CCD camera will offer options ranging from 2 × 2 binning (final image size of 512 × 512) to 8 × 8 binning (final image size of 128 × 128). Binning has the advantage of improving signal levels because two or more pixels are treated as one pixel, effectively pooling the charges accumulated. Binning also has the advantage of reducing the effective readout noise. Although the same field of view is maintained, the image resolution is reduced upon binning due to the fewer pixels available to resolve a given field of view.

Recent development of the electron multiplier CCD (EMCCD) has significantly reduced the readout noise problem. EMCCDs are equipped with an electron multiplier, placed after the pixel shift registers and before the charge amplifier. It amplifies the charge several times so that the readout noise is negligible compared to the actual fluorescence signal.

3.3 Data Analysis and Instrument Calibration

Quantitative analysis of fluorescent images is necessary to understand changes in cell or tissue behavior in response to perturbations. There are certain guidelines to keep in mind before performing a reproducible and accurate measurement using fluorescence microscopy. The first is to always measure the background level using unlabeled (reference) sample (Skala and Ramanujam 2010). Another approach for estimating background signal level is to use areas free from cells/tissues just outside the region of interest. The background image collected with any of these methods can then be subtracted from the fluorescence image of the sample of interest in order to minimize the background signal contribution.

The second and likely most important guideline is the calibration of fluorescence images. Like most measurements, fluorescence from samples can vary between imaging sessions due to changes in the excitation light intensity, zoom, focal plane, and system response. For exogenous fluorescent labels, there also could be possible variations in the stability of the label itself. To compare measurements made on different days or in time-lapse imaging, it is important to establish a calibration standard using a photostable fluorophore with negligible photobleaching. Fluorescence from a stable calibration standard will only vary due to changes in the illumination light intensity and thus allow for correcting any changes in fluorescence that are not due to actual changes in the sample of interest. Most manufacturers of fluorescent labels provide the number of excitation–emission cycles prior to observable signal degradation. For example, rhodamine B is a stable calibration standard and an attractive choice due to its broad absorption spectrum. To calibrate a fluorescence microscope, the fluorescence from a droplet of rhodamine, either dried on a microscope slide (Skala and Ramanujam 2010) or in solution, must be measured as close as possible in time to the actual sample of interest. The same microscope settings (gain, exposure/integration time, excitation light power, and objective) must be used for both the calibration standard and the samples of interest. Fluorescence images of the sample can then be divided by the average intensity from the calibration slide in order to obtain a calibrated image for a significant comparison with other images that may be recorded at different times or physiological state of the cell or tissue. While conducting a time-lapse imaging or comparative cell-treatments experiment, it is advisable to limit changes in settings to the integration or exposure time alone. Quantifying changes in fluorescence intensity images for biological or biomedical significance becomes more complex as the number of experimental variables increases. A sufficiently low concentration of the calibration standard should be imaged at multiple integration times in order to determine linearity with fluorescence intensity and therefore avoiding photobleaching-based artifact. Subsequent experiments should be conducted within these settings thereby ensuring that fluorescence images collected at different integration times can be calibrated and compared. For example, if a change in integration time from 50 to 200 ms results in a consistent fourfold increase in intensity of the calibration standard, samples can be safely imaged at either 50 or 200 ms exposure times. Following imaging, the samples can be divided by the calibration standards that were imaged under the same experimental condition. Establishing a well-defined calibration standard, concentration and range of integration times is critical for quantitative fluorescence imaging and therefore functional imaging of cells or tissues.

3.4 Pitfalls and Remedies

This section describes some of the challenges encountered in fluorescent microscopy and potential solutions toward a genuine functional and quantitative imaging.

3.4.1 Photobleaching

Autofluorescence from biological samples is relatively weak—nearly two to three orders of magnitude lower than a typical intensity of the excitation light used in optical microscopy. Autofluorescence emitters, such as flavins and NADH, have low quantum yields and therefore very low signal level is expected (Weber 1950). One potential solution is to increase the intensity of the excitation light, illuminating the sample. However, this creates a problem that is common in fluorescence experiments—photobleaching, a photophysical process that gradually leads to an irreversible loss of fluorescence. Photobleaching timescale depends on both the fluorophore and surrounding environment; the effect of photobleaching may be reduced (or even avoided entirely) by exposing samples to very short durations (if possible) or low excitation light intensities (Bernas et al. 2004).

3.4.2 Overlap between Multiple Fluorophores

Often, more than one fluorophore is used to label different proteins or organelles in a given sample. In such cases, knowledge of corresponding absorption and emission profiles of each fluorophore becomes important in experimental design to avoid the bleed-through from one fluorophore to the other. Spectral overlap can adversely impact quantitative experiments that are specifically designed for biological studies, such as fluctuations in metabolic states of live cells, by creating a false readout and interpretation.

3.5 Clinical Applications

Autofluorescence imaging of live cells and tissues in the context of oncology has been widely reported in literature. Of all the endogenous fluorophores in tissues (see Chapter 2), NADH and FAD have shown the most promise for discriminating early cancer or dysplasia from normal tissue (Ramanujam 2000). NADH and FAD are important biomarkers for the redox state and a number of metabolic pathways in cells and tissues. Single-photon excited autofluorescence imaging of intracellular FAD has been used to differentiate between normal tissue and cancers of the colon (Fiarman et al. 1995). Carlson et al. (2005) showed that autofluorescence from mitochondrial NADH or FAD could be used for tracking

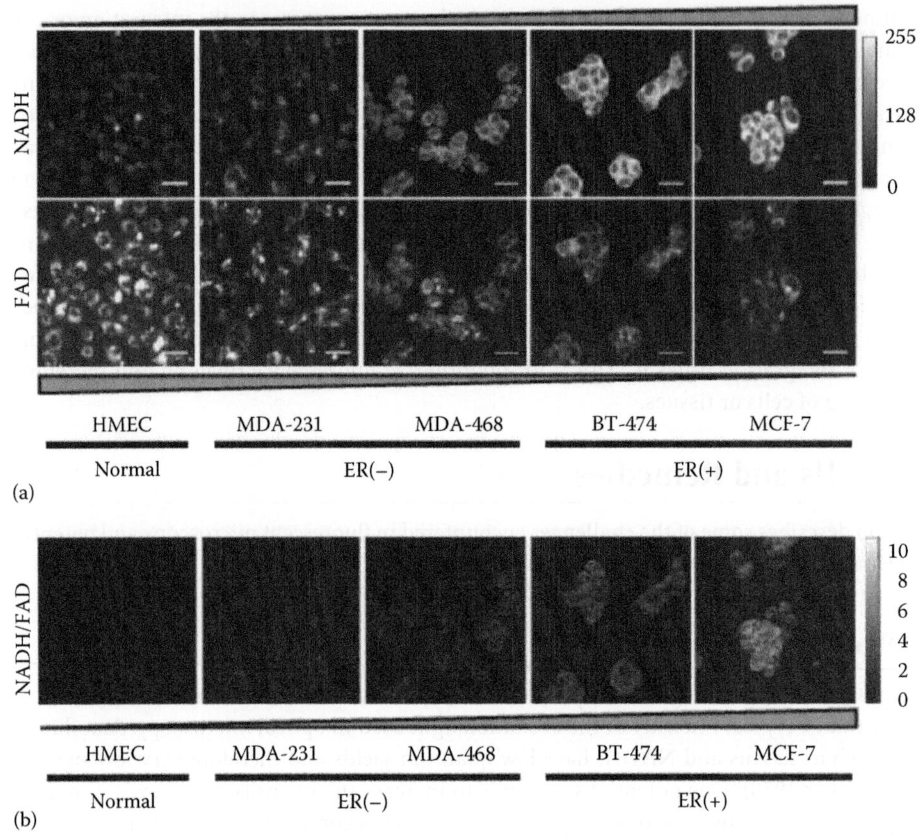

FIGURE 3.4 (a) Representative NADH and FAD images from a panel of breast cancer cell lines stratified by estrogen receptor (ER) status. HMEC indicates normal mammary epithelial cells. (b) Optical redox ratio (NADH/FAD) can differentiate breast cancer cell lines based on ER status. (From Ostrander, J.H. et al., *Cancer Research*, 70(11), 4759, 2010.)

biochemical changes during progression of cervical neoplasia. Ostrander et al. measured the ratio of NADH to FAD fluorescence, also termed the optical redox ratio, using confocal microscopy and found that such ratio could stratify breast cancer cell lines based on estrogen receptor status (Ostrander et al. 2010) (Figure 3.4). The same optical redox ratio was also found to be highly sensitive for discriminating cancer from noncancer in frozen breast biopsies (Xu et al. 2013) as well as indolent from aggressive breast cancer cell lines (Xu et al. 2010).

3.6 Summary

Single-photon microscopy is a versatile laboratory tool for measuring autofluorescence from live cells, frozen tissue sections, and in vivo mouse models. Advances in light sources, detectors, and microscopy technology ensure that imaging of the metabolic fluorophores, NADH and FAD, represents the best opportunity for clinical translation of optical biomarkers for discriminating cancer from normal tissue. In addition, the ability to use low-cost light-emitting diodes and simple phone-based cameras could provide powerful tools in resource-limited settings such as India, Africa, and Haiti for the noninvasive diagnosis of cervical neoplasia and oral cancer (Thekkek and Richards-Kortum 2008; Rahman et al. 2010; Pierce et al. 2012; Quinn et al. 2012). Low-cost fluorescence microscopes using exogenous dye-based labeling are also being developed for other clinical applications such as infectious diseases

(Sia et al. 2004) and tuberculosis (Hänscheid 2008). Single-photon fluorescence also affords a higher signal level compared to multiphoton fluorescence. However, such an advantage comes at significant cost. Fluorescence emitted from the entire sample leads to significant photobleaching and increased background signal levels, thereby reducing contrast from the sample of interest. Although techniques such as confocal microscopy can improve contrast using a variable pinhole, the generation of fluorescent light still occurs; it is merely being rejected by the pinhole. In addition, an important disadvantage of single-photon microscopy is the difficulty in measuring NADH fluorescence without specialized optics for guiding UV light. Special care must also be taken while imaging NADH fluorescence in cells because absorption of UV light at 280–300 nm can cause significant DNA damage (Bliton and Lechleiter 1995). In comparison, two-photon microscopy involves excitation of the sample using wavelengths that are nearly twice the single-photon excitation wavelength. For instance, NADH fluorescence is optimally excited in the 750–800 nm range and completely eliminates the problem of UV-induced phototoxicity. As shown in the following chapters, the advent of multiphoton microscopy has increased the versatility of NADH and FAD as prognostic biomarkers. In addition to imaging fluorescence intensity, fluorescence lifetimes of NADH and FAD are also proving to be capable of discriminating cancer from normal tissue.

References

Bernas, T., M. Zarebki, R. R. Cook, and J. W. Dobrucki. 2004. Minimizing photobleaching during confocal microscopy of fluorescent probes bound to chromatin: Role of anoxia and photon flux. *Journal of Microscopy* 215(3): 281–296.

Bliton, A. C. and J. D. Lechleiter. 1995. Optical considerations at ultraviolet wavelengths in confocal microscopy. In: *Handbook of Biological Confocal Microscopy*, J. B. Pawley (ed.), pp. 431–444. New York: Springer.

Carlson, K., I. Pavlova, T. Collier, M. Descour, M. Follen, and R. Richards-Kortum. 2005. Confocal microscopy: Imaging cervical precancerous lesions. *Gynecologic Oncology* 99(3): S84–S88.

Chance, B., P. Cohen, F. Jobsis, and B. Schoener. 1962. Intracellular oxidation-reduction states in vivo the microfluorometry of pyridine nucleotide gives a continuous measurement of the oxidation state. *Science* 137(3529): 499–508.

Chance, B. and G. R. Williams. 1955a. Respiratory enzymes in oxidative phosphorylation I. Kinetics of oxygen utilization. *Journal of Biological Chemistry* 217(1): 383–394.

Chance, B. and G. R. Williams. 1955b. Respiratory enzymes in oxidative phosphorylation II. Difference spectra. *Journal of Biological Chemistry* 217(1): 395–408.

Chance, B. and G. R. Williams. 1955c. Respiratory enzymes in oxidative phosphorylation III. The steady state. *Journal of Biological Chemistry* 217(1): 409–428.

Fiarman, G. S., M. H. Nathanson, L. I. Deckelbaum, L. Kelly, and C. R. Kapadia. 1995. Differences in laser-induced autofluorescence between adenomatous and hyperplastic polyps and normal colonic mucosa by confocal microscopy. *Digestive Diseases and Sciences* 40(6): 1261–1268.

Hänscheid, T. 2008. The future looks bright: Low-cost fluorescent microscopes for detection of *Mycobacterium tuberculosis* and Coccidiae. *Transactions of the Royal Society of Tropical Medicine and Hygiene* 102(6): 520–521.

Ostrander, J. H., C. M. McMahon, S. Lem et al. 2010. Optical redox ratio differentiates breast cancer cell lines based on estrogen receptor status. *Cancer Research* 70(11): 4759–4766.

Pierce, M. C., Y. Y. Guan, M. K. Quinn et al. 2012. A pilot study of low-cost, high-resolution microendoscopy as a tool for identifying women with cervical precancer. *Cancer Prevention Research* 5(11): 1273–1279.

Quinn, M. K., T. C. Bubi, M. C. Pierce, M. K. Kayembe, D. Ramogola-Masire, and R. Richards-Kortum. 2012. High-resolution microendoscopy for the detection of cervical neoplasia in low-resource settings. *PLoS One* 7(9): e44924.

Rahman, M. S., N. Ingole, D. Roblyer et al. 2010. Research evaluation of a low-cost, portable imaging system for early detection of oral cancer. *Head and Neck Oncology* 2: 10.

Ramanujam, N. 2000. Fluorescence spectroscopy of neoplastic and non-neoplastic tissues. *Neoplasia* 2(1–2): 89–117.

Sia, S. K., V. Linder, B. A. Parviz, A. Siegel, and G. M. Whitesides. 2004. An integrated approach to a portable and low-cost immunoassay for resource-poor settings. *Angewandte Chemie International Edition* 43(4): 498–502.

Skala, M. and N. Ramanujam. 2010. Multiphoton redox ratio imaging for metabolic monitoring in vivo. In: *Advanced Protocols in Oxidative Stress II*, D. Armstrong (ed.), Vol. 594 of Methods in Molecular Biology, J. M. Walker (ed.), pp. 155–162. New York: Springer.

Thekkek, N. and R. Richards-Kortum. 2008. Optical imaging for cervical cancer detection: Solutions for a continuing global problem. *Nature Reviews Cancer* 8(9): 725–731.

Weber, G. 1950. Fluorescence of riboflavin and flavin-adenine dinucleotide. *Biochemical Journal* 47(1): 114–121.

Xu, H. N., S. Nioka, B. Chance, and L. Z. Li. 2013. Imaging the redox states of human breast cancer core biopsies. In: *Oxygen Transport to Tissue XXXIV*, W. J. Welch, F. Palm, D. F. Bruley, and D. K. Harrison (eds.), Vol. 765 of Advances in Experimental Medicine and Biology, I. R. Cohen, A. Lajtha, J. D. Lambris, and R. Paoletti (eds.), pp. 207–213. New York: Springer.

Xu, H. N., S. Nioka, J. D. Glickson, B. Chance, and L. Z. Li. 2010 Quantitative mitochondrial redox imaging of breast cancer metastatic potential. *Journal of Biomedical Optics* 15(3): 036010.

4

Autofluorescence Lifetime Imaging

Michael G. Nichols
Creighton University

Kristina Ward
Creighton University

Lyandysha V. Zholudeva
Creighton University

Heather Jensen Smith
Creighton University

Richard Hallworth
Creighton University

4.1 Introduction: Why FLIM?

Following a light-induced excitation, the fluorescence lifetime is the average residence time that a fluorophore will spend in its excited electronic state prior to returning to the ground state (Figure 4.1). Since the excited state is energetically unstable, this time (usually in the range of picoseconds to microseconds) is sensitive to a number of intramolecular interactions, defined by the fluorophore's structure and the surrounding microenvironment. Such sensitivity makes the fluorescence lifetime an important observable for identifying a molecule and its conformational states (e.g., folding and ligand binding) and sensing its microenvironment (e.g., pH, temperature, and collisional encounters with other molecules or surfaces). While such interactions will also affect the spectral signature of the molecule, these changes are usually minor and often difficult to efficiently resolve due to their characteristically broad emission spectra. In contrast, changes in the fluorescence lifetime can be very pronounced (as much as 20-fold in some cases). In addition, the fluorescence lifetime does not depend on the concentration of the fluorophore under most experimental conditions, which makes fluorescence lifetime probing less sensitive to artifacts produced by photobleaching and scattering. These advantages are utilized in fluorescence lifetime imaging microscopy (FLIM) techniques—a number of methods complimentary to conventional fluorescence intensity measurements, such as wide-field and confocal microscopy (Bastiaens and Squire 1999; Gadella et al. 1993; Lakowicz 2006; Lakowicz and Berndt 1991; Levitt et al. 2009; Periasamy and Clegg 2009; Wang et al. 1989, 1992). While single-point fluorescence lifetime measurements of molecules in different environments have been around for almost a century, the first spatially resolved FLIM image was acquired 20 years ago (Berezin and Achilefu 2010; Wang et al. 1989). Since then, the applications of FLIM techniques have grown exponentially in both basic and applied research. In particular, FLIM of intrinsically

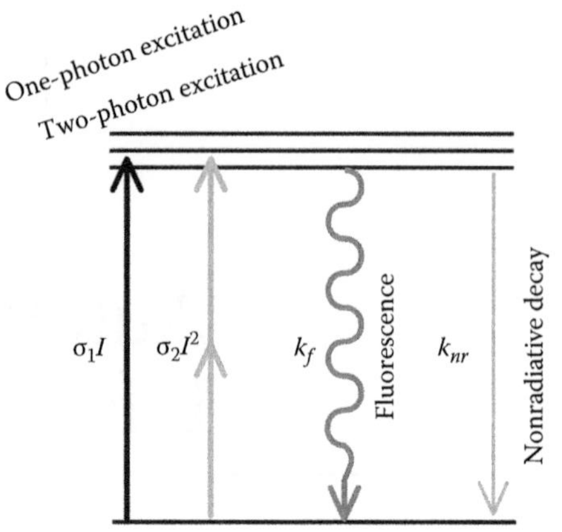

FIGURE 4.1 Energy-level diagram showing single- or two-photon excitation of a fluorophore into an excited vibronic state followed by deexcitation via radiative or nonradiative decay. The excitation rates for single- or two-photon excitation are $\sigma_1 I$ and $\sigma_2 I^2$, respectively. The radiative decay rate (k_r) is a characteristic of the fluorophore, while the nonradiative decay rate (k_{nr}) is governed by interactions that transfer energy from the fluorophore to either other groups within the same molecule or other molecules within the environment of the fluorophore.

fluorescent natural biomarkers of cellular metabolism, such as nicotinamide adenine dinucleotide (NADH) and flavin adenine dinucleotide (FAD), has been applied to clinical research and diagnostics. Over the last two decades, FLIM technology has matured from the design-and-build-it-yourself days of the 1990s into commercially available FLIM upgrade kits for most standard fluorescence microscopes.

A comprehensive review of time-resolved fluorescence techniques is beyond the scope of this chapter. Rather, we will review the main approaches that are actively used by researchers today; introduce their theoretical underpinnings; discuss the type of equipment that is required, pitfalls, and remedies; and briefly highlight some of the interesting applications of cell and tissue autofluorescence lifetime imaging microscopy in energy metabolism studies.

4.2 Theoretical Underpinnings

Figure 4.1 depicts an energy-level diagram and light-induced transitions of a fluorophore. Excitation to different electronic states can be carried out using either single-photon ($p = 1$) or multiphoton ($p > 1$) excitation mechanisms. The p-photon excitation rate at some location, $\vec{r}$, is given by

$$R_A^{(p)}(\vec{r},t) = \sigma_p \int_V N_0(\vec{r}) I^p(\vec{r},t)\, dV \tag{4.1}$$

where

σ_p is the p-photon cross section

N_0 is the ground electronic state number density or concentration of the fluorophore

$I(\vec{r},t)$ is the instantaneous light intensity, and the integral is taken over the spatial profile of the illumination field (Xu and Webb 1996)

Following excitation, the number density of fluorophores in the first excited electronic state, N_1, will decay exponentially:

$$N_1(t) = N_{10} \exp\left(\frac{-t}{\tau}\right) \tag{4.2}$$

where N_{10} is the initial population in the excited state. The fluorescence lifetime τ in Equation 4.2 is determined by the relaxation processes that may compete with fluorescence (e.g., charge transfer, energy transfer, dissociation, and quenching), which can be grouped into nonradiative (k_{nr}) and radiative (k_r) decay rates (Figure 4.1):

$$\tau_f^{-1} = k_{nr} + k_r \tag{4.3}$$

Photobleaching can also affect the fluorescence lifetime and its rate can be grouped with k_{nr}. In practice, the photobleaching rate constant is relatively small in comparison to the other nonradiative relaxation processes. The probability of emitting a fluorescence photon for each excitation photon absorbed, or the fluorescence quantum yield, φ, is given by

$$\varphi = k_r \tau_f \tag{4.4}$$

In general, only a small fraction of the fluorescence is usually detected with an efficiency (η) on the order of a few percent, determined by the collection efficiency of the optical system, the transmission characteristics of the microscope objective and filters, and the sensitivity of the detector. The detected p-photon excited fluorescence decay is given by

$$F^{(p)}(t) = \frac{1}{p}\varphi\eta N_{10} V_{ex} \frac{e^{-t/\tau}}{\tau} \tag{4.5}$$

where V_{ex} is the excitation volume. The time-averaged value of Equation 4.5 represents the fluorescence signal that is recorded by intensity-based techniques:

$$\left\langle F^{(p)}(t)\right\rangle = \frac{1}{p}\varphi\eta N_{10} V_{ex} \tag{4.6}$$

The number density of excited molecules in Equation 4.6 depends on the excitation mechanism (single- or multiphoton) but generally can be approximated as

$$N_{10} \approx N_0\left(g^{(p)}\sigma_p\langle I\rangle^p\right)T \tag{4.7}$$

where T is the time interval between the laser pulses (the reciprocal of the pulse repetition rate) and the pth-order temporal coherence function, $g^{(p)} = \langle I^p\rangle / \langle I\rangle^p$, has been used to obtain a convenient expression in terms of the average light intensity $\langle I\rangle$ (Xu and Webb 2002).* From Equations 4.4 and 4.6, the detected time-averaged fluorescence intensity can be rewritten as

$$\left\langle F^{(p)}(t)\right\rangle \propto N_0\tau \tag{4.8}$$

Equation 4.8 demonstrates that the fluorescence intensity reflects changes in either the fluorophore concentration or the fluorescence lifetime. In FLIM, however, the fluorescence lifetime is independent of the fluorophore concentration under most conditions. While it is tempting to assume the contrast in a fluorescence intensity image reflects the fluorophore concentration, this is not necessarily the case. A bright pixel could indicate either a greater lifetime or a greater concentration. Only when the fluorescence lifetime is uniform is the fluorescence intensity a valid surrogate for the fluorophore concentration.

* For an idealized pulse with constant (peak) intensity I_0 for pulse width τ_{lp}, $g^{(p)} = (T/\tau_{lp})^{p-1}$, which is unity for single-photon excitation or approximately 10^5 for two-photon excitation using a laser that provides one 100 fs pulse every 10 ns.

So far, it has been assumed that all of the fluorophores in the focal volume are in the same microenvironment and thereby exhibit the same fluorescence lifetime. However, under most experimental conditions, the observation volume is likely to include a large number of molecules of different conformations within a heterogeneous microenvironment. For example, estimates of cellular flavin concentrations are on the order of 10–100 μM, while NADH concentrations may be up to 10 times this value (Heikal 2010; Yu and Heikal 2009). If the detection volume is on the order of 1 fL, 10^4–10^6 molecules could contribute to the decay. It is therefore not surprising that fluorescence decays, recorded from cells and tissues by laser-scanning or wide-field microscopy methods, reflect contributions from molecules with several distinct lifetimes.

To measure the fluorescence lifetime, either time- or frequency-domain techniques can be used. In the time domain, the fluorophore is excited by short laser pulses and the resulting fluorescence is measured in time as the fluorophore returns to the ground state. In the frequency domain, the fluorophore is excited by an intensity-modulated light source (either sinusoidal or pulsed), and the relative phase delay and demodulation of the fluorescence signal is used to determine the lifetime. While these approaches may seem fundamentally different, in fact, they are very similar. For example, any time-domain signal can be analyzed in the frequency domain by standard Fourier transform methods, regardless of the temporal profile of the excitation light source. While time-domain methods yield temporal information directly, both approaches can provide the same information regarding the lifetime. There are, however, significant differences in the way these techniques are implemented, the type of equipment that is required, and the applications for which each is best suited.

4.3 Experimental Designs, Data Acquisition, and Analysis

4.3.1 Time-Correlated Single-Photon Counting Technique

In time-correlated single-photon counting (TCSPC)-based FLIM, repetitive excitation with a pulsed laser is performed for each pixel of a raster-scanned image, and the arrival times of the fluorescence photons are accumulated in a large number of discrete time channels to construct the fluorescence decay histogram (Figure 4.2). The laser pulses are monitored via a fast photodiode that generates a synchronization ("sync") pulse to trigger the TCSPC electronics, which are housed and controlled within a separate computer system. This establishes a reference time for the excitation pulse. The time delay between the "sync" pulse and the first detected fluorescence photon is recorded and stored by the TCSPC electronics. This process is repeated for each laser pulse over a predetermined acquisition time to build up the fluorescence decay histogram. During a typical pixel dwell time of 50 μs, 4000 laser pulses may produce 50 fluorescence photons to be recorded. By either binning the signals recorded from neighboring pixels or repeatedly scanning the sample, an accurate histogram of the time of arrival of the first emitted photon following the excitation pulse is assembled (Figure 4.2). The measured fluorescence decay, $F(t)$, is fit by a weighted sum of exponentials:

$$F(t) = F_0 \sum_{i=1}^{n} A_i e^{-t/\tau_i} \quad (4.9)$$

reflecting the possibility of several lifetimes due to either different fluorophores, diverse conformations, or variations in their local environment. In this expression, F_0 is the initial fluorescence, t is the elapsed time following the laser pulse, the A_i coefficients indicate the fraction of the total fluorescence signal associated with fluorophores of lifetime τ_i, and n is the number of exponentials that depends on the chemical structure of the fluorophore(s), environmental heterogeneity, and the signal-to-noise ratio (SNR) of the measured fluorescence decay per pixel.

4.3.1.1 Equipment

A typical imaging setup is shown in Figure 4.2. The light source of choice is a pulsed ultrashort laser with a pulse width in the range of femto- to picoseconds. Mode-locked titanium–sapphire (Ti:S) lasers are

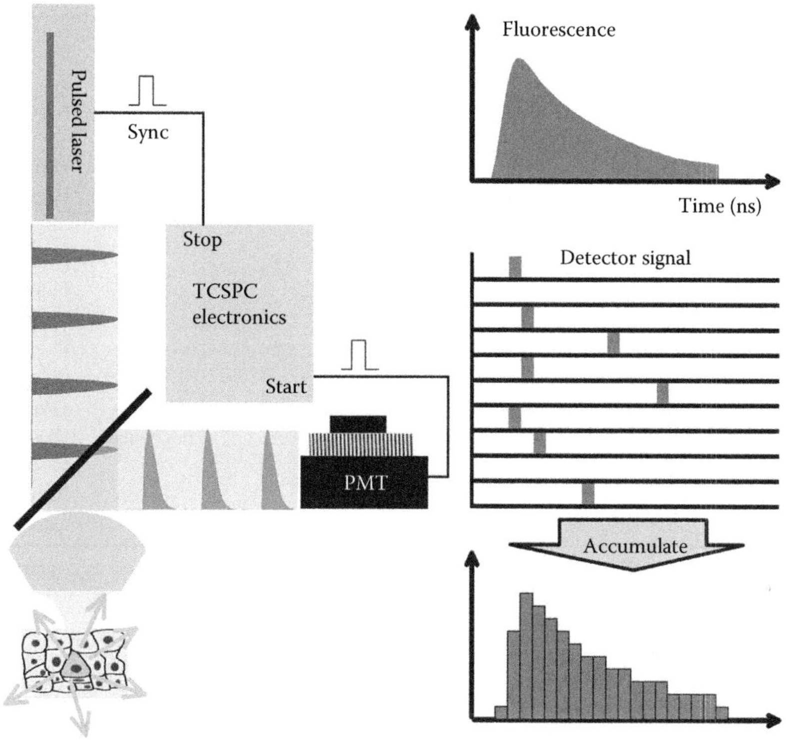

FIGURE 4.2 Schematic diagram showing the TCSPC approach to FLIM. Short pulses of light are focused into living cells and tissue to excite endogenous fluorescence. TCSPC electronics record the time delay between the arrival of the first emitted photon, measured by a PMT and the laser pulse, provided by the "sync" signal from the laser or external photodiode. A histogram of photon arrival times accumulates with each pulse. The resulting decay is fit to a sum of exponentials to obtain the fluorescent lifetimes. The lifetime values collected at each pixel are used to create the lifetime image.

a particularly good choice because of their ultrashort laser pulses (width of 70–150 fs) and wavelength tuning range (680–1050 nm) that enables efficient multiphoton excitation of UV–Vis fluorophores. Moreover, near-infrared pulses penetrate deeper into tissue than UV light and cause less damage to the cells and tissues (Zipfel et al. 2003a,b). High pulse repetition rates of 80 MHz, typical of the Ti:S lasers, are also ideally suited to multiphoton FLIM. Since most fluorophores have lifetimes of about 4 ns or less (see Table 4.1 for examples of endogenous fluorophores), the time interval of 12.5 ns between laser pulses is long enough to allow fluorescence generated by one pulse to fully decay before the arrival of the next. Alternatively, single-photon FLIM is readily accomplished either by using a pulsed diode laser or by frequency doubling (or tripling) a Ti:S or a pulsed neodymium-doped yttrium aluminum garnet (Nd:YAG) laser (Elson et al. 2002; Schneckenburger et al. 2004).

To produce a 2D image, the laser pulses are usually raster scanned across the sample using confocal or two-photon laser-scanning microscopy techniques. The objective lens, used for focusing the laser pulses to the cells or tissues of interest, also collects the fluorescence, and dichroic beam splitters and bandpass filters are used to isolate the signal of interest and direct the fluorescence photons to the detectors, typically photomultiplier tubes (PMTs). From the great variety of the detectors available on the market, the most popular are PMTs with gallium arsenide phosphide (GaAsP) cathodes with high (40%) quantum efficiency, appropriate for use with weak autofluorescence. Moreover, its sensitivity to fluorescence in the range of 300–700 nm permits detection of most popular fluorophores, including endogenous NADH and FAD.

TABLE 4.1 Fluorescence Properties of Intrinsic Fluorophores

Component	Excitation Range (nm)	Emission Range (nm)	Typical Excitation Max (nm)	Typical Emission Max (nm)	Typical Lifetime (ns)	Lifetime References
Proteins/amino acids						
Tryptophan	240–300	300–400	280	350	3.1	Lakowicz (2006)
Tyrosine	260–280	280–340	275	300	3.6	Lakowicz (2006)
Structural proteins						
Collagen	250–360	320–480	325, 360	400, 405	3–4	Marcu et al. (2001) Phipps et al. (2012)
Elastin	300–450	350–500	290, 325	340, 400	2–3	Marcu et al. (2001) Phipps et al. (2012)
Coenzymes						
NAD(P)H	300–400	400–500	340	460	0.4 ns (Free) 1–5 ns (Bound)	Lakowicz et al. (1992) Yu and Heikal (2009) Vergen et al. (2012) Blinova et al. (2008)
FMN/FAD	300–500	500–600	390, 450	535	2.3	Lakowicz (1983)

Excitation and emission peaks and ranges and emission range can be found in reviews such as Wagnieres et al. (1998), Lakowicz (2006), Ramanujam (2000), and Richards-Kortum and Sevick-Muraca (1996).

Since the laser repetition rate is typically a 100-fold greater than the fluorescence detection rate, a more efficient "reversed start–stop configuration" is typically used. In this configuration, the detection of a fluorescent photon serves as the "start" signal. The "sync" pulse is delayed by an appropriately chosen length of cable so that it is detected after the arrival of the fluorescent photon. In this configuration, the "sync" pulse provides the "stop" signal in the measurement in the fluorescence decay time. This is depicted in Figure 4.2. Manufacturers, such as Becker & Hickl GmbH and PicoQuant GmbH (both in Berlin, Germany), provide complete modular TCSPC "upgrade kits" for laser-scanning microscopes.

4.3.1.2 System Response

The time resolution of TCSPC systems is limited primarily by the performance of the detector. The PMT produces a measureable pulse of a large number of electrons following the initial ejection of an electron at the photocathode by a single fluorescence photon. Through the amplification process, variation in the transit time of these electrons as they travel from the photocathode to the anode of the PMT causes the instantaneous detection event to be smeared out in time. This broadening is called the instrument response function (IRF). As a result, the recorded fluorescence decay is convoluted with the IRF. To account for this, the IRF can be recorded from a scattering solution (e.g., urea or even milk droplets in water) or second harmonic generation (SHG) from collagen or artificially synthesized crystals, such as periodically poled lithium niobate for multiphoton excitation. It is also possible to infer the IRF without direct measurement by model fitting to the measured fluorescence decay. Once determined, the IRF temporal profile is then convolved with the model function used for fitting. In principle, this deconvolution process allows for the measurement of fluorescence lifetimes much smaller than the temporal profile of the IRF, though the reliability of this method is still debated.

A more direct way to increase the timing resolution is to use detectors with reduced transit-time spread. Although GaAsP detectors increase the quantum efficiency, they usually broaden the transit-time spread. For example, the Hamamatsu H7422P-40—one of the most popular detectors used in autofluorescence lifetime imaging microscopy—has a transit-time spread of 200–350 ps (Becker 2008). Detectors of more compact design, such as multichannel plate (MCP) PMTs (e.g., Hamamatsu R3809U), have smaller transit-time spread in the range of 28–30 ps (Becker 2008).

4.3.1.3 Data Analysis

Each pixel in a TCSPC FLIM image has a fluorescence decay curve. A typical example is shown on a screenshot of the SPCImage (Becker and Hickl) FLIM data analysis software in Figure 4.3. In this case, the dataset is a multiphoton FLIM image of NADH within an EMT6 tumor spheroid with a necrotic core. The graph below the images shows the recorded fluorescence decay curve for the pixel highlighted by the cursor (blue dots), with the timescale ranging from 0 to 10 ns. The image on the left is the corresponding fluorescence intensity image (i.e., the integrated fluorescence decay per each pixel), while the image on the right shows the average fluorescence lifetime obtained by fitting every pixel to a convolution of the IRF and a double-exponential decay:

$$F(t)=\int_{0}^{t}\left(A_1e^{-t'/\tau_1}+A_2e^{-t'/\tau_2}\right)\cdot \mathrm{IRF}\left(t-t'\right)dt' \tag{4.10}$$

In this particular case, the IRF is inferred from the measured fluorescence decay. The best fit of Equation 4.10 is shown as the red line, while the IRF (in this case for a H7422P-40 PMT) is shown as the green line. The decay parameters obtained from the fit for this pixel are shown in the table on the lower right. Because Equation 4.10 is nonlinear, there is no convenient, analytical solution. The fitting algorithm systematically adjusts the free fitting parameters to produce the best fit, that is, the lowest value of the χ^2

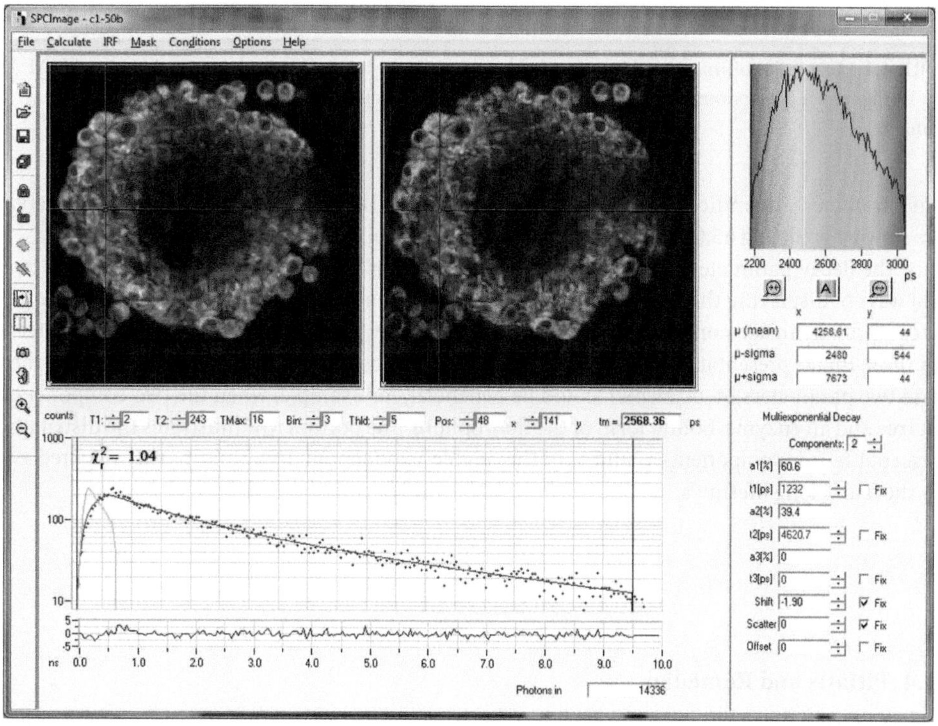

FIGURE 4.3 Screenshot of the Becker and Hickl SPCImage data analysis program used to analyze TCSPC FLIM images. The grayscale image shows the fluorescence intensity, while the pseudocolor image shows the average fluorescence lifetime. The lifetime histogram shows the distribution of lifetimes found in the image and a scale that associates the color of a pixel with the lifetime. The fluorescence decay data for the pixel identified by the blue crosshairs are plotted below the image with the best double-exponential fit to the data indicated by the red line.

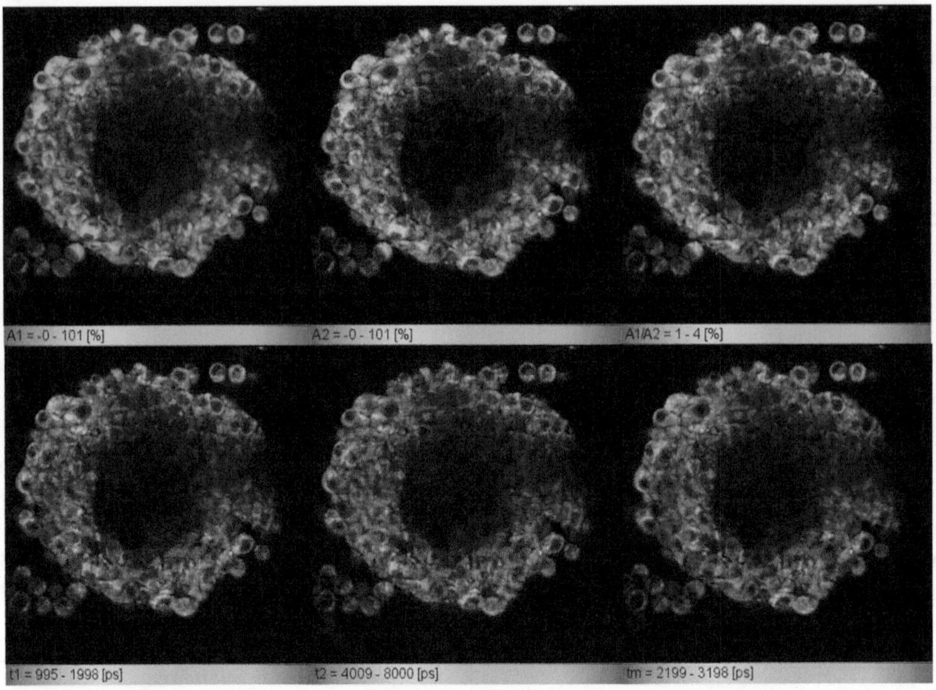

FIGURE 4.4 Examples of image contrast formed by parameters derived from the analysis of the FLIM dataset shown in Figure 4.3. The parameter name and color-coded range of values are indicated on the color bars below each image.

best-fit parameter, also indicated on the graph in Figure 4.3. This is an iterative process, and therefore it can be slow when fitting 65,000 pixels of a typical 256 × 256 pixel image.

Once the decay parameters (A_i, τ_i) have been obtained for each pixel in the FLIM image, there are several ways of displaying the data. Each parameter, or any mathematical manipulation of these parameters, can provide image contrast and be displayed as a color-coded image. Figure 4.4 illustrates just a few of those image presentation possibilities. The top row shows the distribution of A_1, A_2 and the ratio of these two parameters, A_1/A_2, which could be beneficial, for example, when the two components represent free and an enzyme-bound form of the fluorophore. The bottom row illustrates the distribution of fluorescent lifetime components τ_1 and τ_2 of the double-exponential decay and τ_m the weighted average of the short and long lifetimes:

$$\tau_m = \frac{\sum A_i \tau_i}{\sum A_i} \tag{4.11}$$

4.3.1.4 Pitfalls and Remedies

Currently, TCSPC-based FLIM is a popular, high-precision, and mature technology, with common lifetime resolution of ~50 ps or better, depending on the SNR and detectors used in a given system. Unfortunately, high precision comes at the cost of efficiency and acquisition speed. Low fluorescence count rates (~10^6 detected photons/s or less), for example, are required to properly record the delay time between the excitation pulse and the detected fluorescence. The maximum count rate is determined by the dead time set by the electronics of TCSPC modules (i.e., the time between the first fluorescence photon detected and the next excitation–detection cycle). For a TCSPC system with a 125 ns dead time, an 8 MHz count rate would result in 50% conversion efficiency (Becker 2008). The TCSPC count rate is further limited by the so-called "pileup" effect that is inherent in this technique, especially at very high signal levels. To build up an unbiased estimate of the fluorescence decay histogram, TCSPC electronics measure the time of arrival of a single, first-to-arrive photon following the excitation laser pulse (defined by the "sync" pulse). Following the first detected photon in an excitation–detection cycle, any additional fluorescence photons are ignored. This is particularly problematic considering the acquisition time needed to construct an accurate FLIM image. Unfortunately, the likelihood of detecting more than a single photon per cycle increases when operating at high fluorescence count rates, hence the "pileup" effect. As a result, the fluorescence decay curve becomes biased toward shorter lifetimes (Becker 2005). In practice, this limits the maximum count rate to a few percent of the laser repetition rate or about 1 MHz for a typical Ti:S laser. While this can easily be arranged, the requirements for low count rate coupled with serial acquisition by laser beam scanning lead to rather long acquisition times of up to several minutes per image. This can be problematic for measurements in vivo where motion artifacts may be present. To address these issues, fast tandem multifocal multiphoton instrumentation is being developed to simultaneously image many points using multicathode PMTs and parallel TCSPC electronics (Becker et al. 2007; Kumar et al. 2007). Alternatively, if fast acquisition times or a wide-field implementation is required for a given application, time gating or frequency-domain FLIM might be a more appropriate choice.

4.3.2 Rapid Lifetime Determination by Time Gating

Rapid lifetime determination (RLD) employs sequential time gating of the detector to measure the fluorescence intensity within two or more time intervals following the excitation pulse. The lifetime can then be estimated by comparing the integrated fluorescence signal detected during each gate interval. For relatively simple cases, such as single- and double-exponential decay following an ultrashort laser pulse, the lifetime(s) can be directly calculated without the need for time-consuming iterative fitting. This approach can be implemented in either wide-field or laser-scanning microscopes (Buurman et al. 1992; Gerritsen et al. 2002; Periasamy and Clegg 2009; Sytsma et al. 1998).

4.3.2.1 Equipment

As with TCSPC, RLD relies on short pulses of light to excite fluorophores, typically from either a mode-locked or a pulsed diode laser (Figure 4.5). The laser pulse triggers the opening of a series of electronic gates that will stay open for a set period of time after a programmed delay. While the gate is open, the detector(s) accumulate the fluorescence signal for that time period. In wide-field imaging, a gated image intensifier and charge-coupled device (CCD) camera are employed to acquire data for all pixels simultaneously. The signal intensity develops by integrating over a large number of laser pulses (up to 10^6 or so). After acquiring data for a specified gate, the delay time is then adjusted and the process is repeated. For beam scanning, a single PMT is sufficient. In various implementations, the gates are set to open in succession, such that one gate opens as soon as the previous gate closes, or, to increase the number of detected photons, there may be some intentional overlap in the time periods defined by each gate. Unlike TCSPC, where only the first fluorescence photon is detected after the "sync" pulse is recorded, in RLD, all of the photons arriving at the detector within each time-gate interval are used.

The major advantage of the RLD method over TCSPC FLIM is high image acquisition speed and fast data analysis. Moreover, RLD-FLIM can be operated in analog mode, which also facilitates the high acquisition speed. However, photon counting is still used in RLD to help discriminate against detector noise, which is more important for low-signal measurements. In general, greater photon count rates are achievable with this approach because the dead time of the RLD electronics can be very low (<1 ns), making data loss in high repetition rate measurements less of a problem in comparison to TCSPC (Gerritsen et al. 2002).

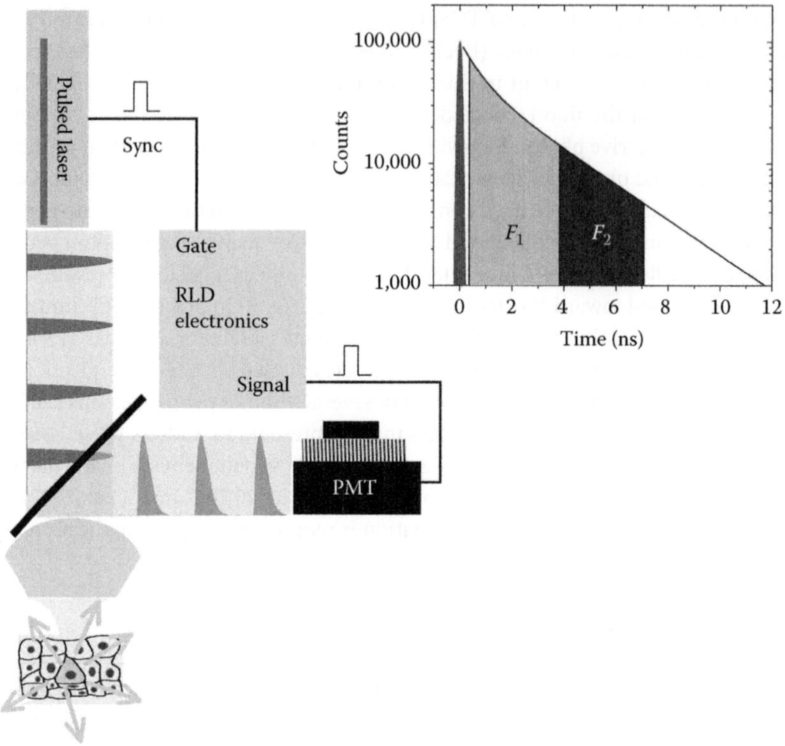

FIGURE 4.5 Schematic diagram showing the RLD approach. As with the TCSPC approach (Figure 4.2), a pulsed laser excites cellular autofluorescence. The pulse "sync" opens an electronic gate and the fluorescence signal (F_1) is measured over a defined time interval before the gate closes. A second gate opens as the first closes, and the fluorescence signal (F_2) is measured. This can be repeated for additional gates (not shown). The signals acquired during each time interval are used to estimate the lifetime (see text for further discussion).

This method also has the advantage of efficiently using nearly all of the detected fluorescence photons to determine the excited-state lifetime. As with other methods discussed in this chapter, the image acquisition time is ultimately limited by the time that it takes to collect enough photons to accurately determine the lifetime. Since the RLD electronics have little dead time, the maximum frame rate is therefore limited by the detector quantum efficiency (and maximum count rate when using single-photon counting detectors). For both laser scanning with conventional PMTs and wide-field imaging employing gated image intensifiers with CCD cameras, fluorescence lifetime images of fluorophores with 1–2 ns lifetimes can be acquired with an acquisition time on the order of a few seconds. For either implementation, the total gate time must be kept less than the repetition rate of the pulsed laser.

4.3.2.2 Data Analysis

When the opening of two successive gates is separated in time by ΔT and the integrated fluorescence intensity during each gate is measured to be F_1 and F_2, respectively, the lifetime and amplitude associated with a single-exponential decay, defined by Equation 4.9, can be directly calculated without fitting:

$$F_1 = \int_{t_0}^{t_0+\Delta T} N_{10} e^{-t/\tau}\, dt \tag{4.12}$$

$$F_2 = \int_{t_0+\Delta T_0}^{t_0+2\Delta T} N_{10} e^{-t/\tau} dt \tag{4.13}$$

$$\tau_1 = \frac{\Delta T}{\ln(F_1/F_2)} \tag{4.14}$$

$$A_1 = \frac{F_1^2 \ln(F_1/F_2)}{(F_1 - F_2)\Delta T} \tag{4.15}$$

To resolve a double-exponential decay, at least four gates are required to determine the four parameters associated with the decay, and formulas for this case have also been determined (Sharman et al. 1999).

In wide-field imaging, Equations 4.12 and 4.13 are collected as individual images. Additional images are required for each gate, and these images will generally be spaced in time by about 1 s. Equations 4.14 and 4.15 (or equivalent expressions for double-exponential decays) are then applied after the acquisition, pixel by pixel, using intensity values drawn from images acquired for each gate. In beam scanning mode, the fluorescence lifetime in each pixel is measured and analyzed sequentially with signals from all of the gates acquired for each pulse for a given pixel. Again, results from a large number of pulses are aggregated to achieve the desired signal level before calculating the lifetime for that pixel. Equations 4.14 and 4.15 can be computed and displayed in real time.

4.3.2.3 Pitfalls and Remedies

While the RLD method is not as precise as TCSPC techniques when analyzing complex fluorescence decays, the accuracy and timing resolution can be sufficient for quantitative imaging provided enough photons are collected during each gating interval to accurately approximate Equations 4.12 and 4.13. Given the exponential nature of the decay, the early gates will have better statistics than the later gates. The optimal choice of the gating time requires a priori knowledge of the fluorescent lifetime. For a single-exponential decay, the optimal gate time should be about two to three times greater than the lifetime. Similarly, for double-exponential decays, the gate time should be at least twice the shortest decay time, though in this case, the accuracy will also depend on the relative values of the two lifetimes as well as the relative amplitudes of each decay (Sharman et al. 1999). Generally, the SNR in the measurement of the integrated fluorescence will improve as $\sqrt{N_i}$ where N_i is the number of photons collected in a given gate. Hence, longer accumulation times or greater laser intensity could improve the accuracy of the measurement, but only if photobleaching is not significant. Photobleaching can be especially problematic in wide-field implementations since each gate is acquired as a separate image, potentially resulting in systematic errors in the lifetime calculation. When using PMTs in beam scanning implementations, longer acquisition times are required to resolve very short lifetime components because of the limitations on the count rate imposed by the PMT. In this case, acquisition times are about 10 s to several minutes, which is comparable to the acquisition time with TCSPC. Given the fact that RLD cannot indicate whether or not a fluorescence decay is simple (single or double exponential) or not and some knowledge of the lifetime is required to optimize gating, an unknown sample should first be characterized using another technique to enable the development of a proper RLD imaging protocol.

4.3.3 Frequency-Domain FLIM

As with RLD, both wide-field excitation and laser scanning are routinely used for frequency-domain FLIM (Gerritsen et al. 2006; Lakowicz and Berndt 1991; Periasamy and Clegg 2009). For either configuration, an intensity-modulated light source is used to excite the sample. While any modulation pattern can be used, it is helpful to think of the incoming light as a sinusoidal wave. The fluorescence signal produced by such a light wave will also vary sinusoidally with the same modulation frequency, but it will be delayed because the fluorophore resides in the excited state for some time before emitting fluorescence. This delay results in a phase shift and a demodulation (i.e., reduction in the amplitude of the oscillation relative to the average value) of the fluorescence signal. Hence, the fluorescence lifetime can be determined by measuring either of these quantities. It is advantageous to be able to measure the response to a wide range of modulation frequencies, approaching the gigahertz range for subnanosecond lifetime measurements needed for fluorophores such as NADH.

4.3.3.1 Equipment

A typical configuration for a frequency-domain FLIM instrument is shown in Figure 4.6. For fluorescence lifetimes of a few nanoseconds, arc lamps and continuous-wave (CW) lasers can be used as the light source. The modulation can be accomplished in a number of ways. For example, the laser beam can be modulated to frequencies up to 200 MHz by an electro-optic modulator (EOM) driven with a sinusoidal voltage or acousto-optic modulator (AOM) (Gadella et al. 1993). Similarly, the supply voltage of bright light-emitting diodes (LEDs) can be directly modulated for wide-field frequency-domain FLIM (Elder et al. 2006; Herman and Vecer 2008). To measure subnanosecond lifetimes, higher modulation

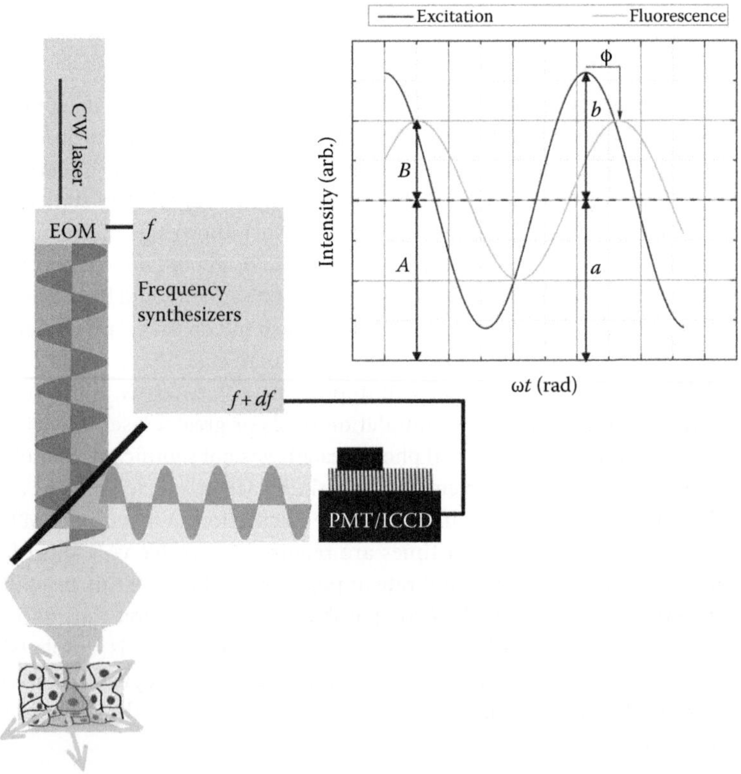

FIGURE 4.6 Schematic diagram showing the frequency-domain approach. The modulated light source is focused into the tissue. The resulting fluorescence emission has the same modulation frequency, but delayed in time by the fluorescence lifetime, and the modulation amplitude is also decreased. Both the time delay (or phase difference) and the demodulation can be used to determine the fluorescence lifetime and create the lifetime image.

frequencies can be obtained from pulsed light sources such as synchronously pumped lasers, mode-locked lasers, or pulsed LEDs. A train of pulses with pulse width of τ_p is equivalent to a superposition of sinusoidally modulated waves containing frequencies up to $f_{max} \sim 1/2\pi\tau_p$, so subnanosecond pulse widths will provide GHz modulation frequencies for frequency-domain FLIM.

On the detection side, both intensified CCD (ICCD) cameras and PMTs can be used as phase-sensitive detectors by sinusoidally modulating their detection gain using frequency synthesizers and digital phase shifters, shown schematically in Figure 4.6. The detector gain is given by

$$G(t) = G_0\left[1 + m_D \sin\left(2\pi\left(f + df\right)t - \phi_D\right)\right] \tag{4.16}$$

where

- G_0 is the time-averaged gain
- m_D is the detector modulation depth
- ϕ_D is the detection phase angle measured relative to the source
- f is the modulation frequency of the light source
- $f + df$ is the modulation frequency of the detector gain, which may be the same as the light source (homodyne, $df = 0$) or at a slightly different frequency (heterodyne, $df \ll f$)

With the heterodyne detection scheme, mixing of the incident fluorescence waveform and the detected waveform at a slightly different frequency results in a beat signal that carries the fluorescence phase shift and demodulation information at the much lower difference frequency, df. Typically, this frequency would be in the kHz range, while the high-frequency modulation is in the GHz range. This reduction in frequency allows the fluorescence signal to be more easily sampled, which has the added benefit of reducing the impact of electronic noise during data acquisition.

As with the time-domain instruments, manufacturers such as Lambert Instruments (the Netherlands) and ISS (Champaign, IL) provide commercial implementations of frequency-domain FLIM as upgrade kits that can be installed on most commercial fluorescence microscopes.

4.3.3.2 Data Analysis

In the frequency domain, we consider the response of a fluorophore to a sinusoidally modulated light source (Figure 4.6):

$$I(t) = a + b\sin(2\pi ft) \tag{4.17}$$

where

- a is a constant offset
- b is the amplitude of the time-dependent variation in the excitation light intensity
- the ratio b/a is the excitation modulation
- f is the modulation frequency (Lakowicz 2006)

Following excitation, the fluorescence emission will be delayed by the characteristic lifetime of the fluorophore, and the phase-delayed and demodulated fluorescence will have the form

$$F(t) = A + B\sin(2\pi ft + \phi) \tag{4.18}$$

where

- A and B are the offset and the amplitude of the time-dependent variation in the fluorescence intensity, respectively
- ϕ is the phase shift of the fluorescence emission measured relative to the light source

For a single-exponential decay, either the phase or the demodulation, m, can be measured to determine the corresponding fluorescence lifetime of a fluorophore:

$$\tan\phi = 2\pi f\,\tau \tag{4.19}$$

$$m = \frac{B/A}{b/a} = \frac{1}{\sqrt{1+(2\pi f\tau)^2}} \tag{4.20}$$

These results can be generalized for any time-dependent fluorescence decay by taking sine (S) and cosine (C) transformations of the fluorescence signal (for details, see Lakowicz 2006) such that

$$\tan\phi = \frac{S}{C} \tag{4.21}$$

$$m = \sqrt{C^2 + S^2} \tag{4.22}$$

For a sum of exponentials, the sine and cosine transforms can be written as

$$S = \frac{2\pi f}{\tau_{avg}} \sum_i \frac{A_i \tau_i^2}{1+(2\pi f\tau_i)^2} \tag{4.23}$$

$$C = \frac{1}{\tau_{avg}} \sum_i \frac{A_i \tau_i}{1+(2\pi f\tau_i)^2} \tag{4.24}$$

High-precision measurements are made by systemically varying the excitation modulation frequency, obtaining experimental measurements for S and C, and then performing nonlinear least-squares fitting using Equations 4.23 and 4.24 to determine the best estimates of the fluorescence lifetimes. While this is routine for point measurements, it is typically not practical to use more than one or a few, well-chosen frequencies for FLIM. The modulation frequencies that produce the most significant phase change or demodulation are given by

$$f_{opt,i} \approx \frac{1}{2\pi\tau_i} \tag{4.25}$$

In this case, modulation frequencies of approximately 40, 160, and 265 MHz are required to optimally detect fluorescence lifetimes on the order of 4, 1, and 0.6 ns, respectively (Elder et al. 2006; Gadella et al. 1993; Lakowicz and Berndt 1991).

In wide-field frequency-domain FLIM, the ICCD is modulated at the same frequency as the light source ($df = 0$), but a constant phase shift is also applied. An image stack is acquired for a set of detection phase angles, ϕ_D, ranging from 0 to 2π rad. The intensity of each pixel recorded by the camera will vary with the detection phase angle according to

$$F(\phi_D) = F_0\left[1 + 0.5mm_D\cos(\phi - \phi_D)\right] \tag{4.26}$$

Each pixel stack in the image can then be extracted and fit to determine the phase shift, demodulation, and hence the lifetime at that pixel. Alternatively, measurements made at a few well-chosen phase angles permit an analytical solution without need for iterative fitting (Elder et al. 2006; Gadella et al. 1993; Lakowicz and Berndt 1991).

Frequency-domain FLIM instruments employing pulsed laser beams that are raster scanned across the sample and detected with a gain-modulated PMT typically use the heterodyne detection scheme. The PMT signal can be digitized at a sample rate that is perhaps 10-fold greater than the difference frequency using standard analog-to-digital data acquisition electronics. The signal from each pixel is then Fourier transformed to calculate the phase shift and demodulation at that pixel. With analog data acquisition in the frequency domain, greater photon count rates become possible without the dead time or pulse-pileup problems that limit the TCSPC technique (Gratton et al. 2003).

4.3.3.3 Pitfalls and Remedies

The data analysis described in the previous section makes the assumption that there is no offset, or dark signal, in the detection system. This should be verified by taking background images while preventing sample fluorescence from entering the detector. When background signals are present, they should be subtracted out prior to the calculation of the phase shift or demodulation. Furthermore, the phase shift and demodulation measurements are always made "relative" to the light source. Unaccounted-for variations in the modulation or phase of the light source during frequency-domain imaging can lead to systematic errors. To prevent this, reference measurements should be made throughout the imaging process. This can be done, for example, by changing optical filters to obtain images of the (attenuated!) scattered laser light instead of the fluorescence signal or by obtaining images of reference fluorophores with known lifetimes.

Reflected light and light scattered from the sample can also produce lifetime artifacts in FLIM images. This is more problematic for wide-field implementations that collect data from all regions simultaneously. Beam scanning implementations are less susceptible because only one region of the sample is illuminated at a time, preventing cross talk with neighboring regions. Of course, for thick samples, photons originating above and below the focal plane can also produce artifacts in the lifetime image. Where possible, illumination and detection strategies such as multiphoton excitation and confocal detection that reduce the detection volume can eliminate this problem.

The most significant source of detection noise in wide-field frequency-domain FLIM comes from the image intensifier, since the photon shot noise is amplified. Reducing the intensifier gain will improve the precision of the measurement, but this will also require longer acquisition times. In fact, this is not just applicable to frequency-domain FLIM. Regardless of the technique, the best way to improve precision is to collect more photons.

4.4 Special Considerations and Approaches in FLIM Data Analysis

Given the variety of data acquisition methods, it is perhaps not surprising that there is a wide variety of data analysis algorithms that have been used to process FLIM datasets. For each technique described earlier, we have briefly indicated how the fluorescence decay parameters are determined. It is important to realize that embedded in each of these descriptions is an implicit model of the fluorophore, its environment, and the interaction between the two. For example, whenever a decision is made to fit a decay curve to two exponentials in TCSPC, or perform RLD with four gates, or use Equations 4.21 through 4.24 to obtain the phase and demodulation of a frequency-domain FLIM experiment, there is an underlying assumption that the fluorescence is actually best represented by a double-exponential decay. However, this assumption can be difficult to test because the number of observed decays depends not only on the molecular system but also on the signal-to-noise level and timing resolution of the measurement.

The SNR in fluorescence imaging experiments scales as the square root of the number of detected photons and generally limits the number of exponential decays that can be reliably determined by the experiment. A factor of 10 improvement in the SNR, for example, requires a 100-fold increase in the number of collected photons. However, the fluorophores themselves are not perfectly stable and there is a limit to the number of photons that a given molecule will emit before photobleaching. A common rule of thumb is that at least 100, 1,000, and 10,000 photons in the peak bin are needed for single-, double-, and triple-exponential fitting in TCSPC experiments (Sun et al. 2011). While this is helpful and easy to remember, the exact requirements depend sensitively on the values of the lifetimes and coefficients. An early statistical analysis by Köllner and Wolfrum (1992) found that a single lifetime could be determined with 10% accuracy from as little as 185 photons detected in the fluorescence decay measurement—less than one photon per time bin, on average. However, a double-exponential decay model with decay parameters of $\tau_1 = 2$ ns ($A_1 = 0.1$) and $\tau_2 = 4$ ns ($A_2 = 90\%$) required at least 400,000 photons or more than 1,500 photons per bin (Köllner and Wolfrum 1992). More recently, Nishimura et al. (2005) have compared different fitting algorithms and found that while conventional least-squares fitting required an average of 100–900 photons per bin for multiexponential fitting, a maximum likelihood estimator, which properly assumes Poisson counting statistics, considerably relaxes this requirement.

Moreover, in the milieu of living cells and tissue, the fluorophores are in a dynamic, heterogeneous environment. Assigning each decay component to a specific molecular conformation or environmental factor is challenging and cannot generally be accomplished without additional independent experiments. Lee et al. (2001) have argued that fitting to stretched exponentials of the form

$$F(t) = F_0 \exp\left[-\left(\frac{t}{\tau}\right)^{1/h}\right] \tag{4.27}$$

where h is a measure of the heterogeneity of the decay, provides a truer representation of the underlying autofluorescence dynamics, since tissue heterogeneity can lead to a continuum of lifetimes (Lee et al. 2001). Performing imaging experiments in such an environment complicates matters, because the need to acquire a signal with good temporal and spatial resolution in a finite period of time can severely limit the number of photons that contribute to an individual measurement. For all of these reasons, relatively simple single- and double-exponential decay models will tend to fit measured decays reasonably well (and provide useful image contrast) regardless of the underlying molecular mechanisms. Therefore, one must be careful not to over interpret the data derived from fluorescence lifetime images.

It is possible to improve the situation by including FLIM as part of a more comprehensive global analysis routine. The goal of this approach is to introduce a priori knowledge from additional experiments into the FLIM analysis. For example, if one or more of the lifetimes are known or otherwise constrained, the effective number of fitting parameters can be reduced, thereby making it possible to test model assumptions (hence functional information can be gained). As an example of this approach, Vishwasrao et al. combined fluorescence lifetime measurements with fluorescence anisotropy measurements to reduce the number of fitting parameters and conclusively identified one free and three bound states of NADH in rat hippocampal slices (Lakowicz 2006; Periasamy and Clegg 2009; Vishwasrao et al. 2005).

Combining FLIM image analysis with computational simulations can also help to explicitly test a given model in silico. Known parameters such as the fluorescence lifetimes and amplitudes are used to create ideal datasets based on the biophysical models. Appropriate amounts of noise are then added to the dataset, consistent with the imaging conditions of a real experiment, and the resulting datasets are then analyzed. If successful, the analysis will reveal the assumed model parameters. By testing competing models in this fashion, it is possible to discern whether or not a given interpretation is appropriate given the limitations of the FLIM imaging procedure.

It is also worth noting that many questions can be answered without an accurate measurement of the lifetime or knowledge of the underlying model for the fluorescence decay process. With appropriate controls, for example, the contrast afforded by FLIM datasets derived from any model may be sufficient to address a given hypothesis without the need for an accurate determination of the lifetime. A relatively new approach for FLIM data analysis is "phasor analysis," a model-free graphical algorithm that simply displays transformed lifetime data in a polar plot (Digman and Gratton 2012; Digman et al. 2008). No fitting is required with this approach. Instead, a straightforward transformation is applied to the measured fluorescence decay at each pixel (i, j) in the image, whether those data were collected in the time domain,

$$g_{i,j}(\omega) = \frac{\int F_{i,j}(t)\cos(\omega t)\,dt}{\int F_{i,j}(t)\,dt} \tag{4.28}$$

$$s_{i,j}(\omega) = \frac{\int F_{i,j}(t)\sin(\omega t)\,dt}{\int F_{i,j}(t)\,dt} \tag{4.29}$$

or in the frequency domain,

$$g_{i,j}(\omega) = m_{i,j}\cos\left(\varphi_{i,j}\right) \tag{4.30}$$

$$s_{i,j}(\omega) = m_{i,j}\sin\left(\varphi_{i,j}\right) \tag{4.31}$$

In this analysis, all pixels are displayed as a 2D histogram, with g values plotted on the horizontal axis and s values plotted on the y-axis. This provides a graphical view of the lifetime distribution for an image, permitting identification of sources of autofluorescence according to their fluorescence lifetimes (Stringari et al. 2011).

4.5 Applications

Autofluorescence lifetime imaging microscopy has been found to be a useful technique for a wide variety of applications ranging from diagnosis of disease states to quantitative analysis of cellular metabolism. The breadth of applications stems from the wide variety of endogenous fluorophores in living cells and tissues that include amino acids (e.g., tryptophan and tyrosine), structural proteins (e.g., collagen and elastin), and key coenzymes related to cellular metabolism, such as NAD(P)H, FMN, and FAD. The relevant optical properties of these are summarized in Table 4.1 and Figures 4.7 and 4.8. While the focus of this volume is cellular metabolism, it is helpful to keep in mind the distinguishing features of all the intrinsic fluorophores when designing experiments or analyzing images that may include nonmetabolic components.

4.5.1 Autofluorescence of Structural Proteins

Collagen and elastin are extracellular structural proteins that are primarily responsible for the strength and elasticity of biological tissues. Collagen provides structural support of all soft tissues, while elastin is more prevalent in elastic tissues, such as skin, arteries, cartilage, lungs, and bladder. While there are 28 classes of collagens that have been categorized, there are 5 primary types of collagen fibers (collagen types I–V), and of these, collagen type I is the most significant. The strength and spectral properties of the structural proteins are due to intermolecular cross-linking. Collagen has a broad absorption spectrum in the range of 300–400 nm and a similarly broad fluorescence emission spectrum that can extend from 400 to 600 nm (Figure 4.7), with a peak that shifts with the excitation wavelength. Collagen emission can overlap with NADH fluorescence and therefore may be seen through the same excitation/emission filters used in a fluorescence microscope. Under 340 nm excitation, for example, collagen has a peak fluorescence emission near 400 nm, but this peak shifts to about 450 nm when the excitation wavelength is

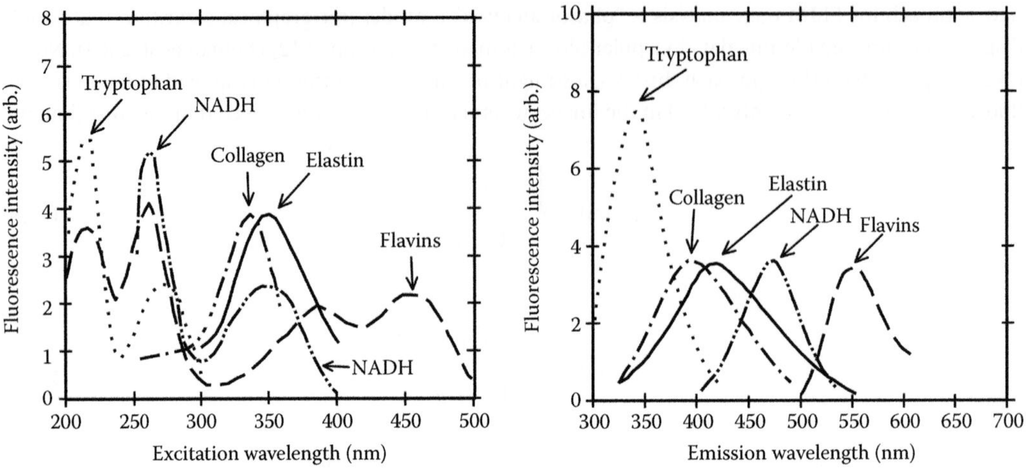

FIGURE 4.7 Autofluorescence excitation and emission spectra of selected endogenous fluorophores discussed in this chapter. (Derived from Monici, M., *Biotechnol. Annu. Rev.* 11, 227, 2005.)

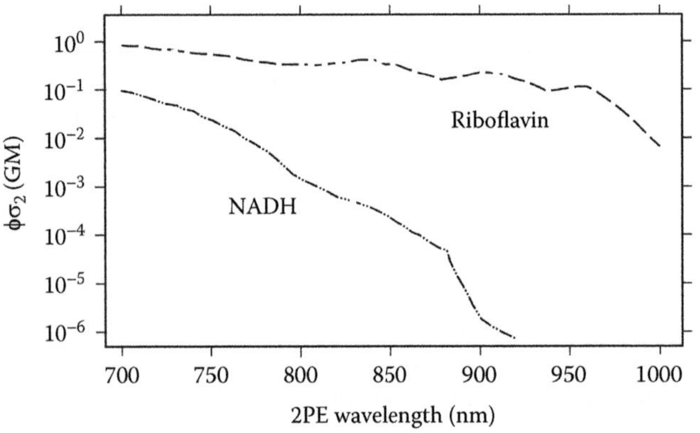

FIGURE 4.8 Two-photon action cross-sectional spectra of NADH and riboflavin. 1 GM = 10^{-50} cm^4 s $photon^{-1}$. (Derived from Zipfel, W.R. et al., *Proc. Natl. Acad. Sci. USA*, 100(12), 7075, 2003a.)

shifted to 360 nm. In the latter case, the emission spectra of collagen and NADH are difficult to distinguish (Georgakoudi et al. 2002). In tissue fluorescence spectroscopy, collagen fluorescence tends to contribute more significantly than NADH to the overall tissue autofluorescence, though the magnitude of both signals have been shown to vary, for example, with disease progression (Drezek et al. 2001).

Spectroscopically, elastin is somewhat similar to collagen, having an absorption peak near 325 nm, an emission peak near 400 nm, and a long wavelength tail extending to 550 nm (Figure 4.7). Again, these features are dictated by molecular cross-linking. Fortunately, these signals can usually be distinguished since collagen and elastin are primarily extracellular, while NADH is intracellular. Note that powdered forms of the structural proteins tend to exhibit additional peaks and different wavelengths than the same proteins in situ. Changes in the fluorescence intensity and lifetimes of both collagen and elastin can reflect changes in tissue architecture. Such changes have been observed with disease progression in cancer as well as in the development of unstable atherosclerotic plaques (Marcu et al. 2001; Ramanujam 2000; Richards-Kortum and Sevick-Muraca 1996).

It is also worthwhile noting that collagen can also be imaged using SHG microscopy due to the intrinsic ordering of collagen fibrils (Zipfel et al. 2003a,b). SHG is a coherent scattering process whereby two photons from the laser field combine to produce a scattered photon of exactly half the incident wavelength (Mertz 2010). Although this is not a fluorescence process, it is important to be aware of the possibility of SHG when designing experiments because the scattered signal can be as significant as fluorescence emission and appears in both fluorescence intensity and lifetime images. While it is not easy to discern SHG from fluorescence via signal intensity, it is easier with FLIM: SHG exhibits no time delay, appearing as an immediate fluorescence decay with a measured lifetime less than the characteristic instrument response time.

4.5.2 Autofluorescence of Metabolic Coenzymes

In addition to the autofluorescence signatures of tissue architecture, living tissues also contain important fluorescent coenzymes that directly reflect the cellular metabolic state. Unfortunately, metabolic profiling based on the fluorescence signature of ADP/ATP is not feasible. Optical absorption (~259 nm) and emission (~370 nm) of ATP is easily masked by tryptophan-containing proteins. Although fluorescent analogs of ATP have been synthesized in order to shift the absorption and emission spectra further into the visible spectrum, addition of the fluorescent tag results in a molecule that is substantially bigger than ATP itself, leading to interference with its normal function (Bagshaw 2001). Fortunately, other endogenous fluorophores such as NADH and flavin coenzymes are more amenable to optical techniques (Heikal 2010).

In free solution, the fluorescence of the nicotinamide ring is quenched by the adenine moiety in the same molecule, resulting in a ~0.40 ns average lifetime, with observations of 0.2, 0.7, and 1.2 ns decays (Vergen et al. 2012; Visser and Hoek 1981; Yu and Heikal 2009). When bound to various enzymes, the NADH molecule adopts a stretched conformation, disrupting the quenching and thereby significantly enhancing the fluorescence intensity and increasing the fluorescence lifetime values to as much as 5 ns (Blinova et al. 2008). Hence, bound NADH is significantly brighter than free NADH (see also the discussion in Section 2.3.1).

The flavin cofactors FMN and FAD, derived from riboflavin (vitamin B_2), are covalently bound in flavoproteins that participate in the Krebs cycle and the electron transport chain (ETC). As outlined in Section 2.3.3, only oxidized form of the flavins, FpOx (FAD or FMN), are fluorescent, while the reduced forms, Fp ($FADH_2$ or $FMNH_2$), are not.

Early studies by Weber demonstrated that when the adenine ring of FAD is in a close, stacked conformation, relative to the isoalloxazine ring, fluorescence is quenched (Weber 1950). The same study showed that riboflavin, similar to FMN, is quenched by the addition of adenine or adenosine in aqueous solutions. Early time-resolved studies showed that free FAD in solution had two lifetime components with 2.8 ns (72%) and 0.3 ns (28%) or an average lifetime of about 2.1 ns, while FMN had a single fluorescence lifetime of 4.7 ns (Visser 1984). Improved timing resolution later revealed the presence of significant, extremely fast decays with time constants of 0.81 and 9 ps in addition to a 2.6 ns decay (Chosrowjan et al. 2003). The fast 9 ps decay was associated with the stacked conformation of FAD where electron transfer from the adenine to the flavin is very efficient. The extremely short timescale of this transition renders it unobservable to traditional lifetime imaging techniques; hence, the stacked conformation is effectively nonfluorescent. Similarly, ultrafast dynamics between the flavin moiety and aromatic amino acid residues within flavoproteins significantly quenches flavin fluorescence to the point where many flavoproteins are nonfluorescent (Kunz and Kunz 1985, Zhong and Zewail 2001). Apart from free flavins, which are not coupled to cellular metabolism, two metabolically active flavoproteins have been identified: lipoamide dehydrogenase (LipDH), a component of both the α-ketoglutarate dehydrogenase complex and the pyruvate dehydrogenase complex, and electron transfer flavoprotein (ETF) (Chance et al. 1979; Huang et al. 2002; Kunz and Gellerich 1993; Rocheleau et al. 2004).

4.5.3 Role of NADH and Flavoproteins in Cellular Metabolism

Given their essential contribution to cellular energetics and the fact that the redox ratios of NADH/NAD^+ and FpOx/Fp can be read out directly via their visible fluorescence, metabolic imaging techniques have focused on these natural biomolecules. Since the fluorescence signal is directly related to both the concentration and the fluorescence lifetime of fluorophores (Equation 4.8), it is helpful to consider the various environments of these coenzymes and reactions in which they participate during cellular metabolism.

As outlined in Chapter 1, during glycolysis, two molecules of NADH are produced in the cytoplasm along with two molecules of pyruvate. To facilitate glycolysis, the cell maintains low levels of NADH within the cytoplasm by transferring the reducing equivalents from cytoplasmic NADH into mitochondria via the malate–aspartate shuttle. In this process, NADH is oxidized to NAD^+ by malate dehydrogenase on the cytoplasmic side, and the reducing equivalents are transported across the inner mitochondrial membrane via malate, which is subsequently oxidized by malate dehydrogenase to reduce mitochondrial NAD^+ to NADH. Within mitochondria, additional NADH is produced by the oxidation of pyruvate to acetyl-CoA by pyruvate dehydrogenase and by the TCA cycle enzymes isocitrate dehydrogenase, α-ketoglutarate dehydrogenase, and malate dehydrogenase.

The energy stored in mitochondrial NADH is ultimately transferred to an electrochemical proton gradient that is necessary to drive the phosphorylation of ADP to ATP. The electron transport process begins when mitochondrial NADH is oxidized to NAD^+ as it donates electrons to reduce FMN to $FMNH_2$ within NADH dehydrogenase (Complex I of the ETC). These electrons are then passed to Fe-S clusters, coenzyme Q, and Complex III and are ultimately used to reduce molecular oxygen to water via Complex IV. Electrons are also injected into the ETC by succinate–ubiquinone oxidoreductase (succinate dehydrogenase) via the TCA cycle. Succinate–ubiquinone oxidoreductase contains a bound FAD cofactor, which is reduced to $FADH_2$ via hydride transfer from succinate. $FADH_2$ is then oxidized, while ubiquinone (also known as coenzyme Q) is reduced to ubiquinol. Again, this reduced form of coenzyme Q then transfers electrons directly to Complex III of the ETC. The electron transport through Complexes I, III, and IV is coupled to the pumping of protons out of the mitochondrial matrix; the resulting proton gradient across the inner mitochondrial membrane is then harnessed by the F_0F_1-ATPase to phosphorylate ADP to ATP.

During respiration, the oxidation of NADH to NAD^+ via the ETC results in a characteristic loss of NADH fluorescence signal, while flavoprotein oxidation leads to an increase in flavoprotein fluorescence. In contrast, blocking electron transport with mitochondrial inhibitors, such as cyanide, produces a characteristic increase in NADH fluorescence and a decrease in flavoprotein fluorescence. Reciprocal changes in fluorescence signals like these indicate that the autofluorescence signal originates from mitochondria.

4.5.4 Metabolic Imaging by Autofluorescence Intensity

While the fluorescence from either NADH or oxidized flavoprotein has been used to characterize the cellular metabolic state as described earlier, Chance et al. (1979) pioneered the use of both signals as a redox ratio, NADH/FpOx, to mitigate the influence of other fluorescence sources and provide a reliable mitochondrial signal. This spectroscopic technique was subsequently extended to an imaging technique and has since been used extensively to quantify the metabolic state of cells and tissues (Masters and Chance 1993).

A hallmark of the disease progression of cancer is the increased use of anaerobic glycolysis for ATP production (also known as Warburg effect). Zhang et al. (2004) developed a near-infrared fluorescent conjugate of 2-deoxyglucose (a glucose sensor), somewhat analogous to the use of positron-emitting fluoro-deoxyglucose for gold-standard cancer detection by positron emission tomography (PET) imaging. They found that the accumulation of this glucose sensor correlated directly with the

NADH/(FpOx + NADH) redox ratio in frozen tumor sections imaged using the filtered output of a mercury arc lamp.

Another approach used to quantify the degree of NADH reduction and flavin oxidation was described by Tiede et al. (2007). To quantify the metabolic state of sensory cells in the intact cochlear organ of Corti, a relative redox scale was devised where the percent reduction of NADH and flavoprotein was calculated from changes in fluorescence when mitochondrial inhibitors and uncouplers were applied. In this study, NADH and flavoprotein were simultaneously two-photon excited using 740 nm pulses of light from a laser. Multiphoton excitation in particular is advantageous for autofluorescence lifetime imaging in cells and tissues because it allows access to the UV-absorption band(s) of those biomolecules using near-infrared ultrashort pulses (Huang et al. 2002; Rocheleau et al. 2004; Zipfel et al. 2003a,b). In addition, multiphoton excitation also restricts photobleaching and photodamage outside the imaging plane (Barth et al. 2005; Tiede and Nichols 2006). Figure 4.8 shows representative two-photon action cross-sectional spectra (i.e., the product of the two-photon absorption cross section and the fluorescence quantum yield) for NADH and riboflavin. NADH fluorescence was primarily detected in a 420–500 nm bandpass, while FpOx fluorescence was primarily observed between 510 and 570 nm, though given the broad emission spectra of these fluorophores (Figure 4.7), these detection channels had to be spectrally unmixed to quantitatively assess NADH and FpOx fluorescence. Maximally reduced NADH was achieved by inhibiting mitochondrial respiration with sodium cyanide, while maximally oxidized NADH was achieved by uncoupling the mitochondria with FCCP (carbonyl cyanide 4-(trifluoromethoxy)-phenylhydrazone). Given these endpoints, the redox state was quantified by the percent reduction of NADH

$$\%R = \left(\frac{F_n - F_{n,u}}{F_{n,i} - F_{n,u}} \right) \cdot 100 \tag{4.32}$$

where

F_n is the average cellular fluorescence obtained from the NADH channel
$F_{n,i}$ is the average cellular NADH fluorescence from the inhibited, NaCN-treated preparation
$F_{n,u}$ is the average NADH fluorescence from the uncoupled, FCCP-treated preparation

Similarly, the amount of FpOx was given by

$$\%O = \left(\frac{F_{fp} - F_{fp,i}}{F_{fp,u} - F_{fp,i}} \right) \cdot 100 \tag{4.33}$$

where again the subscript i denotes the inhibited, cyanide-treated preparation and the subscript u denotes the uncoupled, FCCP-treated preparation (Tiede et al. 2007).

While great progress has been made in elucidating cellular metabolism using assays based on fluorescence intensity alone, these techniques implicitly assume that changes in the fluorescence intensity are due to changes in fluorophore concentration. However, NADH and flavins are found in a wide variety of locations within the cell (cytoplasm, mitochondria, nucleus, and other organelles) and are potentially bound to a wide variety of enzymes. Furthermore, the associations between these cofactors and their binding partners are dynamic. Variations in metabolic state must therefore lead to variations in the microenvironments sampled by the cofactors; it should be no surprise that a wide variety of fluorescence lifetimes have been measured for endogenous fluorophores such as these. At the same time, while there is some chaos in the complexity, there is also a great potential for FLIM to probe in detail the movement and interactions of endogenous fluorophores in response to a number of stimuli as the cell responds to changes in energetic supply or demand.

4.5.5 FLIM of Autofluorescence

The first lifetime images of NADH were obtained over two decades ago by Lakowicz et al. using the frequency-domain method with 355 nm picosecond pulses produced from a synchronously pumped, frequency-doubled dye laser and an ICCD camera that could be modulated at frequencies up to 150 MHz (Lakowicz et al. 1992). Lakowicz and coworkers were able to clearly distinguish NADH bound to malate dehydrogenase from free NADH in solution by its much longer fluorescence lifetime. Their study was also a precursor to imaging of cellular autofluorescence by FLIM. In addition to providing the first illustration of how fluorescence lifetimes could be used to provide an image contrast complimentary to the fluorescence intensity, they also used frequency-domain FLIM to selectively suppress either bound or free NADH in the image.

The development of time-resolved measurements of tissue autofluorescence was initially targeted at the detection of cancer by either autofluorescence or the differential uptake of exogenous fluorescence probes (Wagnieres et al. 1998). Wagnieres et al. followed the frequency-domain approach outlined by Lakowicz and coworkers but implemented it in a standard endoscope. Argon laser light (514 nm) was electro-optically modulated and delivered by fiber optic through the endoscope to excite endogenous fluorophores (presumably flavins) in excised-human bladder tissue as a proof of principle (Wagnieres et al. 1997). The image collected by the endoscope was focused onto a gain-modulated image intensifier for homodyne phase-sensitive imaging. Lifetimes of 2.5 ns were observed in the normal urothelium, whereas lifetimes obtained near an ulceration of the mucous membrane were significantly shorter (0.5 ns). A frame rate of a few Hz was used by acquiring only a few images at well-chosen phase delays, sufficient to provide a useful platform for identifying diseased tissue on the basis of its inherent fluorescence lifetime. Glanzmann et al. (1999) were the first to obtain time- and wavelength-resolved spectra in vivo, using a low-repetition-rate pulsed N_2 pumped dye laser with a streak camera to measure fluorescence decays with subnanosecond resolution. Spectroscopy was performed to provide guidance for the future development of a frequency-domain FLIM for in vivo endoscopic imaging (Glanzmann et al. 1999). Time-resolved autofluorescence spectra were obtained in vivo from the bladder, the bronchi of the lung, and the esophagus (with either 337 or 480 nm excitation). Both subnanosecond and nanosecond fluorescence lifetimes were measured, and variations in the lifetime were sufficient to provide contrast between normal and abnormal tissues. Several endogenous fluorophores were simultaneously excited with 337 nm excitation, and the differences in autofluorescence lifetime between normal and cancer tissues were found to be wavelength dependent. As a result, narrow spectral windows were required to preserve the lifetime contrast. If too many sources of autofluorescence were averaged, the contrast was lost.

This observation highlights an important difficulty encountered with FLIM imaging of autofluorescence. Living tissue has a complex assortment of endogenous fluorophores that are inherently difficult to sort out. The greater the volume of tissue sampled, the greater the number and variety of contributors to the measured fluorescence. Since autofluorescence is relatively weak, sampling a larger volume might improve the signal, but the loss of spatial discrimination also reduces the usefulness of the technique. This is true whether the goal is to discriminate normal tissues from abnormal or quantify changes in cellular metabolism. Furthermore, tissue autofluorescence is dynamic and susceptible to photobleaching. Therefore, care must be taken to avoid photodamage and potential artifacts. This is particularly important since relevant sources of autofluorescence (e.g., the amino acids, collagen, and NADH) are all excited with UV light.

4.5.6 Multiphoton FLIM for Metabolic Imaging

Over the last decade, the potential of FLIM modalities has pushed the development of confocal and multiphoton FLIM imaging techniques to new frontiers. Both multiphoton and single-photon (with confocal detection) excitations reduce the detection volume and, therefore, the ensemble averaging, which in turn allows for optical sectioning and 3D fluorescence lifetime imaging (Elson et al. 2004; Zipfel et al. 2003a,b).

The reduced depth of focus also significantly mitigates interference from fluorophores that are more than a few microns away from the focal volume. These techniques usually employ laser beam scanning with fluorescence detection by point detectors such as PMTs and are amenable to either frequency- or time-domain techniques.

Using two-photon excitation of NADH (at 740 nm) and TCSPC FLIM, Bird et al. reported variations in NADH autofluorescence lifetimes with the density of breast cancer cells in monolayer culture (Bird et al. 2005). Measured lifetimes from a double-exponential fit were approximately 0.4 and 3.0 ns for rapidly dividing cells, consistent with free and enzyme-bound forms of NADH, respectively. Increasing cell density resulted in a 50% reduction of the longer lifetime and a decrease in the free-to-bound ratio for both control and serum-starved cells. They also found that changes in the NADH autofluorescence lifetime were not due to changes in cellular proliferation. Skala et al. also used the multiphoton TCSPC approach to map the progression of disease in the DMBA-treated hamster cheek pouch model of oral carcinogenesis (Skala et al. 2007). To excite NADH and flavins, wavelengths of 780 and 890 nm, respectively, were used, and the blue-green autofluorescence was detected with the BG-39 filter. Short ($\tau_1 = 0.29$ ns) and long ($\tau_2 = 2.03$ ns) lifetime components were consistent with free and bound NADH. Two fluorescence lifetimes of flavins were similarly observed. The short lifetime component ($\tau_1 = 0.15$ ns, 83%) was consistent with a quenched, protein-bound state, while the longer lifetime ($\tau_2 = 2.44$ ns, 17%) was consistent with free flavin. DMBA-induced precancerous cells had similar bound-to-free NADH ratios as normal cells (32%–37%), but there was a significant reduction in the longer lifetime component ($\tau_2 = 1.58$ ns for low-grade precancer and $\tau_2 = 1.83$ for high-grade precancer).

As illustrated by the studies of Bird et al. and Skala et al., very useful observations of changes in cellular metabolism can be obtained from time-domain autofluorescence lifetime measurements when fit to the sum of two exponentials, Equation 4.9, but, given the number of potential binding partners for coenzymes like NADH, it should not be assumed that there are only two pools of NADH within the cell. In fact, several investigators have now confirmed the presence of many bound pools of NADH, as inferred from variations in fluorescence lifetimes as well as measurements of fluorescence anisotropy (Blinova et al. 2005, 2008; Vergen et al. 2012; Vishwasrao et al. 2005). Unfortunately, given the limited number of fluorescence photons that can be collected for a given pixel, fitting more than two exponential decays will not give meaningful results. Furthermore, even with the reduced volume afforded by techniques such as multiphoton FLIM, there are still many fluorophores sampling unique microenvironments within the volume of a voxel, so the fluorescence decay will necessarily be an admixture of a range of fluorescence lifetimes, as discussed earlier. Since the potential spectral shifts of molecules like NADH bound to different proteins are likely to be much smaller than the detection passband, the fluorescence decay is going to be averaged over many molecular conformations and environments. Nevertheless, similar problems have been encountered in related fields, such as super resolution and single-molecule imaging. For example, one approach to tracking molecular position below the diffraction limit was to digitize the point spread function and track the center of the diffraction limited blur (Moerner 2007).

Vergen et al. (2012) took an analogous approach by identifying NADH pools according to the peak of the fluorescence lifetime distribution. Multiphoton TCSPC FLIM images were obtained from rat basophilic leukemia (RBL) cells in monolayer culture, excited at 740 nm with emission between 420 and 500 nm to isolate NADH from other sources of cellular autofluorescence. Fluorescence decays for each pixel were fit to both single- and double-exponential decays, and an F-test was performed to determine whether or not two decays were justified. If so, both best-fit amplitudes and lifetimes were used, otherwise the single-exponential fit was sufficient. The relative concentration associated with a given lifetime was obtained from the amplitudes, and concentration-lifetime histograms were made for all pixels in each image. The histograms were fit with a sum of Gaussian functions according to the number of observed peaks. The location of each Gaussian peak was then used as the mean lifetime for a given pool of NADH. In any given image, at least four lifetime pools were evident. After replicating the experiment many times for uncoupled, routine culture (RC), and inhibited RBL cells, eight lifetime pools could be identified, with the most consistent lifetimes occurring at 0.55, 0.70, 2.6, and 3.2 ns. A 4.0 ns lifetime pool was also occasionally identified. The challenge is now to determine the significance of each lifetime component and, if possible, their assignment to specific molecular conformation or surrounding environment. Additional NADH lifetime pools with lifetimes of approximately 0.25 and 0.87 ns were also observed in some experiments of RC and uncoupled cells, but not for cells inhibited by NaCN. Some NADH lifetimes shifted with metabolic state, while others did not. About 80% of the NADH in untreated cells had a lifetime shorter than 0.75 ns, which increased when respiration was inhibited and decreased when respiration was uncoupled. This was likely due to NADH moving between free and enzyme-bound states with changes in metabolism. In similar experiments with EMT6 adenocarcinoma cells, NADH lifetime distributions were also shown to depend on the extracellular glucose concentration (Vergen et al. 2012). This supports previous findings by Evans et al. (2005), who suggested that NADH FLIM imaging may also be useful as a glucose sensor.

More recently, this approach has been applied to obtain metabolic profiles of sensory cells in the living cochlea. Of the two types of cochlear sensory cells, inner hair cells (IHCs) are significantly more resilient than outer hair cells (OHCs) to acoustic trauma, age-related hearing loss, and antibiotic ototoxicity. To determine whether these differences are related to cellular metabolism, we imaged the sensory cells of acutely cultured cochlear explants obtained from FVB mice (Jensen-Smith et al. 2012). Multiphoton FLIM images of NADH were obtained while maintaining the culture at 32°C throughout the imaging process. Oxidative phosphorylation was inhibited or uncoupled using NaCN and FCCP, respectively. Examples of FLIM images that were obtained are shown in Figure 4.9. Three rows of OHCs and a row of IHCs (from top to bottom) can be identified. Individual IHCs and OHCs were analyzed, following the preceding approach described. Seven NADH pools are routinely observed across all cultures with lifetimes ranging from 0.73 to about 5.0 ns. The distribution of lifetimes varies with both metabolic state and cell type. The ability to identify and develop an understanding of the endogenous differences in metabolic state between these sensory cells will be crucial for the prevention of permanent hearing loss.

4.6 Summary

Fluorescence lifetime imaging of autofluorescence provides a unique view of the dynamic interactions between endogenous fluorophores and their local microenvironment. In this chapter, we have highlighted several reasons why FLIM is an important tool for studying metabolism in living tissue. First, FLIM is a high-resolution imaging technique, capable of distinguishing changes in metabolism in rows of adjacent cells. As an optical imaging modality, FLIM allows for noninvasive monitoring of endogenous fluorophores in situ, without significant tissue damage. This is particularly true when coupled with multiphoton microscopy employing nontoxic, near-infrared pulses of light. When spectral identification is complicated by broad, overlapping excitation and emission spectra, FLIM can help distinguish molecular sources of autofluorescence according to their fluorescence lifetime. Notably, while fluorescence intensity is often assumed to be equivalent to the fluorophores concentration, in fact, this is seldom the case. Since pixel brightness depends on the product of the fluorescence lifetime and the concentration, time-resolved measurements are needed to accurately determine changes in concentration.

Like many other imaging methods, FLIM also has its own disadvantages. To begin with, FLIM is significantly more complicated than intensity-based imaging techniques, both in instrumentation and image analysis. It is also more expensive, requiring both faster detectors and pulsed (or modulated) light sources. FLIM imaging can also be significantly slower, especially with laser-scanning experimental implementations due to the longer imaging times required to collect the data pixel by pixel in order to construct a complete image. As is often the case with technological innovation, improvements on all fronts continue to be made. Indeed, commercial implementations and improvements over the last decade have made FLIM much more practical and accessible.

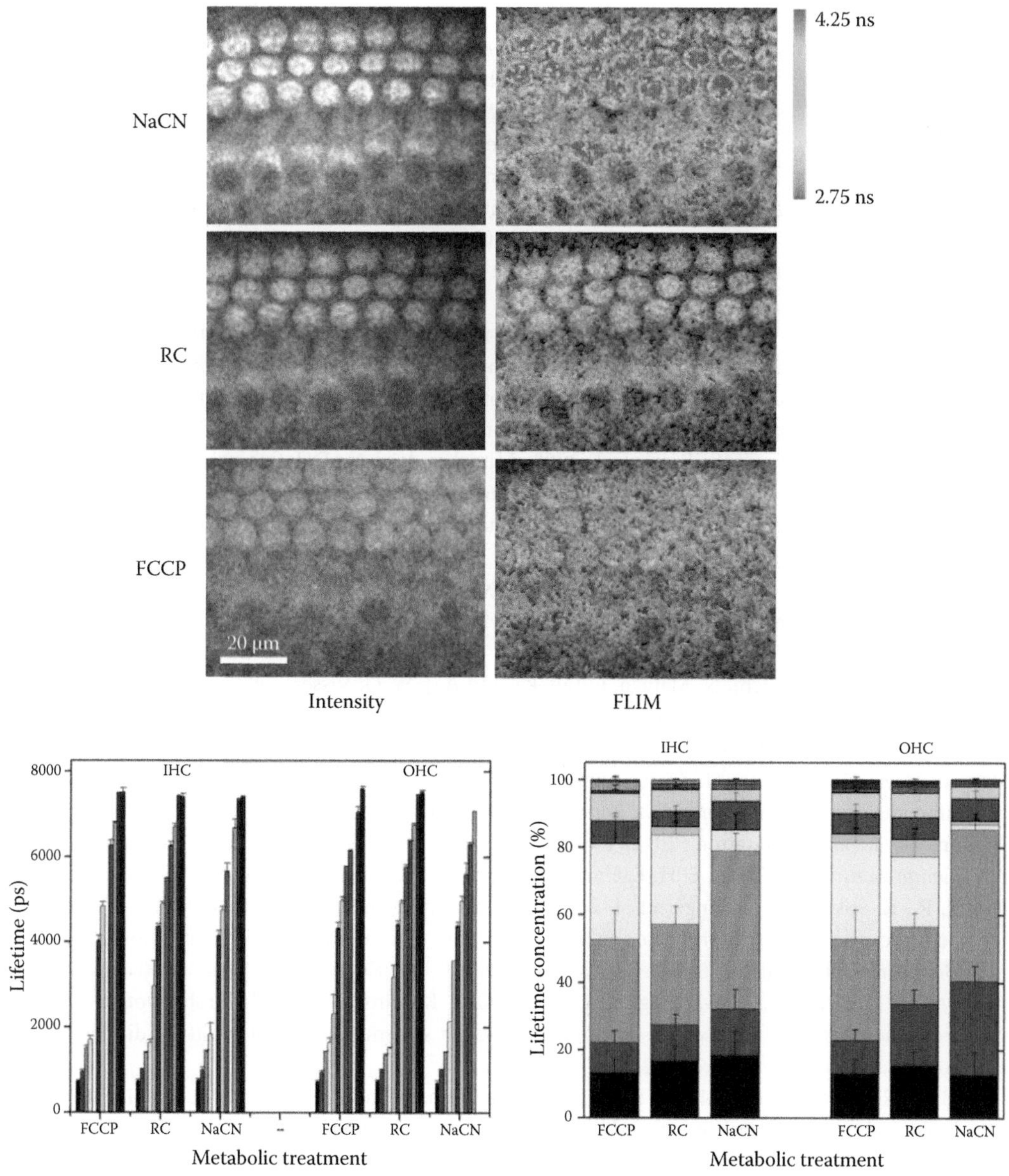

FIGURE 4.9 FLIM NADH images of intact sensory cells from a living cochlear explant. Three rows of OHCs and one row of IHCs can be seen in each image. Application of an inhibitor (NaCN) or uncoupler (FCCP) results in changes in both the intensity and lifetime distribution as compared to the RC. The colors in the FLIM images represent different average lifetimes, τ_m, ranging from 2.75 ns (red) to 4.25 ns (blue). Many NADH pools can be identified by fitting lifetime histograms obtained from a large number of cells (see text for further discussion). Each color shown in the bar graphs represents a unique NADH lifetime and relative concentration that was observed in the HCs.

References

Bagshaw, C. R. 2001. ATP analogues at a glance. *Journal of Cell Science* 114(3): 459–460.

Barth, E. E., R. Hallworth, and M. G. Nichols. 2005. A comparison of the sensitivity of photodamage assays in rat basophilic leukemia cells. *Photochemistry and Photobiology* 81(3): 556–562.

Bastiaens, P. I. H. and A. Squire. 1999. Fluorescence lifetime imaging microscopy: Spatial resolution of biochemical processes in the cell. *Trends in Cell Biology* 9(2): 48–52.

Becker, W. 2005. *Advanced Time-Correlated Single Photon Counting Techniques*. New York: Springer-Verlag.

Becker, W. 2008. *The bh TCSPC Handbook*. Berlin, Germany: Becker & Hickl GmbH.

Becker, W., A. Bergmann, and C. Biskup. 2007. Multispectral fluorescence lifetime imaging by TCSPC. *Microscopy Research and Technique* 70(5): 403–409.

Berezin, M. Y. and S. Achilefu. 2010. Fluorescence lifetime measurements and biological imaging. *Chemical Reviews* 10(5): 2641–2684.

Bird, D. K., L. Yan, K. M. Vrotsos et al. 2005. Metabolic mapping of MCF10A human breast cells via multiphoton fluorescence lifetime imaging of the coenzyme NADH. *Cancer Research* 65(19): 8766–8773.

Blinova, K., S. Carroll, S. Bose et al. 2005. Distribution of mitochondrial NADH fluorescence lifetimes: Steady-state kinetics of matrix NADH interactions. *Biochemistry* 44(7): 2585–2594.

Blinova, K., R. L. Levine, E. S. Boja et al. 2008. Mitochondrial NADH fluorescence is enhanced by complex I binding. *Biochemistry* 47(36): 9636–9645.

Buurman, E. P., R. Sanders, A. Draauer et al. 1992. Fluorescence lifetime imaging using a confocal laser scanning microscope. *Scanning* 14(3): 155–159.

Chance, B., B. Schoener, R. Oshino, F. Itshak, and Y. Nakase. 1979. Oxidation-reduction ratio studies of mitochondria in freeze-trapped samples. NADH and flavoprotein fluorescence signals. *Journal of Biological Chemistry* 254(11): 4764–4771.

Chosrowjan, H., S. Taniguchi, N. Mataga, F. Tanaka, and A. J. W. G. Visser. 2003. The stacked flavin adenine dinucleotide conformation in water is fluorescent on picosecond timescale. *Chemical Physics Letters* 378(3–4): 354–358.

Digman, M. A., V. R. Caiolfa, M. Zamai, and E. Gratton. 2008. The phasor approach to fluorescence lifetime imaging analysis. *Biophysical Journal* 94(2): L14–L16.

Digman, M. A. and E. Gratton. 2012. Fluorescence lifetime microscopy: The phasor approach. In: *Comprehensive Biophysics*, E. H. Egelman (ed.), pp. 24–38. Amsterdam, the Netherlands: Elsevier.

Drezek, R., K. Sokolov, U. Utzinger et al. 2001. Understanding the contributions of NADH and collagen to cervical tissue fluorescence spectra: Modeling, measurements, and implications. *Journal of Biomedical Optics* 6(4): 385–396.

Elder, A. D., J. H. Frank, J. Swartling, X. Dai, and C. F. Kaminski. 2006. Calibration of a wide-field frequency-domain fluorescence lifetime microscopy system using light emitting diodes as light sources. *Journal of Microscopy* 224(2): 166–180.

Elson, D., J. Requejo-Isidro, I. Munro et al. 2004. Time-domain fluorescence lifetime imaging applied to biological tissue. *Photochemical and Photobiological Sciences* 3(8): 795–801.

Elson, D. S., J. Siegel, S. E. D. Webb et al. 2002. Fluorescence lifetime system for microscopy and multiwell plate imaging with a blue picosecond diode laser. *Optics Letters* 27(16): 1409–1411.

Evans, N. D., L. Gnudi, O. J. Rolinski, D. J. S. Birch, and J. C. Pickup. 2005. Glucose-dependent changes in NAD(P)H-related fluorescence lifetime of adipocytes and fibroblasts in vitro: Potential for non-invasive glucose sensing in diabetes mellitus. *Journal of Photochemistry and Photobiology B* 80(2): 122–129.

Gadella Jr., T. W. J., T. M. Jovin, and R. M. Clegg. 1993. Fluorescence lifetime imaging microscopy (FLIM): Spatial resolution of microstructures on the nanosecond time scale. *Biophysical Chemistry* 48(2): 221–239.

Georgakoudi, I., B. C. Jacobson, M. G. Müller et al. 2002. NAD(P)H and collagen as in vivo quantitative fluorescent biomarkers of epithelial precancerous changes. *Cancer Research* 62(3): 682–687.

Gerritsen, H. C., M. A. H. Asselbergs, A. V. Agronskaia, and W. G. J. H. M. Van Sark. 2002. Fluorescence lifetime imaging in scanning microscopes: Acquisition speed, photon economy and lifetime resolution. *Journal of Microscopy* 206(3): 218–224.

Gerritsen, H. C., A. Draaijer, D. J. Van den Heuvel, and A. V. Agronskaia. 2006. Fluorescence lifetime imaging in scanning microscopy. In: *Handbook of Biological Confocal Microscopy*, J. B. Pawley (ed.), pp. 516–534. New York: Springer.

Glanzmann, T., J.-P. Ballini, H. van den Bergh, and G. Wagnières. 1999. Time-resolved spectrofluorometer for clinical tissue characterization during endoscopy. *Review of Scientific Instruments* 70(10): 4067–4077.

Gratton, E., S. Breusegem, J. Sutin, Q. Ruan, and N. Barry. 2003. Fluorescence lifetime imaging for the two-photon microscope: Time-domain and frequency-domain methods. *Journal of Biomedical Optics* 8(3): 381–390.

Heikal, A. A. 2010. Intracellular coenzymes as natural biomarkers for metabolic activities and mitochondrial anomalies. *Biomarkers in Medicine* 4(2): 241–263.

Herman, P. and J. Vecer. 2008. Frequency domain fluorometry with pulsed light-emitting diodes. *Annals of the New York Academy of Sciences* 1130(1): 56–61.

Huang, S., A. A. Heikal, and W. W. Webb. 2002. Two-photon fluorescence spectroscopy and microscopy of NAD(P)H and flavoprotein. *Biophysical Journal* 82(5): 2811–2825.

Jensen-Smith, H. C., R. Hallworth, and M. G. Nichols. 2012. Gentamicin rapidly inhibits mitochondrial metabolism in high-frequency cochlear outer hair cells. *PLoS One* 7(6): e38471.

Köllner, M. and J. Wolfrum. 1992. How many photons are necessary for fluorescence-lifetime measurements? *Chemical Physics Letters* 200(1): 199–204.

Kumar, S., C. Dunsby, P. A. A. De Beule et al. 2007. Multifocal multiphoton excitation and time correlated single photon counting detection for 3-D fluorescence lifetime imaging. *Optics Express* 15(20): 12548–12561.

Kunz, W. S. and F. N. Gellerich. 1993. Quantification of the content of fluorescent flavoproteins in mitochondria from liver, kidney cortex, skeletal muscle, and brain. *Biochemical Medicine and Metabolic Biology* 50(1): 103–110.

Kunz, W. S. and W. Kunz. 1985. Contribution of different enzymes to flavoprotein fluorescence of isolated rat liver mitochondria. *Biochimica et Biophysica Acta (BBA)—General Subjects* 841(3): 237–246.

Lakowicz, J. R., ed. 1983. *Principles of Fluorescence Spectroscopy*. New York: Plenum Press.

Lakowicz, J. R., ed. 2006. *Principles of Fluorescence Spectroscopy*. Boston, MA: Springer.

Lakowicz, J. R. and K. W. Berndt. 1991. Lifetime-selective fluorescence imaging using an rf phase-sensitive camera. *Review of Scientific Instruments* 62(7): 1727–1734.

Lakowicz, J. R., H. Szmacinski, K. Nowaczyk, and M. L. Johnson. 1992. Fluorescence lifetime imaging of free and protein-bound NADH. *Proceedings of the National Academy of Sciences* 89(4): 1271–1275.

Lee, K. C., J. Siegel, S. E. Webb et al. 2001. Application of the stretched exponential function to fluorescence lifetime imaging. *Biophysical Journal* 81(3): 1265–1274.

Levitt, J. A., D. R. Matthews, S. M. Ameer-Beg, and K. Suhling. 2009. Fluorescence lifetime and polarization-resolved imaging in cell biology. *Current Opinion in Biotechnology* 20(1): 28–36.

Marcu, L., M. C. Fishbein, J.-M. I. Maarek, and W. S. Grundfest. 2001. Discrimination of human coronary artery atherosclerotic lipid-rich lesions by time-resolved laser-induced fluorescence spectroscopy. *Arteriosclerosis, Thrombosis, and Vascular Biology* 21(7): 1244–1250.

Masters, B. R. and B. Chance. 1993. Redox confocal imaging: Intrinsic fluorescent probes of cellular metabolism. In: *Fluorescent and Luminescent Probes for Biological Activity*, W. T. Mason (ed.), pp. 44–56. London, U.K.: Academic Press.

Mertz, J. 2010. *Introduction to Optical Microscopy*. Greenwood Village, CO: Roberts and Company Publishers.

Moerner, W. E. 2007. New directions in single-molecule imaging and analysis. *Proceedings of the National Academy of Sciences of the United States of America* 104(31): 12596–12602.

Monici, M. 2005. Cell and tissue autofluorescence research and diagnostic applications. In: *Biotechnology Annual Review*, R. M. El-Gewely (ed.), pp. 227–256. Amsterdam, the Netherlands: Elsevier Science.

Nishimura, G. and M. Tamura. 2005. Artefacts in the analysis of temporal response functions measured by photon counting. *Physics in Medicine and Biology* 50(6): 1327–1342.

Periasamy, A. and R. Clegg, eds. 2009. *FLIM Microscopy in Biology and Medicine.* Boca Raton, FL: CRC Press.

Phipps, J. E., Y. Sun, M. C. Fishbein, and L. Marcu. 2012. A fluorescence lifetime imaging classification method to investigate the collagen to lipid ratio in fibrous caps of atherosclerotic plaque. *Lasers in Surgery and Medicine* 44(7): 564–571.

Ramanujam, N. 2000. Fluorescence spectroscopy of neoplastic and non-neoplastic tissues. *Neoplasia* 2(1–2): 89–117.

Richards-Kortum, R. and E. Sevick-Muraca. 1996. Quantitative optical spectroscopy for tissue diagnosis. *Annual Review of Physical Chemistry* 47(1): 555–606.

Rocheleau, J. V., W. S. Head, and D. W. Piston. 2004. Quantitative NAD(P)H/flavoprotein autofluorescence imaging reveals metabolic mechanisms of pancreatic islet pyruvate response. *The Journal of Biological Chemistry* 279(30): 31780–31787.

Schneckenburger, H., M. Wagner, P. Weber, W. S. L. Strauss, and R. Sailer. 2004. Autofluorescence lifetime imaging of cultivated cells using a UV picosecond laser diode. *Journal of Fluorescence* 14(5): 649–654.

Sharman, K. K., A. Periasamy, H. Ashworth, and J. N. Demas. 1999. Error analysis of the rapid lifetime determination method for double-exponential decays and new windowing schemes. *Analytical Chemistry* 71(5): 947–952.

Skala, M. C., K. M. Riching, D. K. Bird et al. 2007. In vivo multiphoton fluorescence lifetime imaging of protein-bound and free nicotinamide adenine dinucleotide in normal and precancerous epithelia. *Journal of Biomedical Optics* 12(2): 024014.

Stringari, C., A. Cinquin, O. Cinquin, M. A. Digman, P. J. Donovan, and E. Gratton. 2011. Phasor approach to fluorescence lifetime microscopy distinguishes different metabolic states of germ cells in a live tissue. *Proceedings of the National Academy of Sciences of the United States of America* 108(33): 13582–13587.

Sun, Y., R. N. Day, and A. Periasamy. 2011. Investigating protein-protein interactions in living cells using fluorescence lifetime imaging microscopy. *Nature Protocols* 6(9): 1324–1340.

Sytsma, J., J. M. Vroom, C. J. de Grauw, and H. C. Gerritsen. 1998. Time-gated fluorescence lifetime imaging and microvolume spectroscopy using two-photon excitation. *Journal of Microscopy* 191(1): 39–51.

Tiede, L. M. and M. G. Nichols. 2006. Photobleaching of reduced nicotinamide adenine dinucleotide and the development of highly fluorescent lesions in rat basophilic leukemia cells during multiphoton microscopy. *Photochemistry and Photobiology* 82(3): 656–664.

Tiede, L. M., S. M. Rocha-Sanchez, R. Hallworth, M. G. Nichols, and K. Beisel. 2007. Determination of hair cell metabolic state in isolated cochlear preparations by two-photon microscopy. *Journal of Biomedical Optics* 12(2): 021004.

Vergen, J., C. Hecht, L. V. Zholudeva, M. M. Marquardt, R. Hallworth, and M. G. Nichols. 2012. Metabolic imaging using two-photon excited NADH intensity and fluorescence lifetime imaging. *Microscopy and Microanalysis* 18(4): 761–770.

Vishwasrao, H. D., A. A. Heikal, K. A. Kasischke, and W. W. Webb. 2005. Conformational dependence of intracellular NADH on metabolic state revealed by associated fluorescence anisotropy. *Journal of Biological Chemistry* 280(26): 25119–25126.

Visser, A. J. W. G. 1984. Kinetics of stacking interactions in flavin adenine dinucleotide from time-resolved flavin fluorescence. *Photochemistry and Photobiology* 40(6): 703–706.

Visser, A. J. W. G. and A. Van Hoek. 1981. The fluorescence decay of reduced nicotinamides in aqueous solutions after excitation with a UV-mode locked Ar ion laser. *Photochemistry and Photobiology* 33(1): 35–40.

Wagnieres, G., J. Mizeret, A. Studzinski, and H. Van Den Bergh. 1997. Frequency-domain fluorescence lifetime imaging for endoscopic clinical cancer photodetection: Apparatus design and preliminary results. *Journal of Fluorescence* 7(1): 75–83.

Wagnieres, G. A., W. M. Star, and B. C. Wilson. 1998. In vivo fluorescence spectroscopy and imaging for oncological applications. *Photochemistry and Photobiology* 68(5): 603–632.

Wang, X. F., A. Periasamy, B. Herman, and D. M. Coleman. 1992. Fluorescence lifetime imaging microscopy (FLIM): Instrumentation and applications. *Critical Reviews in Analytical Chemistry* 23(5): 369–395.

Wang, X. F., T. Uchida, and S. Minami. 1989. A fluorescence lifetime distribution measurement system based on phase-resolved detection using an image dissector tube. *Applied Spectroscopy* 43(5): 840–845.

Weber, G. 1950. Fluorescence of riboflavin and flavin-adenine dinucleotide. *Biochemical Journal* 47(1): 114–121.

Xu, C. and W. Webb. 1996. Measurement of two-photon excitation cross sections of molecular fluorophores with data from 690 to 1050 nm. *Journal of the Optical Society of America B, Optical Physics* 13(3): 481–491.

Xu, C. and W. Webb. 2002. Multiphoton excitation of molecular fluorophores and nonlinear laser microscopy. In: *Topics in Fluorescence Spectroscopy*, Vol. 5: Nonlinear and Two-Photon-Induced Fluorescence, J. R. Lakowicz (ed.), pp. 471–540. New York: Kluwer Academic Publishers.

Yu, Q. and A. A. Heikal. 2009. Two-photon autofluorescence dynamics imaging reveals sensitivity of intracellular NADH concentration and conformation to cell physiology at the single-cell level. *Journal of Photochemistry and Photobiology B, Biology* 95(1): 46–57.

Zhang, Z., H. Li, Q. Liu et al. 2004. Metabolic imaging of tumors using intrinsic and extrinsic fluorescent markers. *Biosensors and Bioelectronics* 20(3): 643–650.

Zhong, D. and A. H. Zewail. 2001. Femtosecond dynamics of flavoproteins: Charge separation and recombination in riboflavin (vitamin B2)-binding protein and in glucose oxidase enzyme. *Proceedings of the National Academy of Sciences of the United States of America* 98(21): 11867–11872.

Zipfel, W. R., R. M. Williams, R. Christie, A. Y., Nikitin, B. T. Hyman, and W. W. Webb. 2003a. Live tissue intrinsic emission microscopy using multiphoton-excited native fluorescence and second harmonic generation. *Proceedings of the National Academy of Sciences of the United States of America* 100(12): 7075–7080.

Zipfel, W. R., R. M. Williams, and W. W. Webb. 2003b. Nonlinear magic: Multiphoton microscopy in the biosciences. *Nature Biotechnology* 21(11): 1369–1377.

5

Polarization Imaging of Cellular Autofluorescence

Harshad D. Vishwasrao
Columbia University

Qianru Yu
Dartmouth College

Kuravi Hewawasam
University of Minnesota Duluth

Ahmed A. Heikal
University of Minnesota Duluth

5.1 Introduction

Intrinsic fluorophores and their associated biological processes exhibit dynamics on a wide range of timescales throughout the heterogeneous milieu of living cells. Conventional time-lapse autofluorescence intensity imaging is best suited for monitoring slow physiological functions such as changes in cellular morphology, cell migration, and intracellular distribution that take place on a timescale of seconds to minutes. In contrast, ultrafast (10^{-12}–10^{-7} s) time-resolved fluorescence measurements can probe molecular dynamics, such as excited-state processes and rotational dynamics, which are acutely sensitive to the chemical structure, intermolecular interactions, and microenvironment of a given fluorophore. Time-resolved fluorescence and anisotropy measurements are uniquely suited for a detailed biophysical description of intrinsic fluorophores within the context of physiological and pathological changes in living cells or tissues. Consequently, multiparametric autofluorescence detection (i.e., intensity, color, lifetime, and polarization) provides the most complete description of intrinsically fluorescent biomolecules and their role in both the physiology and pathology of cells or tissues.

Using either autofluorescence or analytical biochemistry methods, NADH-based studies of cellular metabolism have traditionally measured the total NADH concentration. However, the reaction velocity of a given intracellular NADH-dehydrogenase binding depends on the concentration of locally available free NADH (Williamson et al. 1967). As a result, the chemical flux in any given NADH-oxidizing pathway (e.g., oxidative phosphorylation) is likely to be more influenced by the local population of free NADH than total NADH. While this argument provides the theoretical basis for the thermodynamic importance of free NADH, recent studies have provided experimental evidence that the free/bound NADH ratio varies significantly as a function of the physiological state of cells (Yu and Heikal 2009; Zheng et al. 2010) and brain tissues (Vishwasrao et al. 2005) as well as with pathological conditions such as cancer (Skala et al. 2007; Yu and Heikal 2009), pre-apoptosis (Yu et al. 2011), and ischemic reperfusion injury (Thorling et al. 2011).

Given the thermodynamic importance of free NADH, and the emerging evidence that the free/enzyme-bound state is sensitive to cellular metabolism and pathological conditions, considerable work has been done to quantify the free/bound state of intracellular NADH. Early estimates of the free and enzyme-bound NADH concentrations came from biochemical techniques such as pyridine nucleotide extraction in tissues (Garofalo et al. 1988; Klaidman et al. 1995) and the metabolite indicator method for resolving the cytoplasmic and mitochondrial free [NADH]/[NAD^+] ratio (Bücher et al. 1972; Merrill and Guynn 1976, 1982). While informative, these methods require destroying the cells/tissue and thereby restricting metabolic studies to a snapshot. Furthermore, these techniques are inherently incapable of resolving the spatial and compartmental variations of the free/bound ratio of NADH (i.e., no morphological context). In contrast, fluorescence techniques (Chance and Baltscheffsky 1958; Gafni and Brand 1976; Hönes et al. 1986; Huang et al. 2002; Kasischke et al. 2004; Paul and Schneckenburger 1996; Salmon et al. 1982; Vishwasrao et al. 2005; Wakita et al. 1995; Yu and Heikal 2009) are nondestructive while preserving the spatial information down to submicron length scales, depending on the imaging modality, within living cells.

Enzyme binding of a small fluorophore like NADH alters its local environment, mobility, and conformation, which affect the fluorophore's photophysical properties. Fluorescence spectroscopy can therefore be used to calculate the partitioning of intracellular NADH between the free and enzyme-bound states. The question then becomes which particular observable offers the most robust assay for estimating the free and enzyme-bound ratio of NADH. As discussed in earlier chapters, enzyme binding induces a spectral shift (up to ~20 nm) in the emission peak of NADH autofluorescence (Chance and Baltscheffsky 1958; Salmon et al. 1982) and significantly enhances its fluorescence lifetime. However, these spectral shifts are small compared to the width of the NADH spectrum (~150 nm). Moreover, the multiexponential decay of bound NADH has shorter components comparable to the decay time of free NADH, making it difficult to assign any individual time component to either the free or bound species of NADH in a mixture (Gafni and Brand 1976; Vishwasrao et al. 2005; Yu and Heikal 2009).

The rotational (or tumbling) time is possibly the most sensitive and direct observable for enzyme binding of NADH (Heikal 2010; Vishwasrao et al. 2005; Yu and Heikal 2009). The molecular size of free NADH (MW = 665 Da) is substantially smaller than that of an NADH-binding enzyme (e.g., MW = 70 kDa for malate dehydrogenase). Since the rotational time is proportional to molecular weight (or size), the rotational time of the NADH–enzyme complex is significantly larger (10–100-fold) than that of free NADH. As a result, time-resolved anisotropy is the technique of choice for measuring the rotational time of a given fluorophore. For example, time-resolved anisotropy has been used for the discrimination between the free and bound states of NADH (Hönes et al. 1986; Vishwasrao et al. 2005; Yu and Heikal 2009).

In this chapter, we discuss various aspects of anisotropy imaging of the cellular autofluorescence associated with both NADH and flavins (i.e., FAD, FMN, and flavoproteins). A theoretical and practical discussion of fluorescence anisotropy imaging will be provided for both steady-state and time-resolved modalities. We will also provide a survey of solution studies of the photophysics of NADH and FAD in different environments. These solution studies provide a reference point that is helpful in correctly interpreting cellular autofluorescence measurements in living cells. Time-resolved anisotropy will require some discussion of fluorescence lifetime imaging, which is covered more thoroughly in Chapter 4. Our objective here is to highlight the potential of the autofluorescence anisotropy imaging of NADH and flavins as powerful analytical tools for studying cell physiology and pathology that can complement other fluorescence-based methods described in this book as well as conventional analytical biochemistry.

5.2 Theoretical Background

5.2.1 Time-Resolved Associated Anisotropy of a Mixture of Fluorophores

In living cells, NADH is a cofactor for a large number of different enzymes (e.g., dehydrogenases) with distinct structures, binding sites, and enzymatic activities. While the enzymes themselves are not fluorescent under NADH excitation conditions, enzyme binding alters both the fluorescence properties (fluorescence lifetime or quantum yield) and the effective hydrodynamic volume of NADH. Therefore, intracellular NADH exists as a mixture of free and enzyme-bound forms that are both fluorescent under selective excitation/detection conditions. The relative contributions of free and bound NADH to the overall autofluorescence signal depend not only on their concentrations but also on their quantum yields or lifetimes. We therefore present the theoretical framework for associated anisotropy, which describes the total fluorescence anisotropy of a mixture of species, each with a distinct fluorescence lifetime and hydrodynamic volume (Figure 5.1).

In any given pixel (x, y) in an autofluorescence image of living cells, consider a mixture of multiple (N) fluorescent species (Figure 5.1a), each of which (e.g., the ith species) has a unique fluorescence lifetime τ_i and a rotational time ϕ_i that is size dependent. The overall fluorescence decay of such a mixture will be given by

$$F(x,y,t)=\sum_{i=1}^{N}\alpha_i(x,y)\cdot\exp\left[\frac{-t}{\tau_i(x,y)}\right] \tag{5.1}$$

where the amplitude (α_i) reflects the fractional population of the ith species in the mixture such that $0\leq\alpha_i\leq 1$ and $\sum_i \alpha_i = 1$. The total fluorescence decay (Equation 5.1) can be directly measured using time-correlated single photon counting (TCSPC) at magic-angle (54.7°) polarization in order to eliminate the rotational (tumbling) effects of the excited fluorophore. The corresponding average fluorescence lifetime for such a mixture is given by

$$\sum_{i=1}^{N}\alpha_i(x,y)\tau_i(x,y) \tag{5.2}$$

In steady-state fluorescence imaging, however, the ith species will contribute a fraction (f_i) of the total, time-averaged fluorescence signal such that

$$f_i(x,y)=\frac{\alpha_i\cdot\tau_i}{\sum_{i=1}^{N}\alpha_i\cdot\tau_i} \tag{5.3}$$

During the lifetime of the excited state, a fluorophore undergoes rotational motion with a characteristic timescale that depends on the hydrodynamic volume and restrictions imposed by the local

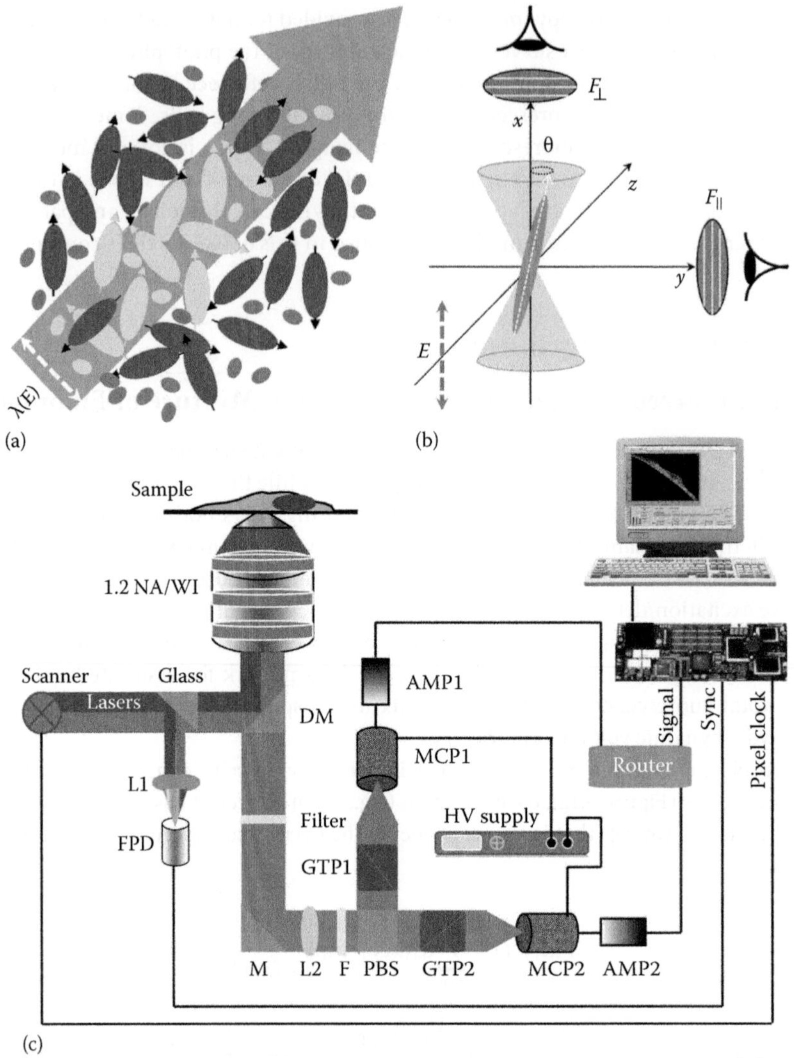

FIGURE 5.1 A sketch of our experimental setup for 2P-FLIM and anisotropy of cellular autofluorescence. (a) In a mixture of two fluorophores with different sizes and quantum yields, polarized laser light selectively excites those fluorophores with absorption dipoles that are parallel to the laser polarization. (b) The parallel and perpendicular polarization components of the fluorescence emission from the rotating dipole moment of the excited fluorophore are then detected simultaneously. (c) A femtosecond laser system (Mira 900F), pumped by a solid-state diode CW laser (Verdi V-10), generates 120 fs pulses at 76 MHz and 700–1000 nm for 2P-FLIM imaging. The laser pulses are conditioned and steered toward a laser scanner before being focused on the sample through a 1.2 NA, 60×, water immersion objective on an inverted microscope (IX80, Olympus). The scanned 2P epifluorescence is separated from the laser using both a dichroic mirror (DM) and appropriate filters (F), before being polarization analyzed and detected by an MCP-PMT. The output signal of the MCP-PMT is then amplified, routed, and then used to build a fluorescence decay histogram for each pixel using an SPC-830 module (Becker & Hickl). This module is synchronized to the laser pulse train using the signal from a fast photodiode (FPD) that detects a small fraction of the laser beam. The pixel-to-pixel signal in the data acquisition software is synchronized with the laser scanning using a pixel clock from the scanner. For single-point, time-resolved fluorescence and anisotropy, the repetition rate is reduced (4.2 MHz) using a pulse picker (Mira 9200, Coherent). The other notations are defined as follows: M, mirror; L, lens for laser/fluorescence conditioning; F, filters; PBS, polarizing beam splitter; GTP, Glan–Thompson polarizer; AMP, amplifier; and HV, high-voltage power supply.

environment. The corresponding rotational motion of a fluorophore (or a mixture of fluorophores) can be characterized using time-resolved fluorescence anisotropy, $r(x,y,t)$, which is calculated from the measured parallel $F_{\parallel}(x,y,t)$ and perpendicularly $F_{\perp}(x,y,t)$ polarized fluorescence signals (Figure 5.1b) such that

$$r(x,y,t) = \frac{F_{\parallel}(x,y,t) - GF_{\perp}(x,y,t)}{F_{\parallel}(x,y,t) + 2GF_{\perp}(x,y,t)} \tag{5.4}$$

The G-factor accounts for a possible relative bias between the two polarization detections and can be determined using a tail-matching method using a reference fluorophore with (1) an emission spectrum that overlaps with the unknown (NADH in this case) and (2) a small size such that the rotational time is much faster than the fluorescence lifetime (Lakowicz 2006). The denominator of Equation 5.4 is basically the same as the total fluorescence lifetime (Equation 5.1), measured using magic-angle detection. For a single species of fluorophore in a given pixel, the anisotropy is typically described by a simple exponential decay:

$$r(x,y,t) = r_0(x,y)\exp\left[\frac{-t}{\phi(x,y)}\right] \tag{5.5}$$

where ϕ is the rotational time of the fluorophore that depends on its hydrodynamic volume (V), or its molecular weight (M), and the local viscosity (η) of the surrounding environment (Lakowicz 2006; Piersma et al. 1998; Yu et al. 2008):

$$\phi = \frac{\eta V}{k_B T} = \frac{\eta M(\upsilon + h)}{N_A RT} \tag{5.6}$$

The specific volume (v) and the degree of hydration (h) relate the hydrodynamic volume of a protein to its molecular weight (Lakowicz 2006). For a protein in an aqueous buffer, the partial specific volume has a value of $v = 0.735$ cm^3 g^{-1} and the degree of hydration is $h \sim 0.2$ cm^3 g^{-1}. The initial anisotropy (r_0) depends on the chemical structure of the fluorophore, the order of excitation, and the presence of ultrafast nonradiative processes that may compete with the fluorescence depopulation of the excited state in its native environment such that (Lakowicz 2006)

$$r_0(x,y) = \frac{2n}{2n+3}\left(\frac{3\cos^2(\theta) - 1}{2}\right) \tag{5.7}$$

where

n is the order of excitation (i.e., $n = 1$ for one-photon [1P] and $n = 2$ for two-photon [2P])
θ is the angle between the absorbing and emitting dipoles of a given fluorophore

For a mixture of fluorophores (or a single species in heterogeneous microenvironments), however, the anisotropy decay may be more complex. The overall time-resolved anisotropy of a mixture of N species can be written in terms of the fluorescence and anisotropy decays of all constituent species such that

$$r(x,y,t) = \frac{\sum_{i=1}^{N} F_i(x,y,t) \cdot r_i(x,y,t)}{\sum_{i=1}^{N} F_i(x,y,t)} \tag{5.8}$$

Assuming that each ith species in a given (x, y) pixel has a distinct fluorescence lifetime (τ_i) and rotational time (ϕ_i), Equation 5.8 can then be rewritten as

$$r(x,y,t)=\frac{\sum_{i=1}^{N}\alpha_i\exp\left(-t/\tau_i\right)\cdot r_{0i}\exp\left(-t/\phi_i\right)}{\sum_{i=1}^{N}\alpha_i\exp\left(-t/\tau_i\right)} \tag{5.9}$$

The apparent initial anisotropy (at $t=0$) of the mixture is $r_0=\sum_{i=1}^{N}\alpha_i\cdot r_{0i}$. Such anisotropy decay is termed "associated anisotropy" since each rotational time is associated with a specific fluorescence lifetime component of a given species in the mixture. While the fluorescence of free aqueous NADH decays as a biexponential, the corresponding fluorescence anisotropy decays predominantly as a single exponential with a rotational time of 290 ps and a 2P initial anisotropy of 0.38. These results indicate that the two free NADH conformations (folded and stretched) are indistinguishable based on their hydrodynamic radii as measured by time-resolved anisotropy. Since intracellular NADH exists in both free and enzyme-bound forms, each of which has a distinct fluorescence lifetime and hydrodynamic radius, we can generalize Equation 5.9 to a mixture of cellular free and enzyme-bound NADH as follows:

$$r(x,y,t)=\frac{\sum_{i=1}^{N_{\text{free}}}\alpha_i\exp\left(-t/\tau_i\right)\cdot r_{0i}\exp\left(-t/\phi_i\right)+\sum_{j=1}^{M_{\text{bound}}}\alpha_j\exp\left(-t/\tau_j\right)\cdot r_{0j}\exp\left(-t/\phi_j\right)}{\sum_{i=1}^{N_{\text{free}}}\alpha_i\exp\left(-t/\tau_i\right)+\sum_{j=1}^{N_{\text{bound}}}\alpha_j\exp\left(-t/\tau_i\right)} \tag{5.10}$$

When NADH binds tightly to a much larger enzyme, its rotational time is proportional to the molecular weight of the NADH–enzyme complex. As a result, the anisotropy decay of bound NADH can, in principle, be used to distinguish between multiple binding states with variable enzymes based on their molecular weights. Practically, however, this is not possible to do in living cells for a number of reasons. First, the accuracy of time-resolved anisotropy decreases as the rotational time (10–100 ns for many NADH dehydrogenases) exceeds the fluorescence lifetime (<4 ns). Second, NADH-binding enzymes are likely to exist in viscous, crowded environments in living cells that would further slow their rotational mobility. Third, NADH binds to a large number of enzymes and, therefore, the contribution of each species to the anisotropy decay may not be very significant. Finally, different fluorescence lifetime components may not necessarily correspond to distinct hydrodynamic volumes that can be resolved using time-resolved anisotropy. Hence, while the discrimination between free and bound NADH species is straightforward because of their vastly different sizes, the discrimination between different enzymes in the complex milieu of living cells is not practically possible. Accordingly, we can approximately assign the intracellular NADH autofluorescence to only two species, free (*subscript f*) and enzyme bound (*subscript b*), such that Equation 5.10 can be simplified further to

$$r(x,y,t)=\frac{\alpha_{\text{f}}\exp\left(-t/\tau_{\text{f}}\right)\cdot r_{0\text{f}}\exp\left(-t/\phi_{\text{f}}\right)+\alpha_{\text{b}}\exp\left(-t/\tau_{\text{b}}\right)\cdot r_{0\text{b}}\exp\left(-t/\phi_{\text{b}}\right)}{\alpha_{\text{f}}\exp\left(-t/\tau_{\text{f}}\right)+\alpha_{\text{b}}\exp\left(-t/\tau_{\text{b}}\right)} \tag{5.11}$$

This simplified equation still contains seven independent parameters (α_f, τ_f, α_b, τ_b, r_{0f}, ϕ_f, r_{0b}, and ϕ_b, where $\alpha_f+\alpha_b=1$) that can make it difficult to attribute these parameters to specific molecular species and biological processes. However, all the parameters that appear in the denominator of Equation 5.11 (i.e., α_f, τ_f, α_b, and τ_b) can be measured independently using the total fluorescence decay at magic-angle detection. The remaining four fitting parameters are associated solely with the anisotropy decay (i.e., β_f, ϕ_f, β_b, and ϕ_b). Moreover, the two rotational times are likely to be vastly different, reflecting the large difference in the hydrodynamic volumes of free and enzyme-bound NADH.

The associated anisotropy curve of the free and bound NADH mixture can alternatively be written as a function of the fluorescence fractions of the two species (Bailey et al. 2001):

$$r(t) = f_{\mathrm{f}}(\alpha,\tau)\cdot r_{0\mathrm{f}}\exp\left(-t/\phi_{\mathrm{f}}\right) + f_{\mathrm{b}}\left(\alpha,\tau\right)\cdot r_{0\mathrm{b}}\exp\left(-t/\phi_{\mathrm{b}}\right) \tag{5.12}$$

Equation 5.12 makes it conceptually easy to see that free NADH, with its shorter lifetime and rapid rotational time, would dominate the associated anisotropy curve at early times (≤500 ps), while the large enzyme-bound NADH, with its longer fluorescence lifetime and slow rotational time, would dominate the anisotropy curve at later times.

5.2.2 Steady-State Anisotropy Imaging

Steady-state autofluorescence anisotropy imaging provides spatial information about the local environment of a given intrinsic fluorophore. However, the polarization-analyzed autofluorescence signal represents a time-averaged intensity in each pixel and therefore sacrifices the temporal resolution inherent in time-resolved anisotropy. For any single species, the parallel, $F_{\|}(x,y)$, and perpendicularly, $F_{\perp}(x,y)$, polarized fluorescence images can be recorded simultaneously in order to calculate the corresponding anisotropy image, $r_{ss}(x,y)$, for each pixel (Ariola et al. 2006; Davey et al. 2007; Yu et al. 2008):

$$r_{ss}(x,y) = \frac{\left(F_{\|}(x,y) - B_{\|}\right) - G\left(F_{\perp}(x,y) - B_{\perp}\right)}{\left(F_{\|}(x,y) - B_{\|}\right) + 2G\left(F_{\perp}(x,y) - B_{\perp}\right)} \tag{5.13}$$

The background in the parallel ($B_{\|}$) and perpendicular ($B_{\perp}$) autofluorescence images can be estimated using an area away from cell bodies. For a single emitting species, the Perrin equation (Lakowicz 2006) relates the measured steady-state anisotropy (r_{ss}) in any pixel to the corresponding initial anisotropy (r_0), rotational time (ϕ_i), and autofluorescence lifetime (τ_i). For a mixture of fluorophores at a given pixel (x, y), however, the corresponding Perrin equation can be rewritten as

$$r_{ss}(x,y) = \sum_{i=1}^{N}\left(\frac{f_i}{1+\tau_i/\phi_i}\right)\cdot r_{0i} \tag{5.14}$$

The steady-state anisotropy becomes the sum of the initial anisotropies of the individual species in the mixture, weighted by both their fractional $\left(f_i = \alpha_i\cdot\tau_i \Big/ \sum_{i=1}^{N}\alpha_i\cdot\tau_i\right)$ contributions to the total fluorescence signal and the τ_i/ϕ_i ratio. Accordingly, the initial anisotropy (r_0) and steady-state anisotropy (r_{ss}) are equal when $\tau_i/\phi_i \ll 1$. This assumption is valid for the enzyme-bound NADH due to the significant enhancement of the rotational time upon enzyme binding in the restrictive cell environment. We should emphasize that this discussion on the relationship between the initial and steady-state anisotropy does not account for the depolarization effect of high NA objectives used in 2P microscopy. As a result, the 2P steady-state anisotropy measured using femtosecond laser pulses, in a microscope setting, may be different from that measured using cw laser excitation in a cuvette with the perpendicular configuration of the fluorescence detection with respect to the excitation laser propagation. The initial anisotropy measured using time-resolved anisotropy may also be lower than the theoretical value in the presence of ultrafast molecular dynamics that may compete with fluorescence.

5.3 Experimental Protocols

5.3.1 Biological Sample Handling

NADH and $FADH_2$ serve as the principle electron donors in energy metabolism and play an important role in cell respiration, differentiation, and apoptosis (Alberts et al. 2003). Hence, it is important to maintain the cells and tissues under investigation in a well-defined physiological state for a meaningful interpretation of their autofluorescence-based results. This entails maintaining the proper ambient temperature, oxygen level, pH, blood flow (in vivo), and concentrations of nutrients (Heikal 2010, 2012).

5.3.1.1 Controlling the Physiological State of Cells/Tissues

It is recommended that the culture medium of mammalian cells should be free of phenol red to avoid background fluorescence. For short-duration imaging of living cells (~30 min), MatTek glass-bottom Petri dishes provide a means for maintaining healthy and functional cells outside the incubator (Yu et al. 2008). During this period, the pH, nutrition, and CO_2 are sufficiently stable for metabolic imaging. Various dish sizes with bottom coverslip sizes and surface coatings (such as collagen) are also commercially available. The Tokai Hit stage-top incubator maintains cell culture for at least 24 h and is excellent for long-duration, time-lapse imaging. It is equipped with a temperature and CO_2 control system using a stage-top water bath and premixed 5% CO_2 for live cell imaging (Furuike et al. 2008; Ueno et al. 2005).

The Bioptechs FCS2 is a closed-flow chamber system that is also used frequently for long-term live cell imaging (Biagini et al. 2006; Strovas et al. 2007). The FCS2 system has a unique perfusion control system with directed medium flow over the cells. The Bioptechs Delta T culture system is an open-dish with thermal control and fast thermal recovery after perfusion. This system is more convenient for experimental manipulations (e.g., microinjection) than the closed FCS2 system. Homemade cell chambers, which take advantage of gravity for media flow, are cost effective and offer more design flexibility.

5.3.1.2 Brain Tissue Preparation, Maintenance, and Incubation in Perfusion Chambers

Thin slices are common ex vivo biological model systems for fluorescence-based studies of many tissue types. Careful preparation of these slices is necessary in order to minimize tissue trauma and ensure biologically relevant results. Moreover, it is necessary to locate target cells within the tissue that are accessible to imaging and yet deep enough to be protected from the surgical trauma. Here, we will focus on the preparation and maintenance of brain slices—one of the most commonly used tissue model systems in neurobiology.

During slice preparation, it is inevitable that cells near the slice surfaces will be physically damaged, leading to cell death or compromised physiology (Bak et al. 1980; Kasischke et al. 2001; Monette et al. 1998). It is therefore imperative that functional metabolic imaging be carried out deeper in the tissue in order to avoid these damaged cell layers. For adult rat and mouse brain slices, this surgically damaged layer typically extends to about 50 μm from the slice surface. However, the extent of damage will depend on many factors including the age of the animals, duration of the experiment, transient metabolic insults, and the method of preparation. It is recommended that each investigator ascertain the extent of such damage in their specific preparation using a cell viability assay such as the LIVE/DEAD Viability/Cytotoxicity Kit (Invitrogen) in which ethidium homodimer-1 stains the nuclei of dead cells and calcein AM stains the cell bodies of healthy cells.

For controlling the environment of tissue samples, a perfusion chamber should allow for continually perfusing the sample with a physiological buffer with an appropriate level of nutrients (e.g., glucose, pyruvate), oxygen (and CO_2), heat, and any required pharmacological agents. Depending on the specific experimental requirements, open or closed perfusion chambers can be used. Open chambers, such as the Large Rectangular Open Bath Chamber (RC-27) from Warner Instruments, maintain a steady flow of buffer over the tissue sample. The buffer is exposed to the atmosphere, which allows for gas exchange and the partial equilibration of the perfusate O_2/CO_2 levels with the atmosphere. These chambers are suitable for experiments in which a precisely defined perfusate oxygen concentration is not critical but direct physical access to the tissue is necessary—for example, when combining metabolic imaging with electrode recording or stimulation. When using the Warner RC-27 chamber to perfuse brain slices, we have found that normoxic artificial cerebrospinal fluid (ACSF) (bubbled with 95% O_2 + 5% CO_2), perfused into one end of the chamber, rapidly loses oxygen to the atmosphere and provides a pO_2 of only ~0.6 atm at the brain slice in the center of the chamber. On the other hand, hypoxic ACSF (95% N_2 + 5% CO_2)

increases its pO_2 from ~0 at the point of entry into the chamber to ~0.2 atm at the slice location. These chambers are therefore not ideal for oxygen-dependence studies of energy metabolism.

Closed perfusion chambers, such as the closed bath imaging chamber series (RC-20, RC-21B) from Warner Instruments, are best suited for experiments in which a well-defined pO_2 of the perfusate is needed. Normoxic and hypoxic buffer solutions are equilibrated with their respective gas mixtures and then perfused into the chamber by separate programmable pumps. The pO_2 in the chamber can then be controlled by the relative flow rates of the normoxic and hypoxic buffer. It is recommended to use a chamber with a small volume, relative to the tissue sample size, for optimized mixing, a fast solution exchange time, a rapid change of pO_2, and sample stability during autofluorescence imaging. A small-volume chamber ensures that any observed spatiotemporal changes in NADH autofluorescence in response to pO_2 manipulation will be solely biological in nature and not due to a slow and inhomogeneous solution exchange.

5.3.1.3 Manipulating the Redox and Metabolic State of Cells

Since the oxidation of NADH in mitochondria is coupled with the phosphorylation of ADP to ATP, the metabolic state of a cell can be measured with the [NADH]/[NAD^+] ratio, also known as the redox ratio. On the other hand, FAD is in equilibrium with the mitochondrial [NADH]/[NAD^+] ratio and can be used for the ratiometric monitoring of cellular metabolic activities (Huang et al. 2002). In optical imaging, the ratio of the fluorescence intensities F_{FAD}/F_{NADH} or $F_{FAD}/(F_{FAD} + F_{NADH})$ can then serve as the redox ratio for a noninvasive, ratiometric probe of mitochondrial energy metabolism (Kunz et al. 2002; Skala et al. 2007).

The cellular redox state can be manipulated by many factors such as the oxygen level (Vishwasrao et al. 2005), cyanide as an inhibitor of the electron transport chain (Huang et al. 2002; Tiede et al. 2007), and carbonyl cyanide 4-(trifluoromethoxy)phenylhydrazone (FCCP) as a mitochondrial uncoupler (Huang et al. 2002; Rocheleau et al. 2004).

In the absence of oxygen (by either hypoxic perfusion or restricted oxygen supply as in a stroke or a tumor), there is no final acceptor for the reducing equivalents taken from NADH in the electron transport chain. As a result, the reaction velocity and the oxidation of NADH in the electron transport chain becomes slow, which leads to the accumulation of the fluorescent reduced form (Alberts et al. 2002; Bird et al. 2005; Vishwasrao et al. 2005). The inhibition of the electron transport chain using cyanide causes a similar change in the redox ratio by similarly preventing NADH oxidation (Bird et al. 2005; Vishwasrao et al. 2005). In contrast, FCCP causes the mitochondrial membrane potential to collapse, which leads to the uncoupling of NADH oxidation from ATP generation in the oxidative phosphorylation pathway. As a result, the increase in the electron transport chain reaction velocity causes a hyperoxidation of NADH and an increase in the nonfluorescent NAD^+ (Huang et al. 2002; Romashko et al. 1998).

5.3.2 Microscope-Based Anisotropy of Cellular Autofluorescence

5.3.2.1 Time-Resolved Fluorescence and Anisotropy Measurements

Our fluorescence microspectroscopy system was used for time-resolved, 2P fluorescence and anisotropy of intracellular NADH and FAD as described elsewhere (Heikal 2012). The experimental setup (Ariola et al. 2006; Yu et al. 2008) consists of an infrared femtosecond laser system that generates pulses (710–990 nm, ~120 fs, 76 MHz), which can be reduced to 4.2 MHz using a pulse picker for single-point measurements with high temporal resolution (Figure 5.1c). For 1P excitation studies, these femtosecond laser pulses pass through a second harmonic generator in order to generate the wavelength range of 350–490 nm. The 2P excitation laser beam is conditioned and steered toward a laser scanner (FV300, Olympus) and an inverted microscope (IX81, Olympus) where it is focused on the sample through a high NA microscope objective (1.2 NA, 60×, water immersion). Due to the optical limitations of our scanner in the UV region, 1P laser pulses are usually steered toward the sample via the back exit port of the microscope for single-point measurements. The epifluorescence is detected via the right exit port of the microscope for a number of microspectroscopy modalities (Figure 5.1c) with minor modifications of the optical setup as described in the following:

1. For 2P fluorescence lifetime imaging microscopy (2P-FLIM), high-repetition-rate laser pulses are usually used to minimize the acquisition time while enhancing the signal-to-noise (S/N) level (see Chapter 4 for more details on FLIM). For a single-fluorophore FLIM image, the 2P epifluorescence is separated from the excitation laser using both a dichroic mirror and optical filters, which are selected based on the fluorophore of interest. The 2P epifluorescence is then steered toward the right exit port where it is filtered and polarization analyzed prior to detection by a microchannel plate photomultiplier tube (MCP-PMT). For fluorescence lifetime measurements, rotational effects on the excited-state dynamics can be removed by using magic-angle (54.7°) polarization detection. A Glan–Thompson polarizer, mounted on an indexed rotational mount, can be used to control the polarization of the detected epifluorescence. Since our experimental setup is integrated with a confocal microscope with DIC capability, it is necessary to remove the DIC polarizer from the excitation/epifluorescence pathway prior to fluorescent lifetime or anisotropy measurements. The signal from the MCP-PMT is then amplified, routed, and registered using an SPC-830 TCSPC module (Becker & Hickl). A small fraction (~5%) of the excitation laser is diverted to a fast photodiode in order to generate a synchronization (SYNC) pulse to trigger the SPC-830 module.

   The S/N of FLIM images can be improved by increasing the acquisition time and/or the excitation power. Care must be taken, however, to minimize photobleaching of the intrinsic fluorophore or photodamage to the cells. In 2P-FLIM, 256 × 256 pixels are used with 256 time bins per pixel, where these parameters may change based on the signal level and photostability of the fluorophore of interest.
2. A pseudo single-point, time-resolved fluorescence modality can be used to enhance both the S/N ratio and temporal resolution. In this modality, the excitation laser is scanned over a region of interest in the sample, but the fluorescence is registered as 1 pixel with 1024 time bins. While this modality lacks the spatial resolution (averaging over the region of interest), it greatly improves the S/N ratio and temporal resolution, while minimizing the cellular photodamage that would occur for a stationary excitation laser on the sample. To implement this single-point mode of FLIM, the magic-angle autofluorescence is detected with the SPC-830 module, which is set at ADC resolution = 1024, time/channel = 12.2 ps, and routing channel x and $y = 1$, in order to generate the fluorescence decay, $F(t)$.
3. The same setup described earlier can be used for single-point, time-resolved 2P fluorescence anisotropy measurements with minor modifications (Figure 5.1c). For example, the 2P autofluorescence, steered toward the right-hand exit port, is split into two channels using a polarizing beam splitter. These orthogonal, time-resolved polarization components are then detected using two MCP-PMTs as described earlier. Both parallel, $F_{\parallel}(t)$, and perpendicular, $F_{\perp}(t)$, polarization components of cellular autofluorescence are then recoded simultaneously in order to minimize any artifacts due to sample movements and laser instability (Ariola et al. 2006; Davey et al. 2007; Yu et al. 2008). The DIC polarizer should be removed for a well-defined polarization of both the excitation and detection of autofluorescence. The polarization-analyzed signals from both detectors are then amplified, routed, and registered using the SPC-830 module, which is set at ADC resolution = 1024, time/channel 12.2 ps, routing channel $X = 2$, and routing channel $Y = 2$. The anisotropy decay, $r(t)$, is then calculated in postprocessing of the parallel and perpendicular fluorescence decays using OriginPro (OriginLab).

5.3.2.2 Steady-State Anisotropy Imaging

To measure the steady-state anisotropy of cells/tissues, the time-averaged parallel, $F_{\parallel}(x,y)$, and perpendicular, $F_{\perp}(x,y)$, polarization fluorescence intensity images (Figure 5.1b) are recorded simultaneously (Ariola et al. 2006; Davey et al. 2007; Kress et al. 2011; Yu et al. 2008). While steady-state anisotropy imaging can be carried out using either 1P confocal microscopy or 2P microscopy, we will focus here on the latter with laser-scanning capability. Following the pulsed laser excitation, the 2P epifluorescence is spectrally isolated using the proper dichroic mirror and filters, split into polarization components using a polarizing beam splitter, and detected simultaneously by two MCP-PMTs. The detector signals are then amplified, routed, and then registered using an SPC-830 module (Becker & Hickl) that is synchronized with both the laser pulses via a fast photodiode SYNC signal (Figure 5.1c) and the scanning unit (Ariola et al. 2006; Yu et al. 2008). While the acquisition protocol is the same as for 2P-FLIM measurements, only the time-averaged, polarization-analyzed fluorescence intensities per pixel (i.e., no temporal resolution with the number of time bins = 1 for each pixel) are recorded for steady-state anisotropy for postprocessing calculation of the anisotropy image. The corresponding background signal and noise level in these measurements in both channels should be estimated accurately and used in the anisotropy image calculations. As shown in the text below, the background signal and noise level will affect the calculated anisotropy image. While enhancing the S/N of polarization-analyzed autofluorescence will help minimize artifacts, care must be taken to avoid photobleaching as well as photodamage to the cells/tissue under investigation.

A conventional two-channel laser-scanning confocal microscope can also be modified for anisotropy imaging (Vishwasrao et al. 2012). In a FluoView 300 scanner (Olympus), for example, there are two mounts for each channel, designed for two fluorophores (i.e., color) imaging. In this scanner, we have replaced one of those filter mounts in each detection channel with a polarizer of the same dimension (Ariola et al. 2006). With geometrical constraints imposed by the filter mount design in the FV300, the alignment and mounting of the polarizers is challenging and can be tested using a weak cw laser with a well-defined polarization. Using the same line of thinking, existing TIRF microscopy setups can be easily modified in order to allow for steady-state anisotropy (or polarization) for single-molecule orientation studies.

5.3.2.3 Data Analysis

For fluorescence lifetime measurements (see Chapter 4 for more details), many software packages, such as SPCImage (Becker & Hickl), SymPhoTime (PicoQuant), and Globals (Laboratory for Fluorescence Dynamics) are already available commercially for single-point curve and FLIM fitting. These algorithms allow for nonlinear least squares fitting of multiexponential decays with deconvolution capability with the measured or calculated instrument response function. It is generally recommended to use the simplest model (i.e., fewest components) to fit the experimental data when possible, and the goodness of the fit is judged by both the χ^2 value and the residuals.

Fitting time-resolved fluorescence anisotropy curves (Equations 5.5 and 5.11) can be carried out using the nonlinear least squares fitting routines in OriginPro 8.1 (OriginLab). Usually, the measured anisotropy decay is not deconvoluted from the system response function, which undermines the accuracy of this approach in determining fast rotational (or tumbling) motion that is comparable to the FWHM of the system response function (~45 ps in our case). Considering the large number of fitting parameters in associated anisotropy (e.g., Equation 5.11), a global fitting approach can be used in order to calculate the fluorescence lifetime and anisotropy parameters (Knutson et al. 1983; Vishwasrao et al. 2005). The number of floating parameters in global analysis can be reduced (i.e., constrained) without compromising the integrity of the fit. Global analysis simultaneously fits two or more curves that have (1) a set of shared parameters and (2) a set of parameters that are solely specific to the individual curves. The goodness of the global fit is evaluated by the global reduced χ^2-value as well as the residuals.

5.3.2.4 Control Experiments and Calibration Standards

It is important to calibrate the equipment used for time-resolved fluorescence and anisotropy measurements in order to ensure reliable results and the proper performance of the experimental setup. Appropriate calibration standards are fluorophores with known fluorescence lifetimes, rotational times, and absorption/emission spectra. In addition, the system response function needs to be measured for later deconvolution-based fitting analysis of the measured fluorescence decays. These system response functions are measured experimentally using the second harmonic signal from monobasic potassium phosphate crystals or collagen bundles from a mouse tail. Alternatively, the system response function can be approximated by the fitting software based on the rise time of a given fluorescence decay. However, this approach of computer-generated system response function will only be valid in the absence of ultrafast processes with negative amplitude (i.e., rise) in the observed decay. See a detailed discussion of the system response measurement in Section 4.3.1.2.

For time-resolved anisotropy measurements, the *G*-factor should also be measured using the tail-matching approach (Lakowicz 2006) to compensate for any potential polarization bias in the fluorescence detections. In this approach, a fluorophore with a relatively fast rotational time (e.g., tens of picoseconds) and longer fluorescence lifetime (a few nanoseconds) is used as a calibration standard. Since a small fluorophore rotates much faster than its excited-state lifetime, the fluorescence decay tail should be completely depolarized. The ratio of parallel to perpendicular fluorescence intensity near the decay tail then equals the *G*-factor. In order to avoid possible wavelength dependence of the polarization bias, the excitation/emission spectra of the selected standard fluorophore should also overlap with that of the intrinsic fluorophore under investigation. Aqueous coumarin (350 nm/460 nm) is usually used as a calibration standard when studying the time-resolved fluorescence and anisotropy of NADH, while fluorescein or rhodamine green (488 nm/525 nm) serves as a good standard for FAD studies.

5.3.2.5 Limitation of Depth Penetration in Tissue Imaging

Generally, fluorescence imaging deep in thick tissues is limited by the attenuation of both the excitation laser and the fluorescence due to light scattering and absorption. Anisotropy measurements in thick tissues contend with the additional challenge of fluorescence depolarization due to scattering in a turbid medium (Bigelow and Foster 2006; Lentz et al. 1979; Teale 1969). For the simplest case of isotropic scattering, it has been predicted that a single scattering event could reduce the anisotropy of polarized light by a factor of 0.7 (Teale 1969). Furthermore, optical parameters, such as the absorption coefficient, the scattering coefficient, and the scattering anisotropy, depend on the tissue type, metabolic state, and pathology of the tissue under investigation (Cheong et al. 1990; Sandell and Zhu 2011).

The autofluorescence anisotropy of normal and malignant human breast tissues has been investigated as a function of tissue thickness using 1P-excitation spectroscopy for intrinsic NADH and FAD (Mohanty et al. 2001). It was found that the fluorescence anisotropy decays rapidly as the thickness of the sample increases. Comparative studies on 10 and 100 μm thick samples revealed that the thick sample exhibited a ~40% lower NADH steady-state anisotropy and a ~25% lower FAD anisotropy in normal tissue. In malignant tissues, however, both the NADH and FAD anisotropies in thin samples were larger than they were in normal tissue but with a much steeper decrease as a function of tissue thickness. These 1P autofluorescence studies clearly demonstrate the sensitivity of autofluorescence anisotropy to variations in tissue type and physiology. For deeper penetration in thick tissues, however, 2P fluorescence microscopy is the method of choice where 730 and 900 nm are used for NADH and FAD excitation, respectively. Since the scatter of infrared light is significantly less than visible light in tissue, the problem of anisotropy imaging in tissue is somewhat ameliorated in 2P microscopy. Nonetheless, to the best of our knowledge, a thorough characterization of the effect of light scattering on the fluorescence anisotropy imaging of intracellular NADH and FAD in tissues is still lacking.

5.4 Environmental Effects on the Fluorescence Properties of NADH and FAD

Intracellular NADH and FAD exist in a complex environment and in different structural conformations. As a result, it is important to consider the binding and environmental parameters which are likely to influence their photophysical and rotational properties. An understanding of these environmental effects will inform our interpretation of the observed associated anisotropy of intracellular autofluorescence. For a given fluorophore, the fluorescence lifetime may also be sensitive to a number of environmental parameters, such as viscosity, enzyme binding, pH, polarity, and oxygen concentration. While temperature is also known to affect both lifetime and rotational time, it can be easily maintained constant during most experiments and therefore can be ignored here. In living cells and tissues, these environmental parameters are potentially sensitive to the metabolic state and may vary among subcellular compartments. Here, we will consider only those environmental parameters that are relevant to cell/tissue autofluorescence-based studies.

5.4.1 Enzyme Binding

There are many different enzymes in cells whose activities and biological functions depend on the presence of coenzymes such as NADH and FAD. The fluorescence lifetime alone has been used to discriminate between free and enzyme-bound NADH (Bird et al. 2005; Skala et al. 2007). The observed biexponential fluorescence decay of free NADH in an aqueous solution (PBS buffer at pH 7.4 and room temperature) suggests the existence of two conformers of NADH with lifetimes of 350 ps (77%) and 760 ps (23%) with an average fluorescence lifetime of 444 ps. The two fluorescence decay components are commonly assigned to folded and stretched NADH, describing the proximity of the nicotinamide ring to the adenine base (Couprie et al. 1994; Gafni and Brand 1976).

When bound to most enzymes, NADH is rigidly held in a stretched conformation that reduces the interaction between the nicotinamide ring and the adenine base. This tight binding also suppresses the nonradiative relaxation rate of the excited state, which leads to an increase in the fluorescence lifetime. The fluorescence lifetime of NADH typically shows a relatively modest enhancement upon enzyme binding. For example, the average lifetime of the NADH bound to mitochondrial malate dehydrogenase (mMDH) is only two times larger (~0.8 ns) than that of free NADH (~0.43 ns) in a buffer. As a coenzyme, NADH can also form binary (enzyme + NADH) or ternary (enzyme + NADH + substrate) complexes. As shown in Table 5.1, the binding of NADH to liver alcohol dehydrogenase (LADH) at 1°C increased the fluorescence lifetime from 0.54 to 2.81 ns (Gafni and Brand 1976), which is significantly shorter than the corresponding lifetime of 5.6 ns for the ternary complex with isobutyramide (IBA) as a substrate. At 20°C, however, the lifetime of this ternary complex (1.02 ns) exhibited a more modest enhancement (Piersma et al. 1998). There is at least one enzyme, glyceraldehyde-3-phosphate dehydrogenase (GPDH), which is known to quench NADH fluorescence (Velick 1958). The weak enhancement of fluorescence lifetime by enzyme binding, combined with the fact that enzyme-bound NADH decays as a multiexponential, makes it difficult to distinguish between free and enzyme-bound NADH in a mixture using only the fluorescence lifetime.

In PBS (pH 7.4) at room temperature, FAD fluorescence decays as a double exponential with $\tau_1 = 2.55$ ns ($\alpha_1 = 0.64$), $\tau_2 = 4.20$ ns ($\alpha_2 = 0.36$), and an average fluorescence lifetime of 3.14 ns (Heikal 2010, 2012;

TABLE 5.1 Effect of Enzyme Binding on the NADH Fluorescence Lifetime

		Lifetimes (ns)			
Enzyme	T (°C)	$\tau_1(a_1)$	$\tau_2(a_2)$	$\tau_3(a_3)$	$\langle\tau\rangle$
None[a]	RT	0.35(0.77)	0.76(0.23)		0.44
mMDH[a]	RT	0.6(0.72)	1.33(0.28)		0.81
L-LDH[b]	RT	0.34(0.22)	1.1(0.67)	2.53(0.11)	1.08
LADH[c]	1	1.8(0.58)	4.2(0.42)		2.81
LADH + IBA[c]	1	2.9(0.29)	6.9(0.71)		5.7
LADH + IBA[d]	20	0.43(0.15)	1.12(0.85)		1.02

[a] Mitochondrial malate dehydrogenase (mMDH) (Vishwasrao et al. 2005).
[b] L-Lactate dehydrogenase (L-LDH) (Yu and Heikal 2009).
[c,d] Liver alcohol dehydrogenase (LADH) and isobutyramide (IBA) (Gafni and Brand 1976; Piersma et al. 1998).

Lutes 2008; Yu 2009). Under the same experimental conditions, however, the fluorescence of a flavoprotein lipoamide dehydrogenase, LipDH, decays as a biexponential with $\tau_1=2.50$ ns ($\alpha_1=0.64$), $\tau_2=4.14$ ns ($\alpha_2=0.36$), and an estimated average fluorescence lifetime of 3.69 ns (Lutes 2008; Yu 2009), which is somewhat slower than free FAD (see Section 5.8 and Figure 5.7).

5.4.2 Viscosity

Like enzyme binding, viscosity also influences both the excited-state lifetime as well as the rotational mobility of NADH. From steady-state spectroscopy, we found that the time-averaged fluorescence of NADH increases with viscosity, with no significant changes in the absorption profile, indicating a viscosity-dependent fluorescence quantum yield (Φ_F). Independent time-resolved fluorescence measurements of NADH also reveal a viscosity-dependent fluorescence lifetime (Figure 5.2).

These results indicate that the fluorescence properties, including the radiative ($k_r=\Phi_F/\langle\tau\rangle$) and nonradiative ($k_{nr}=\langle\tau\rangle^{-1}-k_r$) rate constants of NADH, are viscosity dependent. The observed slight viscosity dependence of the radiative decay rate of NADH (Figure 5.2) is likely due to a change in the environmental refractive index. In contrast, the nonradiative rate (k_{nr}) shows a significant dependence on viscosity where $k_{nr}=a+b\eta^{-c}$ (Wilhelmi 1982), with $a\sim 0$ ($a=0\pm 1.5$ GHz, $b=2.6\pm 1.5$ GHz cP^{-1}, and $c=0.16\pm 0.11$). Importantly, the viscosity effects on the fluorescence lifetime of NADH, in buffer–glycerol mixture, are significantly less pronounced as compared with the effects of enzyme binding.

5.4.3 pH

The metabolic state of a cell is directly related to the mitochondrial membrane potential, which is maintained by a proton gradient. In healthy and functional mitochondria, the mitochondrial matrix

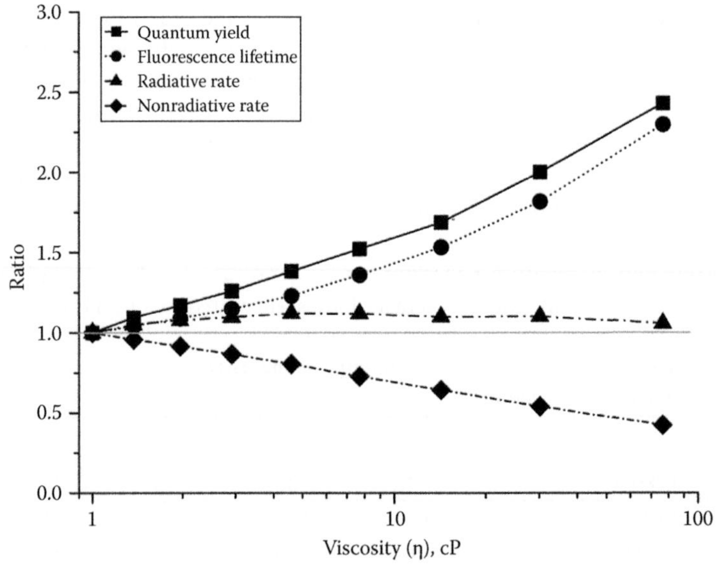

FIGURE 5.2 Viscosity dependence of the fluorescence properties of NADH in a controlled environment. NADH photophysical parameters were measured in buffer–glycerol mixtures as a means to vary the environmental viscosity. All photophysical quantities are normalized to their values in pure buffer ($\eta\sim 0.89$ cP) at room temperature. The fluorescence quantum yield of NADH in each sample of viscosity was measured using steady-state spectroscopy. The average fluorescence lifetime of NADH was measured using TCSPC at magic-angle detection. The radiative and nonradiative rates were then calculated using the fluorescence quantum yield and the corresponding fluorescence lifetime (Lakowicz 2006).

has a pH of 8.0–8.5, while the surrounding cytoplasm has a pH of 7.0–7.5 (Mitchell and Moyle 1969). However, the time-averaged and time-resolved fluorescence intensity of free aqueous NADH reveals no significant change over the pH range of 5–9 (Ogikubo et al. 2011). This lack of pH sensitivity of NADH fluorescence is attributed to the fact that the stacked to extended state transition of the molecule has a pK_a of ~4, far below the physiological pH range (Jardetzky and Wade-Jardetzky 1966). These results indicate that the physiologically normal pH variations in living cells have negligible effects on intracellular NADH autofluorescence.

5.4.4 Oxygen

Cellular metabolism, and, therefore, the concentration and binding state of NADH, is sensitive to the partial pressure of oxygen in the surrounding microenvironment. In addition, the excited state of NADH itself is reactive to oxygen, which leads to an oxygen-dependent quenching of fluorescence. A direct reaction between NADH and oxygen produces NAD^+ (Czochralska et al. 1984) and the superoxide anion O_2^- (Cunningham et al. 1985), which is a factor in inducing oxidative stress. Under aerobic conditions (pO_2 ~ 0.2 atm), the photooxidation quantum yield (Φ_{O_2}) of NADH was found to be 0.02 (Czochralska et al. 1984) and decreased by a factor of 4.3 under anaerobic conditions (pO_2 ~ 0 atm) in an argon atmosphere. Thus, while physiologically relevant variations of pO_2 do greatly affect the photooxidation rate, the quantum yield of this reaction is negligible even under high-pO_2 aerobic conditions. As a result, it is safe to assume that the variation of pO_2 during a typical metabolic experiment has a negligible direct effect on the excited-state dynamics of intracellular NADH.

5.5 Autofluorescence Lifetime Imaging as a Means for Determining the Concentration of Cellular Coenzymes

The concentrations and distributions of intracellular NADH and FAD are key observables relevant to their cellular function. The concentration distribution of an intrinsic fluorophore can be calculated by correcting its fluorescence intensity image for the spatial variation in the average lifetime and comparing this lifetime-corrected intensity to that of a known concentration/lifetime standard. This can be done by using both the fluorescence intensity and lifetime (FLIM) images that are recorded using a calibrated microscope (Yu and Heikal 2009). Using this lifetime-corrected autofluorescence approach, the concentrations of NADH in both living cultured cells (Figure 5.3) and brain tissue have been calculated (Vishwasrao 2004; Yu and Heikal 2009). In cultured normal (Hs578Bst) and cancerous (Hs578T) breast cells, the NADH concentration exhibits a pronounced spatial heterogeneity (Figure 5.3a). The estimated cell-wide average in cancerous cells (168 ± 49 μM) is significantly higher than the corresponding value in normal breast cells (99 ± 37 μM) as shown in Figure 5.3b (Yu and Heikal 2009).

In brain tissues, the average NADH concentration in the dendritic network (stratum radiatum) of the hippocampal CA1 neuronal layer was found to be 119 μM at normal oxygen levels (Vishwasrao 2004). In response to hypoxia, the NADH fluorescence intensity in brain tissues increased by 37%. However, the observed increase in intensity was accompanied by a 17% decrease in the average fluorescence lifetime of NADH, and therefore the lifetime-corrected NADH concentration was calculated to increase by 64% (i.e., to 195 μM) in response to hypoxia.

The same FLIM-based approach was used to quantify the concentration distribution of flavins in the Hs578Bst and Hs578T cell lines at the single-cell level (Figure 5.3b). In these measurements, the microscope detection efficiency was calibrated using a reference solution of free FAD in buffer (90 μM) under the same experimental conditions. While flavin autofluorescence in cancer Hs578T cells is approximately fivefold smaller than that of the normal cells, the concentration difference was actually closer to sevenfold following the lifetime correction. The estimated average concentrations of cellular flavins in Hs578Bst and Hs578T were 144 ± 38 μM ($n = 6$) and 21 ± 10 μM ($n = 5$), respectively (Lutes 2008; Yu 2009).

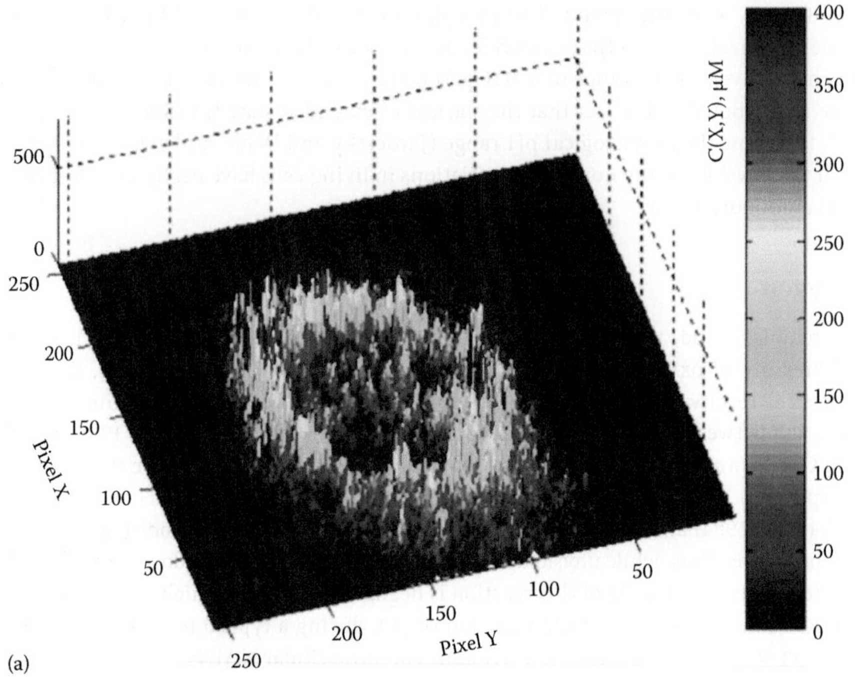

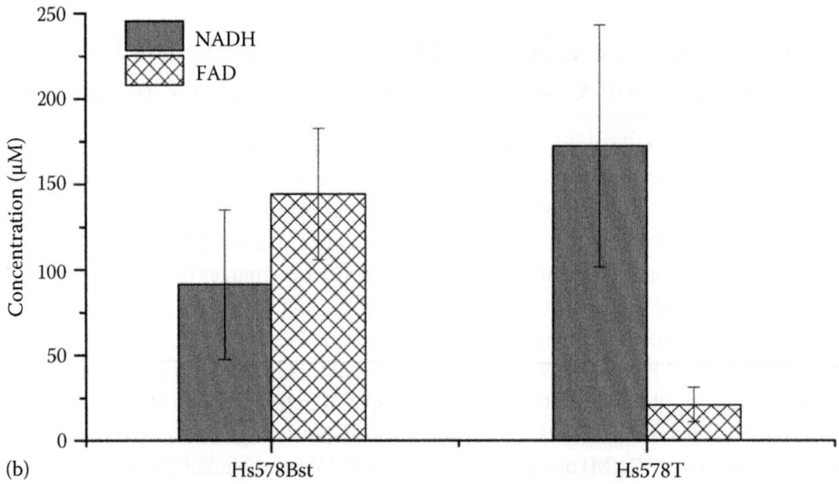

FIGURE 5.3 Determining the concentration of intracellular NADH and flavins using 2P autofluorescence in a calibrated microscope. (a) A typical, autofluorescence-based concentration image of intracellular NADH in cultured Hs578Bst under resting conditions. (b) The estimated NADH concentration in normal Hs578Bst cells is 99 ± 37 μM in culture as compared with 168 ± 49 μM in transformed Hs578T cells (Heikal 2010, 2012; Yu 2009; Yu and Heikal 2009). The estimated concentration of flavins is ~140 μM in cultured Hs578Bst cells as compared with ~24 μM in transformed Hs578Tcells. (From Heikal, A.A., *Biomarkers Med.*, 4(2), 241, 2010; Heikal, A.A., A multiparametric imaging of cellular coenzymes for monitoring metabolic and mitochondrial activities, in *Reviews in Fluorescence*, C.D. Geddes (ed.), vol. 2010 of Reviews in Fluorescence, Springer, New York, 2012, pp. 223–243; Lutes, A.T., Intrinsic flavin as a biomarker for monitoring pathologically-induced changes in cellular energy metabolism, MS thesis, Pennsylvania State University, State College, PA, 2008; Yu, Q., Functional imaging of intracellular metabolic cofactors in human normal and cancer breast cells, PhD dissertation, Pennsylvania State University, State College, PA, 2009.)

These results demonstrate that changes in the autofluorescence intensity alone are not sufficiently accurate to determine the concentration changes. In other words, changes to the lifetime of the fluorophore (due to changes in its conformation and local environment) can have large effect on the measured fluorescence intensity and therefore must be taken into account during autofluorescence-based calculations of concentration. The accuracy of autofluorescence-based concentration imaging will depend on both the reference selection for calibrating the microscope and the assessment of background signal. Binding and self-quenching in the complex milieu of living cells are likely to limit the accuracy of this autofluorescence-based concentration analysis approach of intrinsic coenzymes. The challenge now is to determine the percentage of cellular NADH population that is free or enzyme bound.

5.6 Steady-State Anisotropy Imaging Reveals Environmental Restriction on Intracellular NADH and FAD

The local cellular microenvironments are likely to impose a restriction on the rotational mobility of intrinsic biomolecules such as NADH and FAD. The spatial distribution of these restrictions can be assessed using steady-state autofluorescence anisotropy imaging. As mentioned earlier, the anisotropy in each pixel was calculated using the background-corrected, parallel, and perpendicular polarization images. These images were recorded simultaneously with a high NA objective (1.2 NA, 60×, water immersion), which introduces an additional depolarization effect to the steady-state anisotropy (Yu et al. 2008).

For steady-state anisotropy image analysis, an image-processing algorithm (Figure 5.4a through d) was developed for MATLAB® (The MathWorks) to calculate the anisotropy in each pixel with Equation 5.13 from the simultaneously recorded parallel and perpendicular polarization fluorescence images. This home-built program accounts for the background in each polarization-analyzed image, image size (i.e., number of pixels), the dynamic range associated with 1P or 2P anisotropy, and the *G*-factor prior to the anisotropy calculations in each pixel. The program is also capable of calculating the pixel–anisotropy histogram. We also examined the effects of accurate background signal and electronic noise in the polarization images on the calculated steady-state anisotropy images (Figure 5.4e through h). Using typical polarization-analyzed autofluorescence images of NADH in cultured mouse embryo fibroblast C3H10T1/2 cells as a model system, we calculated the steady-state anisotropy image using 0%, 5%, 10%, and 20% background signal (Figure 5.4). These results highlight the importance of an accurate assessment of the background signal level in constructing the steady-state anisotropy images of cellular autofluorescence. The background level in a given autofluorescence image can be estimated away from the cells in the field of view, which may also be polarization dependent.

Figure 5.5 shows representative steady-state 2P autofluorescence anisotropy images of intrinsic NADH (Figure 5.5a) and flavins (Figure 5.5b) in cultured mouse embryo fibroblast cells (C3H10T1/2) as a model system. The parallel and perpendicular autofluorescence images of NADH and FAD were recorded under 730 and 900 nm excitations, respectively. Because of the difference of both excitation and emission of NADH and FAD, the corresponding *G*-factor was estimated separately using the tail-matching approach (Lakowicz 2006). The broad pixel–anisotropy histogram (Figure 5.5) suggests a spatially heterogeneous microenvironment for intracellular coenzymes with a variable degree of restriction on its rotational freedom. The histogram also suggests that the environmental restriction of flavins seems more pronounced than that of intrinsic NADH (both free and enzyme-bound conformations). It is worth mentioning that the cellular 2P autofluorescence of flavins is relatively low under our experimental conditions as compared with NADH. As a result, care must be taken in analyzing these steady-state anisotropy images in order to avoid an over-reaching interpretation of FAD data.

These averaged values of the steady-state 2P autofluorescence anisotropy are lower than the theoretical maximum, 0.57 (Lakowicz 2006), which suggests a depolarization of the autofluorescence due to the

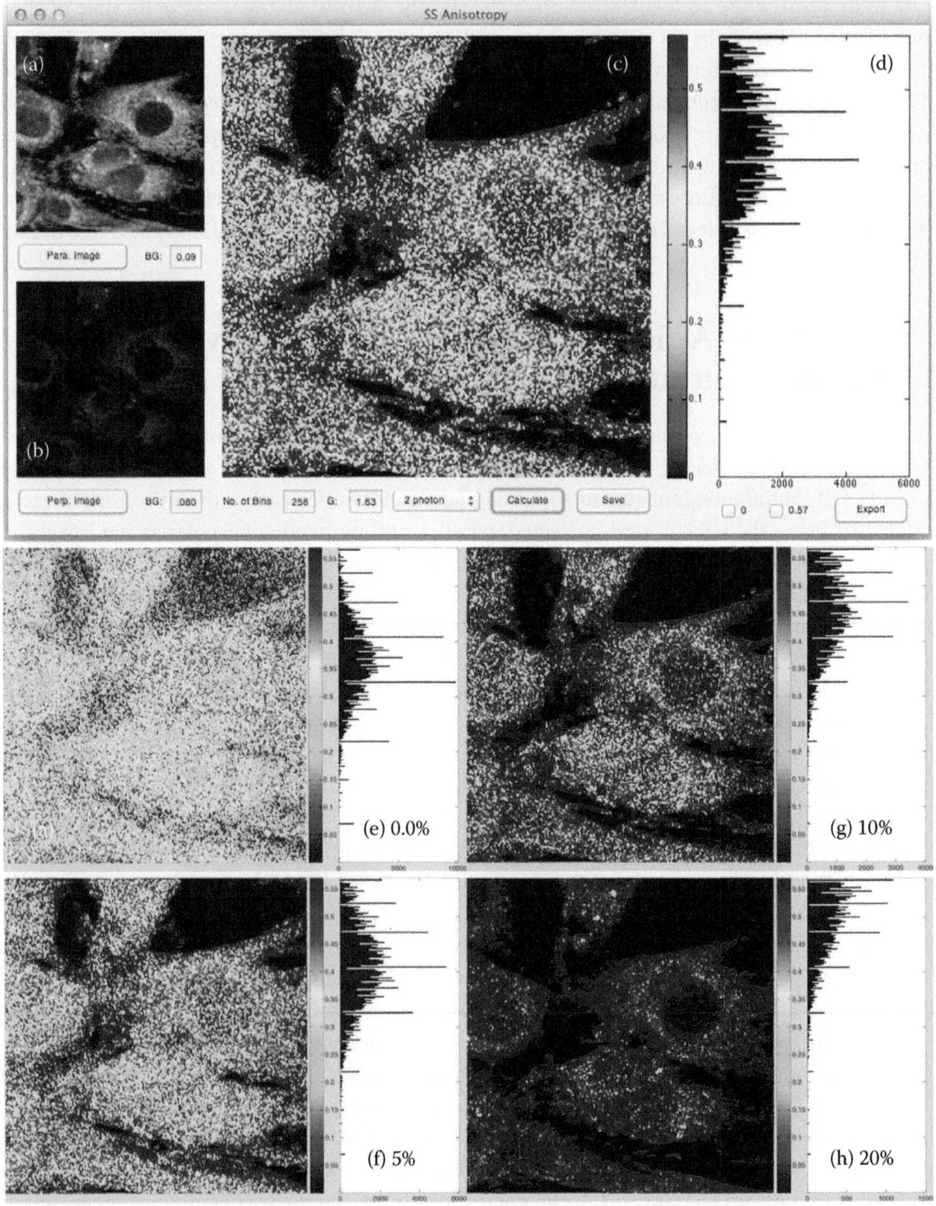

FIGURE 5.4 Steady-state anisotropy image calculations and the effects of background noise. (a–d) A MATLAB®-based image-processing algorithm was developed to calculate the steady-state anisotropy in each pixel (c), Equation 5.13, from the simultaneously recorded parallel (a) and perpendicular (b) polarization fluorescence images. Here, we used the intrinsic NADH and FAD of cultured mouse embryo fibroblast C3H10T1/2 cells as a representative model system. The anisotropy-pixel histogram (c) is also constructed in order to reflect the anisotropy distribution and heterogeneity throughout a given image (or cells). The program allows for changing the background signal, dynamic range (1P or 2P), the image size (here 256×256), and the G-factor ($G = 1.63$ for NADH and $G = 1.72$ for FAD). The calculated steady-state anisotropy image will depend on the background signal level (e: 0.0%, f: 5%, g: 10%, and h: 20%) and, therefore, accurate assessment of the background signal is required for each experimental condition.

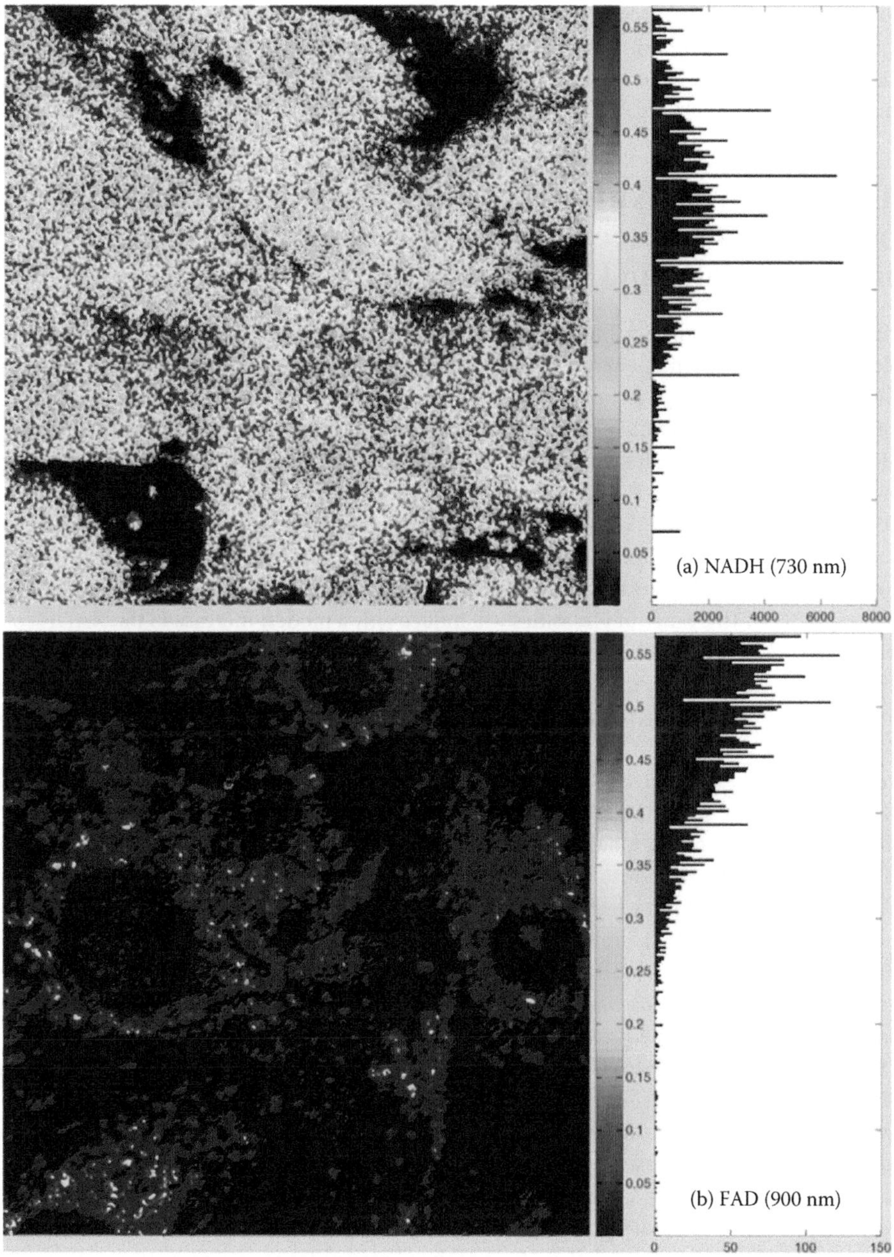

FIGURE 5.5 2P steady-state anisotropy image of intracellular NADH and FAD autofluorescence as a measure of environmental restrictions. (a) 2P anisotropy image of intrinsic NADH in cultured mouse embryo fibroblast C3H10T1/2cells, at room temperature, was recorded under 730 nm excitation in laser-scanning mode. The corresponding 2P anisotropy image of intrinsic flavins in adherent C3H cells, under 900 nm excitation, is also shown (b). It is worth mentioning that the flavins autofluorescence is relatively low. The corresponding anisotropy-pixel histogram for both NADH and FAD is also shown, which suggests that a more restrictive environment for flavins as compared to that of cellular NADH. In these images, the background signal was 8% of the 2P autofluorescence for these two cellular coenzymes. For these steady-state, polarization-analyzed fluorescence, the data acquisition software of SPC-830 was set as follows: ADC Resolution = 1, Routing chan $X = 2$, Routing chan $Y = 1$, Scan pixels $X = 256$, and Scan pixels $Y = 2$.

nonzero angle between the absorption and emission dipoles. In addition to the rotational flexibility of NADH (Kierdaszuk et al. 1996), other depolarization processes may include fluorescence resonance energy transfer (homo-FRET) in cellular microenvironments, and the optical depolarization caused by the high NA objective (Axelrod 1979, 1989).

5.7 Time-Resolved Anisotropy Provides Direct Evidence of Equilibrated Free and Enzyme-Bound NADH in Living Cells and Tissues

According to the Stokes–Einstein model, the rotational time of a given fluorophore is dependent on the molecular size (V), thermal energy ($k_B T$), and the viscosity (η) of the surrounding microenvironment. While enzyme binding increases the rotational time of NADH by increasing its molecular size (V), free NADH can also potentially experience an increase in rotational time due to the heterogeneous local microviscosity. The estimated viscosity in the cytoplasm of living cells (1–1.28 cP) is slightly higher than that of water (Kao et al. 1993; Luby-Phelps et al. 1993). In the mitochondrial matrix, however, the reported viscosity varies, based on the employed method and the fluorescent probe. The photobleaching recovery of mitochondrial AcGFP, for example, yielded a viscosity of 1.5–1.97 cP (Dieteren et al. 2011), which is somewhat higher than the reported value (1.1 cP) using mitochondrial GFP anisotropy (Partikian et al. 1998). This range of viscosities will induce only a modest increase in the NADH rotational time, while enzyme binding will increase the rotational time by two orders of magnitude. As a result, time-resolved anisotropy measurements are best suited for the noninvasive and quantitative analysis of a mixture of fluorophores such as free and enzyme-bound NADH in biological samples (Vishwasrao et al. 2005; Yu and Heikal 2009).

Representative fluorescence decay curves of both free NADH in a buffer and intrinsic NADH in cultured Hs578Bst cells are shown in Figure 5.6a. The corresponding fluorescence lifetime represents the time window during which rotational diffusion is being monitored (Figure 5.6b, c). Below the saturation of LDH binding sites, the time-resolved anisotropy of NADH–LDH mixture (16:1) reveals an associated anisotropy (Figure 5.6b) in a buffered solution at room temperature (Yu and Heikal 2009). Similar results were reported for NADH–mMDH mixture as a function of the enzyme titration (Vishwasrao et al. 2005; Yu and Heikal 2009). The corresponding time-resolved anisotropy of intracellular NADH in cultured Hs578Bstcells (Figure 5.6c) also exhibits a similar type of multiphasic associate anisotropy curve that was observed in an aqueous mixture of LDH and NADH (Figure 5.6b). Similar associated anisotropy curves were also observed in different cell lines such as cultured MCF7 and MCF10A cells (Yu 2009) as well as in brain tissue (Vishwasrao et al. 2005).

Vishwasrao et al. (2005) demonstrated for the first time that NADH autofluorescence in brain tissues exhibits an associated anisotropy using a sample-scanning approach (Vishwasrao et al. 2005), thereby directly demonstrating an equilibrium between free and enzyme-bound intracellular NADH. They were further able to separately resolve changes in the concentrations of free and bound NADH in response to the hypoxia. Using a global analysis of the observed associated anisotropy, the ratio of free to enzyme-bound NADH in brain was determined to be 0.78 ± 0.05. While the ratio did not change significantly in response to hypoxia, the relative amplitudes of the three decay components in NADH autofluorescence were sensitive to hypoxia, suggesting that the enzyme binding profile of intracellular NADH may be altered by hypoxia. These results indicate that the cellular NADH response to hypoxic inhibition of energy metabolism is more complicated than a simple increase in the concentration of NADH (Vishwasrao et al. 2005). Furthermore, the estimated local viscosity seems to decrease from 1.6 ± 0.5 cP (normoxia) to 1.1 ± 0.5 cP (hypoxia), as revealed by the rotational time of the free NADH species. This is likely due to a dilution of the mitochondrial matrix caused by the osmotic swelling of mitochondria (Vishwasrao et al. 2005). Conceptually, these viscosities of the local microenvironment calculated from the rotational time may be different from the bulk, long-range viscosity in the

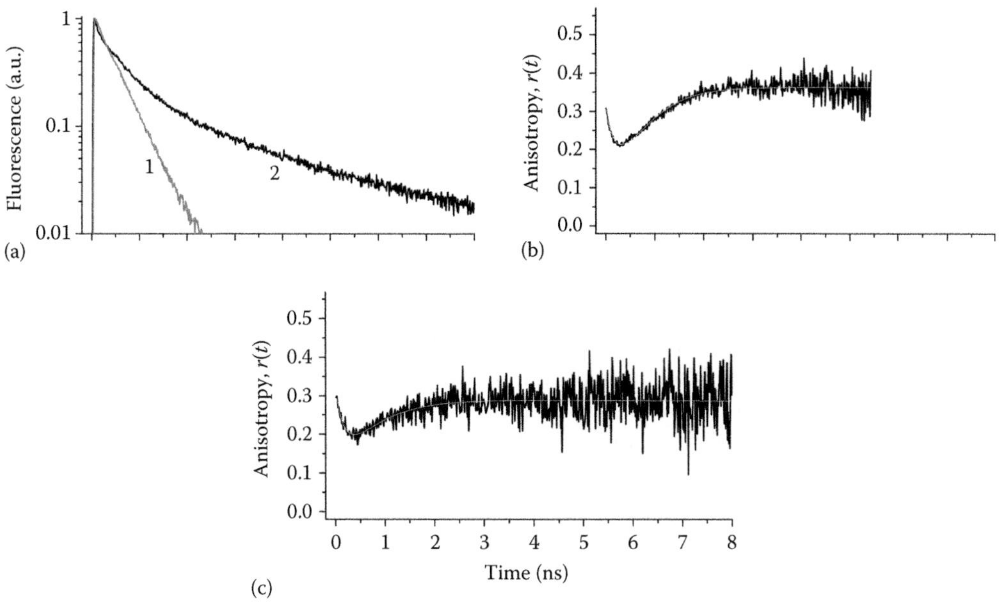

FIGURE 5.6 Time-resolved fluorescence and anisotropy of aqueous and intracellular NADH. (a) Fluorescence decay of NADH in a buffer (PBS, pH 7.4, curve 1) as compared with cellular autofluorescence from cultured Hs578Bst cells excited at 730 nm (curve 2). (b) Time-resolved anisotropy of a mixture of NADH and LDH (16:1) in PBS (pH 7.4, at room temperature) exhibits associated anisotropy decay. (c) Associated anisotropy curve of NADH autofluorescence in cultured Hs578Bst cells, under 730 nm excitation, in pseudo single-point measurements in order to avoid cellular photodamage. Please note that the time-axis in (a), (b), and (c) are the same.

Stokes–Einstein model. We are currently investigating the difference between microenvironment viscosity and the bulk viscosity in a crowded environment as a mimic of the milieu of living cells.

At the single-cell level, the sensitivity of intracellular NADH autofluorescence to cell transformation (normal vs. cancerous) was also examined using time-resolved anisotropy in cultured breast cell lines, Hs578Bst and Hs578T (Yu and Heikal 2009). Using 2P laser-scanning microscopy, the observed associated anisotropy of cellular NADH was described satisfactorily using only two species (free and enzyme-bound NADH), each of which has a distinct fluorescence lifetime and rotational time. According to Equations 5.11 and 5.12, the estimated population fraction of intracellular free NADH (0.18 ± 0.08) in normal breast cells was found to be significantly smaller than the enzyme-bound fraction (0.82 ± 0.08). Moreover, the fraction of free NADH in normal breast cells was found to be lower than that in breast cancer cells (Yu and Heikal 2009). Such enhancement in intracellular free NADH levels in transformed cells was attributed to changes in the activity of oxidative phosphorylation and, perhaps, also of glycolysis. These single-cell studies lend further support to the use of time-resolved associated anisotropy to directly calculate the relative population of free and enzyme-bound NADH in response to physiological variations in the metabolic state of living cells or tissues.

Associated anisotropy studies in a breast cancer cell line (Yu and Heikal 2009) show that the rotational time for intracellular free NADH is 180 ps as compared with 300 ps in the nontransformed cell line. Using the same experimental approach in hippocampal brain slices (Vishwasrao et al. 2005), a rotational time of 318 ps was found in normoxic tissue versus 230 ps in hypoxic tissue. These NADH autofluorescence anisotropy studies indicate that intracellular viscosity is sensitive to cell physiology and pathology. Since intracellular NADH fluorescence predominantly arises from mitochondria (Yu and Heikal 2009), it is possible that the reduction in viscosity reflects a dilution of the mitochondrial matrix due to mitochondrial swelling under pathological conditions (Alirol and Martinou 2006; Bahar et al. 2000; Wilson et al. 2005).

5.8 Time-Resolved Anisotropy of Intracellular Flavins Indicates a Distinct, Predominantly Bound Species in Living Cells

The autofluorescence of cellular flavins was spectrally resolved using 900 nm excitation and detection at 550 ± 75 nm. Under these conditions, the contribution of NADH autofluorescence (730 nm/450 nm) was negligible. In contrast with NADH, the autofluorescence of intrinsic flavins exhibits a multiexponential decay with a shorter average lifetime than that of free FAD in a buffered solution (Figure 5.7a). The pseudo single-point, time-resolved autofluorescence of flavins reveals a triple exponential decay with $\tau_1 = 57 \pm 5$ ps ($\alpha_1 = 0.44 \pm 0.09$), $\tau_2 = 0.76 \pm 0.11$ ns ($\alpha_2 = 0.34 \pm 0.04$), $\tau_3 = 2.8 \pm 0.44$ ns ($\alpha_3 = 0.23 \pm 0.06$), and an estimated average fluorescence lifetime of 0.90 ± 0.17 ns (Heikal 2010, 2012; Lutes 2008).

Figure 5.7b shows the time-resolved anisotropy of free FAD in pure buffer (curve 1) and in the presence of 30% glycerol (curve 2) at room temperature. A buffer with 30% glycerol will have a bulk viscosity of 2–2.5 cP at room temperature as compared with ~1 cP for a pure buffer. This viscosity difference explains the observed slow rotational diffusion of FAD in the presence of 30% glycerol. Time-resolved anisotropy of LipDH in pure buffer decays as a single exponential with a much slower rotational time (Figure 5.7b, curve 3) due to the larger hydrodynamic volume (molecular weight of ~110 kDa) with no significant segmental mobility of the intrinsic FAD moiety. These results demonstrate the significant effect of enzyme binding on the rotational mobility of FAD as compared with the physiologically relevant viscosity.

In cultured Hs578Bst cells, the pseudo single-point, time-resolved anisotropy of intracellular flavin exhibits biexponential anisotropy (Figure 5.7c, curve 2) with a dominant slow component ($\phi_1 = 44 \pm 31$ ns,

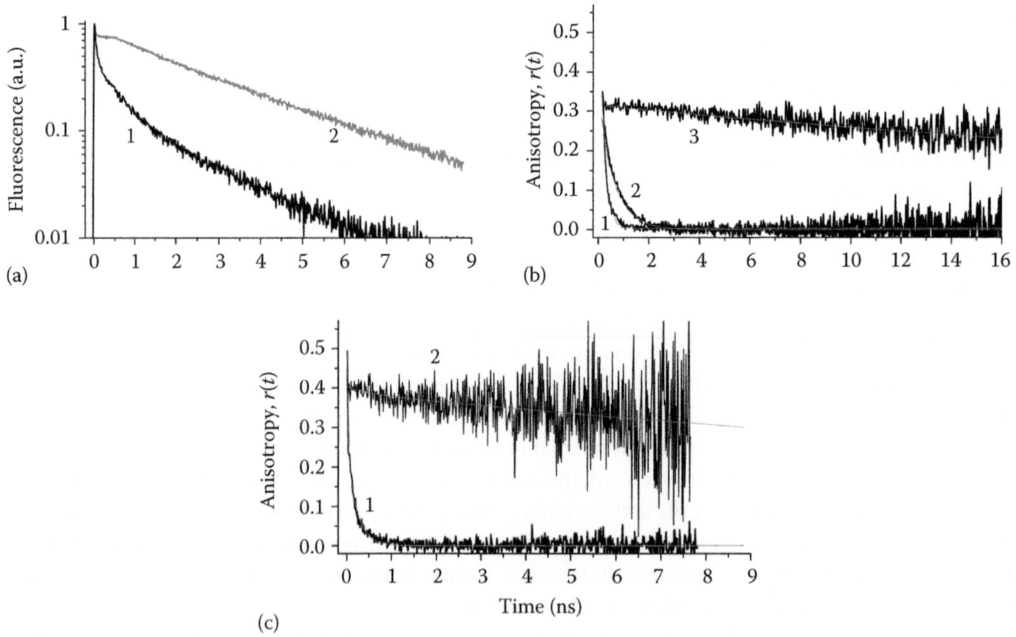

FIGURE 5.7 Time-resolved fluorescence and anisotropy of aqueous and intracellular flavins. (a) Intracellular autofluorescence decay of flavins in cultured Hs578Bst cells (curve 1), excited at 900 nm, as compared with FAD in a buffer (PBS, pH 7.4, curve 2). (b) Time-resolved anisotropy of FAD decays as a single exponential with distinct rotational times in PBS with 0% (curve 1) and 30% glycerol (curve 2), at room temperature. Under the same experimental conditions, however, LipDH anisotropy (PBS, pH7.4) decays with a much slower rotational time (curve 3). (c) Intracellular flavins, excited at 900 nm, exhibit biexponential anisotropy decay (curve 2) with a significantly slower rotational time than that of free, aqueous FAD (curve 1).

$r_{01} = 0.34 \pm 0.02$), which is distinct from the native NADH autofluorescence. A minor ($r_{02} = 0.08 \pm 0.03$) rotational component also decays on a fast timescale ($\phi_2 = 0.7 \pm 0.3$ ns). The comparative rotational diffusion studies in buffer and living Hs578Bst cells indicate that the intracellular flavins predominantly exist in protein-bound conformation. The distinct nature of NADH and FAD rotational dynamics indicates that the associated autofluorescence signals do indeed originate from different intrinsic coenzymes that can be selectively monitored using proper excitation/detection wavelengths (Huang et al. 2002; Yu and Heikal 2009).

5.9 Time-Resolved Anisotropy of NADH and FAD in a Buffered Solution

5.9.1 Viscosity Effects on Rotational Diffusion of NADH and FAD in a Controlled Environment

According to the Stokes–Einstein model, the viscosity also affects the rotational dynamics of a given fluorophore. The viscosity dependence of the rotational time of free NADH in aqueous solution, however, is described well by a modified version of the Debye–Stokes–Einstein relation (Couprie et al. 1994):

$$\phi = \frac{\eta V}{k_B T} + \phi_0 \tag{5.15}$$

The predicted hydrodynamic volume of NADH is 1000 ± 100 Å^3 with an offset (ϕ_0) of 40 ps. Since the viscosity (η) is temperature dependent, it is important to remember that many solution studies are carried out at room temperature (20°C–25°C) as compared with 32°C–35°C for physiological studies. While the corresponding difference in absolute temperatures is only 3%–5%, the difference in water viscosity (1 cP at 20°C vs. 0.77–0.72 cP at 32°C–35°C) is substantial (23%–28%).

Fluorescence anisotropy measurements of free, aqueous NADH indicate a rotational time of 255 ± 35 ps at room temperature (Couprie et al. 1994; Vishwasrao et al. 2005; Yu and Heikal 2009). Using Equation 5.15 along with the temperature-dependent viscosity of water (Korson et al. 1969), the predicted rotational time for free aqueous NADH is ~150 ps at 32°C (Vishwasrao et al. 2005). Similar studies on FAD in aqueous solution reveal a similar sensitivity of the rotational time to the environmental viscosity (Figure 5.7b, curve 2). The time-resolved anisotropy curves are shown for free FAD in an aqueous buffer (Figure 5.7b, curve 1), viscous buffer (30% glycerol, curve 2), and protein-bound complex (LipDH, curve 3). For example, the rotational time of FAD in 30% glycerol is ~620 ps with an $r_0 = 0.275$, which is significantly slower than its rotation in pure buffer (180 ps) at room temperature. The 2P fluorescence lifetime of aqueous free FAD is significantly longer than that of intracellular flavins (Figure 5.7a).

5.9.2 Enzyme-Binding Effects on the Rotational Diffusion of NADH and FAD in a Buffered Solution

There are many different enzymes in cells whose activities and biological functions depend on the presence of coenzymes such as NADH and FAD. The sizes of these enzymes are generally much larger (MW = 10–100 kDa) than the free coenzymes. Mitochondrial malate dehydrogenase (mMDH), for example, has a molecular weight of 70 kDa. In order to assign the observed associated anisotropy of intrinsic NADH in cultured cells and brain tissue, we carried out a controlled time-resolved anisotropy measurements of NADH–enzyme mixtures (Vishwasrao et al. 2005; Yu and Heikal 2009).

NADH binding to mMDH and LDH has been investigated using both time-resolved fluorescence and anisotropy measurements in a PBS buffer, pH 7.4 (Vishwasrao et al. 2005; Yu and Heikal 2009). The fitting parameters for both types of decay were found to depend on the relative concentrations of the enzyme and free NADH (Yu and Heikal 2009). The multiexponential 2P fluorescence decay of NADH (PBS, pH 7.4) as a function of enzyme concentration suggests the presence of multiple conformations

(Deng et al. 2001; Schauerte et al. 1995). Upon complete binding of NADH by mMDH, the fluorescence intensity and average lifetime of NADH–mMDH were twice that of the free cofactor (Labrou et al. 1996; Ovádi et al. 1994). In contrast, the occupied four binding sites in LDH (Deng et al. 2001; Fromm 1963) caused threefold increase in the corresponding average fluorescence lifetime that is attributed to the restriction imposed on NADH by the protein environment (Schauerte et al. 1995; Vishwasrao et al. 2005; Visser and van Hoek 1981).

At saturated binding sites, time-resolved anisotropy of the NADH–mMDH complex (PBS buffer, pH 7.4) exhibits a single exponential decay with a slow rotational time of $\phi = 30$ ns at room temperature. The lack of a fast (<1 ns) rotational component suggests a complete immobilization of NADH within its binding site in mMDH on the nanosecond timescale (Piersma et al. 1998; Vishwasrao et al. 2005; Yu and Heikal 2009; Zheng et al. 2010). While this tight binding is common to most dehydrogenases, crystallographic studies of aldehyde dehydrogenases indicate an exception, where the nicotinamide ring of NADH has some conformational flexibility within the binding site (Hammen et al. 2002). The interaction between NAD and the binding site in aldehyde dehydrogenase seems distinct from other families of dehydrogenases, despite having a similar binding site structure (Liu et al. 1997). The rotational time of NADH–mMDH complex is roughly two orders of magnitude slower than that of free NADH in PBS buffer at room temperature (Vishwasrao et al. 2005; Yu and Heikal 2009). In contrast, the fluorescence lifetime of NADH typically shows a relatively modest enhancement upon enzyme binding. The large difference in the molecular weights of free and bound NADH enables us to unambiguously distinguish between these two states (species) in living cells or tissues using time-resolved associated anisotropy.

The qualitative agreement between the associated anisotropy of intracellular and solution studies indicates that the associated anisotropy of NADH autofluorescence can be attributed to the presence of two equilibrated populations of NADH (free and enzyme bound). For quantitative analysis, one must account for the complex environment of both free and bound NADH in living cells (Yu and Heikal 2009) or brain tissues (Vishwasrao et al. 2005).

Time-resolved anisotropy of free FAD in PBS (pH 7.4) reveals a rotational time of 180 ps ($r_0 = 0.342$) at room temperature (Heikal 2010, 2012; Lutes 2008; Yu 2009). In contrast, the anisotropy of LipDH (PBS, pH 7.4) decays as a single exponential (Figure 5.7b) with a rotational time of 51.63 ns and an initial anisotropy of 0.316 (Lutes 2008).

5.10 Summary and Future Outlook

Intrinsically fluorescent biomolecules, including the coenzymes NADH and FAD, are integral to a complex network of biochemical reactions and metabolic pathways that are essential for the function and survival of cells. In this chapter, we have described noninvasive autofluorescence imaging modalities for quantifying the concentration and the state of binding of the intracellular NADH and FAD. While complementing conventional biochemical techniques, these methods do not require the destruction of cells or tissues and therefore retain the morphological context of these coenzymes. Extrinsic fluorescent dyes are brighter and more photostable than intrinsic fluorophores; however, they may suffer from potential toxicity, nonspecific binding, and interference with the cellular processes associated with the labeling target. In contrast, cellular coenzymes are natural biomolecules that participate in metabolic activities and biochemical reactions in living cells.

The concentrations, distributions, binding states, and local microenvironments of these intrinsically fluorescent coenzymes vary with the metabolic and physiological states of cells and tissues. Fluorescence lifetime and anisotropy imaging of intracellular NADH provide a unique opportunity to probe physiological changes in cellular processes of healthy cells (mitochondrial activity, apoptosis, glycolysis, and oxidative stress) and diseased cells (cancer, aging, and neurodegenerative diseases).

Since the excited-state lifetime is the time window during which rotational time can be measured, the S/N ratio of the anisotropy curve degrades as the rotational time significantly exceeds the fluorescence lifetime (e.g., labeled proteins). As a result, the accuracy in the estimated hydrodynamic volume will decrease as the size of or the environmental restrictions on the fluorophore increase. In order to enhance the S/N ratio in solution measurements, we simply increase the integration time of the detected fluorescence. Working in living cells or tissues, however, is more challenging since increasing the integration time for a better S/N ratio may cause cellular photobleaching/photodamage that would compromise the biological significance of these measurements. This trade-off between signal level and potential photodamage is a critical limiting factor in the use of autofluorescence for live cell diagnostics. Nonetheless, integrated fluorescence microspectroscopy and multiparametric approaches using intracellular coenzymes as natural biomarkers hold a great potential for biological and biomedical research.

Despite the advantages of utilizing these intrinsic fluorophores for biological research and biomedical/clinical applications, some inherent challenges remain (Heikal 2010, 2012). First, assessing the spatial distribution of NADH concentration requires noninvasive, quantitative techniques with high spatial resolution to differentiate between cytosolic (e.g., glycolysis), mitochondrial (e.g., oxidative phosphorylation and the tricarboxylic acid cycle), and nuclear (e.g., transcription) coenzymes. Second, an ability to selectively monitor the concentration of these coenzymes without photobleaching or photodamage is essential for a biologically meaningful readout. Third, these coenzymes are involved in a range of biochemical reactions and are therefore sensitive to many physiological conditions, such as oxygen level, in vivo blood flow, the presence of other metabolites, and chemical stimulation. This functional complexity can complicate the biological interpretation of autofluorescence variations. Fourth, in vivo imaging is essential for clinical diagnostics to realize the full potential of these coenzymes as natural biomarkers for numerous health problems including cancer, neurodegeneration, diabetes, and aging. However, in vivo imaging is challenging due to the limited penetration of light into the turbid biological samples. New advances in endoscopy potentially hold the key for applying in vivo autofluorescence imaging and spectroscopy toward clinical diagnostics and medical applications.

Acknowledgments

We thank John Alfveby for his help with cell culture and imaging. This work was supported in part by the Department of Chemistry and Biochemistry, Swenson College of Science and Engineering, and the Department of Pharmacy Practice and Pharmaceutical Sciences, University of Minnesota Duluth. Additional support was provided by Grant-in-Aid of Research, Artistry and Scholarship (University of Minnesota), NSF (MCB0718741), and NIH (AG030949). Harshad D. Vishwasrao was supported by NIH grant 1R01MH097062-01. Qianru Yu was supported by NIH grants (P20-RR018787/GM103413, P30GM106394, and R01 HL074175; PI: Bruce A. Stanton, Dartmouth College).

References

Alberts, B., A. Johnson, J. Lewis, M. Raff, K. Roberts, and P. Walter. 2002. *Molecular Biology of the Cell.* New York: Garland Science.

Alberts, B., K. Roberts, D. Bray, and J. Lewis. 2003. *Essential Cell Biology.* New York: Garland Science.

Alirol, E. and J. C. Martinou. 2006. Mitochondria and cancer: Is there a morphological connection? *Oncogene* 25:4706–4716.

Ariola, F. S., D. J. Mudaliar, R. P. Walvick, and A. A. Heikal. 2006. Dynamics imaging of lipid phases and lipid-marker interactions in model biomembranes. *Physical Chemistry Chemical Physics* 8(39):4517–4529.

Axelrod, D. 1979. Carbocyanine dye orientation in red cell membrane studied by microscopic fluorescence polarization. *Biophysical Journal* 26(3):557–573.

Axelrod, D. 1989. Fluorescence polarization microscopy. In *Fluorescence Microscopy of Living Cells in Culture B*, D. L. Taylor and Y.-L. Wang (eds.), Vol. 30 of Methods Cell Biology, L. Wilson and P. Tran (eds.), pp. 333–352. London, U.K.: Academic Press.

Bahar, S., D. Fayuk, G. G. Somjen, P. G. Aitken, and D. A. Turner. 2000. Mitochondrial and intrinsic optical signals imaged during hypoxia and spreading depression in rat hippocampal slices. *Journal of Neurophysiology* 84(1):311–324.

Bailey, M. F., E. H. Z. Thompson, and D. P. Millar. 2001. Probing DNA polymerase fidelity mechanisms using time-resolved fluorescence anisotropy. *Methods* 25(1):62–77.

Bak, I. J., U. Misgeld, M. Weiler, and E. Morgan. 1980. The preservation of nerve cells in rat neostriatal slices maintained in vitro: A morphological study. *Brain Research* 197(2):341–353.

Biagini, G. A., P. Viriyavejakul, P. M. O'Neill, P. G. Bray, and S. A. Ward. 2006. Functional characterization and target validation of alternative complex I of *Plasmodium falciparum* mitochondria. *Antimicrobial Agents and Chemotherapy* 50(5):1841–1851.

Bigelow, C. E. and T. H. Foster. 2006. Confocal fluorescence polarization microscopy in turbid media: Effects of scattering-induced depolarization. *Journal of the Optical Society of America A* 23(11):2932–2943.

Bird, D. K., L. Yan, K. M. Vrotsos et al. 2005. Metabolic mapping of MCF 10A human breast cells via multiphoton fluorescence lifetime imaging of the coenzyme NADH. *Cancer Research* 65(19):8766–8773.

Bücher, T., B. Brauser, A. Conze, F. Klein, O. Langguth, and H. Sies. 1972. State of oxidation-reduction and state of binding in cytosolic NADH-system as disclosed by equilibration with extracellular lactate pyruvate in hemoglobin-free perfused rate liver. *European Journal of Biochemistry* 27(12):301–317.

Chance, B. and H. Baltscheffsky. 1958. Respiratory enzymes in oxidative phosphorylation VII. Binding of intramitochondrial reduced pyridine nucleotide. *Journal of Biological Chemistry* 233(3):736–739.

Cheong, W.-F., S. A. Prahl, and A. J. Welch. 1990. A review of the optical properties of biological tissues. *IEEE Journal of Quantum Electronics* 26(12):2166–2185.

Couprie, M. E., F. Merola, P. Tauc et al. 1994. First use of the UV Super-ACO free-electron laser: Fluorescence decays and rotational dynamics of the NADH coenzyme. *Review of Scientific Instruments* 65(5):1485–1495.

Cunningham, M. L., J. S. Johnson, S. M. Giovanazzi, and M. J. Peak. 1985. Photosensitized production of superoxide anion by monochromatic (290–405 nm) ultraviolet irradiation of NADH and NADPH coenzymes. *Photochemistry and Photobiology* 42(2):125–128.

Czochralska, B., W. Kawczynski, G. Bartosz, and D. Shugar. 1984. Oxidation of excited-state NADH and NAD dimer in aqueous medium involvement of O_2^- as a mediator in the presence of oxygen. *Biochimica et Biophysica Acta* 801(3):403–409.

Davey, A. M., R. P. Walvick, Y. Liu, A. A. Heikal, and E. D. Sheets. 2007. Membrane order and molecular dynamics associated with IgE receptor cross-linking in mast cells. *Biophysical Journal* 92(1):343–355.

Deng, H., N. Zhadin, and R. Callender. 2001. Dynamics of protein ligand binding on multiple time scale: NADH binding to lactate dehydrogenase. *Biochemistry* 40(13):3767–3773.

Dieteren, C. E. J., S. C. A. M. Gielen, L. G. J. Nijtmans et al. 2011. Solute diffusion is hindered in the mitochondrial matrix. *Proceedings of the National Academy of Sciences of the United States of America* 108(21):8657–8662.

Fromm, H. J. 1963. Determination of dissociation constants of coenzymes and abortive ternary complexes with rabbit muscle lactate dehydrogenase from fluorescence measurements. *Journal of Biological Chemistry* 238(9):2938–2944.

Furuike, S., K. Adachi, N. Sakaki et al. 2008. Temperature dependence of the rotation and hydrolysis activities of F_1-ATPase. *Biophysical Journal* 95(2):761–770.

Gafni, A. and L. Brand. 1976. Fluorescence decay studies of reduced nicotinamide adenine dinucleotide in solution and bound to liver alcohol dehydrogenase. *Biochemistry* 15(15):3165–3171.

Garofalo, O., D. W. G. Cox, and H. S. Bachelard. 1988. Brain levels of NADH and NAD^+ under hypoxic and hypoglycemic conditions in vitro. *Journal of Neurochemistry* 51(1):172–176.

Hammen, P. K., A. Allali-Hassani, K. Hallenga, T. D. Hurley, and H. Weiner. 2002. Multiple conformations of NAD and NADH when bound to human cytosolic and mitochondrial aldehyde dehydrogenase. *Biochemistry* 41(22):7156–7168.

Heikal, A. A. 2010. Intracellular coenzymes as natural biomarkers for metabolic activities and mitochondrial anomalies. *Biomarkers in Medicine* 4(2):241–263.

Heikal, A. A. 2012. A multiparametric imaging of cellular coenzymes for monitoring metabolic and mitochondrial activities. In: *Reviews in Fluorescence*, C. D. Geddes (ed.), Vol. 2010 of Reviews in Fluorescence, pp. 223–243. New York: Springer.

Hönes, G., J. Hönes, and M. Hauser. 1986. Studies of enzyme-ligand complexes using dynamic fluorescence anisotropy. II. The coenzyme-binding site of malate dehydrogenase. *Biological Chemistry Hoppe-Seyler* 367(1):103–108.

Huang, S., A. A. Heikal, and W. W. Webb. 2002. Two-photon fluorescence spectroscopy and microscopy of NAD(P)H and flavoprotein. *Biophysical Journal* 82(5):2811–2825.

Jardetzky, O. and N. G. Wade-Jardetzky. 1966. The conformation of pyridine dinucleotides in solution. *Journal of Biological Chemistry* 241(1):85–91.

Kao, H. P., J. R. Abney, and A. S. Verkman. 1993. Determinants of the translational mobility of a small solute in cell cytoplasm. *Journal of Cell Biology* 120(1):175–184.

Kasischke, K. A., M. Büchner, A. C. Ludolph, and M. W. Riepe. 2001. Nuclear shrinkage in live mouse hippocampal slices. *Acta Neuropathologica* 101(5):483–490.

Kasischke, K. A., H. D. Vishwasrao, P. J. Fisher, W. R. Zipfel, and W. W. Webb. 2004. Neural activity triggers neuronal oxidative metabolism followed by astrocytic glycolysis. *Science* 305(5680): 99–103.

Kierdaszuk, B., H. Malak, I. Gryczynski, P. Callis, and J. R. Lakowicz. 1996. Fluorescence of reduced nicotinamides using one- and two-photon excitation. *Biophysical Chemistry* 62(1):1–13.

Klaidman, L. K., A. C. Leung, and J. D. Adams. 1995. High-performance liquid-chromatography analysis of oxidised and reduced pyridine dinucleotides in specific brain-regions. *Analytical Biochemistry* 228(2):312–317.

Knutson, J. R., J. M. Beechem, and L. Brand. 1983. Simultaneous analysis of multiple fluorescence decay curves: A global approach. *Chemical Physics Letter* 102(6):501–507.

Korson, L., W. Drost-Hansen, and F. J. Millero. 1969. Viscosity of water at various temperatures. *Journal of Physical Chemistry* 73(1):34–39.

Kress, A., P. Ferrand, H. Rigneault et al. 2011. Probing orientational behavior of MHC class I protein and lipid probes in cell membranes by fluorescence polarization-resolved imaging. *Biophysical Journal* 101(2):468–476.

Kunz, D., K. Winkler, C. E. Elger, and W. S. Kunz. 2002. Functional imaging of mitochondrial redox state. In: *Redox Cell Biology and Genetics A*, C. K. Sen and L. Packer (eds.), Vol. 352 of Methods in Enzymology, pp. 135–150. London, U.K.: Academic Press.

Labrou, N. E., E. Eliopoulos, and Y. D. Clonis. 1996. Dye-affinity labelling of bovine heart mitochondrial malate dehydrogenase and study of the NADH-binding site. *Biochemical Journal* 315:687–693.

Lakowicz, J. R., ed. 2006. *Principles of Fluorescence Spectroscopy.* New York: Springer.

Lentz, B. R., B. M. Moore, and D. A. Barrow. 1979. Light-scattering effects in the measurement of membrane microviscosity with diphenylhexatriene. *Biophysical Journal* 25(3):489–494.

Liu, Z.-J., Y.-J. Sun, J. Rose et al. 1997. The first structure of an aldehyde dehydrogenase reveals novel interactions between NAD and the Rossmann fold. *Nature Structural Biology* 4(4):317–326.

Luby-Phelps, K., S. Mujumdar, R. B. Mujumdar, L. A. Ernst, W. Galbraith, and A. S. Waggoner. 1993. A novel fluorescence ratiometric method confirms the low solvent viscosity of the cytoplasm. *Biophysical Journal* 65(1):236–242.

Lutes, A. T. 2008. Intrinsic flavin as a biomarker for monitoring pathologically-induced changes in cellular energy metabolism. MS thesis, Pennsylvania State University, State College, PA.

Merrill, D. K. and R. W. Guynn. 1976. Electroconvulsive seizure—Investigation into validity of calculating cytoplasmic free NAD^+/NADH ratio from substrate concentrations of brain. *Journal of Neurochemistry* 27(2):459–464.
Merrill, D. K. and R. W. Guynn. 1982. The calculation of the mitochondrial free NAD^+ NADH ratio in brain—Effect of electroconvulsive seizure. *Brain Research* 239(1):71–80.
Mitchell, P. and J. Moyle. 1969. Estimate of membrane potential and pH difference across cristae membrane of rat liver mitochondria. *European Journal of Biochemistry* 7(4):471–484.
Mohanty, S. K., N. Ghosh, S. K. Majumder, and P. K. Gupta. 2001. Depolarization of autofluorescence from malignant and normal human breast tissues. *Applied Optics* 40(7):1147–1154.
Monette, R., D. L. Small, G. Mealing, and P. Morley. 1998. A fluorescence confocal assay to assess neuronal viability in brain slices. *Brain Research Protocols* 2(2):99–108.
Ogikubo, S., T. Nakabayashi, T. Adachi et al. 2011. Intracellular pH sensing using autofluorescence lifetime microscopy. *The Journal of Physical Chemistry B* 115(34):10385–10390.
Ovádi, J., Y. Huang, and H. O. Spivey. 1994. Binding of malate dehydrogenase and NADH channeling to complex I. *Journal of Molecular Recognition* 7(4):265–272.
Partikian, A., B. Ölveczky, R. Swaminathan, Y. Li, and A. S. Verkman. 1998. Rapid diffusion of green fluorescent protein in the mitochondrial matrix. *Journal of Cell Biology* 140(4):821–829.
Paul, R. J. and H. Schneckenburger. 1996. Oxygen concentration and the oxidation-reduction state of yeast: Determination of free/bound NADH and flavins by time-resolved spectroscopy. *Naturwissenschaften* 83(1):32–35.
Piersma, S. R., A. J. W. G. Visser, S. de Vries, and J. A. Duine. 1998. Optical spectroscopy of nicotinoprotein alcohol dehydrogenase from *Amycolatopsis methanolica*: A comparison with horse liver alcohol dehydrogenase and UDP-galactose epimerase. *Biochemistry*37(9):3068–3077.
Rocheleau, J. V., W. S. Head, and D. W. Piston. 2004. Quantitative NAD(P)H/flavoprotein autofluorescence imaging reveals metabolic mechanisms of pancreatic islet pyruvate response. *Journal of Biological Chemistry* 279(30):31780–31787.
Romashko, D. N., E. Marban, and B. O'Rourke. 1998. Subcellular metabolic transients and mitochondrial redox waves in heart cells. *Proceedings of the National Academy of Sciences of the United States of America* 95(4):1618–1623.
Salmon, J.-M., E. Kohen, P. Viallet et al. 1982. Microspectrofluorometric approach to the study of free bound NAD(P)H ratio as metabolic indicator in various cell-types. *Photochemistry and Photobiology* 36(5):585–593.
Sandell, J. L. and T. C. Zhu. 2011. A review of in-vivo optical properties of human tissues and its impact on PDT. *Journal of Biophotonics* 4(11–12):773–787.
Schauerte, J. A., B. D. Schlyer, D. G. Steel, and A. Gafni. 1995. Nanosecond time-resolved circular polarization of fluorescence: Study of NADH bound to horse liver alcohol dehydrogenase. *Proceedings of the National Academy of Sciences of the United States of America* 92(2):569–573.
Skala, M. C., K. M. Riching, A. Gendron-Fitzpatrick et al. 2007. In vivo multiphoton microscopy of NADH and FAD redox states, fluorescence lifetimes, and cellular morphology in precancerous epithelia. *Proceedings of the National Academy of Sciences of the United States of America* 104(49):19494–19499.
Strovas, T. J., L. M. Sauter, X. Guo, and M. E. Lidstrom. 2007. Cell-to-cell heterogeneity in growth rate and gene expression in *Methylobacterium extorquens* AM1. *Journal of Bacteriology* 189(19):7127–7133.
Teale, F. W. J. 1969. Fluorescence depolarization by light-scattering in turbid solutions. *Photochemistry and Photobiology* 10(6):363–374.
Thorling, C. A., X. Liu, F. J. Burczynski, L. M. Fletcher, G. C. Gobe, and M. S. Roberts. 2011. Multiphoton microscopy can visualize zonal damage and decreased cellular metabolic activity in hepatic ischemia-reperfusion injury in rats. *Journal of Biomedical Optics* 16(11):116011.
Tiede, L. M., S. M. Rocha-Sanchez, R. Hallworth, M. G. Nichols, and K. Beisel. 2007. Determination of hair cell metabolic state in isolated cochlear preparations by two-photon microscopy. *Journal of Biomedical Optics* 12(2):021004.

Ueno, H., T. Suzuki, K. Kinosita, and M. Yoshida. 2005. ATP-driven stepwise rotation of F_0F_1-ATP synthase. *Proceedings of the National Academy of Sciences of the United States of America* 102(5):1333–1338.

Velick, S. F. 1958. Fluorescence spectra and polarization of glyceraldehyde-3-phosphate and lactic dehydrogenase coenzyme complexes. *Journal of Biological Chemistry* 233(6):1455–1467.

Vishwasrao, H. D. 2004. Quantitative two-photon redox fluorescence microscopy of neurometabolic dynamics. PhD dissertation, Cornell University, Ithaca, New York.

Vishwasrao, H. D., A. A. Heikal, K. A. Kasischke, and W. W. Webb. 2005. Conformational dependence of intracellular NADH on metabolic state revealed by associated fluorescence anisotropy. *Journal of Biological Chemistry* 280(26):25119–25126.

Vishwasrao, H. D., P. Trifilieff, and E. R. Kandel. 2012. In vivo imaging of the actin polymerization state with two-photon fluorescence anisotropy. *Biophysical Journal* 102(5):1204–1214.

Visser, A. J. W. G. and A. van Hoek. 1981. The fluorescence decay of reduced nicotinamides in aqueous solution after excitation with a UV-mode locked Ar ion laser. *Photochemistry and Photobiology* 33(1):35–40.

Wakita, M., G. Nishimura, and M. Tamura. 1995. Some characteristics of the fluorescence lifetime of reduced pyridine-nucleotides in isolated-mitochondria, isolated hepatocytes, and perfused-rat-liver-in-situ. *Journal of Biochemistry* 118(6):1151–1160.

Wilhelmi, B. 1982. Influence of solvent viscosity on excited state lifetime and fluorescence quantum yield of dye molecules. *Chemical Physics* 66(3):351–355.

Williamson, D. H., P. Lund, and H. A. Krebs. 1967. Redox state of free nicotinamide-adenine dinucleotide in cytoplasm and mitochondria of rate liver. *Biochemical Journal* 103:514–527.

Wilson, J. D., C. E. Bigelow, D. J. Calkins, and T. H. Foster. 2005. Light scattering from intact cells reports oxidative-stress-induced mitochondrial swelling. *Biophysical Journal* 88(4):2929–2938.

Yu, J.-S., H.-W. Guo, C.-H. Wang, Y.-H. Wei, and H.-W. Wang. 2011. Increase of reduced nicotinamide adenine dinucleotide fluorescence lifetime precedes mitochondrial dysfunction in staurosporine-induced apoptosis of HeLa cells. *Journal of Biomedical Optics* 16(3):036008.

Yu, Q. 2009. Functional imaging of intracellular metabolic cofactors in human normal and cancer breast cells. PhD dissertation, Pennsylvania State University, State College, PA.

Yu, Q. and A. A. Heikal. 2009. Two-photon autofluorescence dynamics imaging reveals sensitivity of intracellular NADH concentration and conformation to cell physiology at the single-cell level. *Journal of Photochemistry and Photobiology. B, Biology* 95(1):46–57.

Yu, Q., M. Proia, and A. A. Heikal. 2008. Integrated biophotonics approach for noninvasive and multiscale studies of biomolecular and cellular biophysics. *Journal of Biomedical Optics* 13(4):041315.

Zheng, W., D. Li, and J. Y. Qu. 2010. Monitoring changes of cellular metabolism and microviscosity in vitro based on time-resolved endogenous fluorescence and its anisotropy decay dynamics. *Journal of Biomedical Optics* 15(3):037013.

6

Real-Time In Vivo Monitoring of Cellular Energy Metabolism

Avraham Mayevsky
Bar-Ilan University

Efrat Barbiro-Michaely
Bar-Ilan University

6.1 Introduction

Tissue oxygen or energy positive balance is necessary for the required level of ATP production by mitochondria. The majority of oxygen in a mammalian organism is delivered to cells by erythrocytes: it diffuses along the partial pressure gradient from the lungs (100 mm Hg) to the bloodstream, where it combines chemically to hemoglobin (Hb) in erythrocytes and is carried by convective transport through the large vessels down to arterioles and capillaries. Smaller pressure near the mitochondria (1 mm Hg) favors oxygen diffusion from the microcirculation compartments into the cells (Figure 6.1).

Oxygen supply to the cells in all tissues is determined by the level of Hb saturation, as well as microcirculatory blood flow and volume. The demand for oxygen or energy currencies in different tissues is also dependent upon the physiological and biochemical activities in each tissue or organ (Figure 6.2). Changes in oxygen supply, oxygen demand, or both determine the oxygen balance. In normal tissues, stored energy meets the demand within a certain range. An increase in energy demand, however, will induce an increase in oxygen supply via elevating microcirculatory blood

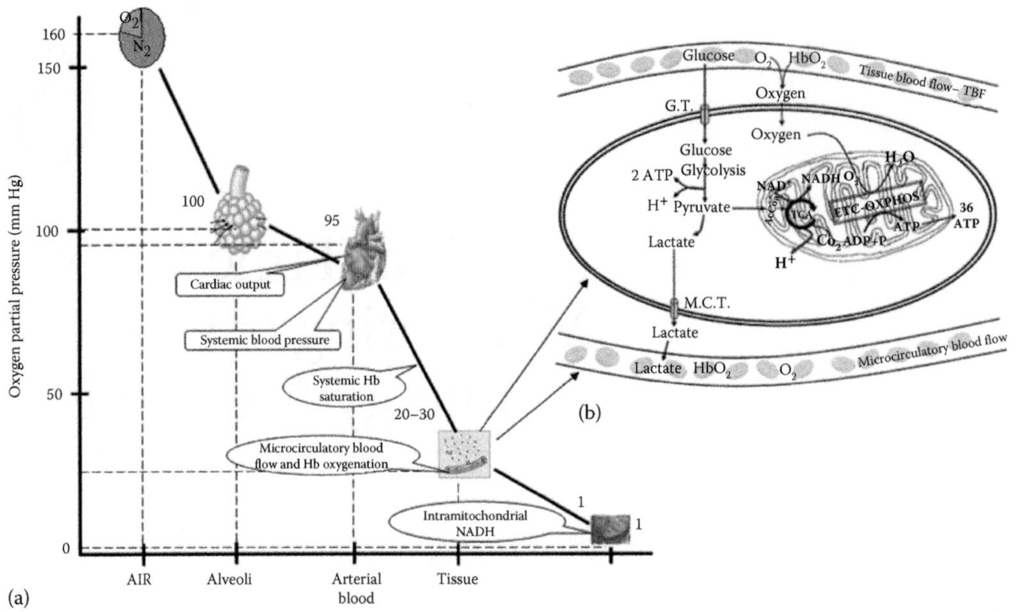

FIGURE 6.1 (a) Oxygen gradient from its high level in the air to its very low level in the mitochondria. (b) All cells and tissues accept oxygen and substrates from the bloodstream. Oxygen is transferred in the blood by the hemoglobin. It is then diffused from the blood to the extracellular space and to the intracellular space. There it serves in the mitochondria to induce ATP.

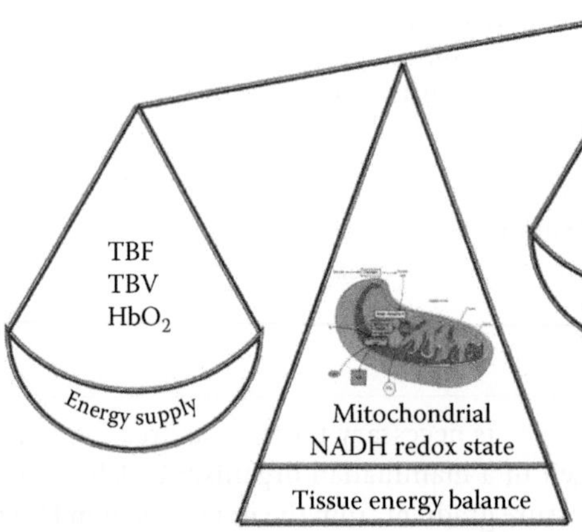

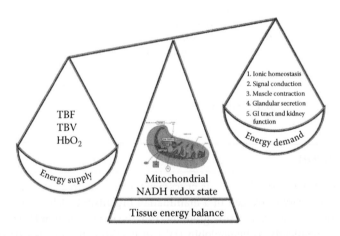

FIGURE 6.2 Mitochondrial NADH redox state level is a good indicator of the balance between tissue energy supply by the bloodstream, which is similar in all tissues, and tissue energy demand, which is specific to each tissue and depends on its activity.

flow, which delivers more oxygenated blood. In contrast, a decrease in oxygen supply will limit oxygen utilization and lead to the development of hypoxic damage in brain and other organs (Mayevsky and Chance 1982).

Evaluation of tissue energy metabolism in vivo can be carried out by monitoring various physiological parameters as shown in Figure 6.3. Given the coupling of ATP production to the availability of oxygen, and thereby to microcirculation, energy metabolism could be evaluated by the levels of

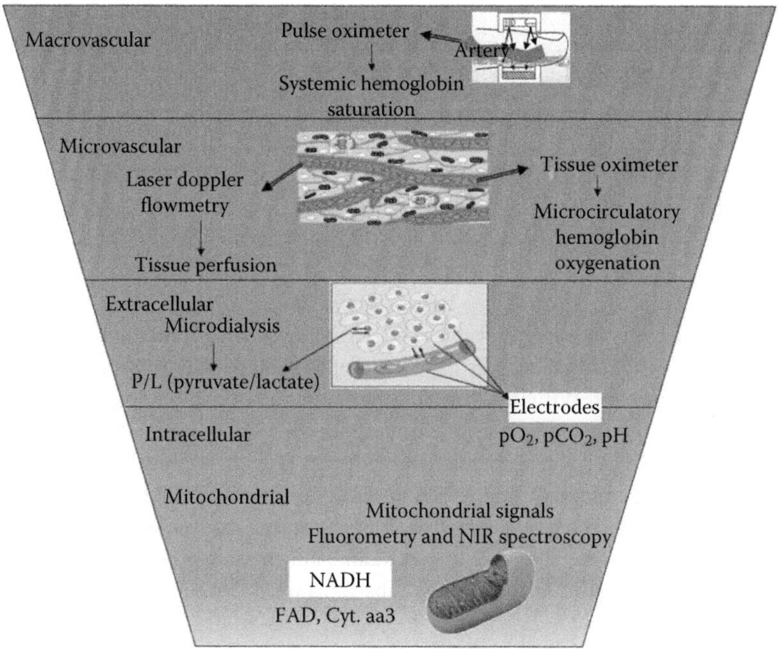

FIGURE 6.3 Tissue oxygenation and energy state are monitored in various methods, which can be divided according to monitoring location. The pulse oximeter monitors the macrovascular bed, whereas laser Doppler flowmetry monitors the microvascular bed and so does the tissue oximeter. Mini- and microelectrodes are used for the monitoring of various chemicals in the extracellular and intracellular spaces as well as in the bloodstream. The most accurate intracellular indicator for tissue energy state in vivo is the mitochondrial NADH redox state. In other studies, flavins and cytochrome aa3 were monitored but the contamination of blood volume and oxygenation was not eliminated.

oxygen in arterioles and capillaries. Most of the current methods, for example, pulse oximeter, which is routinely used in clinics, probe oxygen content only in macrovascular compartments. However, microcirculatory blood flow and Hb oxygenation can still be monitored in arterioles and capillaries (Meirovithz et al. 2007).

Microdialysis is another approach that enables the measurement of metabolite (e.g., pyruvate and lactate) levels in the extracellular space. Such measurement can also be conducted by surface electrodes, providing with the partial pressure of oxygen (pO_2) and carbon dioxide (pCO_2), as well as the hydrogen ion concentration (pH). This method reflects the overall changes in pO_2, pCO_2, and pH levels in intracellular volume, extracellular space, and the small blood vessels and does not distinguish between these compartments. The direct and most accurate approach for in vivo cellular energy metabolism evaluation is based on the optical monitoring of NADH, flavoproteins (Fp), and cytochrome aa3. Of these, only the NADH redox state allows for relatively accurate measurements: as of today, the correction of the Fp and cytochrome aa3 measurements for hemodynamic changes in blood-perfused organs in vivo is almost impossible. The level of NADH autofluorescence was also considered a good indicator for decreased oxygen availability in cellular compartments (Chance et al. 1973) under both in vitro and in vivo conditions.

In this chapter, we will discuss current methods of in vivo NADH monitoring, describe the setups and necessary calibrations, review the pitfalls and remedies, and assess the potential of this approach in the clinical diagnosis of pathologies.

6.2 Brief History of NADH as Natural Biomarker

Over a century of mitochondria research in relation to energy metabolism can be divided into three periods, summarized in Table 6.1 (for a review, see Mayevsky and Rogatsky, 2007). During the first period (1906–1955), NADH was discovered and its absorbance and fluorescence measurements were first performed. The second period (1956–1973) was marked by a transition from monitoring NADH in vitro (by its absorbance) to in vivo (by its fluorescence) in various organs and animal models. During the third period (1974 until today), the number of NADH monitoring methods and applications significantly expanded. This period also highlights a shift from NADH monitoring in animals to its clearance by the Federal Drugs Administration (FDA) for use in clinics.

The breakthrough in the field was achieved by the seminal work by Chance and Williams (1955), who showed for the first time that NADH can be used as an optical biomarker of energy metabolism. Under in vitro conditions, Chance and Williams defined different metabolic states of isolated mitochondria by changing the levels of ADP, substrate (energy-rich compound), and oxygen (Table 6.2). They also measured the redox state of NADH, flavoproteins, and cytochromes through the absorbance of the mitochondrial suspension. At the initial State 1, ~90% of NAD was found to be reduced. Addition of ADP to the mitochondrial suspension led to the exhaustion of the endogenous substrate and full oxidation of NADH (State 2). Under this state, the ADP was available, but the substrate became the limiting factor. Transition to State 3 with the addition of various substrates was marked by a considerable reduction of

TABLE 6.1 Milestones in the Early Period of Mitochondrial NADH Monitoring

Year	Discovery	Author(s)
1905	Involvement of adenine containing nucleotides in yeast fermentation	Harden and Young (1906)
1935	Description of the complete structure of "hydrogen-transferring coenzyme" in erythrocytes	Warburg et al.
1936	Definition of the cofactors NAD (originally denoted diphosphopyridine nucleotide [DPN]) and NADP (originally denoted triphosphopyridine nucleotide [TPN])	Warburg
1951	A shift in the absorption spectrum of NADH with alcohol dehydrogenase	Theorell and Bonnichsen
1952	Development of a rapid and sensitive spectrophotometer	Chance and Legallias
1952	Monitoring of pyridine nucleotide enzymes	Chance
1954	The first detailed study of NADH using fluorescence spectrophotometry	Duysens and Amesz
1958	Measurement of NADH fluorescence in isolated mitochondria	Chance and Baltscheffsky
1959	Measurement of muscle NADH fluorescence in vitro	Chance and Jobsis
1962	In vivo monitoring of NADH fluorescence from the brain and kidney	Chance et al.
1965	Comparison between NADH fluorescence in vivo and enzymatic analysis of tissue NADH	Chance et al.
1968	Monitoring tissue reflectance in addition to NADH fluorescence	Jobsis and Stansby
1971	The first attempt to monitor the human brain during a neurosurgical procedure	Jobsis et al.
1973	The first fiber-optic-based fluorometer–reflectometer used in the brain of an awake animal	Chance et al. Mayevsky and Chance
1982	Simultaneous monitoring of NADH in vivo in four different organs in the body	Mayevsky and Chance
1985	Monitoring of brain NADH together with 31P NMR spectroscopy	Mayevsky et al.
1991	Simultaneous real-time monitoring of NADH, CBF, ECoG, and extracellular ions in experimental animals and in the neurosurgical operating room	Mayevsky et al.
1996	The multiparametric response (including NADH) to cortical spreading depression that is for the first time measured in a comatose patient	Mayevsky et al.
2000	Development of the FDA-approved "tissue spectroscope" medical device for real-time monitoring of NADH and tissue blood flow	Mayevsky et al.
2006	Monitoring of tissue vitality (NADH, TBF, and HbO_2) by a new "CritiView" device	Mayevsky et al.

TABLE 6.2 Metabolic States of Mitochondria In Vitro and Associated Oxidation–Reduction Levels of Respiratory Enzymes

State	$[O_2]$	ADP Level	Substrate Level	Respiration Rate	Rate-Limiting Substance	NADH (%)
1	>0	Low	Low	Slow	ADP	~90
2	>0	High	~0	Slow	Substrate	0
3	>0	High	High	Fast	Respiratory chain	53
4	>0	Low	High	Slow	ADP	99
5	0	High	High	0	Oxygen	~100

NAD^+ (53%–63%). Availability of the substrate also led to the ADP depletion in the suspension through oxidative phosphorylation in State 4 with further reduction of NAD^+ to the extent exceeding 99%. Utilization of ADP resulted in increased oxygen consumption and—very soon after the substrate addition—its depletion in the cuvette holding the mitochondrial suspension. At this period, defined as State 5, NADH was fully reduced.

It is important to note that the metabolic states of mitochondria in vivo are different in their levels of NADH. Accurate determination of the metabolic state of tissues in vivo requires the ability of changing various factors, as has been done by Chance and Williams (1955). In our in vivo experiments, the brain of the awaken rat was exposed to various conditions causing an oxidation or reduction of NADH (Mayevsky and Rogatsky 2007). We aimed to establish the range between the maximum increase and decrease in NADH level, as compared to the normoxic level. Our earlier studies (Mayevsky 1976) described the responses of the brain to an uncoupler, pentachlorophenol (PCP), injected into the lateral ventricle. In order to be certain that the NADH reduction was in fact due to PCP effect, the animal was exposed to an N_2 cycle every few minutes. After each N_2 cycle, which caused a large increase in NADH autofluorescence, the oxidation cycle was recorded as well. The results showed that PCP injection increases the range between maximal and minimal levels of NADH.

6.3 Methods

6.3.1 Principles and Technological Aspects of In Vivo NADH Monitoring

Two main approaches were historically applied in order to monitor NADH autofluorescence (Mayevsky and Rogatsky 2007). The earliest method was based on the measurements of the fluorescence spectrum of NADH (spectral approach). The second approach, established more than 50 years ago by Chance et al. (1962) and dominant for continuous measurements of NADH autofluorescence, relies on measuring the total fluorescence signal integrated into a single intensity using appropriate filters (integrated autofluorescence intensity approach). In addition to the measurement of the fluorescence signal, it is necessary to quantify the changes in tissue reflectance at the excitation wavelength. Tissue reflectance may introduce artifacts in NADH signal quantification and, therefore, affect its biological significance (Chance et al. 1973; Jöbsis et al. 1971; Mayevsky 1984).

As of today, two commercial devices for in vivo NADH monitoring were developed. In order to monitor NADH alone in animal models, a small compact system is available from Prizmatix Ltd. For human and animal studies, a multiparametric monitoring system was developed by CritiSense Ltd. This device is not available yet in the market. Monitoring of NADH redox state was performed by several types of fluorometers that were developed in various laboratories around the world. We highlight in the following two major types of NADH fluorometers.

6.3.2 Direct Current Fluorometer/Reflectometer

As an example of a typical direct current (DC) fluorometer/reflectometer, we present the one used in Mayevsky (1984), Mayevsky and Chance (1982), and Mayevsky et al. (1992). This device includes a metal halide or mercury arc lamps as light source and a Y-shaped light guide (e.g., fiber bundle made of different numbers of fibers and diameter). Appropriate filters and photomultipliers (RCA 931B, Hamamatsu) are used to detect the reflectance and fluorescence signals (Figure 6.4a). Excitation light from the lamp passes through a 366 nm filter toward the tissue via a bundle of one arm of the Y-shaped light guide. The emitted light from the tissue is directed to the fluorometer through the second arm of the light guide and is then split at a 90:10 ratio with 90% of the light passing through a 450 nm filter and used as the fluorescence signal. The remaining 10% passes through the 366 nm filter and is used for the light reflectance measurements. This 90:10 ratio is empirical and provides adequate fluorescence and reflectance signals (Mayevsky et al. 1988). The fluorescence signal needs to be corrected by subtracting the reflectance signal from the fluorescence signal at a 1:1 ratio.

6.3.3 Time-Sharing Fluorometer/Reflectometer

The time-sharing fluorometer/reflectometer (TSFR) enables simultaneous monitoring of mitochondrial NADH redox and microcirculatory Hb oxygenation (Figure 6.4b). In this setup, NADH is monitored using the fluorometric technique described earlier, while two-wavelength reflectance allows for blood oxygenation measurements by comparing the reflectance of oxygenated (HbO_2) and deoxygenated Hb—an approach introduced by Rampil et al. (1992). Although the absorption curves of HbO_2 and Hb are different for the most of the spectra (Prahl 1999), they overlap at some wavelengths (called isosbestic points), where the molar extinction coefficients of both Hb species are the same (Figure 6.4c). Reflectance at an isosbestic wavelength is affected only by blood volume and light scattering. At non-isosbestic wavelengths, however, reflectance depends also on the HbO_2/dHb ratio. When subtracting the reflectance signals at these two wavelengths, the difference is considered as a qualitative representation of the Hb oxygenation levels in small blood vessels.

In these measurements, a mercury arc lamp can be used as the light source. Four sets of excitation/emission filter pairs determine the wavelengths for the measurement of (1) NADH fluorescence (366/450 nm), (2) Hb volumes at an isosbestic point (reflectance, 585/585 nm), (3) Hb oxygenation level (reflectance, 577/577 nm), and (4) correction of the NADH signal (reflectance, 366/366 nm) (Rampil et al. 1992). Filter sets are placed on the circumference of a wheel that rotates at approximately 3000 rpm. The light emitted from the tissue at four wavelengths is transferred to a photodetector, and the digitized signal is then recorded and stored in a special computerized system for further data analysis (LabVIEW A/D software, National Instruments). A laser Doppler flowmeter can also be coupled to the TSFR for measuring the microcirculatory blood flow. The optical fibers of the two devices can be constructed in a single probe, as seen in Figure 6.4b. This combination of laser Doppler flowmeter and TSFR enabled simultaneous monitoring of three parameters (Figure 6.4c).

6.3.4 Preparation of Various Organs for Measurements

In order to measure NADH in vivo, it is necessary to have a constant, tight contact between the tip of the fiber-optic probe and the tissue under investigation. Since NADH autofluorescence signal provides relative units rather than absolute values, any disconnection between the probe and the tissue will require a new calibration after the relocation of the probe to a different region of the tissue. It is important to note that this is the crucial problem in any type of optical monitoring of blood-perfused organs due to the continuous changes in blood volume in response to pathophysiological events. In order to overcome these technical difficulties, we developed various protocols for NADH autofluorescence monitoring in various organs (Figure 6.5).

The simplest way to ensure tight contact between the probe and the tissue for brain investigation is shown in Figure 6.5a (Mayevsky and Chance 1975). In this protocol, a hole of an appropriate size is drilled in the bone and a light guide holder is fixed epidurally by dental acrylic cement

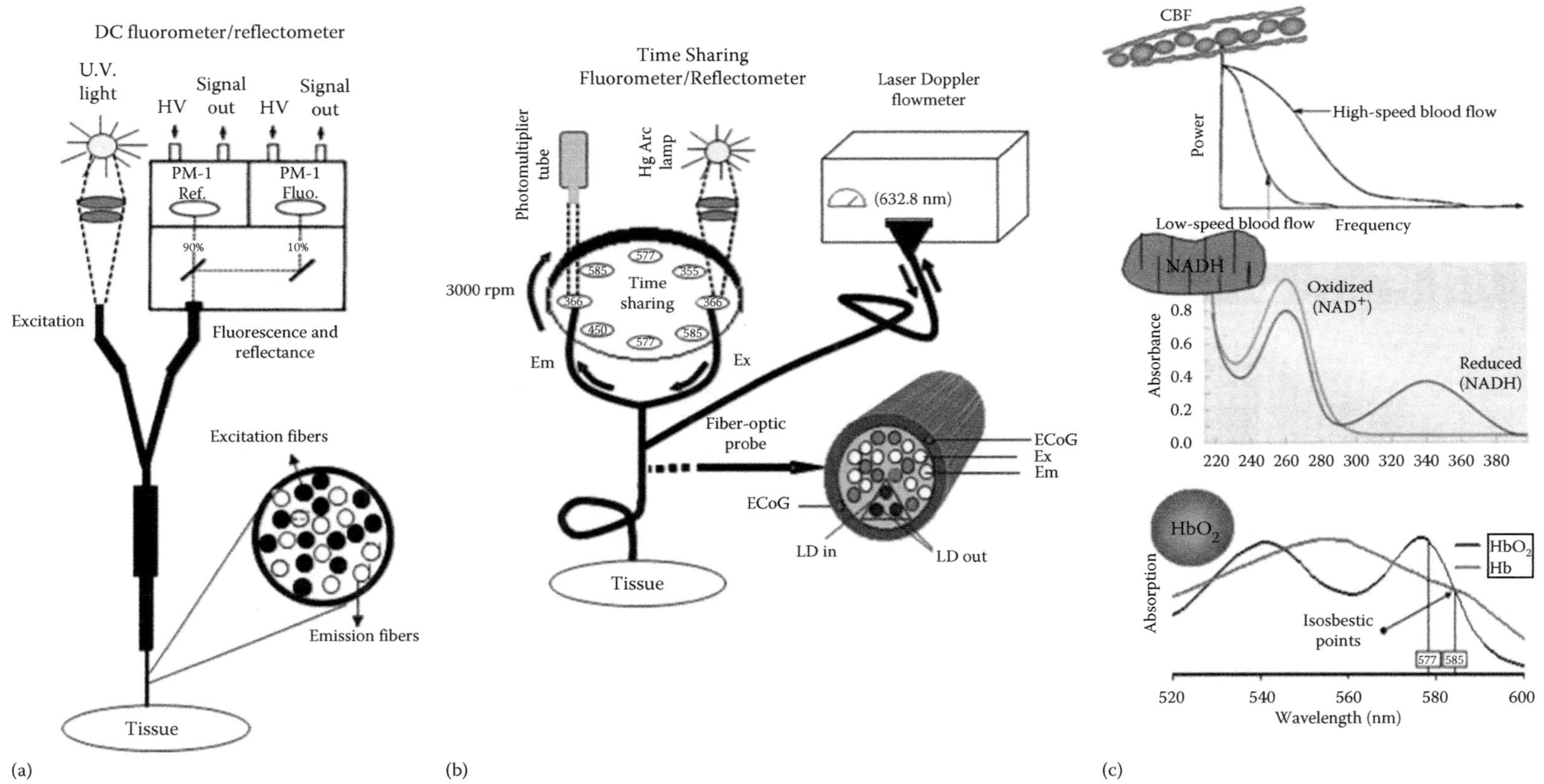

FIGURE 6.4 (a) Schematic presentation of one unit of the DC fluorometer/reflectometer device for NADH fluorometry monitoring. HV, high voltage; PM, photomultiplier. (b) The time-sharing fluorometer–reflectometer device including the laser Doppler flowmeter and electrodes for electrocorticography (ECoG). Excitation and emission fibers for NADH and HbO_2 monitoring, respectively; LD in- and out-optical fibers for blood flow monitoring. The numbers in the spinning disk refer to the wavelength filters. The tissue is connected to the monitoring system via a flexible fiber-optic probe. (c) The principles of monitoring the three parameters CBF, NADH, and HbO_2 in the time-sharing device.

Probe to tissue connection

Connection model	Connection model figure	Tissue/organ
A. Cementation	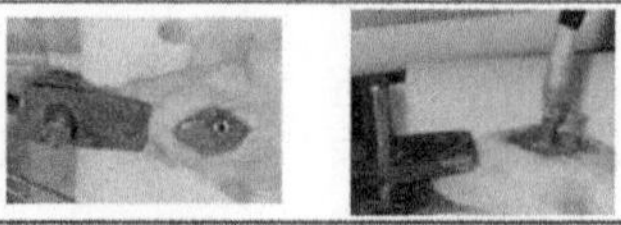	Brain
B. Adhesion	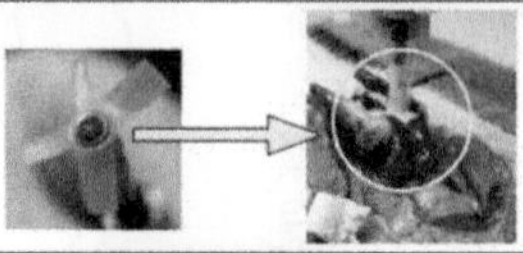	Soft tissue and visceral organs
C. Suturing	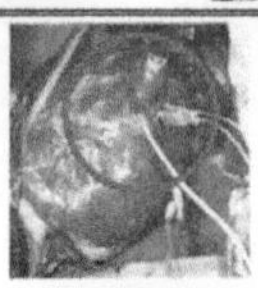	Heart and muscle
D. Micromanipulator		Spinal cord

FIGURE 6.5 Various models for connecting the monitoring probe to the tissue and their application in the tissues and organs.

(Figure 6.5a, left). The fiber-optic probe is introduced into the holder to a preset depth and fixed with screws (Figure 6.5a, right).

The second approach is to use tissue adhesive (cyanoacrylate adhesive) at the fixation point of the probe on the tissue (Figure 6.5b). This approach is suitable for all soft tissues in the body, such as kidneys, small intestine, or liver. In this protocol, a small piece of parafilm paper is winded around the tip of the probe, leaving parafilm remainders that will be glued later on to the tissue (as seen in Figure 6.5b, left). In cases when the tissue under investigation is moving (e.g., beating heart), it is necessary to suture a light guide holder to the muscle tissue (Figure 6.5c). This will allow the contact between the fiber and the tissue to remain stable during data acquisition. We used this approach when a dog heart was monitored using an open-chest model (Kedem et al. 1981). For the spinal cord monitoring, a micromanipulator can be fixed to the animal operating table in order to hold the tip of the optical fiber in contact with the surface of the exposed spinal cord (Simonovich et al. 2008).

6.3.5 Calibration of the Monitored Signals

In order to monitor NADH in vivo in blood-perfused organs, it is necessary to record both fluorescence and reflectance. As mentioned earlier, the corrected NADH level is calculated by the subtraction of the reflectance changes from the fluorescence signal. It is important to note that the correction technology is dependent also on the system used. Therefore, any user that builds a new monitoring system should test and optimize the correction technology for the specific device. In our laboratory, we used the same type of fluorometers since 1972 until 2008. The principle of the calibration procedure was published in several papers (Osbakken and Mayevsky 1996; Osbakken et al. 1989).

6.4 Pitfalls and Remedies

Besides the redox state, other factors also affect the excitation and emission spectra of NADH and may be considered artifacts in fluorescence measurements. Since most fluorometers involve the measurement of total backscattered light at the excitation wavelength, we discuss the artifacts in NADH autofluorescence detection as well as in tissue reflectance recording.

The following factors may affect the two measured signals, namely, reflectance, measured at 366 nm, and fluorescence, excited at 450 nm:

1. Tissue movement
2. Vascular and intravascular events, such as changes in Hb oxygenation level and blood volume due to the autoregulatory vasoconstriction under pathological conditions
3. Extracellular space events, such as volume changes or ion shifts between intra- and extracellular spaces
4. Intracellular space factors, such as O_2 level, ATP turnover rate, substrate availability, and mitochondrial redox state

During the past 40 years, we have used fiber-optic surface fluorometry to monitor organs such as the brain and heart, which were exposed to various physiological conditions. Also, a good correlation was found between mitochondrial NADH and other physiological parameters monitored simultaneously (Mayevsky and Rogatsky 2007).

In the following, we briefly discuss the artifacts and their influence on the in vivo NADH measurements.

6.4.1 Movement-Based Artifacts

In order to obtain a good signal-to-noise ratio and perform reliable and reproducible measurements, good contact between the fibers and the tissue is required during the entire data acquisition period. Pressure on the tissue must be avoided, by using a special light guide holder connected to the tissue as shown in Figure 6.5a and described in Section 6.3.4. In brain measurements, even while the animal was undergoing hyperbaric convulsions or decapitation, only minor changes due to movement-based artifacts were observed, with a negligible effect on NADH autofluorescence. The same approach was applied previously to rats or cats using the "Ultrapac" optics for brain investigation (Jöbsis et al. 1971).

In order to avoid movement-based artifacts in monitoring human patients, we used various approaches. In the neurosurgical intensive care unit, for example, we used a metal holder that was screwed to the skull of comatose patients (Mayevsky et al. 1996). In the operating room, we used a floating light guide probe that was fixed to the head holder in neurosurgical procedures (Mayevsky et al. 1991, 1998, 1999) or a ring used to hold retractors during abdominal operations or kidney transplantations (Mayevsky et al. 2003).

6.4.2 Vascular Events

Vascular events include changes in both blood volume in the microcirculation and blood oxygenation (namely, in the saturation level of HbO_2). Vasodilatation of blood vessels will increase blood volume, while vasoconstriction will decrease blood volume (Figure 6.6a). These changes are the main source of artifacts in monitoring NADH autofluorescence in tissues (Mayevsky and Rogatsky 2007). Since Hb absorbs light strongly at various wavelengths, measurements of NADH autofluorescence are also affected by the level of hemoglobin oxygenation (Figure 6.6b). The influence of blood oxygenation transitions was found to be negligible: a decrease in absorption is compensated by an increase in the transmission of emitted fluorescence (Chance et al. 1973; Mayevsky 1992).

At monitoring a fluorochemically perfused brain preparation (Mayevsky et al. 1981), only negligible changes in reflectance were observed during the anoxic cycle. As shown in Figure 6.6a1, the uncorrected NADH autofluorescence and the corrected fluorescence (CF) had similar kinetics due to a stable reflectance trace. Furthermore, the CF response of the perfused brain to anoxia was similar to that of the

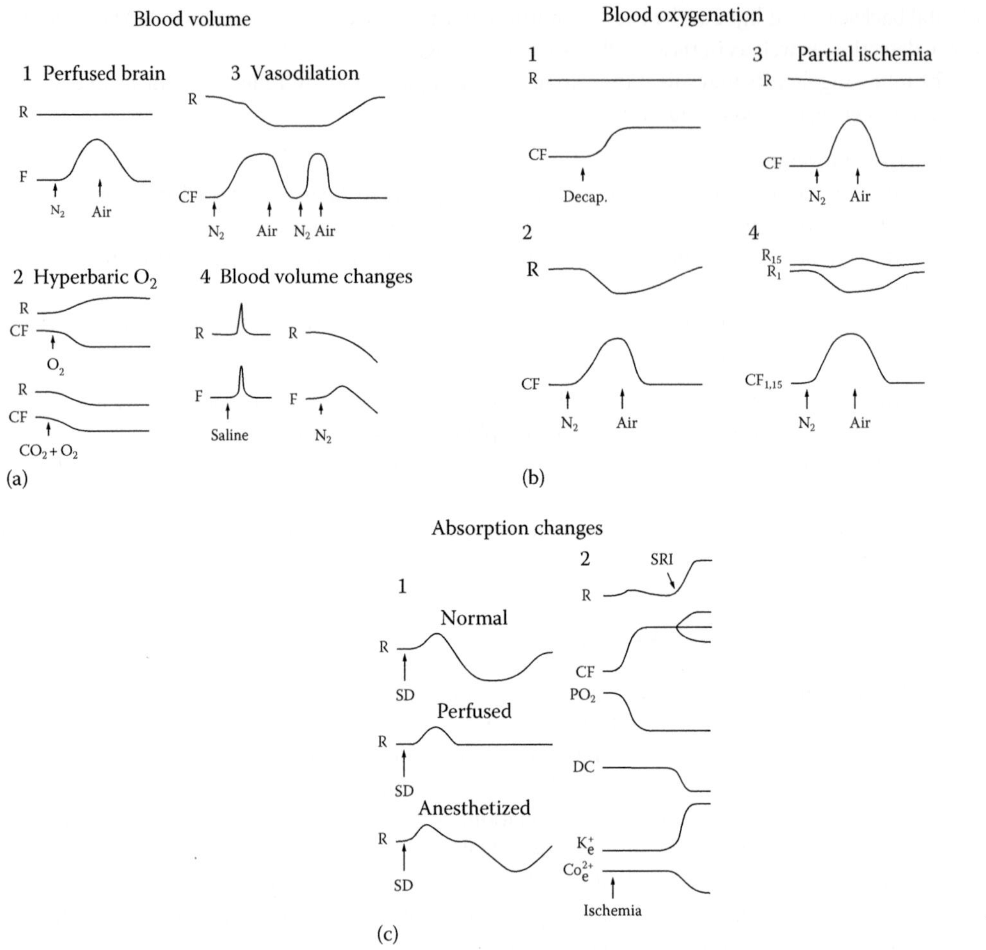

FIGURE 6.6 (a) The effect of changes in blood volume, evaluated by the reflection, on NADH autofluorescence following anoxia (1, 3), hyperbaric oxygenation (2), and saline injection followed by anoxia (4). (b) The effect of blood oxygenation on the reflectance and NADH autofluorescence following decapitation (1) and anoxia (2–4) recorded with the standard DC fluorometer–reflectometer. (c) Effects of spreading depression (1) and ischemia (2) on the response measured from the brain. SD, spreading depression; SRI, secondary reflectance increase. (From Mayevsky, A. and Rogatsky, G.G., *Am. J. Physiol. Cell Physiol.*, 292(2), C615, 2007.)

blood-perfused brain in the same animal before the initiation of perfusion, as shown in Figure 6.6a3. The conclusion is that the blood in the brain affected the fluorescence signal but the reflectance correction approach provided reliable results.

6.4.3 Intra- and Extracellular Space Events

Another potential source of artifacts in NADH autofluorescence measurement is the change in the absorption properties of the tissue at the observation site during various physiological perturbations, such as local ischemia, hypoxia, or spreading depression in the brain. The effect of such fluctuations has mainly been recognized in brain studies.

Very little is known about this factor due to the inability to separate it from other artifacts affecting the recording of NADH fluorescence. For example, when severe ischemia or anoxia is induced for a short time interval, the initial response of the fluorescence and reflectance will provide clear results. But if the event is longer, the reflectance change will be dramatically increasing due to shift of potassium ions to the extracellular space, as seen in Figure 6.6c2. Under these conditions, the correction technique for the fluorescence does not always work properly, and data will be distorted. It seems to us that under physiological or pathological conditions that involve both ions and water movement between the intracellular and the extracellular spaces, this factor may have a greater effect on NADH autofluorescence measurements.

6.5 Applications

This section contains typical results that demonstrate the methodological aspects described in the previous sections.

6.5.1 Effects of Anoxia on Brain

The effect of oxygen deprivation, shown in Figure 6.7, was recorded from the brain of an anesthetized rat. In this animal, NADH was monitored in four sites on the right hemisphere. Four light guides connected to the four-channel DC fluorometer/reflectometer were used for this purpose. Brain exposure to

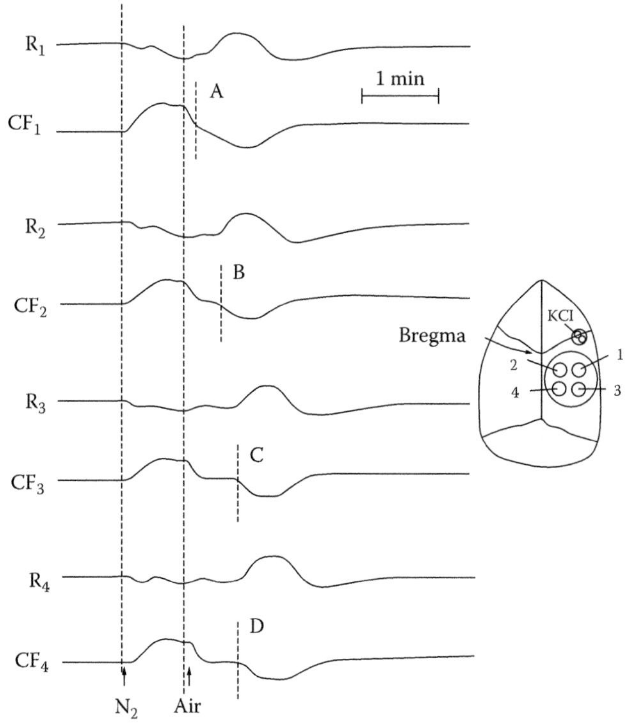

FIGURE 6.7 Metabolic responses of a gerbil brain to anoxia as monitored in four different locations in one hemisphere. The schematic drawing shows the location of the four light guides on the surface of the parietal cortex. The letters A–D mark the point in which spreading depression was observed, demonstrating the propagation of the wave through the hemisphere, as observed also by the increasing distance between the time in which recovery from the anoxic episode was achieved and the point in which SD started. (From Mayevsky, A. and Chance, B., *Science*, 217(4559), 537, 1982.)

anoxia (ventilation with nitrogen) led to a very similar increase in corrected NADH signal at all four points. The NADH reached the baseline immediately after the exposure to air. A wave of cortical spreading depression was developed and led to an oxidation of NADH. This wave propagated from site 1 to site 4, as seen in the location of lines A to D (Mayevsky and Chance 1982).

Application of the TSFR is demonstrated in Figure 6.8. In this system, the laser Doppler flowmeter was added in order to monitor cerebral blood flow (CBF) (Meirovithz et al. 2007). Exposure of the rat to anoxia, similar to the previous experiment, led to a reduction of the HbO_2 and thereby reduced the supply of oxygen to the brain, as revealed by the increase in NADH signal. The microcirculatory blood flow started with a small decrease, followed by a large increase—hyperemia. The reflectance at 366 nm decreased due to a large increase in blood volume induced by the lack of oxygen. The results presented in Figure 6.8 are an average from a group of rats. Statistical analysis indicated that immediately after 100% N_2 inhalation, CBF, reflectance, and HbO_2 levels decreased significantly ($p < 0.01$, $p < 0.001$, and $p < 0.001$, respectively), while mitochondrial NADH significantly increased by about 35% ($p < 0.001$). The maximum decrease in HbO_2 and increase in NADH during the anoxia period were calculated from all experimental groups and were 25.14% ± 1.39% and 42.66% ± 1.12%, respectively. Within 1 min after returning to spontaneous air breathing led to a significant ($p < 0.001$) increase (about 75%) in CBF with further gradual return to the basal levels. Reflectance, NADH redox state, and HbO_2 fully recovered within a few minutes (Meirovithz et al. 2007).

6.5.2 Effects of Adrenaline on the Small Intestine and the Brain

Influence of adrenaline was measured by simultaneous recording from brain and small intestine with two channels of a DC fluorometer/reflectometer and two channels of a laser Doppler flowmeter (Figure 6.9). The mean arterial pressure (MAP) was measured from the tail artery. Adrenaline administration led to large increase in blood pressure due to the sympathetic stimulation and vasoconstriction of the microcirculation in the small intestine, as indicated by the large decrease in intestinal tissue blood flow (TBF) with concurrent increase of blood flow in the brain due to the sparing effect. The change in the perfusion of both organs was also indicated by the NADH responses. In the brain, the NADH level slightly decreased, while the mitochondrial function in the small intestine was inhibited. The corresponding NADH level, measured by its autofluorescence, exhibited a significant increase (Tolmasov et al. 2007).

6.5.3 Effect of Oxygen Deprivation on the Heart

The response of a dog heart to the lack of O_2 was monitored in vivo by a fiber-optic DC fluorometer–reflectometer. The dog was exposed to a short (Figure 6.10a) and long (Figure 6.10b) anoxia period. In both anoxic episodes, the NADH level was elevated significantly. In the second longer anoxia (Figure 6.10b), the heart went into fibrillation and the dog died.

The effect of hypoxia in the same dog model is shown in Figure 6.10c. The dog was artificially ventilated during the entire experiment. Hypoxia was induced by lowering the oxygen level in the respiration mixture. Decrease in inspired oxygen led to an increase in NADH redox state.

The same type of dog model was used in studying the effects of local ischemia (Osbakken and Mayevsky 1996). The left anterior descending artery was isolated for later occlusion. In Figure 6.10d, the coronary artery occlusion led to a dramatic drop in blood flow and a very large increase in NADH, while the change in the reflectance signal was very small. The changes in blood flow measured by a laser Doppler flowmeter were typical for an ischemic event (Osbakken and Mayevsky 1996).

The effect of hypopnea in the dog heart model is shown in Figure 6.10e. Under this condition, the ventilation of the lungs was not optimal and the elevated carbon dioxide in the blood led to an

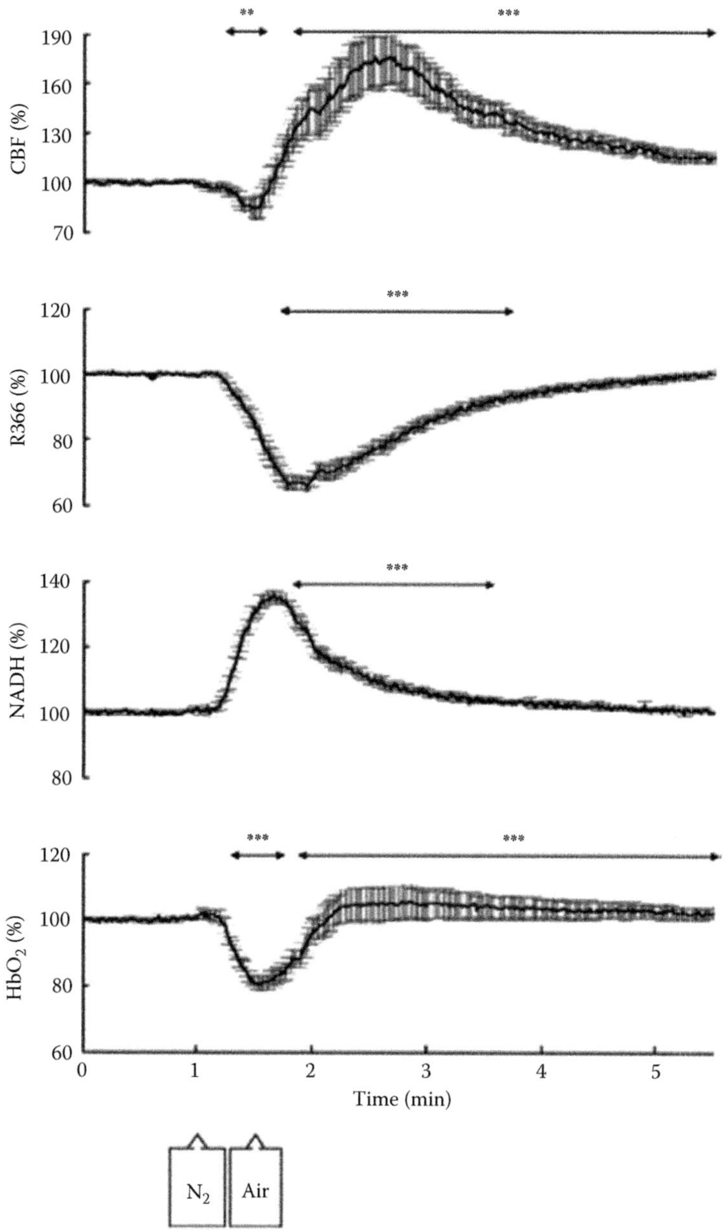

FIGURE 6.8 The effect of anoxia (100% N_2) on CBF, reflectance (R366), mitochondrial NADH redox state (NADH), and hemoglobin oxygen saturation (HbO_2). Values are shown as mean percent values ± SE. $^{**}p < 0.01$, $^{***}p < 0.001$ range of time showing significant differences found by ANOVA repeated measures. Please see Meirovithz et al. (2007).

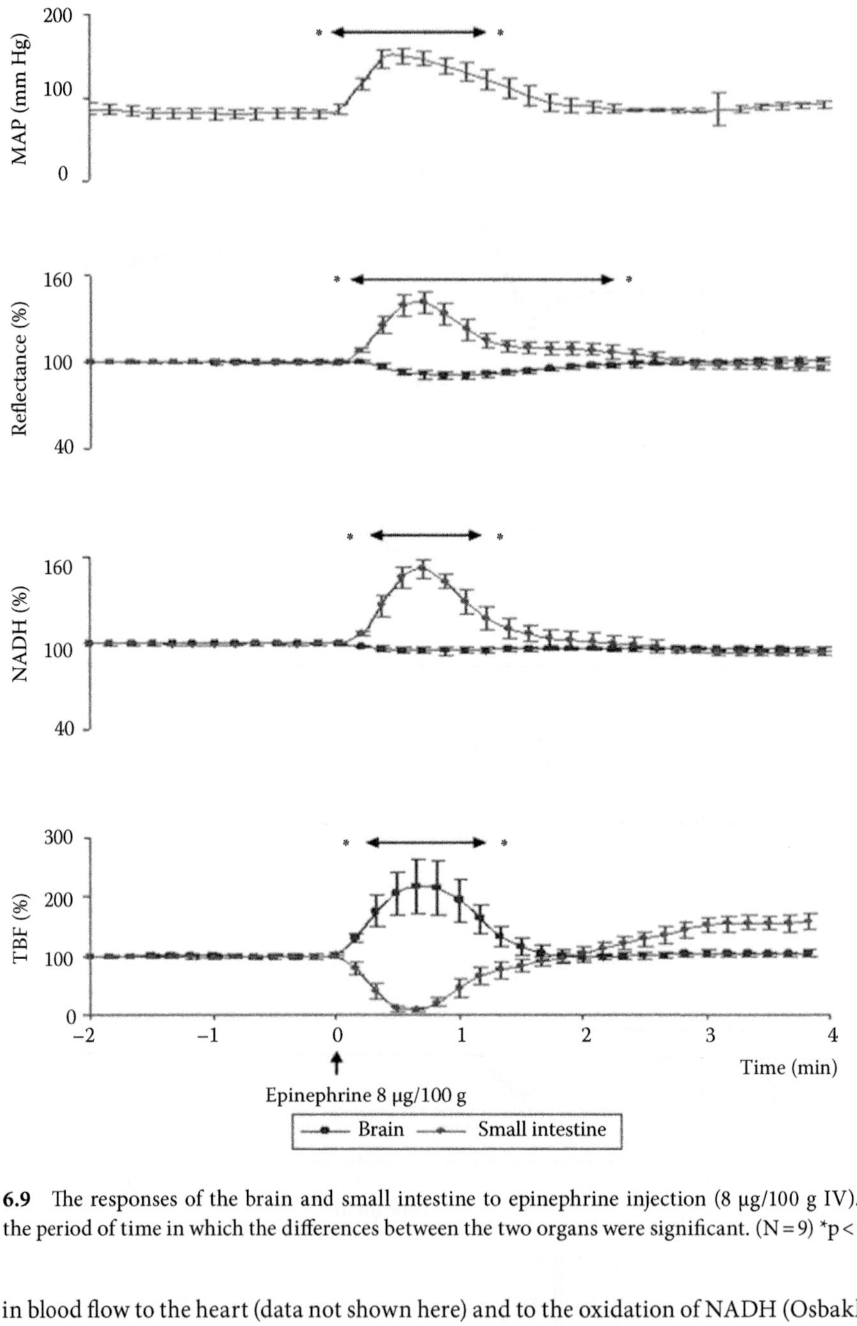

FIGURE 6.9 The responses of the brain and small intestine to epinephrine injection (8 μg/100 g IV). Arrows represent the period of time in which the differences between the two organs were significant. (N = 9) *$p < 0.05$.

increase in blood flow to the heart (data not shown here) and to the oxidation of NADH (Osbakken and Mayevsky 1996).

6.5.4 Responses of the Spinal Cord to Ischemia

Spinal cord ischemia was induced by transient occlusion of the abdominal aorta distal to the left kidney (Figure 6.11). This operation led to a dramatic decrease in spinal cord blood flow to 19.9% ± 6.1% (not shown). This was correlated with a significant rise of the detected signals: by 25% ± 9.3% for tissue

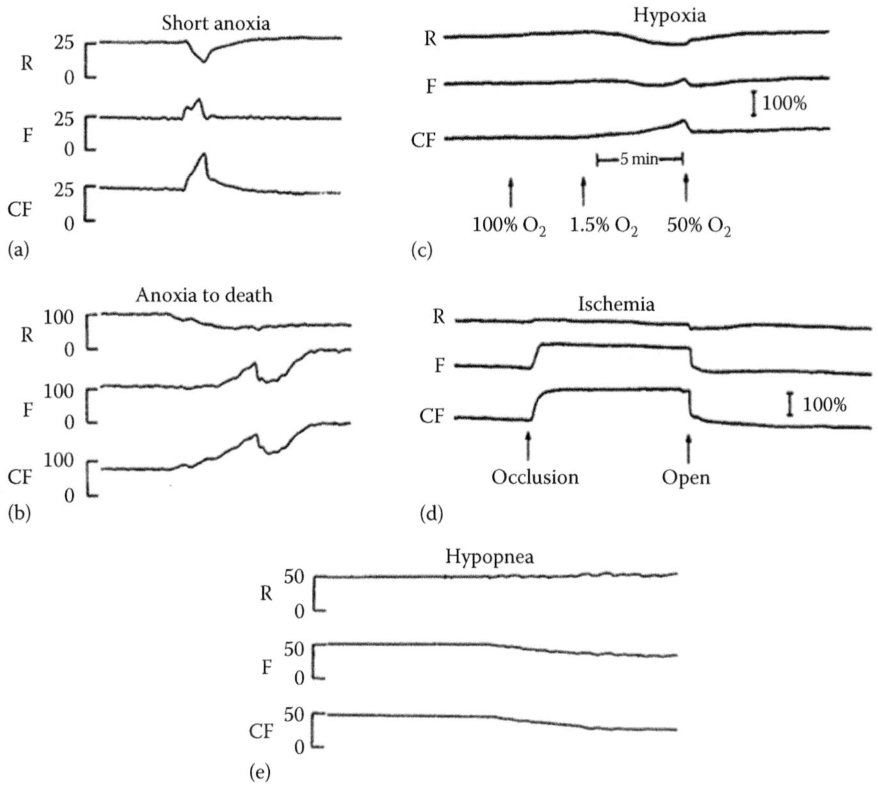

FIGURE 6.10 Responses of the canine heart to various perturbations affecting oxygen supply to the beating heart. Five perturbations are presented including (a and b) short and long anoxia, (c) hypoxia, (d) ischemia, and (e) hypopnea. Please see Osbakken and Mayevsky (1996).

reflectance, by 70% ± 17.2% for tissue autofluorescence, and by 39% ± 11.1% for the corrected NADH levels. These changes were significantly higher than the basal level of each parameter and greater than the control levels ($p < 0.01$).

With the release of the occlusion, the increase in the spinal CBF reached hyperemia levels ($p < 0.01$) within the first 10 min of the reperfusion. The reflectance, fluorescence, and NADH signal also returned gradually to their initial levels. Through the rest of the monitoring period (1.5 h), all parameters stayed stable (Simonovich et al. 2008).

6.5.5 Multiorgan Responses to Hypoxia

Figure 6.12 demonstrates the responses of four different organs to systemic hypoxia that were induced by lowered oxygen level in the air mix supplied for rat ventilation. Four organs were monitored by the multichannel fluorometer/reflectometer: brain, liver, a kidney, and a testis. Levels of oxygen were varied from high (100%) to lower levels of 10% and 5%; pure nitrogen was used to model anoxia. Elevation of NADH levels due to the limited oxygen was proportional to the severity of hypoxia. The responses of all four organs to the systemic hypoxia were similar; however, the amplitude of changes was bigger in the brain, liver, and kidney as compared with the testis, which has low metabolic rate (Mayevsky and Rogatsky 2007).

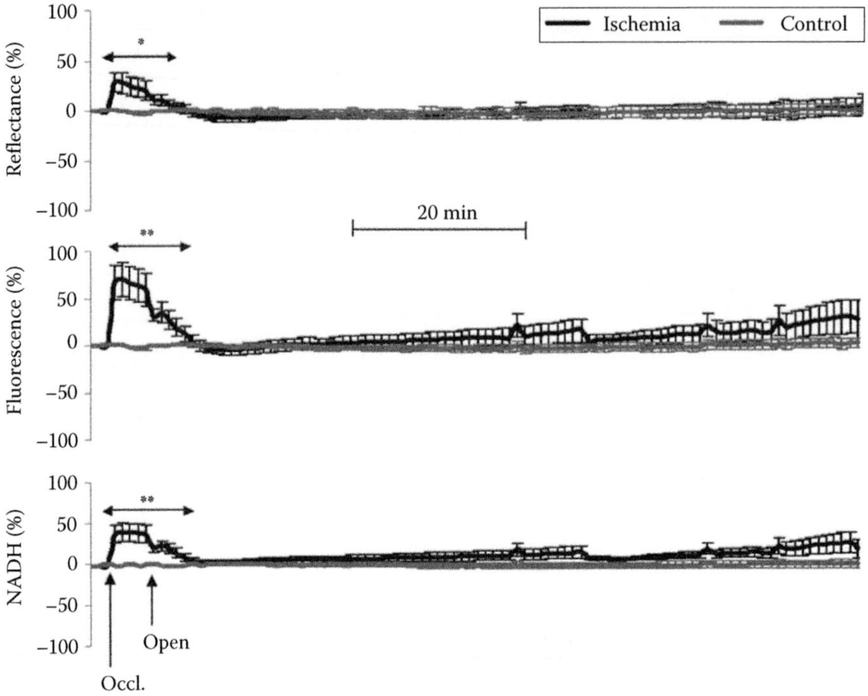

FIGURE 6.11 The responses of the spinal cord to 5 min of ischemia induced by the occlusion of the abdominal aorta (ischemia) versus continuous monitoring (control). Please see Simonovich et al. (2008).

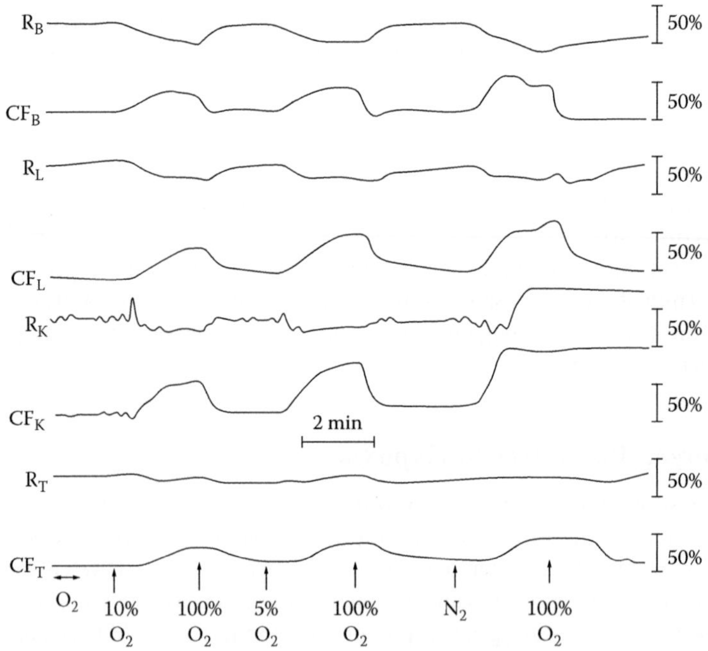

FIGURE 6.12 Typical responses of the tissue reflectance (R) and NADH redox state (CF) simultaneously monitored in four organs to various oxygen concentrations and to anoxia: brain (B), liver (L), kidney (K), and testis (T) (see Mayevsky and Chance 1982).

6.6 Summary

Development of biophotonics in the last 50 years opened up the possibility to monitor energy metabolism of tissues in intact in vivo organs. The main problem in monitoring mitochondrial signals in blood-perfused organs is the hemodynamic artifacts, which require correction in real time for physiological significance.

We found that only NADH autofluorescence could provide relatively a "clean" signal after correcting the hemodynamic-based artifacts. As presented in Figure 6.13, we have applied the multiparametric monitoring technique through the years to various organs and in response to various perturbations under in vivo conditions.

Among those organs that we investigated are brain, spinal cord, kidneys, liver, heart, small intestine, urethra, and testes. The conditions that were tested included models in which oxygen or blood supply is decreased (anoxia, hypoxia, ischemia, and hemorrhage) or increased (hyperoxia and hyperbaria). We also tested the effects of various drugs (norepinephrine [NE], mannitol, etc.) on the various organs. Models of brain activation, such as spreading depression and epilepsy, were also studied. We also run several clinical studies using patients hospitalized in the intensive care units and in operation rooms.

We used the NADH monitoring technology in patients and found results that are basically the same as was found in animal models. In most cases, we combined more physiological parameters to the NADH, and the results showed clinically significance. Two reviews were published recently and the reader can find more details there (Mayevsky and Barbiro-Michaely 2013a,b; Mayevsky et al. 2011).

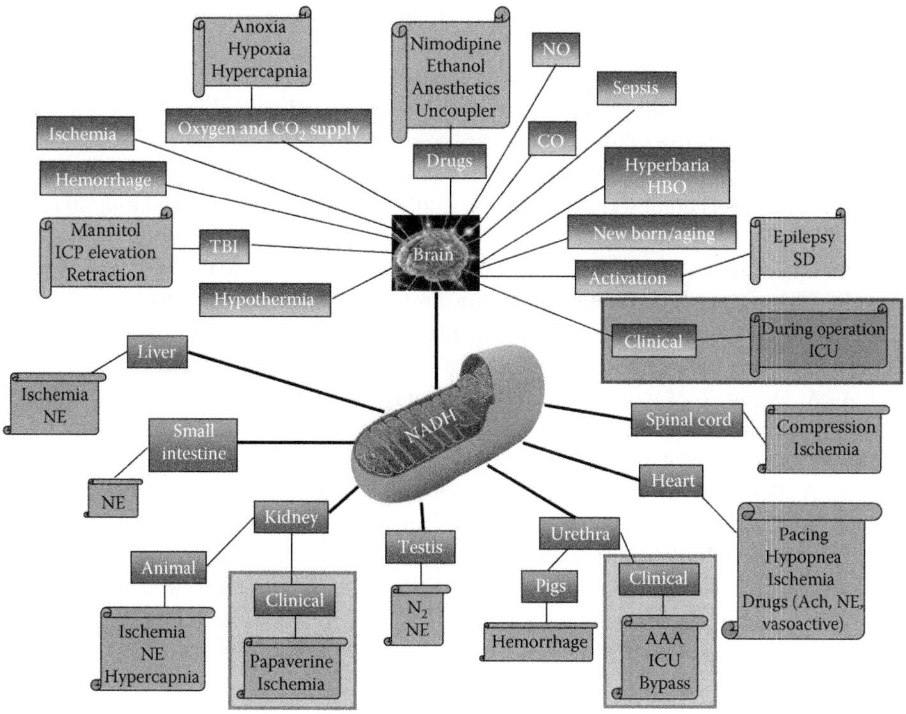

FIGURE 6.13 Summary of 40 years of mitochondrial NADH monitoring in vivo. The monitored organs and perturbations measured by Mayevsky and his collaborators in experimental animals as well as in patients are presented (for details, see Mayevsky and Rogatsky 2007).

References

Chance, B., P. Cohen, F. Jöbsis, and B. Schoener. 1962. Intracellular oxidation-reduction states in vivo the microfluorometry of pyridine nucleotide gives a continuous measurement of the oxidation state. *Science* 137(3529): 499–508.

Chance, B., N. Oshino, T. Sugano, and A. Mayevsky. 1973. Basic principles of tissue oxygen determination from mitochondrial signals. In: *Oxygen Transport to Tissue: Instrumentation, Methods, and Physiology*, H. I. Bicher and D. F. Bruley (eds.), Vol. 37A of Advances in Experimental Medicine and Biology, I. R. Cohen, A. Lajtha, J. D. Lambris, and R. Paoletti (eds.), pp. 277–292. New York: Springer.

Chance, B. and G. R. Williams. 1955. Respiratory enzymes in oxidative phosphorylation. III. The steady state. *Journal of Biological Chemistry* 217(1): 409–427.

Harden, A. and W. J. Young. 1906. The alcoholic ferment of yeast-juice. *Proceedings of the Royal Society of London. Series B, Containing Papers of a Biological Character* 77(519): 405–420.

Jöbsis, F. F., M. O'Connor, A. Vitale, and H. Vreman. 1971. Intracellular redox changes in functioning cerebral cortex. I. Metabolic effects of epileptiform activity. *Journal of Neurophysiology* 34(5): 735–749.

Kedem, J., A. Mayevsky, J. Sonn, and B.-A. Acad. 1981. An experimental approach for evaluation of the O_2 balance in local myocardial regions in vivo. *Quarterly Journal of Experimental Physiology* 66(4): 501–514.

Mayevsky, A. 1976. Brain energy metabolism of the conscious rat exposed to various physiological and pathological situations. *Brain Research* 113(2): 327–338.

Mayevsky, A. 1984. Brain NADH redox state monitored in vivo by fiber optic surface fluorometry. *Brain Research Reviews* 7(1): 49–68.

Mayevsky, A. 1992. Interrelation between intracellular redox state and ion homeostasis in the brain in vivo. In: *Quantitative Spectroscopy in Tissues*, K. Frank and M. Kessler (eds.), pp. 155–168. Frankfurt am Main, Germany: PMI Verlasgruppe.

Mayevsky, A. and E. Barbiro-Michaely. 2013a. Shedding light on mitochondrial function by real time monitoring of NADH fluorescence: I. Basic methodology and animal studies. *Journal of Clinical Monitoring and Computing* 27(1): 1–34.

Mayevsky, A. and E. Barbiro-Michaely. 2013b. Shedding light on mitochondrial function by real time monitoring of NADH fluorescence. II. Human studies. *Journal of Clinical Monitoring and Computing* 27(1): 125–145.

Mayevsky, A. and B. Chance. 1975. Metabolic responses of the awake cerebral cortex to anoxia, hypoxia, spreading depression and epileptiform activity. *Brain Research* 98(1): 149–165.

Mayevsky, A. and B. Chance. 1982. Intracellular oxidation reduction state measured in situ by a multichannel fiber-optic-surface fluorometer. *Science* 217(4559): 537–540.

Mayevsky, A., A. Doron, T. Manor, S. Meilin, N. Zarchin, and G. E. Ouaknine. 1996. Cortical spreading depression recorded from the human brain using a multiparametric monitoring system. *Brain Research* 740(1): 268–274.

Mayevsky, A., A. Doron, S. Meilin, T. Manor, E. Ornstein, and G. E. Ouaknine. 1999. Brain viability and function analyzer: Multiparametric real-time monitoring in neurosurgical patients. In: *Neuromonitoring in Brain Injury*, R. Bullock, A. Marmarou, B. Alessandri, and J. Watson (eds.), Vol. 75 of Acta Neurochirurgica Supplements, H.-J. Steiger (ed.), pp. 63–66. New York: Springer.

Mayevsky, A., E. S. Flamm, W. Pennie, and B. Chance. 1991. Fiber optic based multiprobe system for intraoperative monitoring of brain functions. *Proceedings of SPIE* 1431: 303.

Mayevsky, A., K. Frank, M. Muck, S. Nioka, M. Kessler, and B. Chance. 1992. Multiparametric evaluation of brain functions in the Mongolian gerbil in vivo. *Journal of Basic Clinical Physiology and Pharmacology* 3(4): 323–342.

Mayevsky, A., T. Manor, S. Meilin, A. Doron, and G. E. Ouaknine. 1998. Real-time multiparametric monitoring of the injured human cerebral cortex—A new approach. In: *Intracranial Pressure and Neuromonitoring in Brain Injury*, A. Marmarou et al. (eds.), Vol. 71 of Acta Neurochirurgica Supplements, H.-J. Steiger (ed.), pp. 78–81. New York: Springer.

Mayevsky, A., I. Mizawa, and H. A. Sloviter. 1981. Surface fluorometry and electrical activity of the isolated rat brain perfused with artificial blood. *Neurological Research* 3(4): 307–316.

Mayevsky, A., S. Nioka, and B. Chance. 1988. Fiber optic surface fluorometry/reflectometry and 31 P-NMR for monitoring the intracellular energy state in vivo. In: *Oxygen Transport to Tissue X*, M. Mochizuki (ed.), Vol. 215 of Advances in Experimental Medicine and Biology, I. R. Cohen, A. Lajtha, J. D. Lambris, and R. Paoletti (eds.), pp. 365–374. New York: Springer.

Mayevsky, A. and G. G. Rogatsky. 2007. Mitochondrial function in vivo evaluated by NADH fluorescence: From animal models to human studies. *American Journal of Physiology. Cell Physiology* 292(2): C615–C640.

Mayevsky, A., J. Sonn, M. Luger-Hamer, and R. Nakache. 2003. Real-time assessment of organ vitality during the transplantation procedure. *Transplantation Reviews* 17(2): 96–116.

Mayevsky, A., R. Walden, E. Pewzner et al. 2011. Mitochondrial function and tissue vitality: Bench-to-bedside real-time optical monitoring system. *Journal of Biomedical Optics* 16(6): 067004.

Meirovithz, E., J. Sonn, and A. Mayevsky. 2007. Effect of hyperbaric oxygenation on brain hemodynamics, hemoglobin oxygenation and mitochondrial NADH. *Brain Research Reviews* 54(2): 294–304.

Osbakken, M. and A. Mayevsky. 1996. Multiparameter monitoring and analysis of in vivo ischemic and hypoxic heart. *Journal of Basic and Clinical Physiology and Pharmacology* 7(2): 97–114.

Osbakken, M., A. Mayevsky, I. Ponomarenko, D. Zhang, C. Duska, and B. Chance. 1989. Combined in vivo NADH fluorescence and 31P NMR to evaluate myocardial oxidative phosphorylation. *Journal of Applied Cardiology* 4(5): 305–313.

Prahl, S. 1999. Optical absorption of hemoglobin. http://omlc.ogi.edu/spectra/hemoglobin/. Accessed January 10, 2014.

Rampil, I. J., L. Litt, and A. Mayevsky. 1992. Correlated, simultaneous, multiple-wavelength optical monitoring in vivo of localized cerebrocortical NADH and brain microvessel hemoglobin oxygen saturation. *Journal of Clinical Monitoring* 8(3): 216–225.

Simonovich, M., E. Barbiro-Michaely, and A. Mayevsky. 2008. Real-time monitoring of mitochondrial NADH and microcirculatory blood flow in the spinal cord. *Spine* 33(23): 2495–2502.

Tolmasov, M., E. Barbiro-Michaely, and A. Mayevsky. 2007. Simultaneously multiparametric spectroscopic monitoring of tissue viability in the brain and small intestine. *Proceedings of SPIE* 6434: 64341N.

7

Tryptophan as an Alternative Biomarker for Cellular Energy Metabolism*

Vinod Jyothikumar
University of Virginia

Yuansheng Sun
University of Virginia

Ammasi Periasamy
University of Virginia

7.1 Introduction

The importance of autofluorescence for studying cells and tissues lies mainly in its potential for diagnostic applications (Acuna et al. 2009; Berberan-Santos 2008; Monici 2005) and as a research tool for understanding the underlying mechanisms of molecular interactions and cellular processes under native conditions. As discussed throughout this book, intrinsically fluorescent NADH and flavins are widely utilized as biomarkers for cellular energy metabolism. In this chapter, we show that tryptophan, an essential and least abundant amino acid required to maintain mammalian cell integrity, can also be used as a reporter of cellular metabolic activity.

* This chapter is dedicated to late Professor Robert M. Clegg (1945–2012), one of the pioneers in FLIM–FRET microscopy.

7.2 Amino Acids as Biomarkers

The essential amino acid "L-tryptophan" (Trp) not only serves as a building block for proteins and neurotransmitters, such as serotonin, but is also intricately involved in the regulation of immune responses (Chen and Barkley 1998). In the early 1980s, Pfefferkorn (1984) observed that interferon gamma (IFN-γ), a proinflammatory cytokine that induces Trp degradation, blocks the growth of *Toxoplasma gondii* (Groß and Bohne 2009; Pfefferkorn 1984). Moreover, subsequently, it was suggested that the induction of Trp degradation by inflammatory agents inhibits the growth of pathogens and cancer cells by depriving them of Trp (Taylor and Feng 1991). In the following years, research focused on biostatic Trp depletion as a means of aiding the immune system in fighting infection and neoplasia (Opitz et al. 2007). Alterations in the delicate balance of Trp metabolism were also found among the leading features, underlying the development of such neurodegenerative disorders as Parkinson's, Huntington's, and Alzheimer's diseases (Sas et al. 2007).

As described in detail in Chapter 2, only three amino acids are fluorescent in the range where conventional fluorescence instrumentation can be utilized: phenylalanine, tyrosine, and Trp (Table 7.1). Trp is the least abundant of all fluorescent amino acids: the occurrence frequency of phenylalanine, tyrosine, and Trp is 3.6:3:1, respectively (Brooks et al. 2002). Yet, Trp is the most useful for imaging because of its attractive spectral features—with excitation and emission maxima at 280 and 350 nm, respectively, it has relatively high molar absorptivity (5500 M^{-1} cm^{-1}) and modest quantum yield (Φ) of 0.13 in the 300–400 nm range (Chen 1967). Excited-state fluorescence of Trp decays biexponentially with an average lifetime of 3.03 ns (Figure 7.1). The fluorescence of phenylalanine over the same spectral range suffers from lower Φ(0.024) and molar absorptivity (150 M^{-1} cm^{-1}) (Chen 1967).

Until recently, near-UV spectroscopy and imaging were underexplored, partly because of technical challenges and potential UV-induced damages of live cells and tissues. With the development of novel imaging systems (Urayama et al. 2003), especially two-photon (2P) spectroscopy using supercontinuum lasers (Li et al. 2009), the high sensitivity of Trp fluorescence lifetime to its natural environment (Chen 1967) can be exploited to monitor cellular processes and map cellular organelles.

As outlined throughout this book, concentration, distribution, and redox state of intracellular NADH are important biochemical criteria for many indispensable physiological and pathological events in cellular metabolism (Heikal 2010). As a result, there is a great potential for cellular NADH as a natural biomarker for a range of cellular processes such as apoptosis, redox reactions, and mitochondrial anomalies associated with cancer and neurodegenerative diseases.

Complementing NADH as natural biomarker, both the metabolism and autofluorescence of intrinsic Trp appear to be a promising target for the development of novel therapeutic strategies to treat autoimmune disorders. However, many unresolved questions remain to be elucidated to fully understand this important metabolic pathway and to translate this knowledge into novel therapeutic strategies and diagnostic tools for diseases.

TABLE 7.1 Representative Intrinsic Biomarkers Responsible for Autofluorescence in Cells and Tissues

Fluorophores	Excitation (nm)	Emission (nm)	Lifetime (ns)	References
Phenylalanine	258 (max) 240–270	280 (max)	7.5	McGuinness et al. (2006)
Tyrosine	275 (max) 250–290	300 (max)	2.5	Ashikawa et al. (1982)
Tryptophan	280 (max) 250–310	350 (max)	3.03	Alcala et al. (1987)
NAD(P)H, free	300–380	450–500	0.4	König (2008)
NAD(P)H, protein bound	300–380	450–500	2.0–2.8	König (2008), Koziol et al. (2006)
FAD	420–500	520–570	2.91	
FAD, protein bound	420–500	Weak in 520–570	<0.01	Schweitzer et al. (2007)

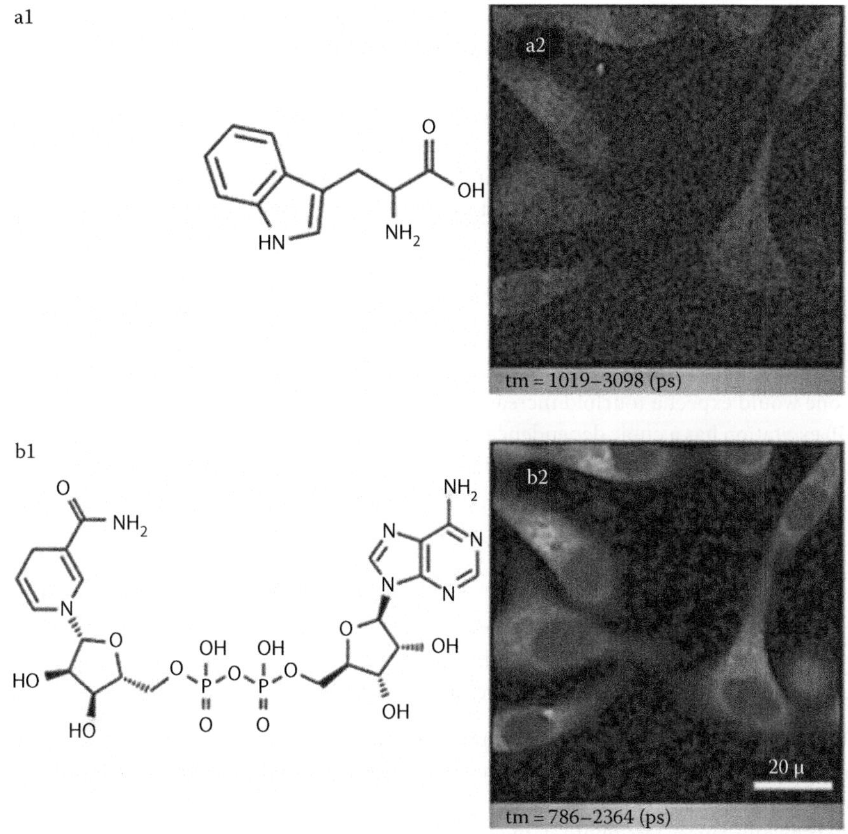

FIGURE 7.1 Chemical structure of Trp and cofactor NADH and their autofluorescence in living cells. (a1) and (b1) show the chemical structures of fluorescent amino acid Trp and cofactor NADH. (a2) and (b2) are their respective fluorescence lifetime images in HeLa cells. Ex. 740 nm.

7.3 Multiphoton Fluorescence Microscopy: Basic Concepts and Advantages for Autofluorescence Imaging in Living Specimens

Two-photon excitation (2PE) was theoretically predicted by Göeppert-Mayer (1931), experimentally proven on a crystal by Kaiser and Garrett in 1961 soon after the invention of lasers (Bayer and Schaack 1970), and first demonstrated on live cells using laser scanning fluorescence microscope by Webb and coworkers in 1990 (Denk et al. 1990). Since then, 2P fluorescence microscopy has been widely used in many areas of the biological and biomedical sciences, including autofluorescence imaging (Denk et al. 1990; Heintzelman et al. 2000; Lakowicz 2009; Li et al. 2009; Pfefferkorn 1984; Taylor and Feng 1991; Urayama et al. 2003; Zipfel et al. 2003).

In a 2P excitation event, two photons, each of which carries approximately half of the total energy that is required for an excitation event, are simultaneously absorbed by a molecule, resulting in the emission of a fluorescent photon from the lowest excited electronic state (Diaspro et al. 2005; Helmchen and Denk 2005; Periasamy and Clegg 2009; Piston 1999; Rubart 2004; So et al. 2000; Svoboda et al. 1997; Svoboda and Yasuda 2006; Venetta 1959; Wang et al. 2010). The probability for the 2P absorption depends on the colocalization of two photons within the absorption cross section of a given fluorophore. Accordingly, the rate of 2P absorption in such a nonlinear optical process is proportional to the square of the instantaneous excitation intensity. Such extremely highly localized photon flux is typically produced by diffraction-limited focusing of femtosecond infrared laser pulses at the specimen plane, which typically corresponds to a 2PE cross section on the order of 10^{-50} to 10^{-49} cm^4 s/photon/molecule. The measured 2P fluorescence intensity depends on the optical detection efficiency (η), the gamma factor (γ) of the laser characteristics (i.e., wavelength, beam waist diameter, pulse width, and repetition rate), square of the average excitation power (P_{av}), and fluorophore's quantum yield (QY), concentration (C), and 2P absorption cross section (σ_{2P}) such that (So et al. 2000; Xu and Webb 1996)

$$F_{2P} = \frac{1}{2}\eta \times \gamma \times QY \times C \times \sigma_{2P} \times P_{av}^2 \tag{7.1}$$

Therefore, by doubling the laser power in the specimen plane while keeping other imaging parameters the same, one would expect a fourfold increase in the detected fluorescence intensity. The rate of three-photon (3P) excitation has a cubic dependence on the instantaneous excitation intensity, where σ_{3P} is the 3P absorption cross section of a given fluorophore (Maiti et al. 1997; Wang and Herman 1996):

$$F_{3P} = \frac{1}{3}\eta \times \gamma \times QY \times C \times \sigma_{3P} \times P_{av}^3 \tag{7.2}$$

Therefore, by tripling the laser power in the specimen plane while keeping other imaging parameters the same, one can expect an eightfold increase in the fluorescence intensity. Compared to single-photon excitation, multiphoton excitation (MPE) offers several advantages in laser scanning fluorescence microscopy:

1. MPE does not require a pinhole for optical sectioning (and therefore descanning) due to the intrinsic diffraction-limited MPE volume (fraction of femtoliter) and thus maximizes the detection efficiency.
2. MPE typically causes less photo damage and photobleaching, since only the fluorophores within the MPE volume are excited. Furthermore, the infrared laser used for MPE is less toxic especially when compared to the deep UV and UV lasers required for the single-photon excitation of Trp and NADH.
3. Because of the longer excitation wavelength, infrared laser beam can penetrate deeper into biological samples and provide unique utilities for imaging tissues and entire living organisms.
4. Fluorescence emission of UV and visible fluorophores can be well separated from the MPE wavelengths by using a short-pass filter. This provides more flexibility in experimental design for MPE emission detection.
5. Due to the MPE nonlinear nature, many fluorophores have broad multiphoton (MP) absorption spectra, and it is therefore possible to select a single MPE wavelength for the simultaneous excitation of multiple-colored fluorophores. Here, we used 740 nm excitation to image both Trp (three-photon excitation [3PE]) and NADH (2PE) in living cells.

7.4 Fluorescence Lifetime Imaging Microscopy and Förster Resonance Energy Transfer: Basic Concepts and Advantages for Investigating Trp–NADH Interactions in Living Specimens

Trp fluorescence is widely used as a tool to monitor changes in proteins and to make inferences regarding local structure and dynamics. In this chapter, we focus on 3P microscopy of intrinsic Trp in live cells. Some advantages of Trp fluorescence over NADH in live cell imaging are as follows:

1. Cells and tissues in apoptotic stages can be detected and isolated.
2. Protein aggregation and granulation within the cell can be studied.
3. Protein synthesis within the cell can be monitored.

The pyridine nucleotides NAD^+ and NADP play vital roles in metabolic conversions as signal transducers and in cellular defense systems. However, intracellular levels of NADH are significantly higher than those of NADPH under physiological conditions. This is because inside the mitochondrial matrix, the tricarboxylic acid (TCA) cycle (see Section 1.3.5) eventually transforms acetyl-CoA, which is derived from glycolytic pyruvate, into carbon dioxide. The TCA cycle produces three NADH molecules and one $FADH_2$, which feed into the respiratory chain. In contrast with yeast, mammals possess two NADP-dependent isocitrate dehydrogenases, a mitochondrial and a cytosolic isoform. Both isoforms are major sources of NADPH supply in mammals. However, these isocitrate dehydrogenase activities only increased during oxidative stress and hypoxia conditions. So, under normal physiological conditions and during glycolysis, we will see more of Trp–NADH interactions than Trp–NADPH.

Fluorescence lifetime is the average time a molecule spends in the excited state before returning to the ground state, typically with the emission of a photon. One of the major applications of fluorescence lifetime imaging microscopy (FLIM) (Chapter 4) is the measurement of the Förster resonance energy transfer (FRET)—nonradiative dipole–dipole interaction between two molecules at which the energy from an excited molecule (the donor) is transferred to another nearby molecule (the acceptor) via a long-range dipole–dipole coupling mechanism (Clegg 1995, 2006; Förster 1946, 1948, 1965). As shown in Equation 7.3, the efficiency of such transfer (E) is dependent on the inverse of the sixth power of the distance (r) separating them. Another parameter that determines E is the distance at which half of the excited-state energy of the donor is transferred to the acceptor (R_0); this characteristic of a given donor–acceptor couple was first described by Theodor Förster in the mid-1940s (Förster 1946, 1948) and named after him the Förster distance. For a given FRET pair, R_0 depends on the orientation factor (κ^2) between the emitting donor and the absorbing acceptor dipoles, the refractive index (n) of the medium, the donor quantum yield (QY_D), the extinction coefficient of the acceptor (ε_A) at its peak absorption wavelength, and the spectral overlap (J) between the donor emission ($f_D(\lambda)$) and the acceptor absorption ($f_A(\lambda)$) spectra (Clegg 1995, 2006; Förster 1946, 1948, 1965):

$$E = \frac{R_0^6}{\left(R_0^6 + r^6\right)} \tag{7.3}$$

$$R_0 = 0.211 \cdot \left\{\kappa^2 \cdot n^{-4} \cdot QY_D \cdot \varepsilon_A \cdot J\right\}^{1/6}, \quad J = \frac{\int_0^\infty f_D(\lambda) f_A(\lambda) \lambda^4 d\lambda}{\int_0^\infty f_D(\lambda) d\lambda} \tag{7.4}$$

Since FRET is usually limited to distances less than ~10 nm, FRET microscopy provides a sensitive tool for investigating a variety of phenomena that produce changes in molecular proximity.

Using FLIM, FRET events can be measured as the reduction in the donor lifetime that results from quenching in the presence of an acceptor. The corresponding energy transfer efficiency (E) can be estimated from the donor lifetimes determined in the absence (τ_D) and the presence (τ_{DA}) of the acceptor:

$$E = 1 - \left(\frac{\tau_{DA}}{\tau_D}\right) \tag{7.5}$$

Compared to steady-state (intensity-based) measurements, FLIM-based FRET offers several advantages:

1. Since only donor signals are measured for determining *E* in FLIM–FRET, this approach does not require corrections for spectral bleed through, which are necessary for intensity-based measurements.
2. Fluorescence lifetime is not sensitive to changes in fluorophore concentration, excitation light intensity, light scattering, and to some extent to photobleaching—the factors that produce artifacts in intensity-based imaging.
3. In addition to the fluorescence lifetime information, time-correlated single-photon counting FLIM (Chapter 4) can also provide the steady-state intensity information at high signal-to-noise ratios through photon counting.

7.4.1 FLIM System Configuration and Data Analysis

The FLIM imaging system, used in the experiments described below, consists of a Bio-Rad Radiance 2100 confocal/MP laser scanning system coupled with Nikon TE300 inverted epifluorescence microscope and controlled with the LaserSharp 2000 software (Carl Zeiss Inc.) (Figure 7.2). An ultrafast (150 fs) tunable 10 W-pumped pulsed Mira 900 laser, operating at 78 MHz and at 740 nm, was used for the MPE of NADH and Trp (Coherent, Inc.). The laser was coupled to the Bio-Rad unit to scan the specimen via an XY raster scanning mechanism using galvo mirrors (Bio-Rad Laboratories, Inc.).

A laser spectrum analyzer (E201, IST Corp.) was used to monitor the MP laser wavelength, and the power was measured at the specimen plane using a power meter (SSIM-VIS and IR, Coherent). A 670UVDCLP dichroic mirror (Chroma Technology) was used to transmit the MP laser beam and reflect the emission light to the external (non-descanned) detection unit in a sealed black box. A mirror in the detection unit routed the emission beam through a short-pass filter to block the infrared light above 650 nm (ET650sp-3P, Chroma Technology) and a filter wheel, holding up to 10 emission filters, to

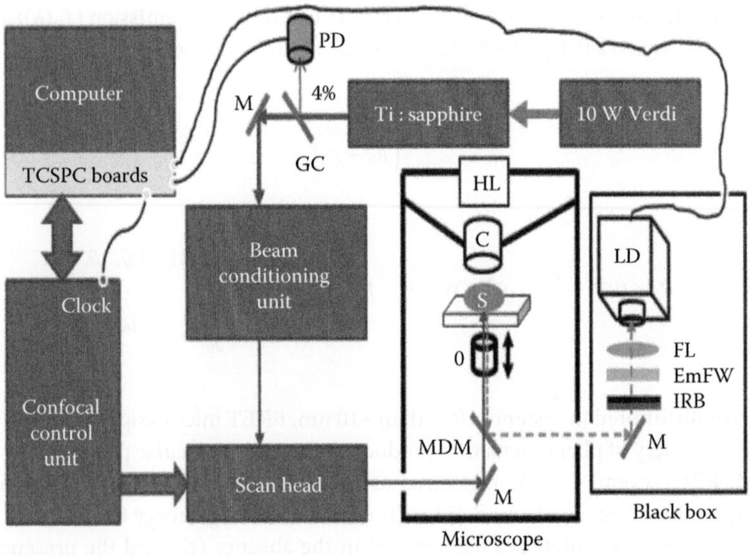

FIGURE 7.2 The basic schematic of the MP TCSPC FLIM system. The excitation source is a 10 W Verdi pumped Ti/sapphire laser, which is coupled to the Bio-Rad Radiance 2100 confocal/MP control unit for scanning the specimen. The laser pulse reference is generated using a GC to reflect 4% of the laser to a PD connected to the TCSPC device. (M, mirror; HL, halogen lamp; C, condenser; S, specimen; O, objective lens; MDM, movable dichroic mirror; IRB, infrared light blocker filter; EmFW, emission filter wheel; FL, focusing lens; LD, lifetime detector.)

a biconvex lens with an antireflective coating for the UV range (Linos G311338322, Qioptiq Photonics GmbH & Co.). The lens focused the light on the bialkali PMH-100-0 detector with the spectral response of 300–600 nm and a full width at half maximum response time of ~150 ps (Becker & Hickl).

The fluorescence decay per pixel was measured using SPC-150 time-correlated single photon counting (TCSPC) board (Becker & Hickl GmbH), and data acquisition was controlled by SPCM software (v. 8.91), set at 256 time bins in each excitation–detection period (39 ps/bin). The TCSPC device synchronizes the lifetime detector to the excitation pulse and the scanning clock and records both the arrival time (relative to the excitation pulse) and the spatial (X, Y, Z) information for each detected photon (Becker 2012). A glass coverslip (GC) reflected ~4% of the MP laser to a photodiode (PD), which generated the reference signal fed to the TCSPC device. Repeated excitation–emission cycles result in a photon count histogram (often termed as a fluorescence decay profile) recorded for each pixel of an image.

A Nikon Plan Fluor 60X/1.2NA WI infrared (IR) objective lens was used to focus the light on the sample and collect the emission (the lens transmission at 340 nm is ~40%). The "donor" Trp and "acceptor" NADH photons were collected using HQ360/40-2p (item no. 230753, Chroma Technologies) and ET480/40-2p (item no. 232292, Chroma Technologies) band-pass filters, respectively. To maintain cell viability, the average excitation power at the specimen plane was kept at 8 mW with the collection time of at least 2 min to collect acceptable photon counts. Photobleaching at this imaging condition was less than 10%, negligible for FLIM analysis.

The decay data were analyzed by the SPCImage software v. 2.8.9 (Becker & Hickl GmbH), which allows single-/multiexponential curve fitting on a pixel-by-pixel basis using a weighted least-squares numerical approach. Cellular NADH lifetime images were analyzed using biexponential model functions to calculate the lifetime components of free and bound NADH and relative contribution each to the decay. Cellular Trp lifetime images were obtained by fitting the data to a single-exponential model function, since double-exponential model did not yield any significant improvement of fitting. Although this could be due to the insufficient photon counts of Trp, the single-exponential fitting results were sufficient in our studies. At the completion of the cellular imaging experiments, the standard solutions were reimaged to ensure that no changes to the instrument had occurred during the experiment (i.e., fluorescence lifetime or intensity). The measured lifetimes of Trp (3.3 ns) and NADH (0.67 ns) in solution were consistent with that reported in literature (Pradhan et al. 1995). The instrument response function (IRF) of the FLIM system was measured using the second harmonic-generated signals from urea crystal and found to be 300 ps. Statistical significance of the data was calculated using unpaired t-test.

7.4.2 Experimental Conditions

7.4.2.1 Standard Solution Preparation

Trp and NADH were obtained from Sigma (item no. T0254 and N4505, respectively) and used without further purification. To prepare 5 mM solution, Trp was dissolved in 25 mM Tris buffer (pH 6.5). NADH was dissolved in Tris buffer (pH 7.4) at 2×10^{-3} M. Solutions were in equilibrium with air. Different concentrations of NADH were measured and the lifetime values were consistent between 0.53 and 0.67 ns.

Representative decays of NADH and Trp in solutions are shown in Figure 7.3. We measured the Trp fluorescence lifetime as a function of its concentration (from solutions buffered at pH 6.5) under conditions identical to live cell imaging and did not observe any significant changes at high concentrations (2.5 mM). Using a spectrofluorometer, we also demonstrated that NADH quenches Trp in solution (see Figure 7.4). The Förster distance (R_0) for the resonance energy transfer (RET) between Trp and NADH was estimated to be 22.66 Å, close to the R_0 value for Trp–NADPH FRET pair (23.4 Å) reported earlier (Torikata et al. 1979a,b).

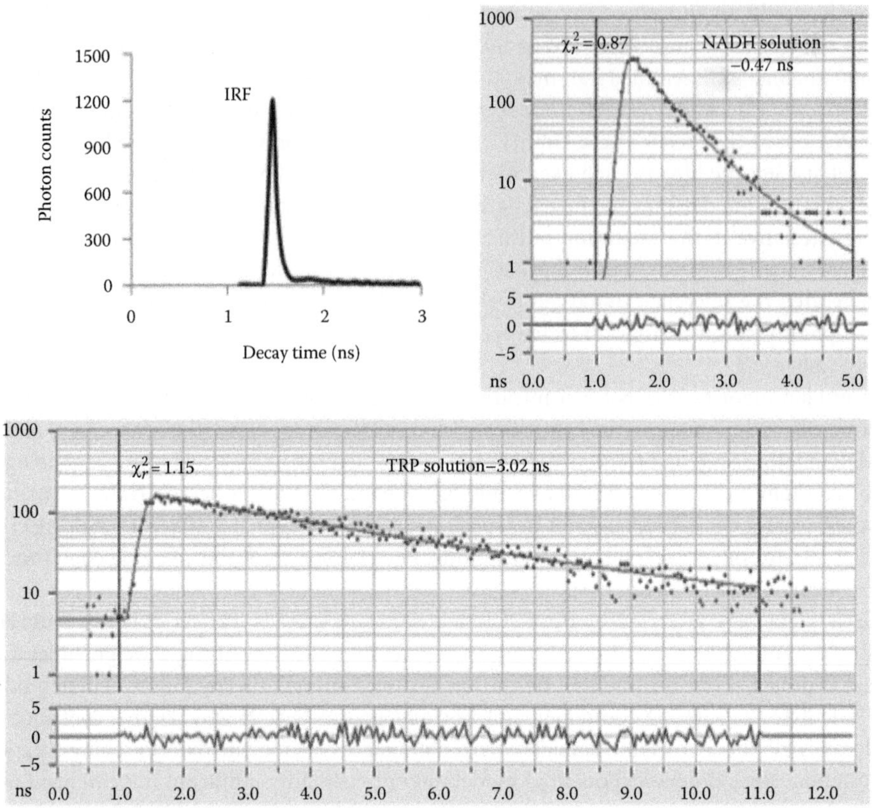

FIGURE 7.3 Representative decays of NADH and Trp in solution. The decay data were analyzed using the IRF, measured from the second harmonic signals of the urea crystal powers. The reduced chi-square (χ_r^2) value and residual indicate the goodness of the fit.

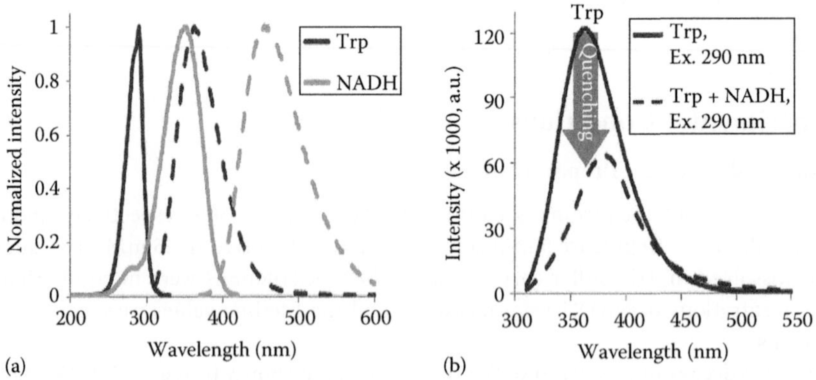

FIGURE 7.4 Trp (Ex. 295 nm, Em. 353 nm)–NADH (Ex. 340 nm, Em. 460 nm) FRET pair (Förster distance, 22.66 Å). The excitation (solid) and emission (dashed) spectra of Trp (0.01 mM) and NADH (0.02 mM) were measured by a spectrofluorometer. (a) The overlap between the Trp emission and NADH excitation spectra is sufficient for FRET. (b) Spectrofluorometer measurements of the mixture solution of 0.1 mM Trp and 0.1 mM NADH clearly shows the Trp quenching (dashed line) in the presence of NADH demonstrating FRET between Trp and NADH.

7.4.2.2 Sample Preparation

Human epithelial breast tissue cell culture MCF10A and human cervical cancer HeLa cells were obtained from the American Type Culture Collection. The cells remained free of mycoplasma and other contaminants and were propagated by adherent culture according to established protocols (Soule et al. 1990). MCF10A cells were grown in Dulbecco's modified Eagle medium (DMEM)-F12 (product no. 11965-118, Life Sciences Corp.) supplemented with 5% horse serum, 20 ng/mL epidermal growth factor, 10 ng/mL insulin (product no. I-1882, Sigma-Aldrich), 1 mg/mL cholera toxin (product no. C-8052, Sigma-Aldrich Co. LLC), and 0.5 ng/mL hydrocortisone (product no. H-0888, Sigma-Aldrich). HeLa cells were suspended in DMEM (product no. 11965-084, Life Technologies) supplemented with 10% fetal bovine serum and 100U penicillin–streptomycin (product no. 15070-063, Life Technologies). Plated cells were stored in an incubator, maintaining 10% CO_2 and 37°C, and were replated every 14–18 h. To prepare the imaging sample, cells were detached from the flasks by trypsinization and washed three times in 10 mL of phosphate buffered saline (PBS). MCF10A and HeLa cells were plated on 25 mm GCs (product no. 12-545-102, Thermo Fisher Scientific), immersed in their standard growth media, and permitted to grow overnight in the incubator. All individual samples were imaged with MP FLIM to complete the FLIM–FRET study.

7.4.2.3 Cellular Treatment and Metabolic Perturbation Study

To increase the metabolism of the MCF10A and HeLa cells, glucose was added to the normal growth medium at concentrations 1, 3, 5, and 7.5 mM and incubated with cells for 30 min. Three independent imaging sessions were conducted with an average nine images per session. A total of 40–50 cells were selected between 7 and 10 different images.

7.4.2.4 Determining the Order of Laser Excitation

The order of excitation was calculated from the fluorescence intensity of the Trp and NADH solutions at different excitation power levels. The intensity (photon counts, *I*) and average laser power levels (mW, *P*), plotted as log(*I*) against log(*P*), clearly show a cubic dependence (slope = 3.1 ± 0.29) for Trp solution and square dependence (slope = 1.89 ± 0.22) for NADH (Figure 7.5). These results confirm that Trp and NADH in pure solutions were 3P and 2P excited, respectively. The demonstration of 2P and 3P process in live cells may not be feasible because of the high laser power at the specimen plane, which can kill the cells.

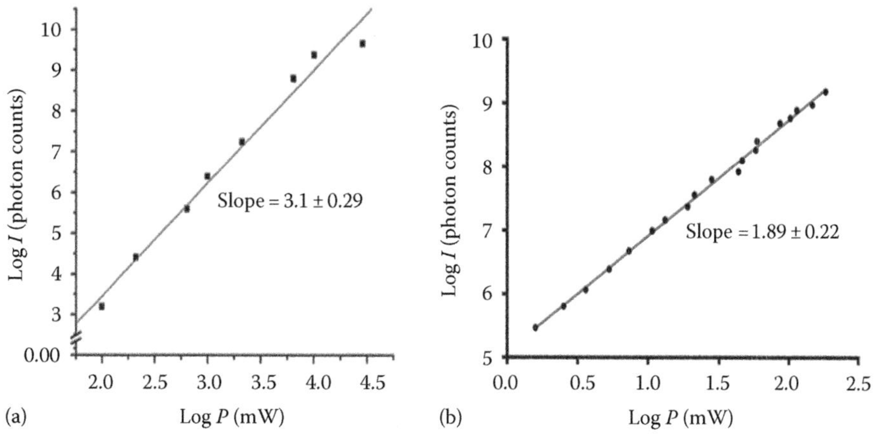

FIGURE 7.5 Identifying the excitation order for Trp and NAD in solution. 3PE of Trp (a) and 2PE of NADH (b) at 740 nm excitation.

7.5 Intracellular Autofluorescence Exhibits a Heterogeneous Fluorescence Lifetime between NADH and Trp in Live Cells

Typical MP laser scanning FLIM images of intracellular NADH (2P) and Trp (3P) in HeLa cells at different glucose concentrations are shown in Figure 7.6. Since the majority of cellular NADH is localized in the mitochondria, autofluorescence 2P-FLIM imaging clearly distinguishes between the cytosol and nucleus. ROIs were selected in both the NADH and Trp FLIM images in order to compare the response of these natural biomarkers to the addition of glucose.

Figure 7.7 shows the average fluorescence lifetime of Trp and protein-bound/free NADH ratio in cytosol–mitochondria and nucleus from cells. The plot clearly illustrates a decrease in the fluorescence

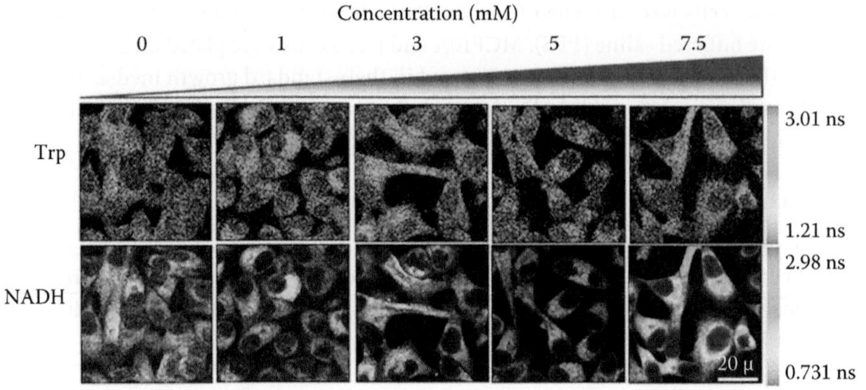

FIGURE 7.6 FLIM images of HeLa cells on gradient glucose concentration. The bottom panel represents FLIM image of 2PE NADH and top panel represents FLIM image of 3PE Trp under 740 nm excitation.

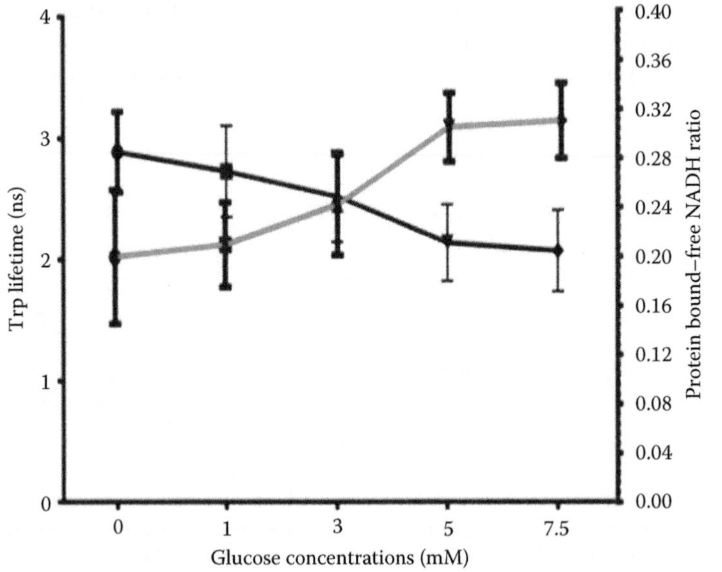

FIGURE 7.7 Effect of glucose perturbation on Trp lifetime and protein-bound/free NADH ratio. The average fluorescence lifetime (with standard deviations) of the Trp fluorescence lifetimes (black line) and the protein-bound/free NADH ratios (gray line) as a function of glucose concentration in live HeLa cells.

lifetime of Trp and an increase in the protein-bound/free NADH ratio at high glucose concentration. Significant reduction of Trp fluorescence lifetime is observed only at the 3 mM glucose threshold, with minimal changes observed over the range of 5–7.5 mM glucose. The average ratio of protein-bound/free NADH in HeLa cells, treated with 7.5 mM glucose, was 0.32 ± 0.06, which is statistically higher than the value of 0.22 ± 0.03 at resting conditions (i.e., no additional supplement of glucose in media). The average Trp lifetime in HeLa cells treated with 7.5 mM glucose was statistically lower 2.1 ± 0.3 than the measured value of 3.0 ± 0.3 at 0 mM glucose.

The result suggests that protein-bound NADH is increased due to oxidative phosphorylation, which markedly reduces the corresponding lifetime of Trp.

Most of the Trp detected by 3P-FLIM of cells is present not in a free form but as a constituent of proteins (Berezin and Achilefu 2010). FLIM images of cellular Trp (Figure 7.6) show broad variations of Trp fluorescence lifetime due to the fluorescence quenching by NADH. Trp fluorescence intensity and lifetime convey the information on protein content and changes in the cellular microenvironment due to metabolic shifts (Ingber 2006; Kilhoffer et al. 1989; Kunwar et al. 2006; Vivian and Callis 2001).

Previous studies have shown that the fluorescence of Trp is quenched by neighboring NADH within protein moiety. The following observation was made by Torikata et al., where the proteins lactate dehydrogenase (LDH) and malate dehydrogenase (MDH) were isolated from pig hearts each containing 5–6 Trp residues in their protein structure (Torikata et al. 1979a,b). Both LDH and MDH are involved in mitochondrial generation of NADH molecules upon the addition of glucose during oxidative phosphorylation. The titration of NADH with proteins like LDH and MDH leads to the quenching of Trp fluorescence and a decreased fluorescence lifetime, which was attributed to energy transfer between Trp and NADH (Torikata et al. 1979a,b). Yet, it is worth noting that many fluorescent Trp molecules in cells do not interact to NADH and therefore may not undergo energy transfer in FRET-based studies. Generally, the Trp fluorescence lifetime is not influenced by protein expression, since it is independent of the fluorophore concentration.

Figure 7.8 shows the estimated energy transfer efficiency, *E* (%), using the cellular Trp lifetime (τ_{DA}) and the fluorescence lifetime of Trp in solution (τ_D). The mean *E* in mitochondria was 8.2 ± 3.7 at no glucose supplement in media as compared with 32.5 ± 5.1 at 7.5 mM of glucose ($p < 0.001$). In the nucleus of HeLa cells, however, the estimated energy transfer efficiency is 7.8 ± 2.3 in the absence of excess glucose as compared with 10.5 ± 4.1 at 7.5 mM of glucose ($p = 0.68$, statistically insignificant difference).

The mean ratio of protein-bound and free NADH in the cytosol–mitochondria of HeLa cells was 0.23 ± 0.03 (at 0 mM glucose) and 0.32 ± 0.06 (at 7.5 mM glucose) ($p < 0.001$). However, the protein-bound

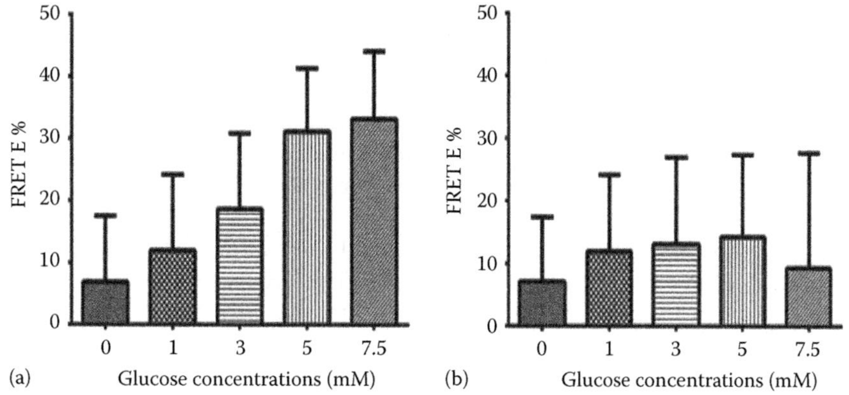

FIGURE 7.8 Effect of glucose perturbation on the energy transfer efficiency, *E* (%), between Trp and NADH. The mean FRET efficiency, *E* (%), as a function of Trp quenching along the gradient glucose concentrations (mM) in HeLa cells between (a) cytosol–mitochondria and (b) nucleus.

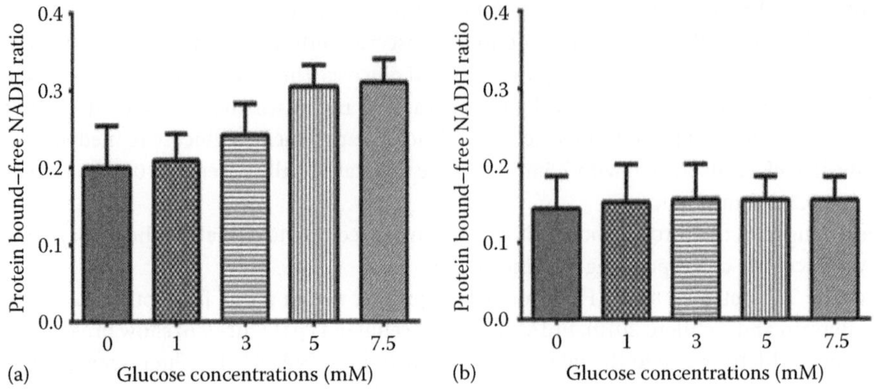

FIGURE 7.9 Effect of glucose perturbation on protein-bound/free NADH ratio in HeLa cells. Mean and standard deviations of the protein-bound/free NADH ratio (A2:A1) along the gradient glucose concentrations (mM) in HeLa cells between (a) cytosol–mitochondria and (b) nucleus.

and free NADH ratios measured in the nuclei, at the same glucose appeared statistically, were indistinguishable (0.21 ± 0.04 at no glucose and 0.22 ± 0.02 at 7.5 mM glucose) (Figure 7.9).

Our results suggest that the fluorescence lifetime of Trp is decreased by NADH at higher bound/free ratios in the cytosol–mitochondria. Unlike solution studies, the fluorescence lifetime of cellular Trp is distributed over a wide range. This indicates that the fluorescence lifetime of Trp may be highly dependent on the metabolic state of the cells and the corresponding NADH levels. Thus, 3P-FLIM–FRET provides an ideal approach for following the Trp quenching pattern at different pathologies, such as cancer, diabetes, and other clinical disease conditions.

7.6 FLIM–FRET Imaging of Trp and NADH for Differentiating Tumorigenic and Nontumorigenic Cells

One of the hallmarks of carcinogenesis is a shift from oxidative phosphorylation to glycolysis for ATP production (Warburg effect) (Bensinger and Christofk 2012; Ferreira 2010; Kim et al. 2009; Upadhyay et al. 2012). Many enzymes bind to NADH in the metabolic pathway, and as favored, metabolic pathways shift with cancer progression; the distribution of NADH binding sites is likely to change. This suggests that changes in metabolism with cancer development can be probed by the lifetime of protein-bound NADH (Bird et al. 2005; Jyothikumar et al. 2013; Skala et al. 2005; Skala and Ramanujam 2010). After establishing that endogenous NADH can serve as the principal acceptor for the excited Trp (as a donor), we now demonstrate how Trp–NADH FRET pair can be used to monitor the metabolic shifts in live cells. In so doing, MP fluorescence lifetime imaging is used to differentiate tumorigenic HeLa cells and nontumorigenic MCF10A cells based on changes in intracellular Trp–NADH fluorescence lifetime (Figure 7.10).

Glucose perturbation study was performed to compare the response of both tumorigenic HeLa and nontumorigenic MCF10A cells using MP FLIM of intrinsic Trp–NADH. Figures 7.11 and 7.12 represent data from 40 to 50 cells with approximately 300 region of interest (ROIs). Preliminary studies indicated that the most dramatic change in Trp lifetime in MCF10A and HeLa cells occurred 30 min after cellular perturbation using 5 mM glucose.

Consequently, we focused our analysis on samples after 30 min of the perturbation studies. We defined the criteria for the selection of ROIs based on the NADH fluorescence lifetime images (intensity > 200 AU) in cells with the presence and absence of glucose. As shown in Figure 7.6, a decrease in Trp fluorescence lifetime was observed during the gradient perturbation studies with glucose in HeLa cells.

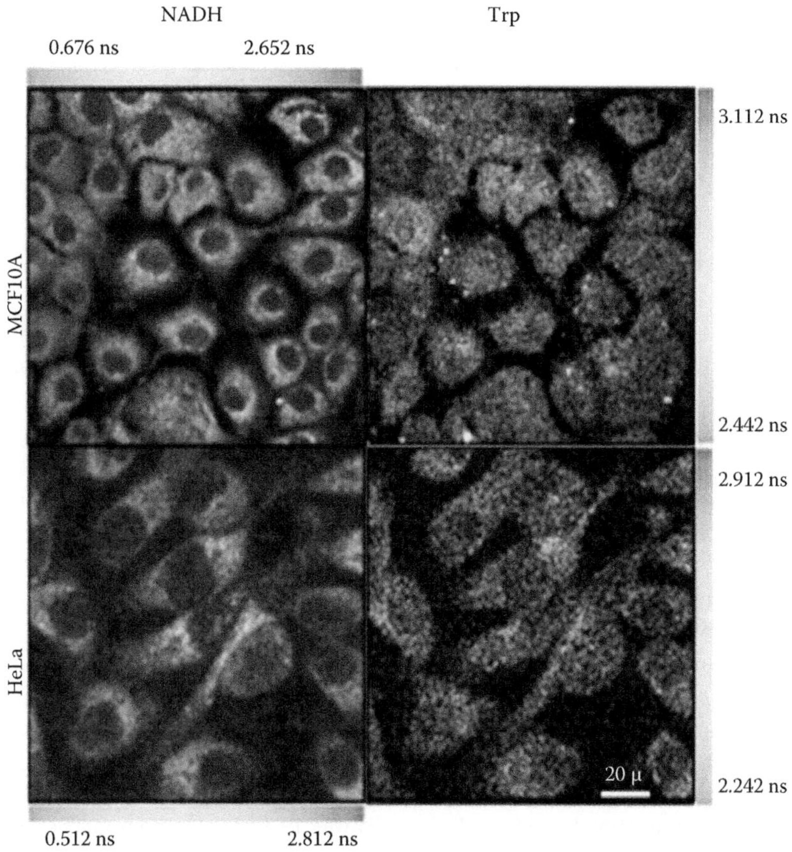

FIGURE 7.10 Fluorescence lifetime images of HeLa and MCF10A cells. Lifetime images from MCF10A and HeLa cells. Trp was imaged using 3PE and NADH in 2PE fluorescence at 740 nm.

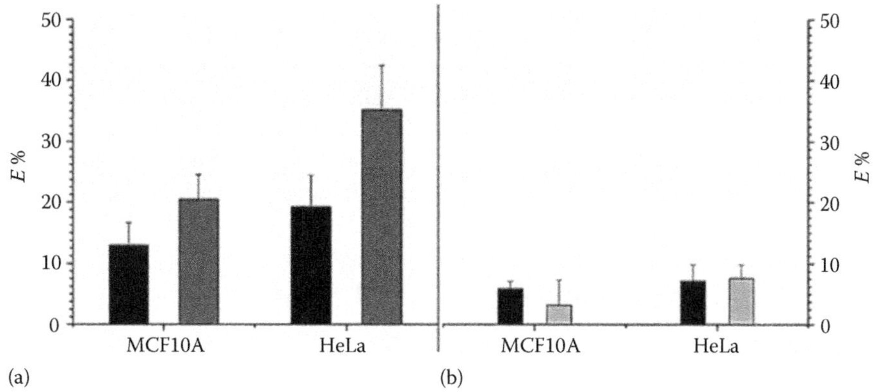

FIGURE 7.11 Comparison of energy transfer efficiency, *E* (%), for Trp–NADH in HeLa and MCF10A cells. Mean and standard deviations of the FRET efficiency, *E* (%), as a function of Trp quenching with (gray) and without (black) glucose in MCF10A and HeLa cells in (a) cytosol–mitochondria and (b) nucleus. (Adapted from Jyothikumar, V. et al., *J. Biomed. Opt.*, 18(6), 060501, 2013.)

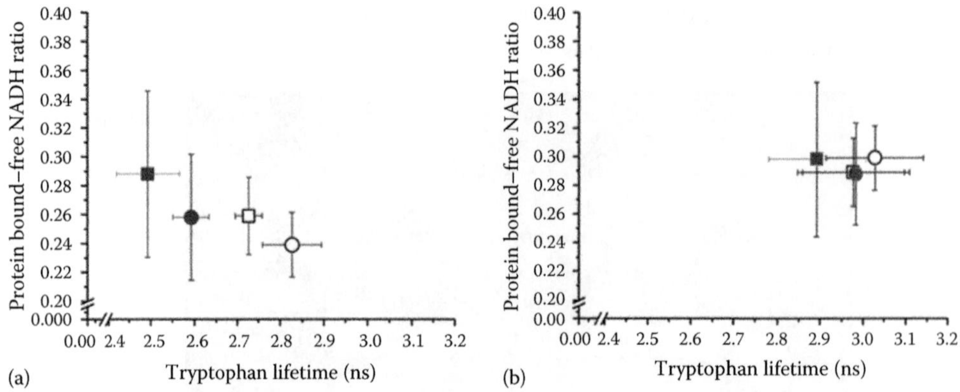

FIGURE 7.12 Correlation between the bound-to-free ratio of NADH and the fluorescence lifetime of Trp in HeLa and MCF10A cells. Mean and standard deviations of protein-bound/free NADH ratio (a2/a1) in relation with Trp quenching in (a) cytosol–mitochondria and (b) nucleus with (gray) and without (white) glucose between MCF10A (○) and HeLa (□) cells. (Adapted from Jyothikumar, V. et al., *J. Biomed. Opt.*, 18(6), 060501, 2013.)

This result suggests that there is an increase in protein-bound NADH levels due to changes in the cellular metabolic state. Our further investigation shows increased quenching of Trp fluorescence in tumorigenic HeLa cells compared to nontumorigenic MCF10A cells (Figure 7.11).

The energy transfer efficiencies, *E* (%), were calculated as described earlier using the fluorescence lifetime of Trp in solution as a reference (τ_D). Figure 7.11 shows the average value of *E* (%) for ROIs selected from cytosol–mitochondria and nucleus of cells as a function of Trp quenching. For MCF10A, the mean cytosol–mitochondria *E* (%) value was 13.0 ± 3.7 (absence of glucose) and 20.5 ± 4.1 (presence of glucose). However, the cytosol–mitochondria *E*% in HeLa cells was 19.3 ± 5.2 (without glucose) and 35.2 ± 7.2 (with glucose) at increased NADH content. Analysis of variance (ANOVA) statistics showed statistically significant differences ($p < 0.05$) of the *E*-values (%) between MCF10A and HeLa cells. The *E*%-values observed with and without glucose in the nuclei of MCF10A (5.9 ± 1.2 and 3.1 ± 1.7) and HeLa (7.2 ± 2.6 and 7.6 ± 2.2) cells were found to be statistically insignificant when compared to the corresponding values observed in cytosol–mitochondria.

Figure 7.12 shows the mean values of the Trp fluorescence lifetimes and the protein-bound/free ratios of NADH for all ROIs combined in cytosol–mitochondria and nucleus, respectively. One-way ANOVA tests between HeLa and MCF10A cells show statistically significant differences ($p < 0.05$) in the Trp fluorescence lifetimes and the ratios of protein-bound and free NADH, both in the presence and absence of glucose. The average ratios of protein-bound/free NADH in HeLa cells treated with glucose were statistically higher than HeLa cells without glucose and MCF10A cells with/without glucose, that is, an increase in the protein-bound form versus free NADH within cells. The result suggests that protein-bound NADH is increased due to glycolysis and oxidative phosphorylation, which markedly quenches Trp lifetime.

7.7 Other Applications for Trp–NADH Studies

7.7.1 Calcium Signaling

Calcium (Ca^{2+}) is the ubiquitous second messenger that plays a crucial role in cerebrovascular processes (i.e., the neurons, astrocytes, and blood vessels in the brain that can regulate different cellular functions and intercellular cross talk) (Berridge et al. 2003; Brenner et al. 2000; Chow et al. 2010). Neurons have a host of Ca^{2+} modulators in the plasma membrane, endoplasmic reticulum (ER), and mitochondria that regulate the intracellular calcium concentration. Simulations of the IP3-mediated pathway of calcium signaling revealed that, in the event of the astrocytes not being connected through gap junctions, intracellular Ca^{2+} concentration $(Ca^{2+})_i$ in individual astrocytes continues to oscillate (Ehrenreich and Schilling 1995; Ghosh and Basu 2012; Höfer et al. 2002; Peers et al. 2004; Salter and Hicks 1995). Delayed coupling of astrocytes tends to dampen these astrocytic Ca^{2+} oscillations and eventually reducing the amplitude of $(Ca^{2+})_i$ to zero. A similar oscillatory response has been suggested for NADH (which reflects mitochondrial activities) (Contreras and Satrústegui 2009; Contreras et al. 2010; D'Andrea and Thorn 1996; Dupont et al. 1990), the key variable for understanding the relationship between Ca^{2+} oscillations and brain activation, established experimentally in mouse hippocampal brain slices (Kann and Kovács 2007). Because cytosolic NADH/NAD^+ ratio can act as a sensor for the coupling of cerebral blood flow (CBF) to the redox state of astrocytes (through lactate–pyruvate cycling in astrocytic mitochondria as a consequence of astrocytic TCA cycle), the NADH/NAD^+ ratio can also serve as a biomarker for the astrocytic energy metabolism (Luciani et al. 2006; Mason et al. 2000; Voronina et al. 2002). Specifically, monitoring the Trp–NADH fluorescence (indicative of the astrocyte's oxidative metabolism) in regions of intense neuronal activation can provide the link between astrocytic metabolism and neuronal activity.

7.7.2 Drug and Ligand Screening

The interactions between lymphohematopoietic cells and drugs have been investigated in relation to the problem of antiblastic drug resistance (Greco et al. 1992; Herner et al. 2013; Rosa et al. 2003). Some classes of antiblastic drugs have fluorogenic properties. Thus, microspectrofluorometry and multispectral imaging autofluorescence microscopy are particularly suitable techniques to perform pharmacokinetic studies on single cells, including uptake, distribution, retention, and efflux of drugs (Adachi et al. 1999; Alberti et al. 1987; Engel et al. 2006; La Schiazza and Bille 2008; Mörtelmaier et al. 2002; Pankow et al. 1995; Spital et al. 1998; Yang et al. 2012). HL60 cells, for example, were treated with doxorubicin in order to investigate the intracellular distribution of the drug and its effects on NADH autofluorescence at the single cell level (Stoya et al. 2002). We propose that the FLIM–FRET method based on the monitoring of the fluorescence lifetime of Trp–NADH as intrinsic probe would be a feasible method for studying biopharmaceutical effects at the intracellular level. With this approach, we will able to identify resistant cells while providing important information for diagnostic, therapeutic, and prognostic evaluations. Ultimately, this would lead to the selection of more effective drugs and substances able to induce the reversion of multidrug resistance.

7.7.3 Diagnostic Applications

Blood cells are among the most studied cells with respect to autofluorescence properties (Monici et al. 1995; Stoya et al. 2002) and the first studies were done on eosinophil granulocytes (Andersson et al. 1998; Li et al. 2010). The fluorescence of their granules is intense enough to be revealed with less sophisticated instrumentation available many years ago. Research on the NADH autofluorescence of cells has been devoted primarily to diagnostic applications (Andersson-Engels et al. 1997; Bader et al. 2011; Bigio and Mourant 1997; Brancaleon et al. 2001; Bremer et al. 2003; Palmer et al. 2010; Pavlova et al. 2012; Sieroń et al. 2008). Studies on the intrinsic fluorescence properties of human white blood cells showed that the leukocyte families differ in both NADH autofluorescence pattern. As a result, it is possible to distinguish one leukocyte family from another using Trp–NADH FLIM–FRET.

7.8 Conclusions and Future Directions

We described in this chapter intrinsic FRET between NADH and Trp in single cells, a FLIM-based FRET approach for autofluorescence monitoring, and potential applications in the biomedical field. Recent advances in laser technology, computational power, and imaging algorithms have extended the application of fluorescence lifetime methods into medical imaging. The lifetime signatures of these biomolecules can be altered by changes in normal cell physiology or under pathological conditions. The use of autofluorescence lifetimes in biological imaging also provides a direct path from cell-based to clinical studies. Because these molecules are often involved in fundamental biological processes, they are significant parameters for checking the metabolic state of cells and tissues. Consequently, analytical techniques based on autofluorescence monitoring have great potential in both basic research and biomedical diagnostics, which explains the growing interest in applying these new analytical tools both in vitro and in vivo applications. The major attraction of autofluorescence-based techniques is to have, in principle, the capability to provide noninvasively and in real time biochemical and morphological information about the native state of the samples without biopsy and harsh preparative treatments needed for standard histological procedures. Numerous applications can be envisioned, from the use for guiding surgical intervention to the use for checking the concentration and effect of drugs at cell and tissue level during pharmacological treatments (e.g., during chemotherapy). Thus, from a theoretical point of view, the autofluorescence-based techniques could give more information with lower costs and lower sampling errors, which often occur in biooptical and preparative procedures.

Acknowledgments

We would like to thank Ms. Kathryn Christopher for the cell preparation and Mr. Horst Wallrabe for his suggestions. We acknowledge funding from National Institutes of Health (PO1HL101871 & OD016446) and University of Virginia.

References

Acuna, A. U., F. Amat-Guerri, P. Morcillo, M. Liras, and B. Rodriguez. 2009. Structure and formation of the fluorescent compound of *Lignum nephriticum*. *Organic Letters* 11 (14): 3020–3023.

Adachi, R., T. Utsui, and K. Furusawa. 1999. Development of the autofluorescence endoscope imaging system. *Diagnostic and Therapeutic Endoscopy* 5 (2): 65–70.

Alberti, S., D. R. Parks, and L. A. Herzenberg. 1987. A single laser method for subtraction of cell autofluorescence in flow cytometry. *Cytometry* 8 (2): 114–119.

Alcala, J. R., E. Gratton, and F. G. Prendergast. 1987. Resolvability of fluorescence lifetime distributions using phase fluorometry. *Biophysical Journal* 51 (4): 587–596.

Andersson, H., T. Baechi, M. Hoechl, and C. Richter. 1998. Autofluorescence of living cells. *Journal of Microscopy* 191 (1): 1–7.

Andersson-Engels, S., C. Klinteberg, K. Svanberg, and S. Svanberg. 1997. In vivo fluorescence imaging for tissue diagnostics. *Physics in Medicine and Biology* 42 (5): 815–824.

Ashikawa, I., Y. Nishimura, M. Tsuboi, K. Watanabe, and I. Koujiro. 1982. Lifetime of tyrosine fluorescence in nucleosome core particles. *Journal of Biochemistry* 91 (6): 2047–2055.

Bader, A. N., A.-M. Pena, C. Johan van Voskuilen et al. 2011. Fast nonlinear spectral microscopy of in vivo human skin. *Biomedical Optics Express* 2 (2): 365–373.

Bayer, E. and G. Schaack. 1970. Two-photon absorption of CaF_2:Eu^{2+}. *Physica Status Solidi* 41 (2): 827–835.

Becker, W. 2012. *The bh TCSPC Handbook*. http://www.beckerhickl.de/literature.htm#handb, accessed June 24, 2014.

Bensinger, S. J. and H. R. Christofk. 2012. New aspects of the Warburg effect in cancer cell biology. *Seminars in Cell and Developmental Biology* 23 (4): 352–361.

Berberan-Santos, M. N., ed. 2008. *Fluorescence of Supermolecules, Polymers, and Nanosystems*. Vol. 4 of Springer Series on Fluorescence. Berlin, Germany: Springer.

Berezin, M. Y. and S. Achilefu. 2010. Fluorescence lifetime measurements and biological imaging. *Chemical Reviews* 110 (5): 2641–2684.

Berridge, M. J., M. D. Bootman, and H. L. Roderick. 2003. Calcium signalling: Dynamics, homeostasis and remodelling. *Nature Reviews Molecular Cell Biology* 4 (7): 517–529.
Bigio, I. J. and J. R. Mourant. 1997. Ultraviolet and visible spectroscopies for tissue diagnostics: Fluorescence spectroscopy and elastic-scattering spectroscopy. *Physics in Medicine and Biology* 42 (5): 803–814.
Bird, D. K., L. Yan, K. M. Vrotsos et al. 2005. Metabolic mapping of MCF10A human breast cells via multi-photon fluorescence lifetime imaging of the coenzyme NADH. *Cancer Research* 65 (19): 8766–8773.
Brancaleon, L., A. J. Durkin, J. H. Tu, G. Menaker, J. D. Fallon, and N. Kollias. 2001. In vivo fluorescence spectroscopy of nonmelanoma skin cancer. *Photochemistry and Photobiology* 73 (2): 178–183.
Bremer, C., V. Ntziachristos, and R. Weissleder. 2003. Optical-based molecular imaging: Contrast agents and potential medical applications. *European Radiology* 13 (2): 231–243.
Brenner, R., G. J. Peréz, A. D. Bonev et al. 2000. Vasoregulation by the β1 subunit of the calcium-activated potassium channel. *Nature* 407 (6806): 870–876.
Brooks, D. J., J. R. Fresco, A. M. Lesk, and M. Singh. 2002. Evolution of amino acid frequencies in proteins over deep time: Inferred order of introduction of amino acids into the genetic code. *Molecular Biology and Evolution* 19 (10): 1645–1655.
Chen, R. F. 1967. Fluorescence quantum yields of tryptophan and tyrosine. *Analytical Letters* 1 (1): 35–42.
Chen, Y. and M. D. Barkley. 1998. Toward understanding tryptophan fluorescence in proteins. *Biochemistry* 37 (28): 9976–9982.
Chow, S.-K., D. Yu, C. L. MacDonald, M. Buibas, and G. A. Silva. 2010. Amyloid beta-peptide directly induces spontaneous calcium transients, delayed intercellular calcium waves and gliosis in rat cortical astrocytes. *ASN Neuro* 2 (1): art:e00026.
Clegg, R. M. 1995. Fluorescence resonance energy transfer. *Current Opinion in Biotechnology* 6 (1): 103–110.
Clegg, R. M. 2006. The history of FRET. In *Reviews in Fluorescence 2006*, C. D. Geddes and J. R. Lakowicz (eds.), pp. 1–45. New York: Springer.
Contreras, L., I. Drago, E. Zampese, and T. Pozzan. 2010. Mitochondria: The calcium connection. *Biochimica et Biophysica Acta (BBA)—Bioenergetics* 1797 (6): 607–618.
Contreras, L. and J. Satrústegui. 2009. Calcium signaling in brain mitochondria interplay of malate aspartate NADH shuttle and calcium uniporter/mitochondrial dehydrogenase pathways. *Journal of Biological Chemistry* 284 (11): 7091–7099.
D'Andrea, P. and P. Thorn. 1996. Ca^{2+} signalling in rat chromaffin cells: Interplay between Ca^{2+} release from intracellular stores and membrane potential. *Cell Calcium* 19 (2): 113–123.
Denk, W., J. H. Strickler, and W. W. Webb. 1990. Two-photon laser scanning fluorescence microscopy. *Science* 248 (4951): 73–76.
Diaspro, A., G. Chirico, and M. Collini. 2005. Two-photon fluorescence excitation and related techniques in biological microscopy. *Quarterly Reviews of Biophysics* 38 (2): 97–166.
Dupont, G., M. J. Berridge, and A. Goldbeter. 1990. Latency correlates with period in a model for signal-induced Ca^{2+} oscillations based on Ca^{2+}-induced Ca^{2+} release. *Cell Regulation* 1 (11): 853–861.
Ehrenreich, H. and L. Schilling. 1995. New developments in the understanding of cerebral vasoregulation and vasospasm: The endothelin-nitric oxide network. *Cleveland Clinic Journal of Medicine* 62 (2): 105–116.
Engel, R., P. J. M. Van Haastert, and A. J. W. G. Visser. 2006. Spectral characterization of *Dictyostelium* autofluorescence. *Microscopy Research and Technique* 69 (3): 168–174.
Ferreira, L. M. R. 2010. Cancer metabolism: The Warburg effect today. *Experimental and Molecular Pathology* 89 (3): 372–380.
Förster, T. 1946. Energy transport and fluorescence [in German]. *Naturwissenschafien* 6: 166–175.
Förster, T. 1948. Zwischenmolekulare energiewanderung und fluoreszenz. *Annalen Der Physik* 437: 55–75.
Förster, T. 1965. Delocalized excitation and excitation transfer. In *Modern Quantum Chemistry. Istanbul Lectures. Part III. Action of Light and Organic Chemistry*, O. Sinanoglu (ed.), pp. 93–137. London, U.K.: Academic Press.

Ghosh, S. and A. Basu. 2012. Calcium signaling in cerebral vasoregulation. In *Calcium Signaling*, Md. Shahidul Islam (ed.). Vol. 740 of Advances in Experimental Medicine and Biology, I. R. Cohen, A. Lajtha, J. D. Lambris, and R. Paoletti (eds.), pp. 833–858. New York: Springer.

Göppert-Mayer, M. 1931. Über elementarakte mit zwei quantensprüngen. *Annalen der Physik* 401 (3): 273–294.

Greco, F., L. De Palma, N. Specchia, S. Jacobelli, and C. Gaggini. 1992. Polymethylmethacrylate-antiblastic drug compounds: An in vitro study assessing the cytotoxic effect in cancer cell lines—A new method for local chemotherapy of bone metastasis. *Orthopedics* 15 (2): 189.

Groß, U. and W. Bohne. 2009. Type II NADH dehydrogenase inhibitor 1-hydroxy-2-dodecyl-4 (1H) quinolone leads to collapse of mitochondrial inner-membrane potential and ATP depletion in *Toxoplasma gondii*. *Eukaryotic Cell* 8 (6): 877–887.

Heikal, A. A. 2010. Intracellular coenzymes as natural biomarkers for metabolic activities and mitochondrial anomalies. *Biomarkers in Medicine* 4 (2): 241–263.

Heintzelman, D. L., R. Lotan, and R. R. Richards-Kortum. 2000. Characterization of the autofluorescence of polymorphonuclear leukocytes, mononuclear leukocytes and cervical epithelial cancer cells for improved spectroscopic discrimination of inflammation from dysplasia. *Photochemistry and Photobiology* 71 (3): 327–332.

Helmchen, F. and W. Denk. 2005. Deep tissue two-photon microscopy. *Nature Methods* 2 (12): 932–940.

Herner, A., I. Nikić, M. Kállay, E. A. Lemke, and P. Kele. 2013. A new family of bioorthogonally applicable fluorogenic labels. *Organic and Biomolecular Chem*istry 11 (20): 3297–3306.

Höfer, T., L. Venance, and C. Giaume. 2002. Control and plasticity of intercellular calcium waves in astrocytes: A modeling approach. *The Journal of Neuroscience* 22 (12): 4850–4859.

Ingber, D. E. 2006. Cellular mechanotransduction: Putting all the pieces together again. *The FASEB Journal* 20 (7): 811–827.

Jyothikumar, V., Y. Sun, and A. Periasamy. 2013. Investigation of tryptophan–NADH interactions in live human cells using three-photon fluorescence lifetime imaging and Förster resonance energy transfer microscopy. *Journal of Biomedical Optics* 18 (6): 060501.

Kann, O. and R. Kovács. 2007. Mitochondria and neuronal activity. *American Journal of Physiology—Cell Physiology* 292 (2): C641–C657.

Kilhoffer, M. C., D. M. Roberts, A. Adibi, D. M. Watterson, and J. Haiech. 1989. Fluorescence characterization of VU-9 calmodulin, an engineered calmodulin with one tryptophan in calcium binding domain III. *Biochemistry* 28 (14): 6086–6092.

Kim, H. H., H. Joo, T. Kim et al. 2009. The mitochondrial Warburg effect: A cancer enigma. *Interdisciplinary Bio Central* 1: 1–7.

König, K. 2008. Clinical multiphoton tomography. *Journal of Biophotonics* 1 (1): 13–23.

Koziol, B., M. Markowicz, J. Kruk, and B. Plytycz. 2006. Riboflavin as a source of autofluorescence in *Eisenia fetida* coelomocytes. *Photochemistry and Photobiology* 82 (2): 570–573.

Kunwar, A., A. Barik, R. Pandey, and K. I. Priyadarsini. 2006. Transport of liposomal and albumin loaded curcumin to living cells: An absorption and fluorescence spectroscopic study. *Biochimica et Biophysica Acta (BBA)—General Subjects* 1760 (10): 1513–1520.

La Schiazza, O. and J. F. Bille. 2008. High-speed two-photon excited autofluorescence imaging of ex vivo human retinal pigment epithelial cells toward age-related macular degeneration diagnostic. *Journal of Biomedical Optics* 13 (6): 064008.

Lakowicz, J. R., ed. 2009. *Principles of Fluorescence Spectroscopy*. New York: Springer.

Li, C., C. Pitsillides, J. M. Runnels, D. Côté, and C. P. Lin. 2010. Multiphoton microscopy of live tissues with ultraviolet autofluorescence. *IEEE Journal of Selected Topics in Quantum Electronics* 16 (3): 516–523.

Li, D., W. Zheng, and J. Y. Qu. 2009. Two-photon autofluorescence microscopy of multicolor excitation. *Optics Letters* 34 (2): 202–204.

Luciani, D. S., S. Misler, and K. S. Polonsky. 2006. Ca^{2+} controls slow NAD(P)H oscillations in glucose-stimulated mouse pancreatic islets. *The Journal of Physiology* 572 (2): 379–392.

Maiti, S., J. B. Shear, R. M. Williams, W. R. Zipfel, and W. W. Webb. 1997. Measuring serotonin distribution in live cells with three-photon excitation. *Science* 275 (5299): 530–532.

Mason, M. J., J. F. Hussain, and M. P. Mahaut-Smith. 2000. A novel role for membrane potential in the modulation of intracellular Ca^{2+} oscillations in rat megakaryocytes. *The Journal of Physiology* 524 (2): 437–446.

McGuinness, C. D., A. M. Macmillan, K. Sagoo, D. McLoskey, and D. J. S. Birch. 2006. Excitation of fluorescence decay using a 265 nm pulsed light-emitting diode: Evidence for aqueous phenylalanine rotamers. *Applied Physics Letters* 89 (6): 063901.

Monici, M. 2005. Cell and tissue autofluorescence research and diagnostic applications. *Biotechnology Annual Review* 11: 227–256.

Monici, M., R. Pratesi, P. A. Bernabei et al. 1995. Natural fluorescence of white blood cells: Spectroscopic and imaging study. *Journal of Photochemistry and Photobiology B: Biology* 30 (1): 29–37.

Mörtelmaier, M., E. J. Kögler, J. Hesse, M. Sonnleitner, L. A. Huber, and G. J. Schütz. 2002. Single molecule microscopy in living cells: Subtraction of autofluorescence based on two color recording. *Single Molecules* 3 (4): 225–231.

Opitz, C. A., W. Wick, L. Steinman, and M. Platten. 2007. Tryptophan degradation in autoimmune diseases. *Cellular and Molecular Life Sciences* 64 (19–20): 2542–2563.

Palmer, G. M., R. J. Boruta, B. L. Viglianti, L. Lan, I. Spasojevic, and M. W. Dewhirst. 2010. Non-invasive monitoring of intra-tumor drug concentration and therapeutic response using optical spectroscopy. *Journal of Controlled Release* 142 (3): 457–464.

Pankow, W., K. Neumann, J. Rüschoff, and P. Von Wichert. 1995. Human alveolar macrophages: Comparison of cell size, autofluorescence, and HLA-DR antigen expression in smokers and non-smokers. *Cancer Detection and Prevention* 19 (3): 268–273.

Pavlova, I., K. R. Hume, S. A. Yazinski et al. 2012. Multiphoton microscopy and microspectroscopy for diagnostics of inflammatory and neoplastic lung. *Journal of Biomedical Optics* 17 (3): 0360141–0360149.

Peers, C., I. F. Smith, J. P. Boyle, and H. A. Pearson. 2004. Remodelling of Ca^{2+} homeostasis in type I cortical astrocytes by hypoxia: Evidence for association with Alzheimer's disease. *Biological Chemistry* 385 (3–4): 285–289.

Periasamy, A. and R. M. Clegg, eds. 2009. *FLIM Microscopy in Biology and Medicine*. Boca Raton, FL: CRC Press.

Pfefferkorn, E. R. 1984. Interferon gamma blocks the growth of *Toxoplasma gondii* in human fibroblasts by inducing the host cells to degrade tryptophan. *Proceedings of the National Academy of Sciences of the United States of America* 81 (3): 908–912.

Piston, D. W. 1999. Imaging living cells and tissues by two-photon excitation microscopy. *Trends in Cell Biology* 9 (2): 66–69.

Pradhan, A., P. Pal, G. Durocher et al. 1995. Steady state and time-resolved fluorescence properties of metastatic and non-metastatic malignant cells from different species. *Journal of Photochemistry and Photobiology B: Biology* 31 (3): 101–112.

Rosa, M. A., G. Maccauro, A. Sgambato, R. Ardito, G. Falcone, V. De Santis, and F. Muratori. 2003. Acrylic cement added with antiblastics in the treatment of bone metastases ultrastructural and in vitro analysis. *Journal of Bone & Joint Surgery* [*Br*] 85 (5): 712–716.

Rubart, M. 2004. Two-photon microscopy of cells and tissue. *Circulation Research* 95 (12): 1154–1166.

Salter, M. W. and J. L. Hicks. 1995. ATP causes release of intracellular Ca^{2+} via the phospholipase C beta/IP3 pathway in astrocytes from the dorsal spinal cord. *The Journal of Neuroscience* 15 (4): 2961–2971.

Sas, K., H. Robotka, J. Toldi, and L. Vécsei. 2007. Mitochondria, metabolic disturbances, oxidative stress and the kynurenine system, with focus on neurodegenerative disorders. *Journal of the Neurological Sciences* 257 (1): 221–239.

Schweitzer, D., S. Schenke, M. Hammer et al. 2007. Towards metabolic mapping of the human retina. *Microscopy Research and Technique* 70 (5): 410–419.

Sieroń, A., A. Kościarz-Grzesiok, J. Waśkowska et al. 2008. The role of autofluorescence diagnostics in the oral mucosa diseases. *Photodiagnosis and Photodynamic Therapy* 5 (3): 182–186.

Skala, M. and N. Ramanujam. 2010. Multiphoton redox ratio imaging for metabolic monitoring in vivo. In *Advanced Protocols in Oxidative Stress II*, D. Armstrong (ed.). Vol. 594 of Methods in Molecular Biology, J. M. Walker (ed.), pp. 155–162. New York: Humana Press.

Skala, M. C., J. M. Squirrell, K. M. Vrotsos et al. 2005. Multiphoton microscopy of endogenous fluorescence differentiates normal, precancerous, and cancerous squamous epithelial tissues. *Cancer Research* 65 (4): 1180–1186.

So, P. T. C., C. Y. Dong, B. R. Masters, and K. M. Berland. 2000. Two-photon excitation fluorescence microscopy. *Annual Review of Biomedical Engineering* 2 (1): 399–429.

Soule, H. D., T. M. Maloney, S. R. Wolman et al. 1990. Isolation and characterization of a spontaneously immortalized human breast epithelial cell line, MCF-10. *Cancer Research* 50 (18): 6075–6086.

Spital, G., M. Radermacher, C. Müller, G. Brumm, A. Lommatzsch, and D. Pauleikhoff. 1998. Autofluorescence characteristics of lipofuscin components in different forms of late senile macular degeneration [in German]. *Klinische Monatsblätter für Augenheilkunde* 213 (1): 23.

Stoya, G., A. Klemm, E. Baumann, H. Vogelsang, U. Ott, W. Linss, and G. Stein. 2002. Determination of autofluorescence of red blood cells (RbCs) in uremic patients as a marker of oxidative damage. *Clinical Nephrology* 58 (3): 198–204.

Svoboda, K., W. Denk, D. Kleinfeld, and D. W. Tank. 1997. In vivo dendritic calcium dynamics in neocortical pyramidal neurons. *Nature* 385 (6612): 161–165.

Svoboda, K. and R. Yasuda. 2006. Principles of two-photon excitation microscopy and its applications to neuroscience. *Neuron* 50 (6): 823–839.

Taylor, M. W. and G. S. Feng. 1991. Relationship between interferon-gamma, indoleamine 2, 3-dioxygenase, and tryptophan catabolism. *The FASEB Journal* 5 (11): 2516–2522.

Torikata, T., L. S. Forster, R. E. Johnson, and J. A. Rupley. 1979a. Lifetimes and NADH quenching of tryptophan fluorescence in pig heart cytoplasmic malate dehydrogenase. *Journal of Biological Chemistry* 254 (9): 3516–3520.

Torikata, T., L. S. Forster, C. C. O'Neal Jr., and J. A. Rupley. 1979b. Lifetimes and NADH quenching of tryptophan fluorescence in pig heart lactate dehydrogenase. *Biochemistry* 18 (2): 385–390.

Upadhyay, M., J. Samal, M. Kandpal, O. Vir Singh, and P. Vivekanandan. 2012. The Warburg effect: Insights from the past decade. *Pharmacology & Therapeutics* 137 (3): 318–330.

Urayama, P., W. Zhong, J. A. Beamish et al. 2003. A UV-visible-NIR fluorescence lifetime imaging microscope for laser-based biological sensing with picosecond resolution. *Applied Physics B* 76 (5): 483–496.

Venetta, B. D. 1959. Microscope phase fluorometer for determining the fluorescence lifetimes of fluorochromes. *Review of Scientific Instruments* 30 (6): 450–457.

Vivian, J. T. and P. R. Callis. 2001. Mechanisms of tryptophan fluorescence shifts in proteins. *Biophysical Journal* 80 (5): 2093–2109.

Voronina, S., T. Sukhomlin, P. R. Johnson, G. Erdemli, O. H. Petersen, and A. Tepikin. 2002. Correlation of NADH and Ca^{2+} signals in mouse pancreatic acinar cells. *The Journal of Physiology* 539 (1): 41–52.

Wang, B.-G., K. König, and K.-J. Halbhuber. 2010. Two-photon microscopy of deep intravital tissues and its merits in clinical research. *Journal of Microscopy* 238 (1): 1–20.

Wang, X. F. and B. Herman, eds. 1996. *Fluorescence Imaging Spectroscopy and Microscopy*. New York: Wiley.

Xu, C. and W. W. Webb. 1996. Measurement of two-photon excitation cross sections of molecular fluorophores with data from 690 to 1050 nm. *Journal of the Optical Society of America B* 13 (3): 481–491.

Yang, L., Y. Zhou, S. Zhu, T. Huang, L. Wu, and X. Yan. 2012. Detection and quantification of bacterial autofluorescence at the single-cell level by a laboratory-built high-sensitivity flow cytometer. *Analytical Chemistry* 84 (3): 1526–1532.

Zipfel, W. R., R. M. Williams, R. Christie, A. Y. Nikitin, B. T. Hyman, and W. W. Webb. 2003. Live tissue intrinsic emission microscopy using multiphoton-excited native fluorescence and second harmonic generation. *Proceedings of the National Academy of Sciences of the United States of America* 100 (12): 7075–7080.

8

Alternative Approaches to Optical Sensing of the Redox State

Yi Yang
East China University of Science and Technology

8.1 Introduction

Oxidation–reduction (redox) reactions are of extreme importance for all organisms from bacteria to mammals, regulating cellular processes and providing cells with energy and building blocks. In order to maintain various cellular functions, redox statuses in a cell are kept balanced by a set of mechanisms, in which pyridine nucleotides and thiol redox buffer systems are the key players. Pyridine nucleotides, reduced and oxidized forms of both nicotinamide adenine dinucleotide (NADH/NAD^+) and its phosphate (NADPH/$NADP^+$), are essential in the redox reactions of various metabolites. NAD^+ receives electrons to form NADH during the oxidation of various nutrients, such as sugars and fats; NADH is then oxidized by the electron transfer chain associated with adenosine triphosphate (ATP) production (see Chapter 1). NADPH provides the reducing equivalents for biosynthetic reactions of anabolic pathways and antioxidant defense in the control of cellular oxidative stress. Finally, appropriate redox state of proteins is maintained by thiols via the redox buffer system comprised of glutathione and thioredoxin (Trx). Glutathione is the most abundant free thiol in cells and the major player in redox homeostasis and function of cellular proteins through interaction with redox proteins (e.g., Trx, glutaredoxin, peroxiredoxin, and protein disulfide isomerase). It also serves as the reducing substrate for antioxidant enzymes,

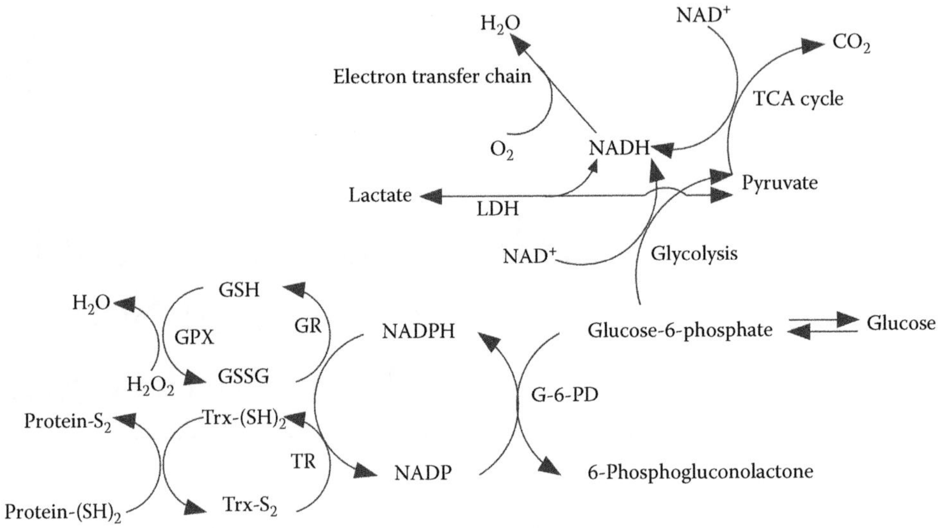

FIGURE 8.1 Simplified scheme for intracellular redox buffering systems. NADH is produced during glycolysis and TCA cycle and consumed by the electron transfer chain. NADPH is generated by the catalysis of glucose-6-phosphate dehydrogenase (G-6-PD) and utilized to reduce the oxidized glutathione (GSSG) by glutathione reductase (GR) or the oxidized Trx by thioredoxin reductase (TR).

such as glutathione peroxidase. Both reduced glutathione and Trx are regenerated by reductases that consume NADPH (Figure 8.1).

The most widespread form of thiol modification by oxidation is the formation of disulfide bonds, mainly considered as abundant structural element important for folding, transport, and function of secreted and surface-membrane proteins (Lee et al. 2004; Sevier and Kaiser 2002). The reaction of a protein thiol with reactive oxygen species (ROS) can lead to different posttranslational modifications, including glutathionylation, formation of a sulfenic, sulfinic, or sulfonic acid, a disulfide, or an S-nitrosothiol (Beltrán et al. 2000; Jacob et al. 2004; Mukhopadhyay et al. 2004). In case of toxic levels of ROS, oxidative modification represents an extreme event that can lead to deleterious consequences, such as apoptosis and loss of function (Simbula et al. 2007). However, if initiated by subtoxic ROS production, thiols may act as redox-sensitive molecular switches (Moran et al. 2001) and thereby provide a common trigger for a variety of ROS-mediated cellular processes that include gene expression, energy metabolism, phosphorylation, and protein translocation.

In eukaryotic cells, redox buffering systems are confined within different cellular compartments. For example, NADH is highly enriched in the mitochondria, and reduced thiols are localized in the cytosol and oxidized thiols in endoplasmic reticulum. In addition, these redox systems are sensitive to environmental changes and perturbations, such as changes in nutrient levels, hyper- or hypoxia, and stress conditions.

An ideal assay to study cellular redox states should be rapid, specific, and effective either in situ or in vivo and maintain the redox signal during the sample preparation and measurement. Although various in vitro biochemical detection methods are widely used for analyzing NADH, NADPH, glutathione, and protein thiols, optical sensing of these cellular redox systems in situ or in vivo can provide valuable information of their localization and dynamic changes in real time. Currently, fluorescence detection is the most widely used method in molecular imaging because of its high sensitivity and selectivity, sufficient temporal and spatial resolutions, and noninvasive nature (Wang et al. 2009). For a long time, researchers have utilized direct imaging of weak endogenous fluorescence of intracellular NADH by single- or two-photon excitation (Kasischke et al. 2004; Patterson et al. 2000; Skala et al. 2007)—an approach extensively reviewed in other chapters of this book. Thiols and their derivatives do not have intrinsic fluorescence; thus, their redox states have to be imaged indirectly by a chemical staining or immunostaining method. Several fluorescent reagents, such as monochlorobimane, have been proposed for determining cellular levels of glutathione (Mandavilli and Janes 2010; Rice et al. 1986; Tauskela et al. 2000); a few antibodies raised against modified thiols have been used in immunoblotting and immunochemistry of protein S-nitrosation, glutathionylation, and sulfenic acid formation (Gow et al. 2002; Seo and Carroll 2009). However, lack of specificity of the antibodies and instability of such thiol modifications limit their application. Thanks to the recent developments of thiol modification-specific labeling methods (Jaffrey et al. 2001; Leichert and Jakob 2004; Saurin et al. 2004; Yang and Loscalzo 2005; Yang et al. 2007b), researchers can now visualize the protein thiol modifications in the cell and collect information related to their distribution and dynamic changes. Thus, S-nitrosoproteins and disulfide-containing proteins can be imaged in situ via indirect chemical labeling, which involves free thiol alkylation followed by reductive generation of thiols from S-nitrosothiols or disulfides with further labeling of the resultant thiols with a fluorescent derivative of methanethiosulfonate (Yang and Loscalzo 2005; Yang et al. 2007b). Immunochemical approaches allow imaging of sulfenic acids in fixed cells by labeling them with dimedone to form a nonfluorescent antigen, which is then recognized by a specific antibody with a fluorescent marker (Maller et al. 2011; Seo and Carroll 2009). For live cell imaging, we recently designed and synthesized a naphthalimide-based fluorescent probe useful for the observation of vicinal dithiol proteins in live cells (Huang et al. 2011).

Although widely used, optical sensing suffers from some limitations for studying cellular redox systems. For example, NADH and NADPH, which have distinct functions in cells, cannot be distinguished by imaging of endogenous autofluorescence as they have similar absorption and emission spectra, and thereby the signal from the corresponding spectral band is collectively denoted as NAD(P)H. It is also hard to obtain dynamic information of redox fluctuations in live cells by fluorescence imaging of cellular protein thiol modifications, which, to a great extent, are restricted to fixed cell studies (Maller et al. 2011; Seo and Carroll 2009; Yang and Loscalzo 2005; Yang et al. 2007b). Precautions must be taken for the determination of glutathione by thiol-reactive sensors due to severe interference of these sensors with cellular redox homeostasis and difficulties for quantification. The reason is that these probes depend on the activity of cellular glutathione transferase and may react with intracellular thiols other than glutathione (e.g., proteins), depleting the cellular thiol pool (Mandavilli and Janes 2010; Tauskela et al. 2000).

8.2 Genetically Encoded Fluorescent Sensors

The issues of cellular activity perturbation and lack of specificity of many optical approaches may be resolved by the next-generation sensors based on fluorescent proteins, that is, genetically encoded sensors. In many biological labs worldwide, fluorescent proteins are currently used as a benchmark to track proteins, determine their cellular localization, and report gene expression activity (Shaner et al. 2005). Fluorescent proteins can be engineered to make sensors for live cell imaging, in particular, for reporting levels of metabolites and cellular activities by changes of their fluorescent properties (Frommer et al. 2009; Gross and Piwnica-Worms 2005; Knöpfel et al. 2006). Since the genetically encoded proteins are self-sufficient to form intrinsic fluorophore without extraneous chemicals, it is possible to target these sensors to specific type of cells and even to different subcellular organelles. Such labeling specificity can be achieved via either changing the promoter of the encoding gene or fusing them with different subcellular organelle-targeting signal peptides, thus allowing accurate spatiotemporal imaging of live cell activities.

There are a few efficient genetically encoded sensors optimized for probing cellular redox states, such as fluorescent proteins directly reacting with superoxide radicals (Bou-Abdallah et al. 2006; Wang et al. 2008b). Belousov et al. developed the HyPer probe, capable of detecting the hydrogen peroxide (H_2O_2) via fusing the prokaryotic H_2O_2-sensing protein, OxyR, to a circularly permuted yellow fluorescent protein (cpYFP) (Belousov et al. 2006). The sensor allowed monitoring of H_2O_2 production during physiological stimulation and apoptosis. Introduction of a cysteine pair on the adjacent surface-exposed β-strands of the green fluorescent protein (GFP) derived its reduction-oxidation sensitive variant, roGFP (Cannon and Remington 2009; Dooley et al. 2004; Hanson et al. 2004; Meyer and Dick 2010; Østergaard et al. 2001). Oxidation of these cysteines to a disulfide caused structural rearrangements of the fluorophore and enhancement of the minor excitation peak at 400 nm, making this molecule suitable for monitoring redox dynamics of intracellular thiols. The roGFP was further modified to vary its response rate (Cannon and Remington 2006), midpoint potentials (Lohman and Remington 2008), and sensitivity to cellular oxidized glutathione via fusion with glutaredoxin (Björnberg et al. 2006; Gutscher et al. 2008). Redox-sensitive fluorescent proteins have been successfully applied in yeasts, plants, fruit flies, and mice (Albrecht et al. 2011; Merksamer et al. 2008; Rosenwasser et al. 2010; Xu et al. 2011).

Recently, our group and Yellen et al. independently developed genetically encoded sensors for monitoring cellular NADH/NAD^+ redox states (Hung et al. 2011; Zhao et al. 2011b). In the following sections, we will focus on live cell imaging and measurement of intracellular NADH level using genetically encoded sensors.

8.3 Genetically Encoded Fluorescent Sensors for NADH

8.3.1 Natural Protein Sensors of NADH and NADPH

As NAD^+ is a major electron acceptor during the oxidation of fuel molecules and NADH is a critical indicator of cellular metabolism, it is no surprise that cells of different organisms from bacteria to mammals develop regulatory proteins and transcription factors that directly sense intracellular NADH concentrations (McLaughlin et al. 2010; Rutter et al. 2001; Zhang et al. 2002, 2006).

The first redox-controlled transcript factor that binds with NADH was identified during the study of circadian rhythms. It is known for a long time that cellular gene expression and metabolism, as well as NADH and NADPH levels, oscillate with about a 24 h cycle, regulated by light or food (Brody and Harris 1973; O'Neill and Reddy 2011). By utilizing an in vitro assay, Rutter et al. showed that NADH and NADPH significantly increased the DNA-binding affinity of the CLOCK protein and its analog NPAS2, both of which are major players of mammalian circadian rhythms (Rutter et al. 2001). Interestingly, CLOCK-SIRT1, which is inhibited by NADH, was found to regulate the NAD^+ salvage pathway, making intracellular NAD^+ concentration oscillate with a 24 h rhythm (Nakahata et al. 2009). Soon after Rutter et al.'s discovery, another transcript factor, the C-terminal-binding protein (CtBP) corepressor, was found to be also regulated by intracellular NADH (Zhang et al. 2002). In both vertebrates and invertebrates, the CtBP family proteins play important roles during cell cycle, development, oncogenesis, and aging, functioning as transcriptional corepressors (Chen et al. 2009; Chinnadurai 2002). They were the first proteins reported to bind to the C-terminal region of the human adenovirus E1A proteins and inhibit oncogenic transformation (Boyd et al. 1993; Schaeper et al. 1995). It is interesting that these proteins, highly conserved across phyla, not only share a striking degree of amino acid homology with NAD-dependent 2-hydroxy acid dehydrogenases but also possess functional dehydrogenase activity (Kumar et al. 2002). NADH binding leads to conformational changes of CtBP, enhancing its binding to various transcriptional repressors. Furthermore, CtBP-mediated repression increases when cells are treated with reagents that enhance NADH level (Zhang et al. 2002). Subsequent studies found that when cellular NADH level elevates under hypoxia conditions, CtBP is recruited to E-cadherin promoter and represses E-cadherin gene expression, thus increasing tumor cell migration (Zhang et al. 2006). The results of the studies comparing CtBP binding to NADH and NAD^+ are controversial: in vitro protein interaction assays showed that NADH has more than 100-fold higher affinity for CtBP than NAD^+ (Fjeld et al. 2003; Kim et al. 2005; Zhang et al. 2002), while other studies suggested that NAD^+ and NADH are equally effective in stimulating CtBP binding to its partners (Balasubramanian et al. 2003; Kumar et al. 2002).

In addition to these NADH-sensing proteins, eukaryotic cells also have sensors for phosphorylated pyridine dinucleotides—$NADP^+$ and NADPH (Lamb et al. 2008). The NADPH-binding HSCARG protein was shown to associate with argininosuccinate synthetase and downregulate its activity along with nitric oxide production when cellular NADPH level decreased (Zhao et al. 2008; Zheng et al. 2007). Yeast galactose regulon and its Gal4–Gal80 transcriptional axis have been well studied and used as a model for studying transcriptional activation in eukaryotes during the past few decades. It is rather surprising that $NADP^+$-binding site was identified in the recently resolved crystal structure of Gal80, and the Gal4–Gal80 transcription system was also found to be regulated by intracellular $NADP^+$ level (Kumar et al. 2008).

Although the structures of these eukaryotic pyridine nucleotide-sensing proteins have been determined (Dai et al. 2009; Kumar et al. 2008, Nardini et al. 2003; Sanders et al. 2007; Zheng et al. 2007), there is not enough information to clearly demonstrate the underlying mechanism of how they function as signaling molecules to regulate gene transcription or protein activity.

To date, the most well-studied transcription factors that sense NADH are the Rex family proteins, which act as transcriptional regulators of the central carbon and energy metabolisms (Brekasis and Paget 2003; Pagels et al. 2010; Ravcheev et al. 2012). Rex was discovered as a novel redox sensing repressor of cydABCD operon (Brekasis and Paget 2003), widely distributed in Gram-positive bacteria, including *Bacillus* and *Streptomyces* genera, and quite a few human pathogens, such as *Streptococcus pneumoniae* and *Clostridium tetani*. This operon encodes both subunits of the cytochrome *bd* terminal oxidase complex, which is highly induced during oxygen limitation or respiration inhibition. Because of high affinity of Rex to NADH and NADPH, its ability to bind to the operator sequence decreases with increasing concentration of these reduced pyridine nucleotides as shown by electromobility shift and surface plasmon resonance (SPR)-based assays (Brekasis and Paget 2003; Sickmier et al. 2005; Wang et al. 2008a). At this, NADH was shown to bind with at least two orders of magnitude more affinity with the reported K_d range of 5–100 nM (Pagels et al. 2010; Wang et al. 2008a), similar to that of the eukaryotic sensor ctBP (Fjeld et al. 2003). NAD^+ competes with NADH for binding and retains (or even enhances) the DNA-binding activity of Rex, however, with at least 1000-fold less affinity (McLaughlin et al. 2010; Pagels et al. 2010; Sickmier et al. 2005; Wang et al. 2008a). Thus, the repressor function reflects the NAD^+/NADH poise (Wang et al. 2008a), which is more physiologically relevant to the cellular redox states than the absolute concentration of NADH.

X-ray crystallography allowed to understand the nature of NAD^+/NADH poise sensing of Rex. The crystal structures of Rex homologs from *Thermus thermophilus* and *Bacillus subtilis* showed a homodimer protein, where each subunit has an N-terminal domain that adopts the winged helix-turn-helix fold. This fold interacts with DNA and a C-terminal NADH-binding domain, which has the Rossmann fold motif, common for pyridine nucleotide-dependent dehydrogenases (McLaughlin et al. 2010; Sickmier et al. 2005; Wang et al. 2008a). NADH binds in an extended conformation to the C-terminal domain near the dimer interface (Sickmier et al. 2005) with its nicotinamide moiety deeply buried inside a hydrophobic pocket (Figure 8.2a). However, in the NAD^+-bound structure, shown in Figure 8.2b, the oxidized nicotinamide ring adopts a *syn* conformation, which rotates 180° relative to the *anti* conformation of bound NADH (McLaughlin et al. 2010).

A comparison of NAD^+- and NADH-bound Rex structures shows that NADH binding induces a significant conformational change of the Rex dimer (Figure 8.2c and d). Upon NADH binding, a 40° closure between the dimeric subunits of Rex occurred so that the dimer of Rex can no longer bind to DNA.

8.3.2 Generation and Optimization of Sensors for NADH

Genetically encoded fluorescent sensors are usually fusions of specific sensing proteins and one or two fluorescent proteins. Förster resonance energy transfer (FRET)-based sensors have two fluorescent proteins of different colors, and their sensitivity is achieved via changes in the FRET efficiency between these two fluorescent proteins, while single fluorescent protein–based sensors have one circularly permuted fluorescent protein (cpFP) fused to sensing domain. In these cpFPs, the amino- and carboxy-termini

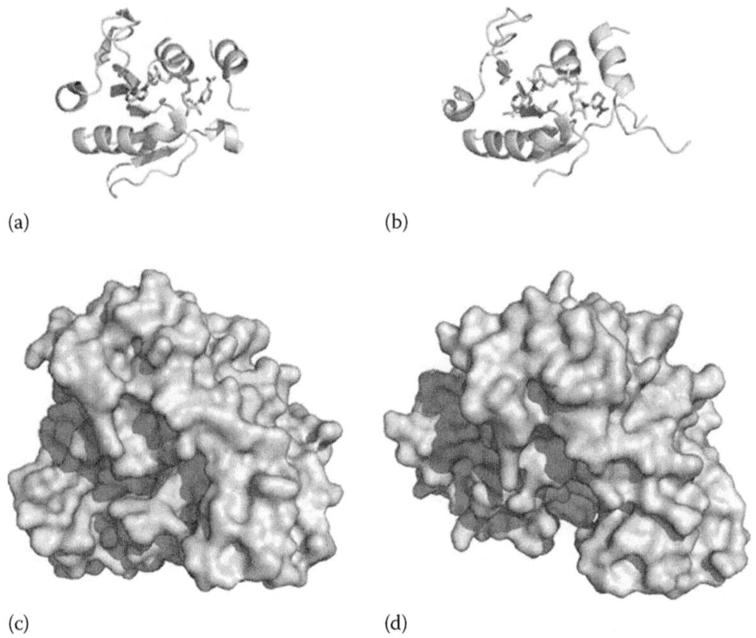

FIGURE 8.2 Structural changes of Rex when the ligand switches from NAD^+ to NADH. (a) *anti* conformation of NADH bound to Rex and (b) *syn* conformation of NAD^+ bound to Rex. (c, d) Surface representation of Rex dimer with NAD^+ or with NADH based on Protein Data Bank files 3IKT and 1XCB.

of the parent fluorescent proteins are fused with a polypeptide linker, and new termini are formed close to the fluorophore, enhancing its fluorescence sensitivity to the surrounding microenvironment. This approach usually exhibits more significant changes of fluorescent properties upon binding with an analyte molecule than the sensors based on the FRET, the most popular fluorescence technique for binding assays. Circularly permuted design was first used to make a Ca^{2+} sensor with an expanded dynamic range as compared to FRET-based sensors (Nagai et al. 2004; Nakai et al. 2001). A few other highly responsive genetically encoded fluorescent biosensors were developed by fusing sensitive protein domains with cpFPs for monitoring various cellular molecules, such as H_2O_2 and ATP (Belousov et al. 2006; Berg et al. 2009; Simen Zhao et al. 2010). Most of these studies utilized green or yellow fluorescent proteins; however, expansion of cpFP-based sensor spectra to the blue and red regions has been reported recently (Niino et al. 2010; Zhao et al. 2011a).

We reasoned that the conformational changes of Rex proteins induced by NADH may be transduced to the cpFPs altering their fluorescence properties (Zhao et al. 2011b). We first constructed three chimeras in which cpYFP molecules were inserted between two *B. subtilis* Rex (B-Rex) subunits with or without the DNA-binding domain. One of the chimera proteins, named NS2 (NADH sensor 2), which was a product of the fusion of a complete Rex monomer, cpYFP, and the NADH-binding domain of Rex (Figure 8.3a), showed changes in the ratio of fluorescence emission intensity values under 390 and 485 nm excitations in the presence of NADH. A series of truncated variants, targeting residues involved in the linker between Rex and cpYFP, were created to maximize the dynamic range of the NS2 fluorescence in response to NADH binding. One of the truncated variants, termed C3, had a dramatic increase, and another one, C8, decrease in the fluorescence intensity in response to NADH when excited at 485 nm, but not at 390 nm. High affinity of the native B-Rex to NADH ($K_d = 24$ nM) suggests that these sensors could be easily saturated at the physiological level of NADH. Furthermore, since Rex protein also binds to NAD^+ and NADPH, albeit at much lower affinity as compared to that of NADH, the sensor efficiency may be abrogated by high concentrations of NAD^+ in the cell.

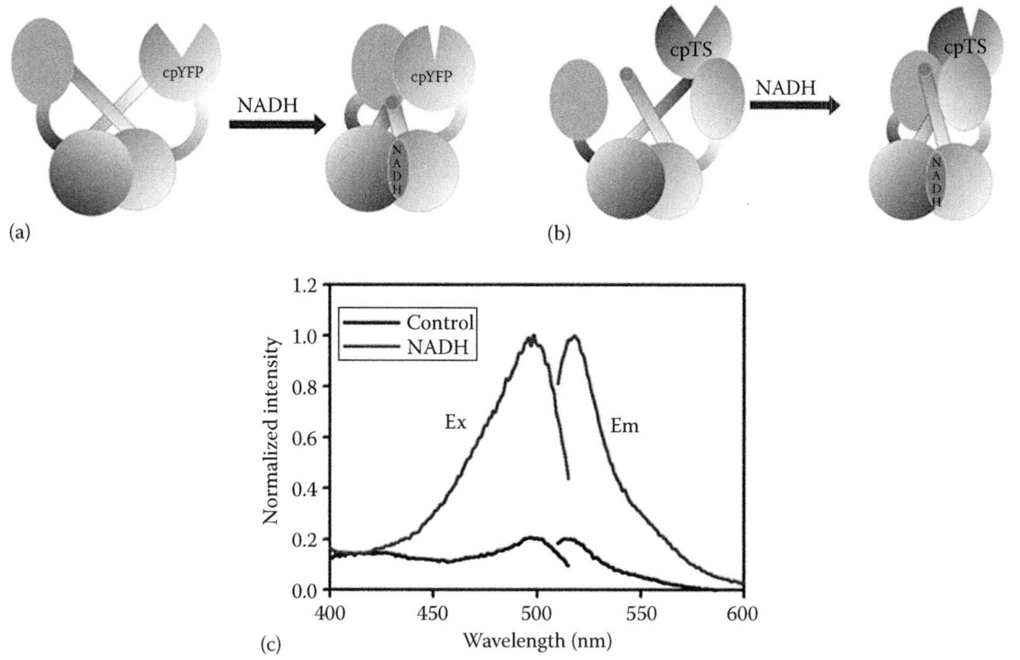

FIGURE 8.3 Genetically encoded NADH sensor. (a) Design of Frex, which is a fusion of a complete Rex monomer, cpYFP, and the NADH-binding domain of a second Rex molecule. (b) Design of Peredox, which is a fusion of two complete Rex monomer and a circular permuted T-Sapphire in between. (c) Fluorescence spectra of purified Frex in the presence (upper curve) or absence (lower curve) of NADH. Excitation spectrum, recorded at 530 nm emission, has two maxima at 420 and 500 nm.

Indeed, we observed fluorescence fluctuations of the C3 and C8 sensors in the presence of NAD^+. In order to enhance the sensor sensitivity and selectivity to NADH, we mutated single amino acid residues around the NADH-binding site. Among these site-directed mutations, C3L194E and C8N120E exhibited an eightfold increase and threefold decrease of fluorescence intensity upon NADH binding, respectively (Figure 8.3c).

Like other cpYFP-based genetically engineered sensors, these variants have one emission peak at 518 nm and two excitation maxima around 421 and 490 nm. Addition of NADH affected their fluorescence only under excitation at 490 nm, allowing for ratiometric imaging (Figure 8.3c). Both C3L194E and C8N120E mutants, termed as fluorescent Rex (Frex) and Frex of high affinity (FrexH), respectively, exhibited high selectivity toward NADH, showing no apparent fluorescence changes in the presence of NADH analogues, including NAD^+, NADPH, or $NADP^+$.

In the same issue of the journal *Cell Metabolism* where we reported Frex and FrexH, Hung et al. introduced yet another genetically encoded sensor for probing intracellular NADH–NAD^+ redox states (Hung et al. 2011). The design of this sensor (Figure 8.3b) was remarkably similar to Frex in that it was also created by inserting a cpFP into the tandem dimer of Rex. However, in contrast to Frex, the new sensor was based on the circular permuted GFP variant T-Sapphire (Zapata-Hommer and Griesbeck 2003) as a reporter and the Rex protein from *Thermus aquaticus* (T-Rex) as a sensor.

There were two major incentives for choosing these variants as the sensor and the reporter. First, as a pH-resistant GFP variant, the circular permuted T-Sapphire and its fusion with sensor proteins should also retain its pH-resistant properties. Second, T-Rex binds with NADH much tighter ($K_d < 5$ nM) than B-Rex ($K_d = 25$ nM) or Rex from *Staphylococcus aureus* ($K_d = 100$ nM), suggesting higher sensitivity of T-Rex and ability to detect lower levels of free NADH as compared to its analogues. The fluorescence of the initial construct of Hung et al. termed P0 responded to NADH, while NAD^+ yielded only minimal change in fluorescence intensity. However, increasing NAD^+ concentration effectively lowered the sensor's apparent affinity for NADH, suggesting that the sensor could be used to monitor NADH/NAD^+ redox poise. Subsequent studies showed that the binding of P0 to NADH/NAD^+ was sensitive to pH. Such sensitivity was largely eliminated in a P0 variant that had Tyr98Trp and Phe189Ile mutations in the first subunit and Tyr98Trp mutation in the second subunit. This variant, named Peredox, was used for cytosolic NADH–NAD^+ redox sensing.

8.4 Spatiotemporal Measurements of NADH in Living Cells

8.4.1 NADH in Subcellular Organelles

Measuring NADH concentrations in living cells is critical for the monitoring of metabolic state fluctuation. The total content of NAD^+–NADH pool in the mitochondria, measured by optical and biochemical methods, varied from 360 μM to 3 mM depending on the cell type (Joubert et al. 2004; Yang et al. 2007a), while the mitochondrial NAD^+/NADH ratio values reported by different authors varied in the range of 2–16 (Henley and Laughrey 1970; Kasimova et al. 2006; Williamson et al. 1967). Significant variation—from 0:1 (Wakita et al. 1995) and 1.5:1 (Blinova et al. 2005) to 1:4 (Yu and Heikal 2009)—was also reported for the ratio between the protein-bound and free NADH species in the mitochondria, which can be measured by a number of imaging techniques, such as fluorescence lifetime (see Chapter 4), fluorescence anisotropy (reviewed in Chapter 5), and fluorescence spectral decomposition analysis (Blinova et al. 2005; Kasimova et al. 2006; Vishwasrao et al. 2005; Yu and Heikal 2009; Zhang et al. 2002). It is interesting that the free NADH concentration in plant mitochondria was reported to be constant under different metabolic conditions (Kasimova et al. 2006). Such variety may reflect different metabolic conditions in the mitochondria of different species and tissues; it may be also due to the limitations of the employed techniques.

Quantification of free NADH level in the cytosol or nucleus using traditional endogenous NAD(P)H fluorescence methods (Mayevsky 2009) is complicated and challenging not only due to very low NADH concentration in these cellular compartments but also because of the major contribution of cytosolic NADPH to the fluorescence detected—the NADPH/NADH ratio reported was as high as ~4 (Jones 1981; Shigemori et al. 1996).

Frex sensor-based ratiometric fluorescence measurement allows quantification of free intracellular NADH level in different subcellular compartments. We performed measurements in mammalian cells through Frex and FrexH using a microplate reader and found extremely low cytosolic NADH levels ranging from 120 to 130 nM (Table 8.1). Ratiometric fluorescence imaging of free NADH in the mitochondria showed that 70%–75% of the mitochondria-targeted Frex (Frex-Mit) sensors were bound to NADH in the mitochondrial matrix of 293FT cells, determining free NADH level in the mitochondrial matrix as 33 ± 9 μM. A slightly lower value of free NADH content (27 ± 5 μM) was measured by a low-affinity sensor.

TABLE 8.1 Occupancy of Genetically Encoded NADH Sensors in Subcellular Compartments

Sensors	Cell	Subcellular Location	Detection Method	Glucose in Medium (mM)	Sensor Occupancy (%)	NADH Levels (μM)
FrexH-Cyt	293FT	Cytosol	Microplate	25	~76	~0.13
Frex-Cyt	293FT	Cytosol	Microplate	25	~3.3	~0.12
Frex-Mit	293FT	Mitochondria	Imaging	25	~70–75	~33
Frex-Mit	293FT	Mitochondria	Microplate	25	~70–75	~26
C3L194K	293FT	Mitochondria	Imaging	25	~35	~27
Peredox	Neuro-2a	Cytosol	Imaging	5	~100	N.D.
Peredox	Neuro-2a	Cytosol	Imaging	0.16	~50	N.D.

Here, C3L194K served as a control assay of the mitochondria with high matrix NADH levels, as it is much less occupied by NADH compared to Frex sensor. Using traditional biochemical assays in isolated mitochondria (Corcoran et al. 2009; Szabo et al. 1996; Thornburg et al. 2008; Ying et al. 2001) and taking as a basis that 1.5 μL of the mitochondria is associated with 1 mg of mitochondrial protein (Schwerzmann et al. 1986), we found that the mitochondria of 293FT cells contain 345 ± 28 μM NAD^+ and 91 ± 8 μM bound NADH (Zhao and Yang 2012). Thus, the mitochondrial free/bound NADH ratio measured in our experiments was 1:2–1:3, which is in accordance with the range previously reported by other authors (Bird et al. 2005; Yu and Heikal 2009). Results from the Frex fluorescence measurements also showed that the concentration of free NADH in the mitochondria is several hundredfold higher than that in the cytosol. Considering that the mitochondrial volume in a cell ranges from 5% to 50%, most intracellular NADH should be localized in the mitochondria, again consistent with the results of the endogenous fluorescence measurements of NADH (Mayevsky 2009).

8.4.2 Dynamic Tracing of NADH in Live Cells

Frex and Peredox sensors are much brighter than the fluorescence of endogenous NADH. As a result, these sensors are more suitable for the dynamic tracing of NADH, especially in subcellular compartments like the cytosol and nucleus, where the endogenous NAD(P)H fluorescence is especially weak and originates mainly from NADPH.

It is known that in the cytosol, NADH and NAD^+ are buffered by pyruvate and lactate via lactate dehydrogenase. Both Frex and Peredox sensors expressed in the cytosol responded rapidly to the addition of lactate or pyruvate into the cell culture medium, reporting an elevation of NADH content by the extraneous lactate and decrease of NADH level by the pyruvate (Hung et al. 2011; Zhao et al. 2011b). Both sensors also exhibited high sensitivity to glucose (Hung et al. 2011; Zhao et al. 2011b), which increases cytosolic and nuclear NADH level through the glycolysis pathway (Eto et al. 1999; Patterson et al. 2000). For these assays, Peredox sensor performed very well under very low glucose conditions (i.e., 0–1 mM), while Frex sensors proved to be more suited for glucose concentrations closer to the physiological range (1–20 mM). Higher spatial and temporal resolution, achieved by simultaneous measurement of the fluorescence from cytosolic and mitochondria-targeted Frex sensors, revealed that the mitochondrial NADH level was more sensitive to the extracellular glucose concentration: half-maximal response of Frex-Mit was detected at 0.16 mM of glucose, while for the Frex sensor expressed in the cytosol, this value was one order of magnitude bigger. Overall, cytosolic NADH level responded to these metabolites much quicker than mitochondrial NADH level. These results suggest that cytosolic and mitochondrial NADH levels measured by Frex could be a sensitive indicator for the status of cellular metabolism.

We then tested Frex as a probe to monitor NADH transport in cells. Glycolysis-generated NADH is transported from the cytosol into the mitochondria via the malate–aspartate shuttle. Blocking the shuttle with aminooxyacetate decreased the Frex-Mit fluorescence intensity, indicative of the reduction of mitochondrial NADH content, whereas no changes in NAD(P)H autofluorescence were detected. Such decrease is widely common in different mammalian cells (Zhao et al. 2011b).

For a long time, it was commonly believed that NAD^+ and NADH cannot be transported across plasma membranes of any cell type (Ying 2008); however, recent studies identified molecular complexes enabling such transport. In particular, Lu et al. reported that P2X7 receptors mediated the transport of NADH across the plasma membrane in astrocytes under millimolar NADH concentrations (Lu et al. 2007). By monitoring Frex fluorescence, we unexpectedly found that even micromolar concentrations of exogenous NADH induced an immediate, dose-dependent, and saturable increase in intracellular NADH in different cells (Zhao et al. 2011b), including the P2X7 receptor-deficient HEK293 cells. Furthermore, the P2X7 receptor inhibitor pyridoxalphosphate-6-azophenyl-20,40-disulphonic acid (PPADS) did not affect the entry of NADH into cells. According to these data, it appears that Frex displayed superior specificity and sensitivity for NADH translocation studies in living cells as compared to the endogenous NAD(P)H fluorescence measurement, which failed to report the transport of NADH across the plasma in the P2X7 receptor-deficient HEK293 cells, or in the presence of micromolar concentration of exogenous NADH (Lu et al. 2007).

8.5 Methods for the Quantification of Cellular NADH Utilizing Genetically Encoded Fluorescent Sensors

8.5.1 Preparation of Frex Sensor–Expressing Cells

The coding sequences of Frex and FrexH were subcloned into pcDNA3.1 Hygro($^+$) (Invitrogen) behind a Kozak sequence for mammalian expression. The threefold nuclear localization signal (3xNLS) DPKKKRKVDPKKKRKVDPKKKRKV was added to the C-terminus for the nuclear targeting of Frex; the mitochondrial localization signal MRKMLAAVSRVLSGASQKPASRVLVASRNFANDATF was inserted at the N-terminus for the corresponding mitochondrial targeting. 293FT cells (Invitrogen, United States) were maintained in Dulbecco's modified eagle medium (DMEM) (high glucose) supplemented with 10% fetal bovine serum (FBS) (Bovogen), 0.1 mM modified eagle medium (MEM) nonessential amino acids (Invitrogen), 6 mM L-glutamine (Invitrogen), and 1 mM sodium pyruvate (Invitrogen) at 37°C in a humidified atmosphere of 95% air and 5% CO_2. Cells were plated in antibiotic-free high-glucose DMEM supplemented with 10% FBS for 16 h before transfection. We typically used 0.8 μg endotoxin-free plasmids with 3.2 μL Lipofectamine 2000 (Invitrogen) for each well of a 12-well plate according to the manufacturer's protocol. Cells were plated on 35 mm glass bottom dishes (NEST Biotechnology) or glass bottom microplates (Matrical Bioscience) with phenol red-free fresh growth medium for fluorescence microscopy measurements. Frex was expressed in different subcellular compartments by tagging with organelle-specific signal peptides and monitored for 24–30 h posttransfection.

8.5.2 Equipment

A Zeiss LSM 710 system based on a Zeiss Axio Observer Z1 inverted microscope stand was used for confocal microscopy. Images were acquired with a Plan Apo 63 × 1.4 NA oil immersion objective from cells maintained at 37°C in a humidified atmosphere using a CO_2 incubator (PECON). Frex sensors were excited sequentially (line-mode scanning) using the 405 nm diode laser (1.2% laser power) and 488 nm line of a multiline Argon laser (2.6% laser power); the emission was detected at 500–550 nm. The line-mode scanning (switching between two excitation sources in every line of the acquired image) allows for accurate measurements of the fluorescence ratio from each pixel while minimizing the effects of vibration and subcellular particle movements. Images were acquired at 1024 × 1024 pixel digital resolution, 12 bit depth, line average of 2, and pinhole size of 3.0 airy units.

To assess the Frex and C3L194K sensor occupancy by NADH in the mitochondria, we performed ratiometric fluorescence imaging of living cells and purified recombinant sensor protein solution with and without NADH, using a high-performance fluorescence microscopy system, based on a Nikon Eclipse Ti-E automated microscope, Plan Apo VC 60x/1.2 NA water immersion objective, Evolve 512 EMCCD (Photometrics), highly stable Lambda XL light source, and Lambda 10-XL filter wheel (Sutter Instruments). Excitation filters (410 BP20, 480 BP30 band pass) were switched for dual-excitation ratiometric imaging. Emission filter was 535 BP40 band pass. Images were captured using 512 × 512 pixel digital resolution, 16 bit depth, and 100 ms exposure for both channels.

For high-throughput measurement of intracellular NADH dynamics, a Synergy 2 Multi-Mode Microplate Reader (Biotek) was used to read the fluorescence of the cells suspended on glass bottom microplates. To this end, 293FT cells were trypsinized at 24–48 h after transfection and then counted, washed, and suspended in phosphate-buffered saline. Aliquots of cells were then incubated at 37°C with specific chemicals during the measurement. Dual-excitation ratios were obtained with 410 and 500 nm excitations, while the fluorescence emission was detected at 528 nm.

8.5.3 Pitfalls and Remedies

Similar to many other cpFP-based genetically encoded sensors (Belousov et al. 2006; Berg et al. 2009; Dooley et al. 2004; Hanson et al. 2004; Imamura et al. 2009; Nagai et al. 2001, 2004; Wang et al. 2008b), Frex is also sensitive to pH. Such sensitivity may represent a limitation for many, if not all, genetically encoded sensors due to its effect on both the fluorescence properties and the sensing domain. Fluorescence of the cpYFP domain, excited at 500 nm, increased significantly with the increase in pH, while the 420 nm–excited emission remained largely unaffected. As a result, pH fluctuations effectively changed the ratio of the fluorescence signals acquired from Frex at two different excitation wavelengths, making the quantification of free intracellular NADH level rather challenging. Frex sensor had an apparent K_d for NADH of ~3.7 μM at pH 7.4, which increased to ~11 μM at pH 8.0. Such effect of pH on the affinity of ligands or the redox potential of the sensing domain has been reported for other sensors as well (Berg et al. 2009; Dooley et al. 2004; Hanson et al. 2004; Imamura et al. 2009). Fortunately, the sensors' fluorescence was found not to be altered by the small pH variation existing in the cytosol (Belousov et al. 2006; Berg et al. 2009; Nakai et al. 2001; Zhao et al. 2011b). Care should be taken to maintain a constant pH for either the buffers and culture medium for consistent results with minimum artifacts; significant changes in the pH may occur even as the cells transfer in or out of the CO_2 incubator. When studying the effects of chemicals that regulate cellular metabolism by microplate reading, we usually suspended the cells in PBS buffer rather than in a culture medium. In addition, we adjusted the pH of different solutions investigated. For fluorescence measurements of sensors in live cells inside a microscope incubator, we added 100 mM HEPES (pH 7.4) to phenol red-free complete culture medium to increase its buffering capacity. Cells were kept in the incubator for 10 min before imaging to balance the dissolved CO_2. Since some added chemical solutions could have shifted the intracellular pH level, we measured the fluorescence from cpYFP-expressing cells in parallel as a control: cpYFP fluorescence responds to the changes in pH, but not in NADH levels. For example, Figure 8.4 shows a two-phase increase in Frex-Mit fluorescence during 1 h time course in glucose-starving cells after glucose supplement. We found that slower phase of increase was an artifact caused by slow pH changes in the mitochondrial matrix after glucose treatment and mitochondria recharge. To correct this artifact, Frex-Mit

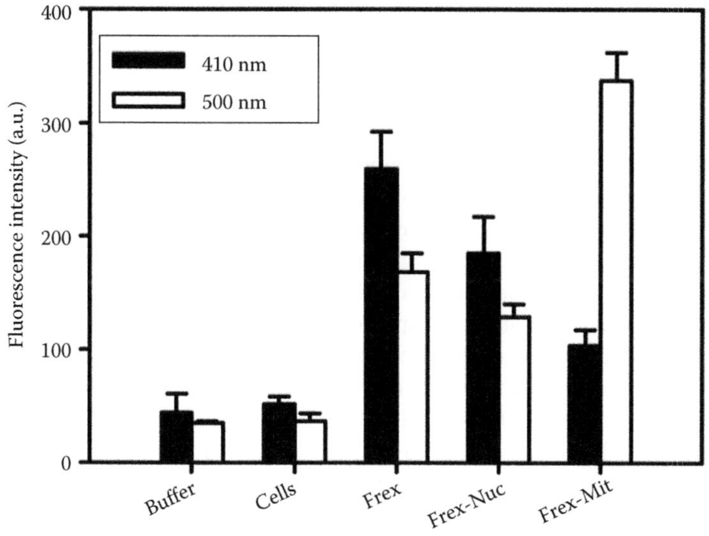

FIGURE 8.4 pH effect correction of Frex fluorescence. Kinetics of fluorescence response of Frex-Mit and cpYFP-Mit after glucose treatment. The solid symbol represents the fluorescence response of Frex-Mit corrected for pH effects.

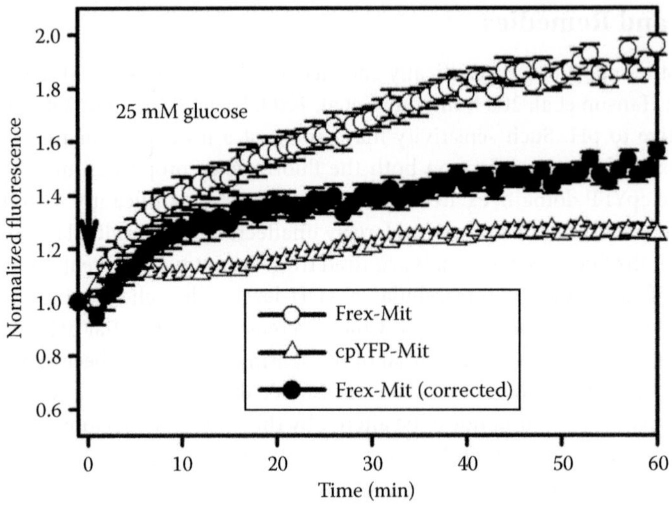

FIGURE 8.5 Typical value of the fluorescence of Frex-expressing living cells. Frex was expressed in different subcellular compartments. Fluorescence was excited at 410 nm (black bar) or 500 nm (white bar). The data show that cell autofluorescence from FAD did not significantly interfere with Frex fluorescence assay.

fluorescence was normalized by cpYFP-Mit fluorescence emission; the corrected Frex-Mit fluorescence reached its maximum within 10 min after glucose supplement, consistent with previous reports using other cell lines (Eto et al. 1999).

Another concern is the detection of cellular autofluorescence that may spectrally overlap with Frex sensors. Intracellular FAD is a major source of autofluorescence in living cells with excitation maximum at 450 nm and emission peak at 535 nm. Thus, background FAD signal may adversely affect the quantification of Frex fluorescence. However, for the acquisition settings described earlier, we found that FAD emission was not significant during the confocal and wide-field measurements (Figure 8.5). Strong autofluorescence from plasticware should also be taken into consideration in microplate reading measurements during high-throughput assays—usage of the clear polystyrene plates should be avoided due to their high background fluorescence. To minimize the autofluorescence of the microplate, we used 96-well glass bottom MatriPlate from Matrical Bioscience (also available from Whatman and Greiner).

Figure 8.5 shows a typical fluorescence reading of Frex-expressing HEK293 cells. It is worth noting that the main contributor of the overall background fluorescence under 410 and 500 nm excitations was the plate itself, while the autofluorescence from untransfected cells was negligible. We estimated that cell autofluorescence was less than 2% of 500 nm–excited Frex fluorescence and ~4% of 410 nm–excited Frex emission. In addition, no significant changes in cell autofluorescence were observed under different metabolic conditions (i.e., in the presence of mitochondrial inhibitors, uncoupler, aminooxyacetic acid [AOA], or hydrogen peroxide). In order to improve the accuracy of our quantifications, we eliminated even this negligible contribution from cell autofluorescence and plasticware by measuring the signal from untransfected (control) cells and subtracting the values obtained from the fluorescence intensity values registered from Frex-expressing cells. For Frex-Mit, it was difficult to accurately measure the weak fluorescence signal by the microplate reader under 421 nm excitation because of the system's inherent noise. As Frex sensors are largely saturated in the mitochondria, a small error in the 410 nm–excited fluorescence may lead to a large error in free NADH estimation. To avoid that, we measured the free NADH in the mitochondria using ratiometric fluorescence imaging. The mitochondrial pH is higher than that of the cytosol, and therefore, the free mitochondrial NADH fluorescence, detected at 485 nm, can be measured by microplate reading in comparison with the Frex emission in living cells (with cell lysates in 0.3% digitonin and 0.3% digitonin with NADH). These two approaches yielded very similar results, suggesting that 70%–75% of the Frex-Mit sensors were bound to mitochondrial NADH in 293FT cells as summarized in Table 8.1.

Additional concerns may arise as to the potential interference of the Frex protein with physiological functions of NADH buffering and homeostasis. There are several buffering systems for NADH in cells, which include several NADH-binding proteins in the cytosol and in the mitochondria. Furthermore, NADH and NAD^+ are also buffered by pyruvate–lactate through lactate dehydrogenase in the cytosol as well as by oxoglutarate–glutamate through β-hydroxybutyrate dehydrogenase. As a reminder, the concentrations of these metabolites are in the millimolar range (Williamson et al. 1967). Some investigators believe that most cellular NADH in the mitochondria is protein bound (Wakita et al. 1995; Yu and Heikal 2009; Zhang et al. 2002); others report similar amounts of free and bound NADH (Blinova et al. 2005; Vishwasrao et al. 2005). Expression of exogenous NADH-binding protein such as the Frex sensor will certainly contribute to the buffering systems for NADH. We estimated that Frex protein concentration expressed in the cytosol of cells is in the range of 4–13 μM, depending on the cell type and transfection efficiency. Consider the following scenario: In resting cells, the free/bound ratio of NADH is 1:3 (Wakita et al. 1995; Yu and Heikal 2009; Zhang et al. 2002), and then cytosolic free and protein-bound NADH concentrations can be taken as 130 and 390 nM, respectively. We assume that the concentration of Frex sensor in the cytosol is 10 μM. Importantly, however, only 1% of these sensors are bound to NADH in resting cells, which translate into 100 nM of Frex-bound NADH in the cytosol. Therefore, Frex-bound NADH is much less than the amount of NADH bound to endogenous proteins affecting only 19% of the total cytosolic NADH. Based on these calculations, it is unlikely that the Frex sensors have a major impact on total NADH concentration and metabolic status of the cells. This conclusion is substantiated with the fact that no significant increase in endogenous NADH fluorescence was observed in the Frex-expressing cells. Immediate response of cytosolic Frex fluorescence to exogenous NADH, lactate, pyruvate, and glucose (Hung et al. 2011; Zhao et al. 2011b) also indicates that the temporal buffering effects by the Frex sensor are negligible.

8.5.4 Data Analysis and Calibration of Intracellular NADH Levels

For microplate reading, samples containing equal numbers of Frex-expressing and untransfected (control) cells were both measured. Fluorescence values were background corrected by subtracting the intensity of 293FT cells that do not express Frex. Unlike the measurement of endogenous fluorescence or the Peredox sensor, it is possible to quantify and compare free NADH concentration in different cells and different organelles after calibration of Frex fluorescence in live cells. For the calibration of cytosolic NADH level, Frex-expressing cells were resuspended with PBS buffer (pH 7.4) that contains 0.001% digitonin. The fluorescence of cells or cell lysate under the excitation at 485 nm was then measured in either the presence or absence of 100 μM NADH.

Raw imaging data were exported as 12-bit tagged image format (TIF) to ImageJ software for analysis. The pixel-by-pixel ratio of the 488 nm/405 nm–excited images of the same cell was used to pseudocolor the images in *hsb* color space. In this protocol, magenta with RGB (255, 0, 255) and red with RGB (255, 0, 0) represent the lowest and highest ratio correspondingly, while color brightness is proportional to the fluorescent signals. In order to assess the Frex sensor occupancy by NADH in the mitochondria, a calibration standard of sensor protein in the presence or absence of NADH should be obtained using the same microscope setting. For calibration, Frex protein was diluted into 100 mM potassium phosphate (KP) buffer (0.1% bovine serum albumin, pH 8.0) to final concentrations of 0.1 and 0.3 μM, respectively.

8.6 Comparison of NADH Sensing Techniques

Compared with conventional measurements of the weak endogenous NAD(P)H autofluorescence, genetically encoded NADH sensors are on average 100-fold brighter and have high specificity, which make these sensors superior for real-time tracking of intracellular NADH levels, especially in

TABLE 8.2 Properties of Genetically Encoded Sensor for NADH or NADH/NAD Ratio

Sensors	Sensor Origin	Fluorescent Protein	Dynamic Changes (%)	Protein Size (kD)	K_d for NADH	Ratiometric Sensing	pH Sensitivity[a]	References
FrexH	B-Rex	cpYFP	800	65	40 nM	Yes	Sensitive	Zhao et al. (2011b)
Frex	B-Rex	cpYFP	200	65	3.7 μM 11 μM	Yes	Sensitive	
C3L194K	B-Rex	cpYFP	300	65	50 μM	Yes	Sensitive	
P0	T-Rex	cpT-Sapphire and mCherry	100	74	<5 nM	No	Sensitive	Hung et al. (2011)
Peredox	T-Rex	cpT-Sapphire	150	74	N.D.	No	Resistant	
Peredox-mCherry	T-Rex	cpT-Sapphire and mCherry	150	101	N.D.	Yes	Resistant	

[a] Measured at pH 8.0.

cytosolic and nuclear compartments. The sensors can be detected in living cells using instruments that are common in laboratories—fluorescence microplate readers, flow cytometers, and wide-field and confocal fluorescence microscopes. Being genetically encoded, these sensors can be specifically targeted at different cells and various subcellular organelles, including the mitochondria and nucleus, allowing for spatiotemporal imaging of cellular activities at an unprecedented level (Hung et al. 2011; Zhao and Yang 2012; Zhao et al. 2011b).

Frex and Peredox sensors both have their corresponding advantages and disadvantages as summarized in Table 8.2. The increase in Frex fluorescence intensity of up to 800% upon NADH binding makes it one of the most responsive genetic sensors currently available. On the other hand, Peredox exhibits up to 150% increase in fluorescence upon NADH binding. The higher response of Frex sensors may be useful for resolving more subtle differences in NADH levels among different cells during biological processes. Frex sensors are intrinsically ratiometric with two excitation wavelengths, as compared to the single excitation wavelength of Peredox. Thus, readouts of Frex sensors are irrelevant to the sensor concentration in the cells, while ratiometric imaging of Peredox can only be achieved by tagging the sensor with mCherry, a red fluorescent protein, which increases the overall size of the sensor and sometimes causes the aggregation of the fusion protein (Hung et al. 2011). The high binding affinity of Peredox toward NADH prohibits its usage in the mitochondria, which have much higher NADH concentrations than the cytosol. The high NADH affinity of Peredox may also complicate the cytosolic NADH measurement and data interpretation. The cytosol- or nucleus-targeted Peredox sensors were largely saturated under normal physiological conditions, that is, with 1–5 mM glucose supplement in the medium or in vivo (Hung et al. 2011). Therefore, the sensor should be calibrated with pyruvate and lactate without glucose during each measurement; due to the same reasons, usage of Peredox sensor for in vivo imaging can be very challenging. The affinity of Frex sensors was fine-tuned, which can be used for cytosolic or mitochondrial NADH measuring under physiological conditions. The major disadvantage of Frex sensors is their sensitivity to intracellular pH level, which requires a pH calibration using the cpYFP emission as a control under the same experimental conditions. Peredox sensors are relatively resistant to fluctuations in the pH level and, therefore, are more suitable for cellular studies with fluctuating pH. However, pH effect is not fully eliminated for Peredox (Hung et al. 2011); thus, caution should be exercised when measuring subtle changes in NADH.

Finally, Frex sensors are specific for NADH. Thus, they will not permit the measurement of the NADH/NAD^+ ratio. The NAD^+ and NADH pools of the cytosol and mitochondria are well separated (Yang et al. 2007b), and it is unlikely that either of these pools would change in the timescale of many imaging studies. Thus, the NADH/NAD^+ ratio will predominantly depend on the level of NADH under those conditions. Peredox sensors are known to sense the cytosolic NADH/NAD$^+$ redox state, as NAD$^+$ competes with NADH during sensor binding. As Peredox sensor has lower affinities for NAD$^+$ and NADH compared to the P0 construct, it does not strictly report NADH/NAD$^+$ ratio. Threefold change in the NAD$^+$ pool size in the physiological range produced a twofold change in the sensor midpoint for NADH/NAD$^+$ ratio; thus, the sensor's output is affected by not only NADH/NAD$^+$ ratio but also NAD$^+$ pool (Hung et al. 2011).

8.7 Summary and Future Outlook

The genetically encoded NADH sensors Frex and Peredox provide very good alternatives to the measurement of the endogenous fluorescence of intracellular NAD(P)H. These sensors are highly specific, have a large change in fluorescence upon NADH binding, and can be targeted to different subcellular compartments. The ability to image cytosolic and nuclear NADH levels in real time could be very useful for better understanding of cellular signaling pathways involving NADH since many signal transduction and transcription regulation events involve NADH (Rutter et al. 2001; Zhang et al. 2002, 2006). Yet these sensors have their own limitations, and therefore, a highly responsive, pH-insensitive, and purely NADH/NAD$^+$ ratio–specific sensor remains to be desirable for the comparison of the redox potentials of different cells and long-term living cell monitoring, without corrections or normalizations. We are currently working on the improved version of NADH/NAD$^+$ sensors, in the hope to use these sensors together with Frex to measure all three critical parameters simultaneously, that is, NADH, NAD$^+$, and their ratio.

8.A Appendix: Sequence Information

8.A.1 Amino Acid Sequence of Frex

MNKDQSKIPQATAKRLPLYYRFLKNLHASGKQRVSSAELSDAVKVDSATIRRDFSYFGALGKKG
YGYNVDYLLSFFRKTLDQDEMTDVILIGVGNLGTAFLHYNFTKNNNTKISMAFDINESKI
GTEVGGVPVYNLDDLEQHVKDESVAILTVPAVAAQSITDRLVALGIKGILNFTPARLNVPEHIR
IHHIDEAVELQSLVYFLKHYYNSDNVYIMADKQKNGIKANFKIRHNVEDGSVQL
ADHYQQNTPIGDGPVLLPDNHYLSFQSVLSKDPNEKRDHMVLLEFVTAAGITLGMDELYNV
DGGSGGTGSKGEELFTGVVPILVELDGDVNGHKFSVSGEGEGDATYGKLTLKLIC
TTGKLPVPWPTLVTTLGYGLKCFARYPDHMKQHDFFKSAMPEGYVQERTIFF
KDDGNYKTRAEVKFEGDTLVNRIELKGIGFKEDGNILGHKLEYNGTMTDVILIGVGNLGT
AFLHYNFTKNNNTKISMAFDINESKIGTEVGGVPVYNLDDLEQHVKDESVAILTVP
AVAAQSITDRLVALGIKGILNFTPARLNVPEHIRIHHIDEAVELQSLVYFLKHYSVLEEIE.

8.A.2 Amino Acid Sequence of FrexH

MNKDQSKIPQATAKRLPLYYRFLKNLHASGKQRVSSAELSDAVKVDSATIRRDFSYFGALGKKG
YGYNVDYLLSFFRKTLDQDEMTDVILIGVGNLGTAFLHYNFTKNNNTKISMAFDIEESKI
GTEVGGVPVYNLDDLEQHVKDESVAILTVPAVAAQSITDRLVALGIKGILNFTPARLNVPEHIR
IHHIDLAVELQSLVYFLKHYSVLEEYNSDNVYIMADKQKNGIKANFKIRHNVEDGSVQL
ADHYQQNTPIGDGPVLLPDNHYLSFQSVLSKDPNEKRDHMVLLEFVTAAGITLGMDELYNV
DGGSGGTGSKGEELFTGVVPILVELDGDVNGHKFSVSGEGEGDATYGKLTLKLIC
TTGKLPVPWPTLVTTLGYGLKCFARYPDHMKQHDFFKSAMPEGYVQERTIFF
KDDGNYKTRAEVKFEGDTLVNRIELKGIGFKEDGNILGHKLEYNGTMTDVILIGVGNLGT
AFLHYNFTKNNNTKISMAFDIEESKIGTEVGGVPVYNLDDLEQHVKDESVAILTVP
AVAAQSITDRLVALGIKGILNFTPARLNVPEHIRIHHIDLAVELQSLVYFLKHYSVLEEIE.

References

Albrecht, S. C., A. G. Barata, J. Großhans, A. A. Teleman, and T. P. Dick. 2011. In vivo mapping of hydrogen peroxide and oxidized glutathione reveals chemical and regional specificity of redox homeostasis. *Cell Metabolism* 14 (6): 819–829.

Balasubramanian, P., L.-J. Zhao, and G. Chinnadurai. 2003. Nicotinamide adenine dinucleotide stimulates oligomerization, interaction with adenovirus E1A and an intrinsic dehydrogenase activity of CtBP. *FEBS Letters* 537 (1): 157–160.

Belousov, V. V., A. F. Fradkov, K. A. Lukyanov et al. 2006. Genetically encoded fluorescent indicator for intracellular hydrogen peroxide. *Nature Methods* 3 (4): 281–286.

Beltrán, B., A. Orsi, E. Clementi, and S. Moncada. 2000. Oxidative stress and S^-nitrosylation of proteins in cells. *British Journal of Pharmacology* 129 (5): 953–960.

Berg, J., Y. P. Hung, and G. Yellen. 2009. A genetically encoded fluorescent reporter of ATP:ADP ratio. *Nature Methods* 6 (2): 161–166.

Bird, D. K., L. Yan, K. M. Vrotsos et al. 2005. Metabolic mapping of MCF10A human breast cells via multiphoton fluorescence lifetime imaging of the coenzyme NADH. *Cancer Research* 65 (19): 8766–8773.

Björnberg, O., H. Østergaard, and J. R. Winther. 2006. Mechanistic insight provided by glutaredoxin within a fusion to redox-sensitive yellow fluorescent protein. *Biochemistry* 45 (7): 2362–2371.

Blinova, K., S. Carroll, S. Bose et al. 2005. Distribution of mitochondrial NADH fluorescence lifetimes: Steady-state kinetics of matrix NADH interactions. *Biochemistry* 44 (7): 2585–2594.

Bou-Abdallah, F., N. D. Chasteen, and M. P. Lesser. 2006. Quenching of superoxide radicals by green fluorescent protein. *Biochimica et Biophysica Acta (BBA)—General Subjects* 1760 (11): 1690–1695.

Boyd, J. M., T. Subramanian, U. Schaeper, M. La Regina, S. Bayley, and G. Chinnadurai. 1993. A region in the C-terminus of adenovirus 2/5 E1a protein is required for association with a cellular phosphoprotein and important for the negative modulation of T24-ras mediated transformation, tumorigenesis and metastasis. *The EMBO Journal* 12 (2): 469–478.

Brekasis, D. and M. S. B. Paget. 2003. A novel sensor of NADH/NAD^+ redox poise in *Streptomyces coelicolor* A3 (2). *The EMBO Journal* 22 (18): 4856–4865.

Brody, S. and S. Harris. 1973. Circadian rhythms in *Neurospora*: Spatial differences in pyridine nucleotide levels. *Science* 180 (4085): 498–500.

Cannon, M. B. and S. J. Remington. 2006. Re-engineering redox-sensitive green fluorescent protein for improved response rate. *Protein Science* 15 (1): 45–57.

Cannon, M. B. and S. J. Remington. 2009. Redox-sensitive green fluorescent protein: Probes for dynamic intracellular redox responses. A review. In *Redox-Mediated Signal Transduction*, J. T. Hancock (ed.). Vol. 476 of Methods in Molecular Biology, J. M. Walker (ed.), pp. 50–64. New York: Humana Press.

Chen, S., J. R. Whetstine, S. Ghosh et al. 2009. The conserved NAD (H)-dependent corepressor CTBP-1 regulates *Caenorhabditis elegans* life span. *Proceedings of the National Academy of Sciences of the United States of America* 106 (5): 1496–1501.

Chinnadurai, G. 2002. CtBP, an unconventional transcriptional corepressor in development and oncogenesis. *Molecular Cell* 9 (2): 213–224.

Corcoran, J. A., H. A. Saffran, B. A. Duguay, and J. R. Smiley. 2009. Herpes simplex virus UL12.5 targets mitochondria through a mitochondrial localization sequence proximal to the N terminus. *Journal of Virology* 83 (6): 2601–2610.

Dai, X., Y. Li, G. Meng et al. 2009. NADPH is an allosteric regulator of HSCARG. *Journal of Molecular Biology* 387 (5): 1277–1285.

Dooley, C. T., T. M. Dore, G. T. Hanson, W. C. Jackson, S. J. Remington, and R. Y. Tsien. 2004. Imaging dynamic redox changes in mammalian cells with green fluorescent protein indicators. *Journal of Biological Chemistry* 279 (21): 22284–22293.

Eto, K., Y. Tsubamoto, Y. Terauchi et al. 1999. Role of NADH shuttle system in glucose-induced activation of mitochondrial metabolism and insulin secretion. *Science* 283 (5404): 981–985.

Fjeld, C. C., W. T. Birdsong, and R. H. Goodman. 2003. Differential binding of NAD^+ and NADH allows the transcriptional corepressor carboxyl-terminal binding protein to serve as a metabolic sensor. *Proceedings of the National Academy of Sciences of the United States of America* 100 (16): 9202–9207.

Frommer, W. B., M. W. Davidson, and R. E. Campbell. 2009. Genetically encoded biosensors based on engineered fluorescent proteins. *Chemical Society Reviews* 38 (10): 2833–2841.

Gow, A. J., Q. Chen, D. T. Hess, B. J. Day, H. Ischiropoulos, and J. S. Stamler. 2002. Basal and stimulated protein S-nitrosylation in multiple cell types and tissues. *Journal of Biological Chemistry* 277 (12): 9637–9640.

Gross, S. and D. Piwnica-Worms. 2005. Spying on cancer: Molecular imaging in vivo with genetically encoded reporters. *Cancer Cell* 7 (1): 5–15.

Gutscher, M., A.-L. Pauleau, L. Marty et al. 2008. Real-time imaging of the intracellular glutathione redox potential. *Nature Methods* 5 (6): 553–559.

Hanson, G. T., R. Aggeler, D. Oglesbee et al. 2004. Investigating mitochondrial redox potential with redox-sensitive green fluorescent protein indicators. *Journal of Biological Chemistry* 279 (13): 13044–13053.

Henley, K. S. and E. G. Laughrey. 1970. The redox state of the mitochondrial NAD system in cirrhosis of the liver and in chronic quantitative undernutrition in the rat. *Biochimica et Biophysica Acta (BBA)—General Subjects* 201 (1): 9–12.

Huang, C., Q. Yin, W. Zhu et al. 2011. Highly selective fluorescent probe for vicinal-dithiol-containing proteins and in situ imaging in living cells. *Angewandte Chemie International Edition* 50 (33): 7551–7556.

Hung, Y. P., J. G. Albeck, M. Tantama, and G. Yellen. 2011. Imaging cytosolic NADH-NAD^+ redox state with a genetically encoded fluorescent biosensor. *Cell Metabolism* 14 (4): 545–554.

Imamura, H., K. P. H. Nhat, H. Togawa et al. 2009. Visualization of ATP levels inside single living cells with fluorescence resonance energy transfer-based genetically encoded indicators. *Proceedings of the National Academy of Sciences of the United States of America* 106 (37): 15651–15656.

Jacob, C., A. L. Holme, and F. H. Fry. 2004. The sulfinic acid switch in proteins. *Organic and Biomolecular Chemistry* 2 (14): 1953–1956.

Jaffrey, S. R., H. Erdjument-Bromage, C. D. Ferris, P. Tempst, and S. H. Snyder. 2001. Protein S-nitrosylation: A physiological signal for neuronal nitric oxide. *Nature Cell Biology* 3 (2): 193–197.

Jones, D. P. 1981. Determination of pyridine dinucleotides in cell extracts by high-performance liquid chromatography. *Journal of Chromatography B: Biomedical Sciences and Applications* 225 (2): 446–449.

Joubert, F., H. M. Fales, H. Wen, C. A. Combs, and R. S. Balaban. 2004. NADH enzyme-dependent fluorescence recovery after photobleaching (ED-FRAP): Applications to enzyme and mitochondrial reaction kinetics, in vitro. *Biophysical Journal* 86 (1): 629–645.

Kasimova, M. R., J. Grigiene, K. Krab et al. 2006. The free NADH concentration is kept constant in plant mitochondria under different metabolic conditions. *The Plant Cell Online* 18 (3): 688–698.

Kasischke, K. A., H. D. Vishwasrao, P. J. Fisher, W. R. Zipfel, and W. W. Webb. 2004. Neural activity triggers neuronal oxidative metabolism followed by astrocytic glycolysis. *Science* 305 (5680): 99–103.

Kim, J.-H., E.-J. Cho, S.-T. Kim, and H.-D. Youn. 2005. CtBP represses p300-mediated transcriptional activation by direct association with its bromodomain. *Nature Structural and Molecular Biology* 12 (5): 423–428.

Knöpfel, T., J. Díez-García, and W. Akemann. 2006. Optical probing of neuronal circuit dynamics: Genetically encoded versus classical fluorescent sensors. *Trends in Neurosciences* 29 (3): 160–166. Kumar, P. R., Y. Yu, R. Sternglanz, S. A. Johnston, and L. Joshua-Tor. 2008. NADP regulates the yeast GAL induction system. *Science* 319 (5866): 1090–1092.

Kumar, V., J. E. Carlson, K. A. Ohgi et al. 2002. Transcription corepressor CtBP is an NAD^+-regulated dehydrogenase. *Molecular Cell* 10 (4): 857–869.

Lamb, H. K., D. K. Stammers, and A. R. Hawkins. 2008. Dinucleotide-sensing proteins: Linking signaling networks and regulating transcription. *Science Signaling* 1 (33): pe38.

Lee, K., J. Lee, Y. Kim et al. 2004. Defining the plant disulfide proteome. *Electrophoresis* 25 (3): 532–541.
Leichert, L. I. and U. Jakob. 2004. Protein thiol modifications visualized in vivo. *PLoS Biology* 2 (11): e333.
Lohman, J. R. and S. J. Remington. 2008. Development of a family of redox-sensitive green fluorescent protein indicators for use in relatively oxidizing subcellular environments. *Biochemistry* 47 (33): 8678–8688.
Lu, H., D. Burns, P. Garnier, G. Wei, K. Zhu, and W. Ying. 2007. $P2X_7$ receptors mediate NADH transport across the plasma membranes of astrocytes. *Biochemical and Biophysical Research Communications* 362 (4): 946–950.
Maller, C., E. Schröder, and P. Eaton. 2011. Glyceraldehyde 3-phosphate dehydrogenase is unlikely to mediate hydrogen peroxide signaling: Studies with a novel anti-dimedone sulfenic acid antibody. *Antioxidants and Redox Signaling* 14 (1): 49–60.
Mandavilli, B. S. and M. S. Janes. 2010. Detection of intracellular glutathione using ThiolTracker violet stain and fluorescence microscopy. *Current Protocols in Cytometry* 53(9.35): 9.35.1–9.35.8.
Mayevsky, A. 2009. Mitochondrial function and energy metabolism in cancer cells: Past overview and future perspectives. *Mitochondrion* 9 (3): 165–179.
McLaughlin, K. J., C. M. Strain-Damerell, K. Xie et al. 2010. Structural basis for NADH/NAD^+ redox sensing by a rex family repressor. *Molecular Cell* 38 (4): 563–575.
Merksamer, P. I., A. Trusina, and F. R. Papa. 2008. Real-time redox measurements during endoplasmic reticulum stress reveal interlinked protein folding functions. *Cell* 135 (5): 933–947.
Meyer, A. J. and T. P. Dick. 2010. Fluorescent protein-based redox probes. *Antioxidants and Redox Signaling* 13 (5): 621–650.
Moran, L. K., J. M. C. Gutteridge, and G. J. Quinlan. 2001. Thiols in cellular redox signalling and control. *Current Medicinal Chemistry* 8 (7): 763–772.
Mukhopadhyay, P., M. Zheng, L. A. Bedzyk, R. A. LaRossa, and G. Storz. 2004. Prominent roles of the NorR and Fur regulators in the *Escherichia coli* transcriptional response to reactive nitrogen species. *Proceedings of the National Academy of Sciences of the United States of America* 101 (3): 745–750.
Nagai, T., A. Sawano, E. Sun Park, and A. Miyawaki. 2001. Circularly permuted green fluorescent proteins engineered to sense Ca^{2+}. *Proceedings of the National Academy of Sciences of the United States of America* 98 (6): 3197–3202.
Nagai, T., S. Yamada, T. Tominaga, M. Ichikawa, and A. Miyawaki. 2004. Expanded dynamic range of fluorescent indicators for Ca^{2+} by circularly permuted yellow fluorescent proteins. *Proceedings of the National Academy of Sciences of the United States of America* 101 (29): 10554–10559.
Nakahata, Y., S. Sahar, G. Astarita, M. Kaluzova, and P. Sassone-Corsi. 2009. Circadian control of the NAD^+ salvage pathway by CLOCK-SIRT1. *Science* 324 (5927): 654–657.
Nakai, J., M. Ohkura, and K. Imoto. 2001. A high signal-to-noise Ca^{2+} probe composed of a single green fluorescent protein. *Nature Biotechnology* 19 (2): 137–141.
Nardini, M., S. Spanò, C. Cericola et al. 2003. CtBP/BARS: A dual-function protein involved in transcription co-repression and Golgi membrane fission. *The EMBO Journal* 22 (12): 3122–3130.
Niino, Y., K. Hotta, and K. Oka. 2010. Blue fluorescent cGMP sensor for multiparameter fluorescence imaging. *PloS One* 5 (2): e9164.
O'Neill, J. S. and A. B. Reddy. 2011. Circadian clocks in human red blood cells. *Nature* 469 (7331): 498–503.
Østergaard, H., A. Henriksen, F. G. Hansen, and J. R. Winther. 2001. Shedding light on disulfide bond formation: Engineering a redox switch in green fluorescent protein. *The EMBO Journal* 20 (21): 5853–5862.
Pagels, M., S. Fuchs, J. Pané-Farré et al. 2010. Redox sensing by a Rex-family repressor is involved in the regulation of anaerobic gene expression in *Staphylococcus aureus*. *Molecular Microbiology* 76 (5): 1142–1161.
Patterson, G. H., S. M. Knobel, P. Arkhammar, O. Thastrup, and D. W. Piston. 2000. Separation of the glucose-stimulated cytoplasmic and mitochondrial NAD(P)H responses in pancreatic islet β cells. *Proceedings of the National Academy of Sciences of the United States of America* 97 (10): 5203–5207.
Ravcheev, D. A., X. Li, H. Latif et al. 2012. Transcriptional regulation of central carbon and energy metabolism in bacteria by redox-responsive repressor Rex. *Journal of Bacteriology* 194 (5): 1145–1157.

Rice, G. C., E. A. Bump, D. C. Shrieve, W. Lee, and M. Kovacs. 1986. Quantitative analysis of cellular glutathione by flow cytometry utilizing monochlorobimane: Some applications to radiation and drug resistance in vitro and in vivo. *Cancer Research* 46 (12): 6105–6110.

Rosenwasser, S., I. Rot, A. J. Meyer, L. Feldman, K. Jiang, and H. Friedman. 2010. A fluorometer-based method for monitoring oxidation of redox‾sensitive GFP (roGFP) during development and extended dark stress. *Physiologia Plantarum* 138 (4): 493–502.

Rutter, J., M. Reick, L. C. Wu, and S. L. McKnight. 2001. Regulation of clock and NPAS2 DNA binding by the redox state of NAD cofactors. *Science* 293 (5529): 510–514.

Sanders, B. D., K. Zhao, J. T. Slama, and R. Marmorstein. 2007. Structural basis for nicotinamide inhibition and base exchange in Sir2 enzymes. *Molecular Cell* 25 (3): 463–472.

Saurin, A. T., H. Neubert, J. P. Brennan, and P. Eaton. 2004 Widespread sulfenic acid formation in tissues in response to hydrogen peroxide. *Proceedings of the National Academy of Sciences of the United States of America* 101 (52): 17982–17987.

Schaeper, U., J. M. Boyd, S. Verma, E. Uhlmann, T. Subramanian, and G. Chinnadurai. 1995. Molecular cloning and characterization of a cellular phosphoprotein that interacts with a conserved C-terminal domain of adenovirus E1A involved in negative modulation of oncogenic transformation. *Proceedings of the National Academy of Sciences of the United States of America* 92 (23): 10467–10471.

Schwerzmann, K., L. M. Cruz-Orive, R. Eggman, A. Sänger, and E. R. Weibel. 1986. Molecular architecture of the inner membrane of mitochondria from rat liver: A combined biochemical and stereological study. *The Journal of Cell Biology* 102 (1): 97–103.

Seo, Y. H. and K. S. Carroll. 2009. Profiling protein thiol oxidation in tumor cells using sulfenic acid-specific antibodies. *Proceedings of the National Academy of Sciences of the United States of America* 106 (38): 16163–16168.

Sevier, C. S. and C. A. Kaiser. 2002. Formation and transfer of disulphide bonds in living cells. *Nature Reviews Molecular Cell Biology* 3 (11): 836–847.

Shaner, N. C., P. A. Steinbach, and R. Y. Tsien. 2005. A guide to choosing fluorescent proteins. *Nature Methods* 2 (12): 905–909.

Shigemori, K., T. Ishizaki, S. Matsukawa, A. Sakai, T. Nakai, and S. Miyabo. 1996. Adenine nucleotides via activation of ATP-sensitive K^+ channels modulate hypoxic response in rat pulmonary artery. *American Journal of Physiology—Lung Cellular and Molecular Physiology* 270 (5): L803–L809.

Sickmier, E. A., D. Brekasis, S. Paranawithana et al. 2005. X-ray structure of a Rex-family repressor/NADH complex insights into the mechanism of redox sensing. *Structure* 13 (1): 43–54.

Simbula, G., A. Columbano, G. M. Ledda-Columbano et al. 2007. Increased ROS generation and p53 activation in α-lipoic acid-induced apoptosis of hepatoma cells. *Apoptosis* 12 (1): 113–123.

Simen Zhao, B., Y. Liang, Y. Song, C. Zheng, Z. Hao, and P. R. Chen. 2010. A highly selective fluorescent probe for visualization of organic hydroperoxides in living cells. *Journal of the American Chemical Society* 132 (48): 17065–17067.

Skala, M. C., K. M. Riching, A. Gendron-Fitzpatrick et al. 2007. In vivo multiphoton microscopy of NADH and FAD redox states, fluorescence lifetimes, and cellular morphology in precancerous epithelia. *Proceedings of the National Academy of Sciences of the United States of America* 104 (49): 19494–19499.

Szabo, C., B. Zingarelli, M. O'Connor, and A. L. Salzman. 1996. DNA strand breakage, activation of poly (ADP-ribose) synthetase, and cellular energy depletion are involved in the cytotoxicity of macrophages and smooth muscle cells exposed to peroxynitrite. *Proceedings of the National Academy of Sciences of the United States of America* 93 (5): 1753–1758.

Tauskela, J. S., K. Hewitt, L.‾P. Kang et al. 2000. Evaluation of glutathione‾sensitive fluorescent dyes in cortical culture. *Glia* 30 (4): 329–341.

Thornburg, J., K. Nelson, B. Clem et al. 2008. Targeting aspartate aminotransferase in breast cancer. *Breast Cancer Research* 10 (5): R84.

Vishwasrao, H. D., A. A. Heikal, K. A. Kasischke, and W. W. Webb. 2005. Conformational dependence of intracellular NADH on metabolic state revealed by associated fluorescence anisotropy. *Journal of Biological Chemistry* 280 (26): 25119–25126.

Wakita, M., G. Nishimura, and M. Tamura. 1995. Some characteristics of the fluorescence lifetime of reduced pyridine nucleotides in isolated mitochondria, isolated hepatocytes, and perfused rat liver in situ. *Journal of Biochemistry* 118 (6): 1151–1160.

Wang, E., M. C. Bauer, A. Rogstam, S. Linse, D. T. Logan, and C. Von Wachenfeldt. 2008a. Structure and functional properties of the *Bacillus subtilis* transcriptional repressor Rex. *Molecular Microbiology* 69 (2): 466–478.Wang, H., E. Nakata, and I. Hamachi. 2009. Recent progress in strategies for the creation of protein-based fluorescent biosensors. *ChemBioChem* 10 (16): 2560–2577.

Wang, W., H. Fang, L. Groom et al. 2008b. Superoxide flashes in single mitochondria. *Cell* 134 (2): 279–290.

Williamson, D. H., P. Lund, and H. A. Krebs. 1967. The redox state of free nicotinamide-adenine dinucleotide in the cytoplasm and mitochondria of rat liver. *Biochemical Journal* 103: 514–527.

Xu, X., K. von Löhneysen, K. Soldau, D. Noack, A. Vu, and J. S. Friedman. 2011. A novel approach for in vivo measurement of mouse red cell redox status. *Blood* 118 (13): 3694–3697.

Yang, H., T. Yang, J. A. Baur et al. 2007a. Nutrient-sensitive mitochondrial NAD^+ levels dictate cell survival. *Cell* 130 (6): 1095–1107.

Yang, Y. and J. Loscalzo. 2005. S-nitrosoprotein formation and localization in endothelial cells. *Proceedings of the National Academy of Sciences of the United States of America* 102 (1): 117–122.

Yang, Y., Y. Song, and J. Loscalzo. 2007b. Regulation of the protein disulfide proteome by mitochondria in mammalian cells. *Proceedings of the National Academy of Sciences of the United States of America* 104 (26): 10813–10817.

Ying, W. 2008. NAD^+/NADH and $NADP^+$/NADPH in cellular functions and cell death: Regulation and biological consequences. *Antioxidants and Redox Signaling* 10 (2): 179–206.

Ying, W., M. B. Sevigny, Y. Chen, and R. A. Swanson. 2001. Poly (ADP-ribose) glycohydrolase mediates oxidative and excitotoxic neuronal death. *Proceedings of the National Academy of Sciences of the United States of America* 98 (21): 12227–12232.

Yu, Q. and A. A. Heikal. 2009. Two-photon autofluorescence dynamics imaging reveals sensitivity of intracellular NADH concentration and conformation to cell physiology at the single-cell level. *Journal of Photochemistry and Photobiology B: Biology* 95 (1): 46–57.

Zapata-Hommer, O. and O. Griesbeck. 2003. Efficiently folding and circularly permuted variants of the Sapphire mutant of GFP. *BMC Biotechnology* 3 (1): 5.

Zhang, Q., D. W. Piston, and R. H. Goodman. 2002. Regulation of corepressor function by nuclear NADH. *Science* 295 (5561): 1895–1897.

Zhang, Q., S.-Y. Wang, A. C. Nottke, J. V. Rocheleau, D. W. Piston, and R. H. Goodman. 2006. Redox sensor CtBP mediates hypoxia-induced tumor cell migration. *Proceedings of the National Academy of Sciences of the United States of America* 103 (24): 9029–9033.

Zhao, Y., S. Araki, J. Wu et al. 2011a. An expanded palette of genetically encoded Ca^{2+} indicators. *Science* 333 (6051): 1888–1891.

Zhao, Y., J. Jin, Q. Hu et al. 2011b. Genetically encoded fluorescent sensors for intracellular NADH detection. *Cell Metabolism* 14 (4): 555–566.

Zhao, Y. and Y. Yang. 2012. Frex and FrexH: Indicators of metabolic states in living cells. *Bioengineered Bugs* 3 (3): 183–190.

Zhao, Y., J. Zhang, H. Li, Y. Li, J. Ren, M. Luo, and X. Zheng. 2008. An NADPH sensor protein (HSCARG) down-regulates nitric oxide synthesis by association with argininosuccinate synthetase and is essential for epithelial cell viability. *Journal of Biological Chemistry* 283 (16): 11004–11013.

Zheng, X., X. Dai, Y. Zhao et al. 2007. Restructuring of the dinucleotide-binding fold in an NADP(H) sensor protein. *Proceedings of the National Academy of Sciences of the United States of America* 104 (21): 8809–8814.

III

Natural Biomarkers for Biochemical and Biological Studies

9

Spatiotemporal Detection of NADH-Linked Enzyme Activities in Single Cell Metabolism

V. Krishnan Ramanujan
Cedars-Sinai Medical Center

9.1 Introduction

Living systems orchestrate an ensemble of structural and functional hierarchy of metabolic networks in order to regulate homeostasis (DeBerardinis et al. 2008; De la Fuente et al. 2006; Digman et al. 2009; Havlin et al. 1999; Ramanujan et al. 2008; Stanley et al. 1999; Wallace 1999). A useful guide to understand this complex labyrinth of energy landscapes is to exploit steady-state approximation, which states that under homeostatic conditions, the rate of generation of any excess metabolite/cofactor is equal to the rate of such perturbation removal. In this paradigm, the function of external/internal stimuli is to enable living systems to explore new steady states by means of transient perturbations. Unfavorable conditions during this exploration process may affect the bioenergetics of the living system reversibly or irreversibly. These changes form the basis of onset of disease processes and various pathophysiology. In the realm of spatial dimensions, a complete theoretical understanding of cellular metabolism is not a realistic goal due to the multitude of players and interactions among them. Experimental investigation is also not straightforward for the same reason. Figure 9.1 illustrates a simple example of NADH oxidation that has an influence on a relatively intricate relationship between the most fundamental bioenergetic pathways in a cell. Historically, most of the enzyme activities have been studied in isolated conditions; these cell-free systems offered an analytical perspective of the various enzymes studied (e.g., binding or dissociation rates and saturation concentration). Most of the prevailing methods of interrogating energy metabolism in living systems have

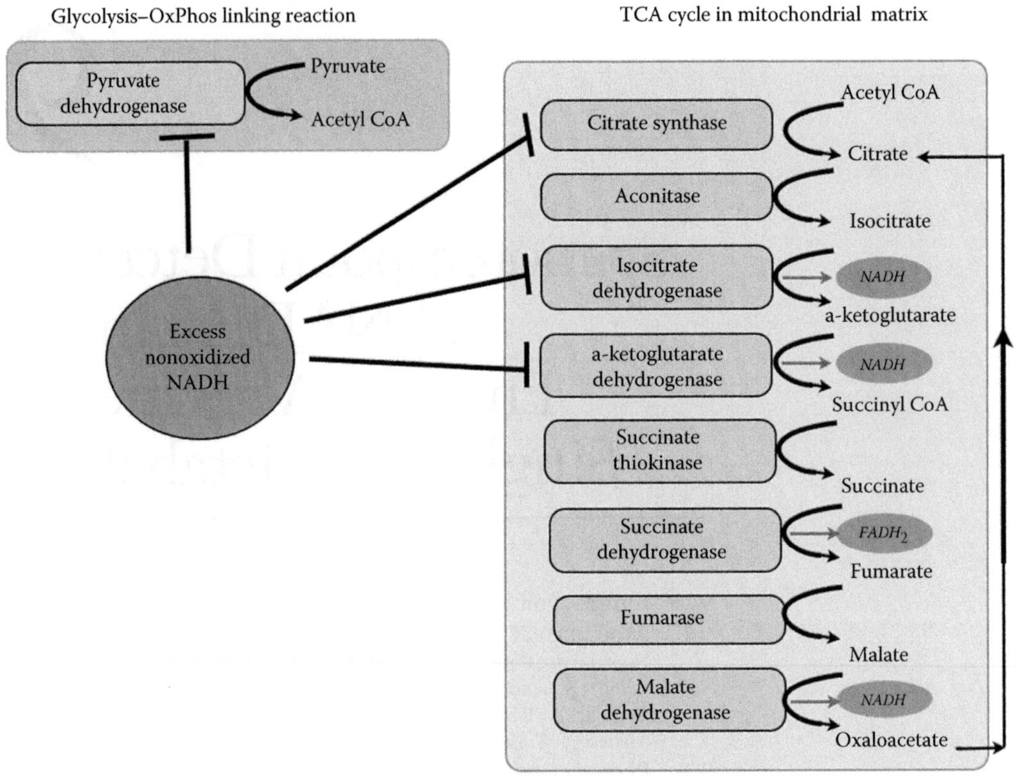

FIGURE 9.1 TCA cycle, also known as tricarboxylic acid cycle, generates electron carriers, NADH, and $FADH_2$. All the enzymes are regulated by substrate availability and product inhibition. In situations where NADH is unable to be oxidized (a variety of pathological states including cancer and other metabolic dysfunctions), the excess nonoxidized NADH inhibits the key TCA cycle enzymes (via product inhibition) as shown here thereby triggering the metabolic switch toward preferential glycolysis even in the presence of oxygen. This aerobic glycolysis or Warburg effect further sets the platform for the increased glycolytic dependence commonly observed in many cancer phenotypes.

a common reductionist approach where any particular dysfunction of the organism is traced back to one or more proteins. Experimentally, individual cell components (e.g., in biopsy specimens) are isolated, and their activity is tested in vitro, often in nonphysiological conditions. Through systematic progress, we have come a long way from the cell-free systems to cell lysates to whole-cell assays and spatially resolved single cell imaging assays (Chance 1964, 2004; Chance and Schoener 1963; Chance et al. 1979). Besides these advancements in spatial domain, technological advancements and sophisticated strategies for data analysis have greatly enabled a significant leap in understanding the cellular enzyme activities at various timescales. Despite the success of such a reductionist approach, a comprehensive understanding of energy metabolism at the level of whole organism is still lacking. Conventional biochemical assays implicitly assume steady-state approximation at all spatial and timescales and therefore yield a steady-state, time-averaged picture of the living systems rather than a dynamic model with nonlinear dynamical or chaotic behaviors governing a variety of regulatory processes. In order to understand the synergetic roles of the interacting partners and the emergent behavior of the various networks, it is imperative to investigate the response of the networks as a whole in addition to conventional methods. Our lab has been involved in addressing these pivotal questions by developing high-resolution imaging approaches as well as novel analysis methods to unravel single cell metabolic networks. In this chapter, we will present one of the exciting aspects of single cell metabolism where we recently identified nonlinear dynamical behavior in NADH-linked enzyme activities in living cells.

9.2 Experimental Methods

9.2.1 Fluorescence Contrast in Monitoring Enzyme Activity in Cellular Systems

Traditionally, enzyme activities have been monitored by absorption spectroscopy; addition of fluorescence modalities increased specificity and sensitivity of these assays. Over the years, our understanding of enzyme activities has greatly improved owing to the steady realization of techniques that allowed monitoring cellular enzyme activities not only in isolated conditions but also in intact cells thereby adding value to the authenticity of the data. Toward this direction, fluorescence imaging methods have revolutionized the visualization of intracellular enzyme activities with unprecedented specificity. Figure 9.2 illustrates a typical scenario where moderate dysfunction in mitochondrial complex I can affect other metabolic routes thereby exacerbating the mitochondrial dysfunction as a whole. The earliest redox fluorimetry measurements in tumor models, pioneered by Britton Chance, focused on monitoring the intensity ratio of oxidized flavoprotein and reduced pyridine nucleotides, which are known to be metabolic indicators in living cells and tissues (Chance et al. 1962). As recognized by Chance and others, fluorescence imaging systems possess (1) greater flexibility in design and implementation and (2) higher spatial resolution and specificity as compared to absorption spectroscopy. Another significant development in redox fluorimetry, a low-temperature high-resolution scanner, was employed in analyzing the intensity ratio of FAD/(FAD + NADH) (redox ratio) from rapidly frozen breast tissues. The authors could find significant differences between normal and severely dysplastic tissues (Xu et al. 2013). Despite the increased sensitivity and good spatial resolution (~50 μm), this method required maintaining the tissues at liquid nitrogen temperature (–196°C) and thereby could not be viable for monitoring tumors in vivo in a live animal.

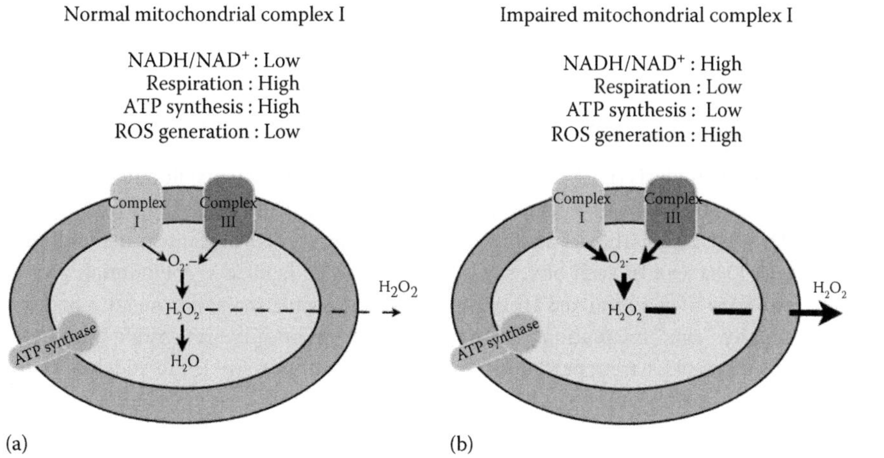

FIGURE 9.2 Mitochondrial complex I is the entry site of the electron transport chain on the inner mitochondrial membrane. This enzyme catalyzes the reaction $NADH + H^+ + CoQ + 4H^+_{in} \rightarrow NAD^+ + CoQH^2 + 4H^+_{out}$. (a) In healthy mitochondria, electron transfer across complex I and complex III produces negligible electron leak and hence very low superoxide (O_2^-) and hydrogen peroxide (H_2O_2). (b) Under conditions of complex I deficiency/dysfunction, $NADH/NAD^+$ ratio becomes high, which in turn leads to a significant increase in reactive oxygen species (ROS).

It is well known that not a single technique could provide a complete picture of tumor metabolism in vivo. For instance, in the context of redox imaging, intensity imaging methods are simpler to implement but are prone to spectral artifacts, such as overlap between NADH and flavoprotein emission spectra, and concentration artifacts, which might obscure the final analysis of redox ratios in thick tissues (Mayevsky and Chance 2007). This could be taken care of by implementing a spectrally resolved data acquisition; however, this modality cannot discriminate between spectrally similar redox species, such as free and protein-bound NAD(P)H. Fluorescence lifetime imaging, described in details in Chapter 4, can be valuable in discriminating spectrally similar redox species that have different lifetimes. Steady-state intensity imaging is insufficient to gain insight into submillisecond enzyme regulation of tumors, which brings in the necessity of kinetic contrast that is being proposed in this project. We therefore believe that a comprehensive metabolic imaging platform that combines all the aforementioned approaches will be a valuable tool in understanding glucose-induced tumor metabolism in vivo. Other chapters in this book are devoted to addressing these various fluorescence methods in a greater detail, and therefore we will focus here on a novel method of monitoring enzyme activity by means of high-resolution time-series analysis of NAD(P)H fluctuations and spatially resolved imaging of NAD(P)H oxidation kinetics in live cells.

9.2.2 Nonlinear Dynamics and Scaling Analysis Approach

Living cells are open systems that operate far from equilibrium (DeBerardinis et al. 2008; De la Fuente et al. 2006; Digman et al. 2009; Havlin et al. 1999; Stanley et al. 1999; Wallace 1999). Experimental studies reporting various manifestations of nonlinear dynamics in clinical pathology have revolutionized our perspectives on health and disease. Despite this realization, a fundamental understanding of these nonequilibrium processes at the level of a single cell is still lacking. Demands of cellular homeostasis under such nonequilibrium conditions require coordinated response of many regulatory networks so as to maintain constant levels of metabolites and cofactors. These networks are comprised of transcription factors and regulatory proteins and form the hub of cellular decision-making process under both normal and stressed conditions. In order to understand the synergetic roles of mechanisms that govern regulatory networks at the single cell level, it is important to develop innovative strategies for monitoring specific in vivo responses in real time. We describe two methods for analyzing the time-series data obtained from cells.

9.2.3 Time-Series Analysis: Detrended Fluctuation Analysis

Detrended fluctuation analysis (DFA) is a modified root-mean-square (rms) fluctuation analysis of random walk originally developed by Peng et al. to quantify statistical correlations in time-series signal (Peng et al. 1992, 1994, 1995). The original time series (of length N) is first integrated and then divided into boxes of equal size (n). In each box, the integrated profile is fit to a polynomial, providing with a local trend, which is then subtracted from the integrated profile in each box—an operation termed "detrending." Finally, "rms" fluctuation $F(n)$ is calculated from both the integrated and detrended signals in each box. These steps are repeated for different values of box size (n) to generate $F(n)$ for broad range of scale sizes n. Intuitively, $F(n)$ will increase as n increases; for scale-invariant signals with power law correlations, there is a scaling relationship $F(n) = n^{\alpha}$. In other words, DFA dissects the original time-series data into many windows with progressively increasing scale size n and calculates the "rms" fluctuations $F(n)$ for every scale size n to yield a scaling function

$$F(n) = n^{\alpha} \tag{9.1}$$

The scaling exponent α is a quantitative measure of extent of correlations in the signal—it can characterize randomness (e.g., $\alpha = 0.5$ for white noise; $\alpha = 1.5$ for Brownian noise) or correlations ($\alpha < 0.5$

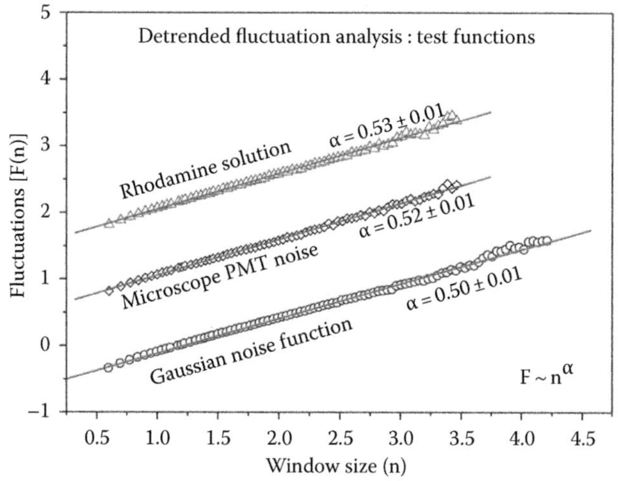

FIGURE 9.3 DFA plots for simulated Gaussian noise function, fluorescent solution (Rhodamine 6G), and instrumental PMT noise. All these yielded a scaling exponent $\alpha = 0.5$ corresponding to uncorrelated randomness.

for anticorrelations and $0.5 < \alpha < 1.5$ for persistent power law–like correlations) regardless of the nature or source of the fluctuations. In a sense, the scaling function is a universal that can be used to analyze fluctuations in any time-series signal. Similar crossover in exponents have been noted in earlier studies of physiological signals (Stanley et al. 1999). An important advantage of the DFA algorithm over other time-series analysis methods is that it can also be reliably used for nonstationary signals since local detrending eliminates the errors associated with nonstationarity. Although this algorithm has been successfully used in quantifying correlations in physiological signals, such as heart beat dynamics and human gait dynamics (Stanley et al. 1994), to the best of our knowledge, there is no systematic application of time-series analysis in single cell dynamics as presented here. Figure 9.3 summarizes the DFA plots for simulated noise functions, microscope photomultiplier tube (PMT) noise, and fluorescent solution specimens (Rhodamine 6G and 100 μM NADH solution in water) to demonstrate that the observed correlated dynamics is not due to instrument artifacts. More details about this algorithm can be found in Ramanujan et al. (2006), and source code can be obtained from the webpage of National Research source (Peng et al. 1992, 1994).

9.2.4 Time-Series Analysis: Fast Fourier Transform

Fast Fourier transform (FFT) is one of the well-established power spectrum analysis methods to quantify correlations in a time-series signal. The basic idea is to convert the signal in time domain (s) to frequency domain (Hz) and compute the different frequency components in the signal. A periodic wave (e.g., a sine wave) oscillating with a period T in time domain will have a single frequency component $f = 1/T$ in FFT amplitude output. Usually, time-series analysis output is represented as power spectral density (PSD), which is the square of the FFT amplitude. For an uncorrelated noise input, the PSD will be flat for the entire frequency scale implying that all frequencies have almost equal amplitudes. In other words, no particular frequency is preferred in noise. This flat PSD output is the source of the name "white noise" for uncorrelated randomness. To quantify correlations in an unknown signal, the FFT PSD is plotted as a function of frequency (log–log scale). PSD follows a scaling relationship

$$P(f) = \frac{1}{f^{\beta}} \tag{9.2}$$

where the exponent β characterizes randomness (β = 0 for white noise and β = 2 for Brownian noise). Figure 9.5c shows the representative PSD plots for control hepatocytes indicating long-range correlations. FFT plots were obtained by using FFT tool in Origin 8.0 (OriginLab). These plots clearly indicate long-range correlations in redox fluctuations and further substantiate the results obtained independently from DFA. Recently, more exhaustive exploration of these ideas has been well documented in literature (Amaral et al. 2000; Aon et al. 2003, 2009; De la Fuente et al. 2008, 2009).

9.3 Representative Case Studies

Recent studies indicate the critical roles that mitochondria play in a variety of metabolic syndromes, including cancer and aging process. It will be intriguing to see how our scaling analysis approach can be exploited to offer insights into the modifications of regulatory dynamics during aging and other disease processes as described in Sections 9.3.1 and 9.3.2.

9.3.1 Time-Series Analysis of NAD(P)H Fluctuations in Living Cells

As a first example demonstrating the utility of the time-series analysis described earlier, we present our recent work, where we applied these statistical correlation analysis tools to gain insight into the enzyme kinetics of a classical mitochondrial enzyme network, the tricarboxylic acid (TCA) cycle. The rationale for the choice of this system is based on two observations: (1) generation of ATP by mitochondria occurs in a highly controlled manner (supply on demand) determined by the cytosolic ADP levels and mitochondrial substrate availability and (2) regulation of key enzymes in the TCA cycle is governed by substrate availability and product inhibition. The mitochondrial electron transport chain couples these two processes at the biochemical level by feeding the output of the TCA cycle (NADH and reduced flavin adenine dinucleotide [$FADH_2$]) to generate ATP and by utilizing the ATP levels to activate/inhibit the TCA cycle. As can be seen, this is a classical situation of nonlinear feedback regulation where the output information (in this case, ADP/ATP ratio) is fed as the input to regulate the main process of electron transport mediated by the TCA cycle enzymes. Under normal circumstances, NADH level determines both ADP/ATP ratio as well as the activity of the TCA cycle enzymes (product inhibition). We therefore reasoned that monitoring statistical correlations in redox (NADH/NAD^+) fluctuations would be a promising experimental approach for understanding the regulatory dynamics of the TCA cycle enzymes. We measured NAD(P)H fluorescence in living primary hepatocytes (liver cells) isolated from a young (5 months old) mouse, as well as in intact mitochondria isolated from these hepatocytes. The latter allowed us to investigate the effects of eliminating nonmitochondrial NAD(P)H contributions in the observed signal fluctuations.

9.3.1.1 Methods: Redox Time-Series Data Acquisition

NAD(P)H fluorescence in cultured hepatocytes or mitochondrial suspensions were monitored by a homebuilt two-photon fluorescence imaging system, based on Olympus IX-71 microscope, coupled to a titanium–sapphire femtosecond laser delivering pulses at the 76 MHz repetition rate (Mira 900, Coherent). Images were acquired with a water immersion 63×/1.2 NA objective (Olympus Inc., IR Corrected). For achieving optimal two-photon imaging of NAD(P)H, all the experiments were carried out by exciting the specimen at 730 nm and detecting the fluorescence in a non-descanned layout (PMT R1894, Hamamatsu Photonics) through a 480/40 nm emission filter (item no. 31044v2, Chroma Technology Corp., Rockingham, VT). Time-series data for monitoring redox fluctuations in mitochondrial suspensions and intact hepatocytes were acquired as follows: a field of view with ~4–6 cells was chosen before measurements. In the case of isolated mitochondrial suspensions, line scans were chosen to represent fields of view with uniform fluorescence intensity to minimize artifacts due to the aggregates of mitochondria. For achieving high time resolution, acquisition was performed in a line scan mode with the line drawn in a way to span all the cells or nearly 50–100 mitochondria. Typically, time-series data were acquired for $N = 8000$ time points (~30 ms/line; ~58 μs/pixel). We employed minimal laser power (0.5%–1% of the maximum 20 mW) at the back aperture of the objective to minimize photobleaching. For spatially resolved NAD(P)H oxidation kinetics imaging, the aforementioned microscope platform was used. However, instead of line (*xt*) scans, 2D (*xy*) images (512 × 512 pixels) were obtained in a time-lapse mode (typical time interval ~10 s) for the duration shown in the figures. NAD(P)H fluctuations were measured in *xt* line scan mode (time-series data; 512 × *T*; ~30 ms/scan). All measurements were performed in 10 mM 4-(2-hydroxyethyl)-1-piperazineethanesulfonic acid (HEPES) buffered Hank's balanced salt solution (phenol free) at room temperature.

9.3.1.2 Results

Figure 9.4 shows representative steady-state fluorescence images of individual mitochondria and intact liver cells. With the microscope configuration and the excitation/emission settings described earlier, these images show clearly the fluorescence signal originating from NAD(P)H in the mitochondrial compartments since $NAD(P)^+$ is known to be nonfluorescent. In the case of isolated mitochondrial suspensions, we also observed regions with stronger fluorescence signal than the rest of the field of view. These regions are speculated to be mitochondrial. Unlike the rounded, punctuate mitochondria in suspensions, NAD(P)H signals from intact cells displayed the classical, wormlike mitochondrial morphology, which further is an indication of good cellular health. In both cases, NAD(P)H signals were significantly above the instrument background.

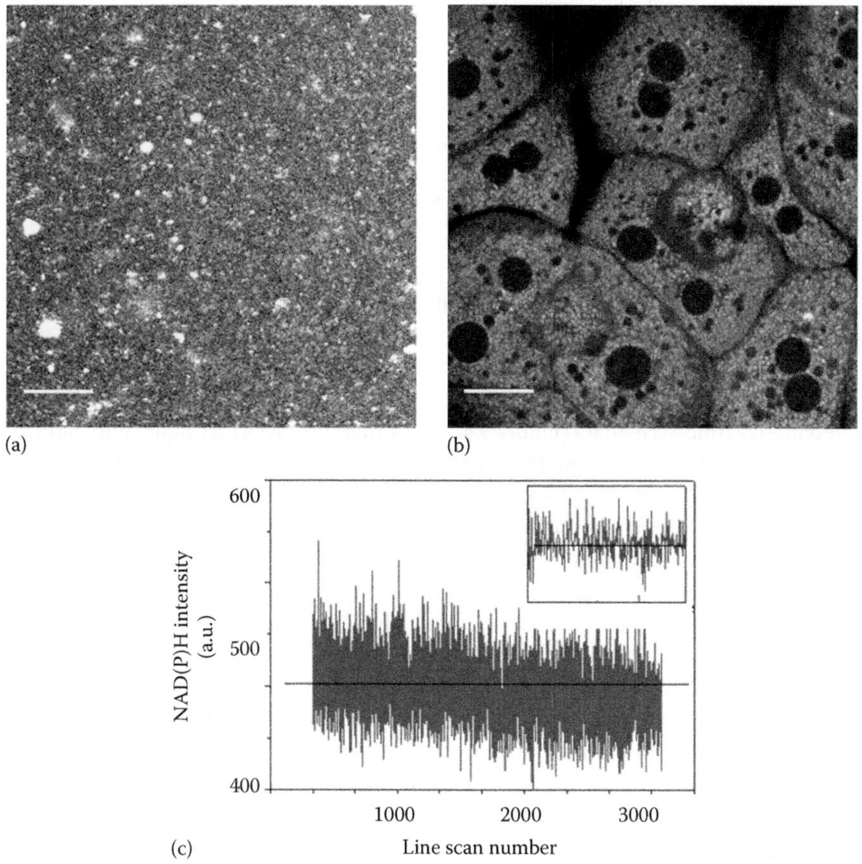

FIGURE 9.4 Steady-state multiphoton NAD(P)H images of (a) isolated mitochondria and (b) living hepatocytes. Scale bars = 20 μm. (c) Representative NAD(P)H intensity fluctuations plotted from the time-series data. The inset shows an enlarged view of shorter time scales.

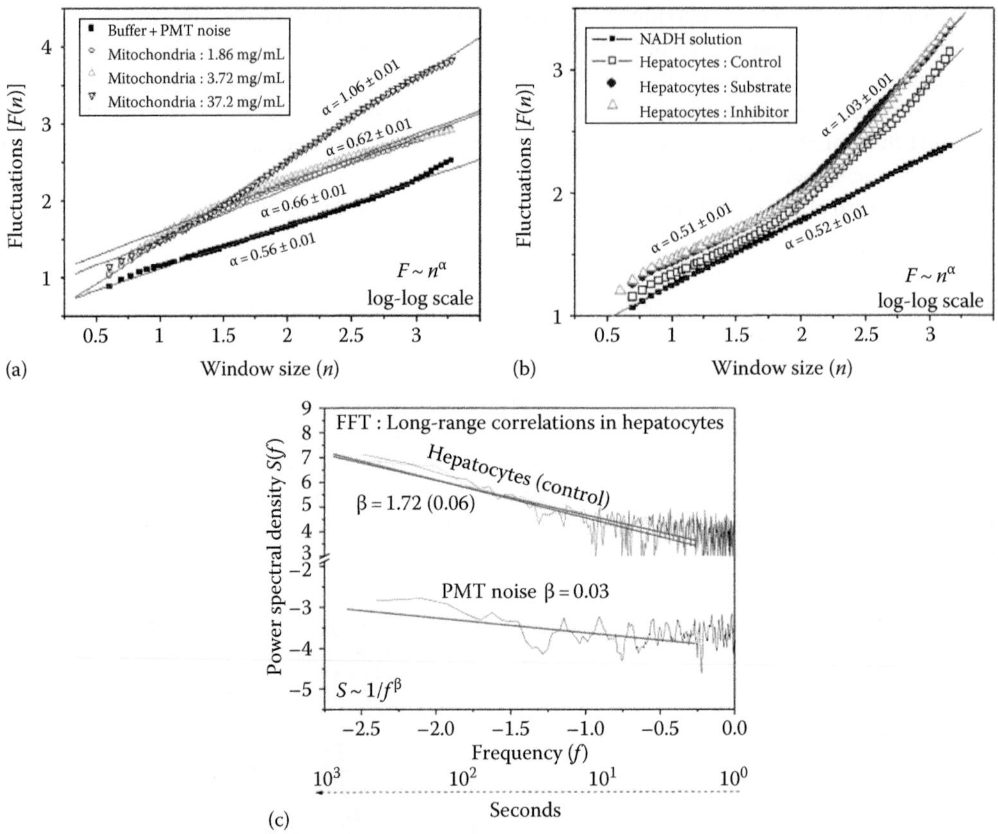

FIGURE 9.5 (a) Log–log plot of fluctuations (F) and scale size (n) obtained from the DFA analysis of the time series recorded from isolated mitochondria in three different concentrations 1.86, 3.72, and 37.2 mg/mL, respectively. (b) Scaling exponents of the DFA analysis of the time series recorded from hepatocytes. (c) Representative FFT plots for hepatocytes showing nonrandom, positive correlations.

The representative time-series data (Figure 9.4c) show that the NAD(P)H signal fluctuates around the mean value. Figure 9.5 shows representative DFA and FFT plots obtained from the time series recorded from isolated mitochondria and hepatocytes. A concentration-dependent increase in positive correlations can be seen on the DFA plots of mitochondria suspension in Figure 9.5a ($\alpha \sim 0.62$, 0.66, and 1.09 corresponding to different mitochondrial concentrations, which are expressed in terms of protein concentrations of 1.86, 3.72, and 37.2 mg/mL, respectively). This correlation suggests that viable mitochondria display nonrandom, non-Gaussian correlations. The DFA plot calculated for the buffer medium with no mitochondria is also shown for comparison ($\alpha \sim 0.56$, white noise). Lack of correlation in the buffer sample further confirms that the scaling exponents obtained for isolated mitochondria indeed correspond to the underlying correlations and are not affected by instrumental or other artifacts. Figure 9.5b shows DFA plots obtained in intact hepatocytes. For shorter scale sizes ($0.75 < \log n < 1.75$), the scaling exponent $\alpha \sim 0.5$ is evident resembling the white noise similar to the control measurements acquired from the 100 μM NADH solution. However, for larger scales ($1.75 < \log n < 3.25$), there is a significant deviation of α that assumes a value of ~1 (power law correlations). Conventional FFT methods showed similar positive correlations (Figure 9.5c). Also plotted in Figure 9.5b are the scaling functions obtained from the time-series data in hepatocytes while they were metabolizing the mitochondrial substrate (5 mM pyruvate/glutamate) or when the activity of complex I of the electron transport chain was inhibited by 10 μM rotenone. The fact that the scaling exponents in these two cases are similar to the one observed in control cells confirms that there is a tight regulation in mitochondrial metabolism even during acute metabolic perturbations.

Digman et al. analyzed images obtained in scanning fluorescence correlation spectroscopy experiments and interpreted the observed spatial correlations in terms of molecular diffusion processes in various timescales (Digman et al. 2005). Our results are consistent with similar diffusion mechanisms in isolated mitochondria suspended in buffer medium; the dependence of scaling exponent on mitochondrial density (protein concentration) may be reasoned as due to diffusion-limited mitochondrial aggregations. However, in contrast to mitochondria, intact hepatocytes also exhibited a crossover in scaling exponent from uncorrelated randomness ($\alpha \sim 0.5$) to persistent positive correlations ($\alpha \sim 1.0$). Although this observation needs more detailed analysis, we speculate that this difference may have arisen from additional sources of time correlations stemming from regulatory dynamics of enzyme network in intact cells. Since the cellular redox poise NADH/NAD^+ has both mitochondrial and non-mitochondrial influences (see Chapter 1), intact cells are anticipated to display more complex dynamics than isolated mitochondria. It is possible that in intact cells, there exists a hierarchical organization of redox regulation that may modulate mitochondrial redox status in an interdependent manner. Owing to the fact that mitochondria in intact cells are dynamically processing substrates shuttling between the cytoplasmic and mitochondrial compartments, the various NADH-linked enzyme activities can be spatially and temporally coordinated to yield an overall optimum response to the enzymatic activity demands. Considering the fact that the equilibrium constants of NAD^+-linked dehydrogenases in liver tissues are $\sim 8 \times 10^{-2}$ mM (Williamson et al. 1967), it is conceivable that intact cells possess more coupled activity of these dehydrogenases than isolated mitochondria. We speculate that such coordinated, coupled activity of multiple NADH-linked enzyme activities might be significantly reduced or even abolished in isolated mitochondria stemming from deficits in substrate availability, enzymatic feedback regulation, etc. Experimentally, we surmise that such differences may result in the drastic differences in scaling exponents as shown in our analysis. Similar crossover in scaling exponent α has earlier been reported in heartbeat dynamics where it was attributed to multiscale fractality (Stanley et al. 1994, 1999). Regardless, the scaling behavior observed in mitochondrial redox fluctuations provides a quantitative basis for understanding metabolic networks in living cells. A logical next step will be to explore the modifications of scaling behavior by selectively knocking out the genes of regulatory enzymes (e.g., by siRNA interference) to determine the specific role every member plays in the overall adaptive performance of the regulatory network. Extending these ideas to the whole organ or even organisms, we can surmise that such nonlinear dynamical behaviors observed at the single cell level could propagate over multiple spatial scales that in turn could regulate the overall performance of the living tissues. Beyond the specific applications in NADH fluctuation analysis, we have identified similar scaling behaviors in free radical fluctuations in living hepatocytes, and these scaling exponents were further shown to signify age-related changes in cellular metabolism (Ramanujan and Herman 2007).

9.3.2 Cancer Metabolism

In living cells, NADH/NAD^+ is a vital redox pair that facilitates a number of electron transfer reactions including mitochondrial electron transport chain (Figure 9.2). Typical concentration of this redox pair is about 0.3–2 mM in eukaryotic cells. In the context of health and disease states, the most relevant component of the mitochondrial electron transport chain is the 45-subunit enzyme complex, NADH-ubiquinone oxidoreductase, or complex I (see Section 1.2.2.1). Mitochondrial complex I is the gatekeeper between glycolytic and mitochondrial pathway, and its dysfunction has been implicated in a number of neurological disorders including Parkinson's disease and Alzheimer's disease. Since the mitochondrial complex I catalyzes the NADH $\rightarrow$ NAD^+ oxidation, one can obtain information on the status of this complex activity by monitoring the conversion of fluorescent NADH to nonfluorescent NAD^+. From the measurement point of view, the conventional approach has been to monitor the spectrophotometric absorbance (maximum absorbance at 350 nm, UV region) or fluorescence of NADH (maximum

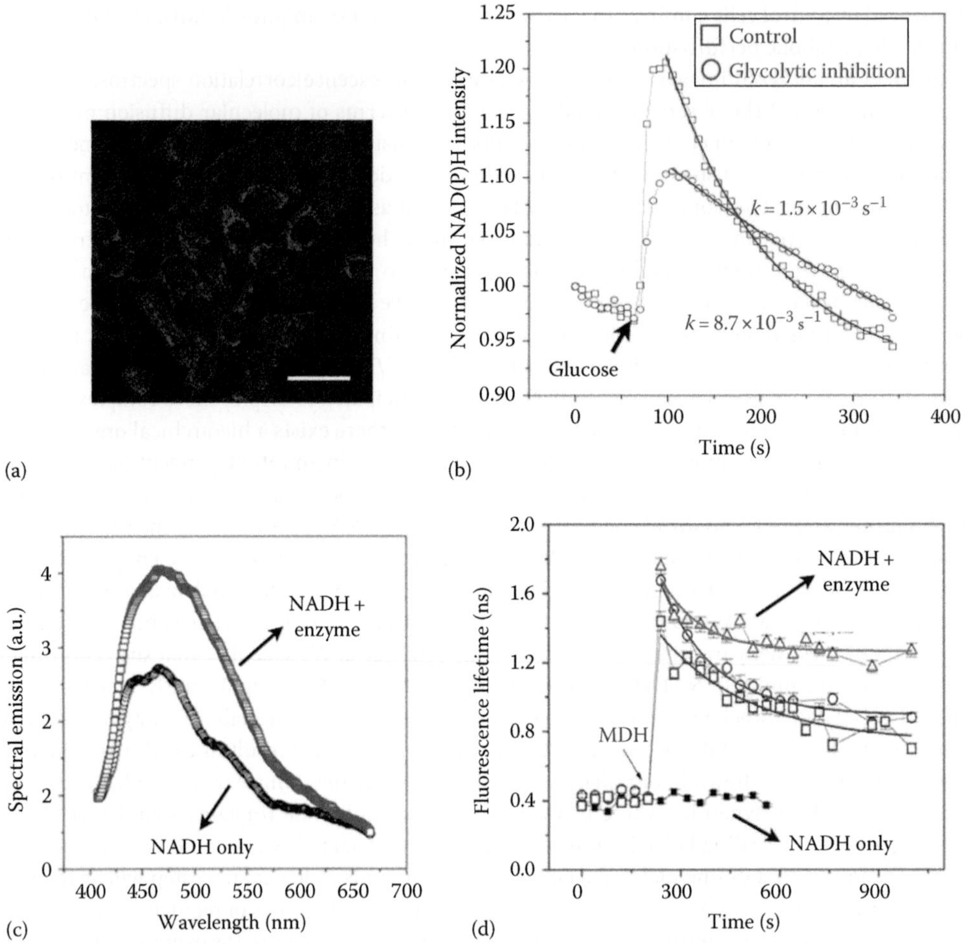

FIGURE 9.6 (a) Two-photon excited NAD(P)H imaging in HeLa Cells. (b) Glycolytic inhibition with 2-deoxyglucose modulates NAD(P)H generation and oxidation. (c) Spectral emission of NADH solutions with and without an NADH-binding enzyme (malate dehydrogenase) shows that there is no spectral profile difference between free and protein-bound NADH. However, the fluorescence lifetime imaging (d) reveals the difference between free and protein-bound NADH populations.

emission at 450 nm) in cell lysates or in isolated mitochondria. There are two major drawbacks in this paradigm: (1) UV radiation is highly cytotoxic to living cells and thus can introduce artifacts in complex I activity measurements by reducing cell viability and (2) shorter wavelengths (350 nm) have poor penetration in tissue thereby limiting access to mitochondrial complex I for monitoring its activity in living animal tissues. Our laboratory has been utilizing two-photon imaging strategies (near-infrared excitation with deep tissue imaging capabilities) to overcome the aforementioned critical bottlenecks in mitochondrial complex I activity measurements. More specifically, we have been focusing on spatially resolved single cell redox imaging to monitor real-time differences in NAD(P)H oxidation rates under various metabolic perturbations. Figure 9.6a shows live HeLa cells (cervical cancer cell line) during glucose metabolism as measured by time-lapse NAD(P)H imaging in a two-photon fluorescence microscope; NAD(P)H signals from multiple mitochondrial regions were analyzed, and Figure 9.6b shows the summary of this analysis during control and metabolically perturbed preconditioning with 20 mM 2-deoxy-glucose that is known to competitively inhibit glycolytic steps within the cells. The advantage of this approach over cuvette-based enzyme activity analysis is that one can obtain high-content information on the spatial distribution of enzyme activity within single cells. Further, by augmenting this with an online image analysis module that can compute the rate constants (as demonstrated in Figure 9.6b) on a pixel-by-pixel basis, these time-lapse image stacks can be readily converted to a "rate constant image" thereby offering visual maps of NADH-linked enzyme activity in real time. Appropriate controls need to be chosen for validation of such an instrument.

This approach will be particularly valuable in situations where the availability of cells/sample is severely limited as in clinical biopsy specimens and primary cells from tissues. The aforementioned strategy involves interrogating cellular redox status by means of the "binary readout" (i.e., fluorescent NADH and nonfluorescent NAD^+) in any given experiment. Although this is a significant component of NADH-linked enzyme activities in cells, differential enzyme activities can also arise from the modulation of binding between the various enzymes and the cofactor NADH. However, since free NADH and enzyme-bound NADH have identical spectral profiles (Figure 9.6c), it is practically impossible to determine the various enzyme-bound states of NADH within the cellular context. A number of earlier studies, including our own publications, have shown that free and enzyme-bound NADH could be distinguished by monitoring the fluorescence lifetime of NADH (free NADH lifetime ~0.4 ns, whereas enzyme-bound NADH lifetime ~2.5–5 ns).

A recent publication from our laboratory demonstrated the utility of such combined modality for monitoring enzyme-linked NAD(P)H oxidation rates in the context of critical biological events (Ramanujan et al. 2008). In this study, we showed that spatially resolved fluorescence lifetime maps of NAD(P)H distribution in primary liver cells isolated from young and aged mouse models could shed light on the effects of aging process at the single cell level. We further demonstrated, in the presence of pharmacological inhibitors, that young and aged cells have significant differences in mitochondrial complex I activity. We envision an interesting applicability of this approach in cancer research as well since a number of recent studies have pointed out that both cancer metabolism and aging metabolism display mitochondrial dysfunction as a common denominator. Figure 9.7 demonstrates this connection in cancer cells where glucose-stimulated NAD(P)H oxidation rate is significantly different in the cells that were preconditioned with mitochondrial complex I inhibitor, rotenone. By clearly showing the connection between glucose metabolism (a cytosolic activity in the cells) and complex I inhibition (a mitochondrial activity in the cells), these data further illustrate the advantages of monitoring enzyme activity in single cell imaging in contrast to isolated enzyme activity measurements. Owing to the endogenous nature of NAD(P)H fluorescence signals, such high-content monitoring of enzyme activities has a great potential in cancer diagnosis and chemotherapy assessment in the clinical arena as well. In the context of preclinical research tools, one can multiplex the measurement modality by using fluorogenic probes and/or fluorescent proteins as reporters of physiological and metabolic status in single cells in addition to monitoring endogenous NAD(P)H signals as was reported in one of our earlier studies. In this particular study, we demonstrated that by combining high-resolution fluorescence lifetime imaging and time-series analysis approaches, the scaling exponents have further utility in delineating the various metabolic pathways that distinguish cancer cells from the noncancerous cells as demonstrated earlier (Ramanujam et al. 2008). In summary, with a strategic combination of imaging instrumentation and novel image analysis modules, probing single cell metabolism at an unprecedented speed and with high resolution can offer great potential to unravel novel information from intact living cells.

9.4 Future Perspectives

Metabolism is a network of highly coordinated chemical reactions in each living cell, and the spatiotemporal propagation of these individual cellular metabolic networks ensures proper regulation of life at the organismal level. Enzymes form the structural and functional nodes of the metabolic hub, and coenzymes such as NADH facilitate accuracy in enzyme-linked metabolic reactions. NADH, as one of

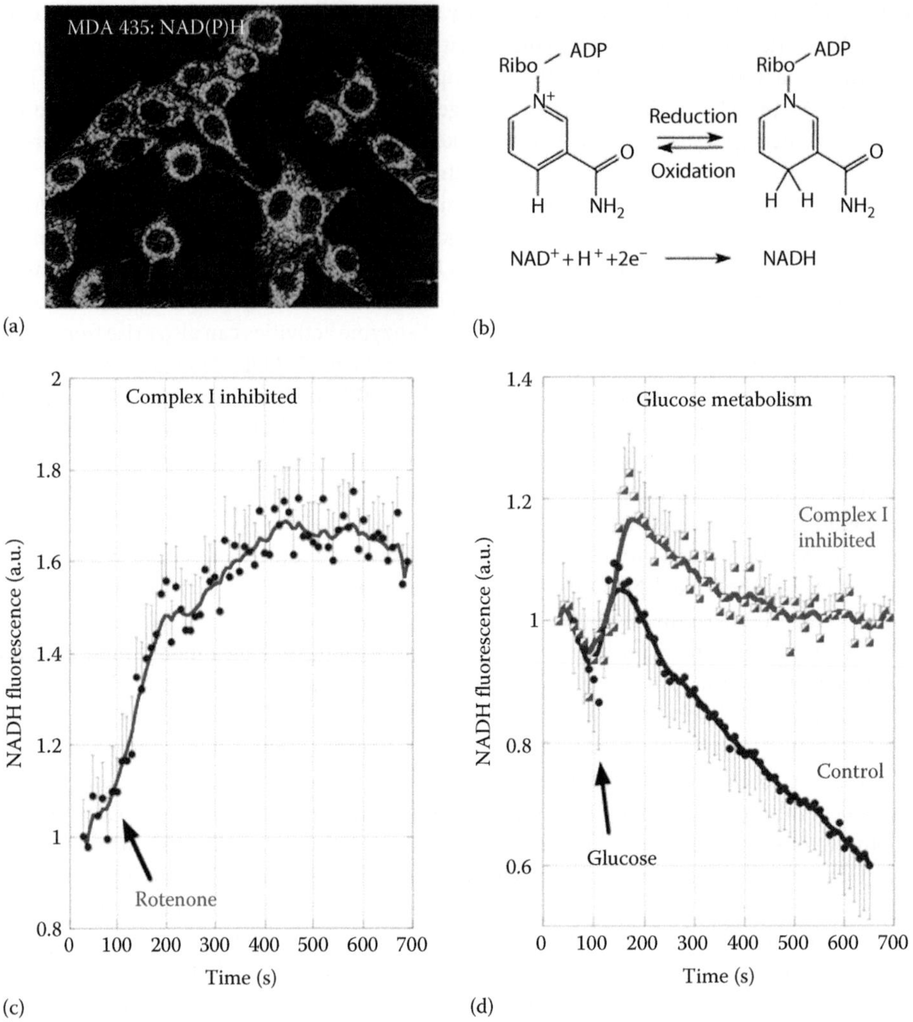

FIGURE 9.7 (a) Two-photon excited NAD(P)H imaging in MDA435 breast cancer cells (730 nm exc; 450/80 nm emission). (b) Schematic of redox equilibrium between NADH (reduced) and NAD^+ (oxidized). (c) Real-time kinetic changes in NAD(P)H levels in MDA435 breast cancer cells during acute complex I inhibition (1 μM rotenone, 15 min). Rotenone irreversibly inhibits NAD(P)H oxidation thereby disturbing the mitochondrial complex I activity. (d) Representative real-time kinetic imaging in MDA435 breast cancer cells reveals that mitochondrial complex I inhibition drastically alters the NAD(P)H oxidation (complex I activity) during glucose metabolism.

the very few compounds in the cell with very high reduction potentials, is particularly significant in maintaining the redox status of cells as well as in regulating multiple electron transfer reactions within the cells. As described earlier, technological advancements for noninvasive monitoring of such redox status have the potential to revolutionize our understanding of metabolism at the single cell level with unprecedented resolution. Differential metabolic activities of cells with metabolic dysfunction (e.g., cancer) can be easily discriminated with appropriate detection system, which then can be translated to novel disease markers and/or diagnosis methods. From the clinical translation point of view, one could conceive fiber-optic imaging modalities in conjunction with the currently utilized endoscopic hardware. This would enable adding new layer of information based on metabolic activities of dysfunctional tissues. Expanding the scope of these approaches, one can therefore envision a repertoire of novel biomarkers and detection approaches based on NADH and other critical coenzymes and their distinct interactions with the cellular enzymes—not only for cancer and aging models discussed here but also for various disease pathology models.

References

Amaral, L. A. N., A. Scala, M. Barthélémy, and H. E. Stanley. 2000. Classes of small-world networks. *Proceedings of the National Academy of Sciences of the United States of America* 97 (21): 11149–11152.

Aon, M. A., S. Cortassa, F. G. Akar, D. A. Brown, L. Zhou, and B. O'Rourke. 2009. From mitochondrial dynamics to arrhythmias. *The International Journal of Biochemistry and Cell Biology* 41 (10): 1940–1948.

Aon, M. A., S. Cortassa, E. Marbán, and B. O'Rourke. 2003. Synchronized whole cell oscillations in mitochondrial metabolism triggered by a local release of reactive oxygen species in cardiac myocytes. *Journal of Biological Chemistry* 278 (45): 44735–44744.

Aon, M. A., S. Cortassa, and B. O'Rourke. 2010. Redox-optimized ROS balance: A unifying hypothesis. *Biochimica et Biophysica Acta* 1797: 865–877.

Chance, B. 1964. Feedback control of metabolism in ascites tumor cells. *Acta—Unio Internationalis Contra Cancrum* 20: 1028–1032.

Chance, B. 2004. Mitochondrial NADH redox state, monitoring discovery and deployment in tissue. *Methods in Enzymology* 385: 361–370.

Chance, B., P. Cohen, F. Jobsis, and B. Schoener. 1962. Localized fluorometry of oxidation-reduction states of intracellular pyridine nucleotide in brain and kidney cortex of the anesthetized rat. *Science* 136 (3513): 325.

Chance, B. and B. Schoener. 1963. Control of oxidation-reduction state of NADH in the liver of anesthetized rats. *Advances in Enzyme Regulation* 1: 169–181.

Chance, B., B. Schoener, R. Oshino, F. Itshak, and Y. Nakase. 1979. Oxidation-reduction ratio studies of mitochondria in freeze-trapped samples. NADH and flavoprotein fluorescence signals. *Journal of Biological Chemistry* 254 (11): 4764–4771.

DeBerardinis, R. J., N. Sayed, D. Ditsworth, and C. B. Thompson. 2008. Brick by brick: Metabolism and tumor cell growth. *Current Opinion in Genetics and Development* 18 (1): 54–61.

De la Fuente, I. M., L. Martínez, A. L. Pérez-Samartín, L. Ormaetxea, C. Amezaga, and A. Vera-López. 2008. Global self-organization of the cellular metabolic structure. *PLos ONE* 3 (8): e3100.

De la Fuente, I. M., A. L. Perez-Samartin, L. Martínez, M. A. García, and A. Vera-López. 2006. Long-range correlations in rabbit brain neural activity. *Annals of Biomedical Engineering* 34 (2): 295–299.

De la Fuente, I. M., F. Vadillo, M.-B. Pérez-Pinilla, A. Vera-López, and J. Veguillas. 2009. The number of catalytic elements is crucial for the emergence of metabolic cores. *PLoS ONE* 4 (10): e7510.

Digman, M. A., P. Sengupta, P. W. Wiseman, C. M. Brown, A. R. Horwitz, and E. Gratton. 2005. Fluctuation correlation spectroscopy with a laser-scanning microscope: Exploiting the hidden time structure. *Biophysical Journal* 88 (5): L33–L36.

Digman, M. A., P. W. Wiseman, C. Choi, A. R. Horwitz, and E. Gratton. 2009. Stoichiometry of molecular complexes at adhesions in living cells. *Proceedings of the National Academy of Sciences of the United States of America* 106 (7): 2170–2175.

Havlin, S., L. A. Nunes Amaral, Y. Ashkenazy et al. 1999. Application of statistical physics to heartbeat diagnosis. *Physica A: Statistical Mechanics and Its Applications* 274 (1): 99–110.

Lin, S.-J., E. Ford, M. Haigis, G. Liszt, and L. Guarente. 2004. Calorie restriction extends yeast life span by lowering the level of NADH. *Genes and Development* 18 (1): 12–16.

Mayevsky, A. and B. Chance. 2007. Oxidation–reduction states of NADH in vivo: From animals to clinical use. *Mitochondrion* 7 (5): 330–339.

Peng, C.-K., S. V. Buldyrev, A. L. Goldberger et al. 1992. Long-range correlations in nucleotide sequences. *Nature* 356 (6365): 168–170.

Peng, C.-K., S. V. Buldyrev, J. M. Hausdorff et al. 1994. Non-equilibrium dynamics as an indispensable characteristic of a healthy biological system. *Integrative Physiological and Behavioral Science* 29 (3): 283–293.

Peng, C.-K., S. Havlin, J. M. Hausdorff, J. E. Mietus, H. E. Stanley, and A. L. Goldberger. 1995. Fractal mechanisms and heart rate dynamics: Long-range correlations and their breakdown with disease. *Journal of Electrocardiology* 28: 59–65.

Ramanujan, V. K., G. Biener, and B. A. Herman. 2006. Scaling behavior in mitochondrial redox fluctuations. *Biophysical Journal* 90 (10): L70–L72.

Ramanujan, V. K. and B. A. Herman. 2007. Aging process modulates nonlinear dynamics in liver cell metabolism. *Journal of Biological Chemistry* 282 (26): 19217–19226.

Ramanujan, V. K. and B. A. Herman. 2008. Nonlinear scaling analysis of glucose metabolism in normal and cancer cells. *Journal of Biomedical Optics* 13 (3): 031219.

Ramanujan, V. K., J. A. Jo, G. Cantu, and B. A. Herman. 2008. Spatially resolved fluorescence lifetime mapping of enzyme kinetics in living cells. *Journal of Microscopy* 230 (3): 329–338.

Stanley, H. E., S. V. Buldyrev, A. L. Goldberger et al. 1994. Statistical mechanics in biology: How ubiquitous are long-range correlations? *Physica A: Statistical Mechanics and Its Applications* 205 (1): 214–253.

Stanley, H. E., L. A. Nunes Amaral, A. L. Goldberger, S. Havlin, P. Ch. Ivanov, and C.-K. Peng. 1999. Statistical physics and physiology: Monofractal and multifractal approaches. *Physica A: Statistical Mechanics and Its Applications* 270 (1): 309–324.

Wallace, D. C. 1999. Mitochondrial diseases in man and mouse. *Science* 283 (5407): 1482–1488.

Williamson, D. H., P. Lund, and H. A. Krebs. 1967. The redox state of free nicotinamide-adenine dinucleotide in the cytoplasm and mitochondria of rat liver. *Biochemical Journal* 103: 514–527.

Xu, H. N., J. Tchou, B. Chance, and L. Z. Li. 2013. Imaging the redox states of human breast cancer core biopsies. In *Oxygen Transport to Tissue XXXIV*, W. Welch, F. Palm, D. F. Bruley, and D. K. Harrison (eds.). Vol. 765 of Advances in Experimental Medicine and Biology, I. R. Cohen, A. Lajtha, R. Paoletti, and J. D. Lambris (eds.), pp. 343–349. New York: Springer.

10

NAD(P)H and FAD as Biomarkers for Programmed Cell Death

Hsing-Wen Wang
University of Maryland

10.1 Introduction

Mitochondria play an important role in the process of cell death, which is now recognized as a complex continuum of mechanisms that include program cell death (apoptosis), necrosis, and autophagy (Kim et al. 2006). Apoptosis is a highly regulated and energy-dependent multistep biological process (Hetts 1998). Dysregulation of apoptosis—either abnormal initiation of the genetic self-termination program or failure to undergo apoptosis—causes various diseases, such as autoimmunity, neurodegeneration, heart disease, and cancer. Apoptosis generally operates via two pathways: (1) mitochondria- and (2) receptor-mediated. In the mitochondria-mediated pathway, an external insult acts on mitochondria with or without the action of proapoptotic proteins, to cause cytochrome *c* release from the mitochondria to the cytoplasm, which is accompanied by the loss of the mitochondrial membrane potential. Released cytochrome *c* interacts with apoptotic protease-activating factor 1 (Apaf-1), ATP, and procaspase 9 to form an apoptosome. Apoptosome then activates a cascade of cellular destruction events, beginning with the activation of death-execution effector caspases, such as caspase 3, and then the activation of downstream caspases, ultimately resulting in the hallmark of apoptosis, including condensation of nuclear and cytoplasmic contents, fragmentation of nuclear DNA, and membrane blebbing (Blankenberg 2008; Schon and Manfredi 2003).

Preclinical and clinical studies have shown that early detection of a triggered cell death pathway can potentially be used to evaluate the effectiveness of a therapy, provide prognostic information, and guide the course of a given treatment (Buchholz et al. 2003; Dubray et al. 1998; Neves and Brindle 2006). Current methods for in vivo detection of cell death include magnetic resonance imaging (MRI), magnetic resonance spectroscopy, nuclear imaging (e.g., single-photon emission computed tomography and positron emission tomography), and optical imaging (Blankenberg 2008, 2009). These methods rely on the use of exogenous molecular markers, such as the 40 kDa vesicle-associated protein, annexin V, and reporter probes that can be specifically cleaved by caspase 3. In contrast, ultrasound (Czarnota et al. 1999) and diffusion MRI (dMRI) (Hamstra et al. 2005) do not require exogenous molecules and provide an indirect means to evaluate cell death. In particular, ultrasound detection is based on the changes in tissue scattering, caused by nuclear condensation at the late stage of apoptotic process. Changes in water diffusion, which is measured by dMRI, are caused by cell shrinkage during apoptosis and cell swelling during necrosis (Moffat et al. 2005). As a result, ultrasound or dMRI methods can be utilized for monitoring morphological changes of cells or tissues at the later stage of cell death, but are not suitable for early apoptosis or necrosis detection.

This chapter aims at providing a summary about the studies on the changes of intracellular NADH fluorescence (both lifetime and intensity) during different cell death pathways and the capacity of this approach to discriminate between apoptosis and necrosis at the early stages of the cell death onset. We discuss the possible underlying mechanisms for the observed NADH autofluorescence changes and perspectives of this approach for clinical applications.

10.2 Monitoring Cell Death Using NADH and FAD Autofluorescence

Several studies have established that NADH and FAD fluorescence intensity and/or redox ratio can be used to monitor mitochondrial dysfunction during cell death in vitro and in vivo (reviewed in Wang et al. 2009). In those studies, programmed cell death was induced using a wide range of stimuli. Following the induction of apoptosis, both increase and decrease of NADH and FAD fluorescence intensities were found in different cell types with different stimuli and at various time points. Several in vitro studies reported an increase and then decrease of NADH and FAD fluorescence intensity during apoptosis. These changes appeared prior to nuclear DNA fragmentation and were found to correlate with mitochondrial membrane potential change. The observed increase in NADH autofluorescence during apoptosis was attributed to protein binding, based on spectral blueshift (Morbidelli et al. 2005). A monotonic decrease in the NADH autofluorescence intensity was also reported in correlation with the mitochondrial membrane potential dissipation (Lemar et al. 2007; Petit et al. 2001), elevated intracellular ROS levels, and lower rate of oxygen consumption (Lemar et al. 2007). A significant decrease in FAD autofluorescence during necrosis was found in the same studies. The in vitro measurements revealed a more consistent trend of increased redox ratio after the induction of cell death (Brewer et al. 2002; Levitt et al. 2006; Morbidelli et al. 2005).

The in vivo studies of intrinsic NADH and FAD autofluorescence did not exhibit the monotonic changes (increase or decrease) in intensity after the induction of apoptosis or necrosis, but the redox ratio consistently increased (Brewer et al. 2002; Mokrý et al. 2007; Zhang et al. 2004). However, the observed trends of redox ratio changes do not allow for differentiating between the apoptotic and necrotic forms of cell death.

The fluorescence lifetime measurement of intrinsic NADH (see Chapter 4) in addition to NADH fluorescence intensity seems to offer a sensitive discrimination of apoptosis and necrosis. For example, the autofluorescence lifetime of NADH increased at the early phase of both mitochondrial and poly(ADP-ribose) polymerase-1 (PARP-1)-mediated cell deaths, which were induced by staurosporine (STS) (Wang et al. 2008a; Yu et al. 2011) and *N*-methyl-*N′*-nitro-*N*-nitrosoguanidine (MNNG) (Guo et al. 2011), respectively. STS-induced apoptosis involves the release of cytochrome *c*, reduction of the mitochondrial membrane potential $\Delta\Psi$, activation of caspase 3, and fragmentation of nuclear DNA. PARP-1-mediated cell death involves significant cytosolic NAD^+ consumption to cause ATP depletion, glycolysis inhibition, and PAR overexpression to trigger apoptosis-inducible factor (AIF) translocation from mitochondria to the nucleus and eventually energy deficiency that leads to cell death. During STS-induced apoptosis in HeLa cells, an initial increase (up to 2 ns) followed by a decrease in the mean autofluorescence lifetime of NADH was

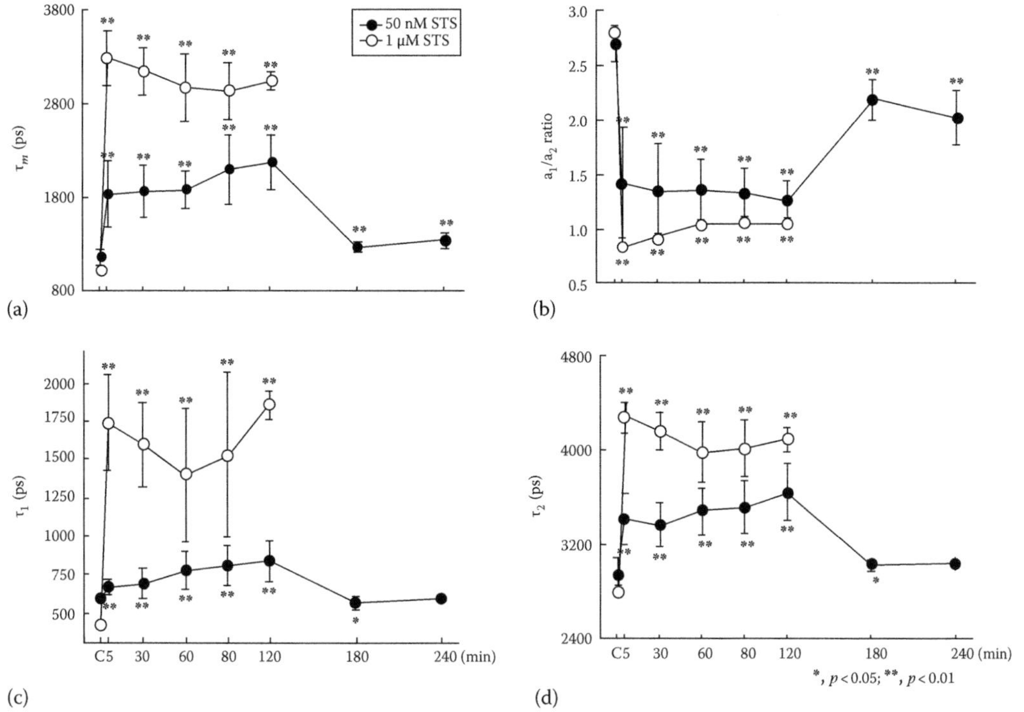

FIGURE 10.1 Mean and standard deviation of (a) the τ_m, (b) a_1/a_2 ratio, (c) τ_1, (d) and τ_2 over 3–12 fields of view as the function of time in cells treated with 50 nM (black circles) and 1 μM (white circles) of STS. "C" on the time axis represents control or the time point prior to the treatment. (Adapted from Yu, J.-S. et al., *Journal of Biomedical Optics*, 16(3), 036008, 2011. With permission.)

observed in dose-dependent manner, with sixfold increase in the corresponding autofluorescence intensity (Wang et al. 2008a; Yu et al. 2011). The observed changes in free and protein-bound NADH lifetime (τ_1 and τ_2) and the corresponding reduction of the a_1/a_2 ratio suggest an increase in protein-bound NADH during STS-induced apoptosis (Figure 10.1).

During MNNG-induced cell death in HeLa cells, a similar initial increase in the mean autofluorescence lifetime of NADH was observed within the 90 min of MNNG induction in dose-dependent manner, with a 60% drop in the corresponding autofluorescence intensity (Guo et al. 2011). Interestingly, no major changes were observed in the fluorescence lifetime of intrinsic NADH under H_2O_2-induced necrosis in both HeLa and 143B cell lines (Wang et al. 2008a). The observed increase in the lifetime of NADH autofluorescence during STS-induced apoptosis and in MNNG-induced cell death was attributed predominantly to the changes in protein-binding state rather than an increase in the cellular content of NADH.

10.3 Correlation between Intrinsic NADH Autofluorescence and Mitochondrial Functions

To understand if NADH fluorescence signal changes directly reflect the mitochondrial function alteration, we further studied the relationship of NADH fluorescence lifetime and mitochondrial function during mitochondria-mediated (through STS induction) and PARP-1-mediated (through MNNG induction) cell death (Guo et al. 2011; Yu et al. 2011). In STS dose-dependent studies, we investigated the time–course relationship between the autofluorescence lifetime of NADH, mitochondrial membrane

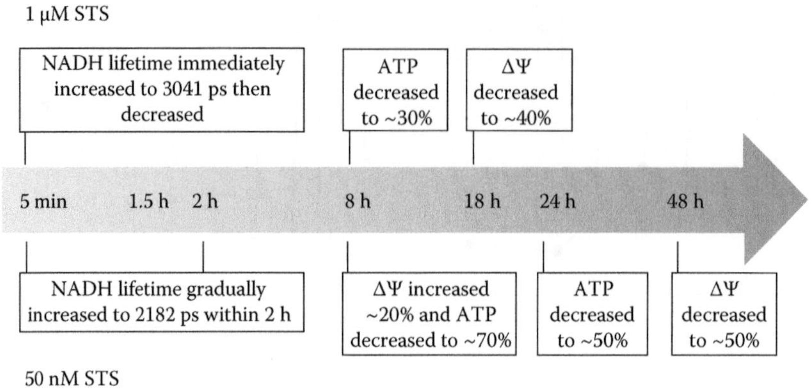

FIGURE 10.2 Time relationship between the NADH fluorescence lifetime, mitochondrial membrane potential (ΔΨ), and ATP content change in HeLa cells treated with 1 μM and 50 nM STS. (Adapted from Yu, J.-S. et al., *Journal of Biomedical Optics*, 16(3), 036008, 2011. With permission.)

potential ΔΨ, ATP level, and caspase 3 activities in HeLa cells (Yu et al. 2011). In brief, at low STS dose (50 nM), a delayed increase of both NADH fluorescence lifetime and the activities of caspase 3 was observed, as well as a reduction of ΔΨ and ATP level as compared to a higher STS dose of 1 μM (Figure 10.2). However, the time course of NADH fluorescence lifetime changes did not seem to associate with that of either ATP or ΔΨ.

In PARP-1-mediated cell death (Guo et al. 2011), mitochondrial membrane potential ΔΨ increased for up to 90 min at a higher MNNG dose (100 μM), while the lifetime of NADH autofluorescence increased. When the cells were incubated with 100 μM MNNG for 30 min, the ATP level was significantly depleted immediately to less than 20%. The concurrent reduction of the oxygen consumption rates suggests an interruption of the electron transport chain. Our time-dependent studies of mitochondrial functions reveal an increase in the fluorescence lifetime of intrinsic NADH prior to the reduction of ΔΨ and mitochondrial uncoupling. Treatment with pyruvate, a glycolytic product that does not require cytosolic NAD^+ for energy metabolism but forces mitochondria to resume the electron transport chain and generate ATP, also affected the autofluorescence lifetime, but not intensity, of cellular NADH (Figure 10.3). In addition, the ATP level, oxygen consumption rate, and cell viability were recovered upon pyruvate (but not glucose) treatment after MNNG was washed out. Pyruvate also prevented the increase of ΔΨ indicating the resumption of the electron transport chain.

The oxygen consumption rate and ATP level are indicators of both the electron transport chain activity and the efficiency of mitochondria-generated ATP in cells. Our studies suggest that the observed increase in the autofluorescence lifetime of NADH is not simply due to changes in cellular metabolism upon the induction of apoptosis. Other mechanisms must be considered because both oxygen consumption rates and ATP levels decreased, instead of increased, during these death pathways.

10.4 Implications to Photodynamic Therapy

Photodynamic therapy (PDT) for cancer treatment involves a photosensitizer drug that interacts with light and oxygen with the generation of singlet oxygen, which triggers the cell death and tumor destruction (Dolmans et al. 2003). The generated singlet oxygen is widely believed to cause the biological damage associated with most current photosensitizers and treatment doses in PDT. Clinical PDT dosimetry is rather complex due to the several treatment factors involved (i.e., light dose, drug dose, light-drug duration, and oxygen), which vary both dynamically and interdependently. Generally, the

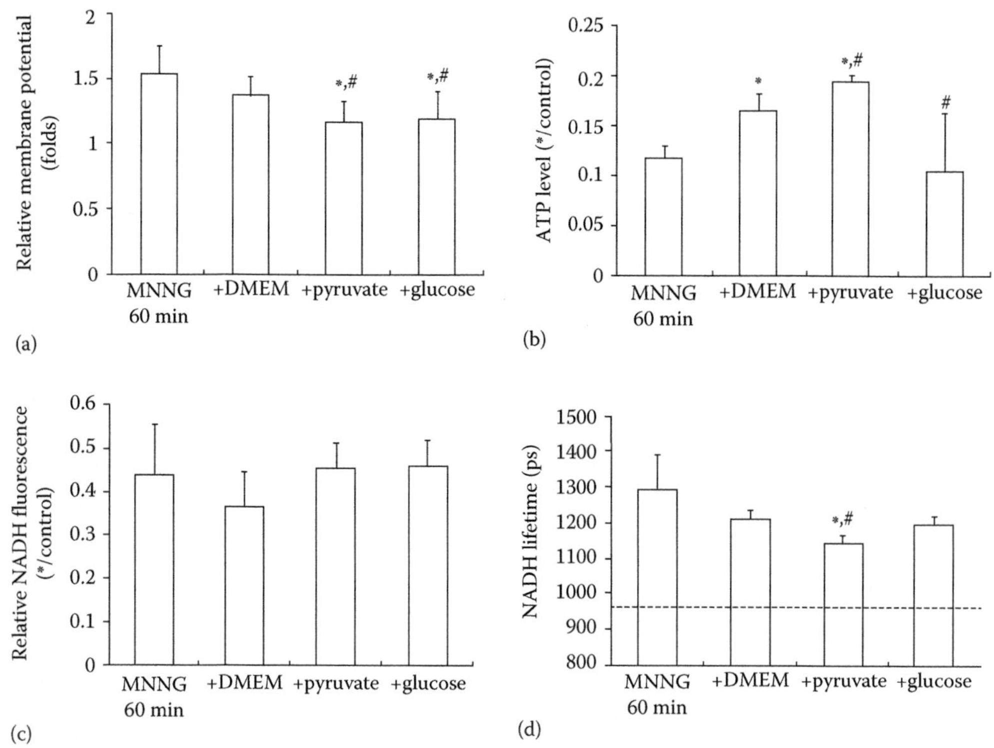

*$p < 0.05$ compared with 60 min MNNG treatment

#$p < 0.05$ compared with 30 min MNNG plus 30 min DMEM treatment

FIGURE 10.3 The change of (a) mitochondrial membrane potential, (b) ATP level, (c) NADH fluorescence intensity, and (d) NADH fluorescence lifetime after 60 min 100 μM MNNG treatment only, or 30 min 100 μM MNNG treatment followed by either 30 min pyruvate, glucose, or Dulbecco's modified eagle medium (DMEM) treatment. Symbols # and * indicate that the value is significantly different ($p < 0.05$) from the value of either the 60 min MNNG treatment only or 30 min MNNG treatment followed by 30 min DMEM treatment. The dotted line shown on (d) indicates the NADH fluorescence lifetime of HeLa cells before the treatment. (Adapted from Guo, H.-W. et al., *Journal of Biomedical Optics*, 16(6), 068001, 2011. With permission.)

following four strategies are used in PDT dosimetry: explicit, implicit, and direct dosimetry, as well as monitoring the response of biological tissues (Wilson et al. 1997). Explicit dosimetry measures the three PDT components (light, drug, and oxygen) and sometimes incorporates a dose-based model to simulate the outcome (Wang et al. 2007). Implicit dosimetry intends to measure a single metric, such as photosensitizer photobleaching, that is predictive of the biological damage or outcome under certain circumstances. In direct dosimetry, however, the singlet oxygen is monitored (Jarvi et al. 2006) without the need for any dose model (Niedre et al. 2003). Yet, studies have shown that singlet-oxygen monitoring approach may fail to predict the tumor response (Wang et al. 2008b), and reporters of biological response to therapy become necessary. Noninvasive in vivo monitoring of tumor perfusion (e.g., blood flow and tissue blood oxygenation) is currently being tested as a means for predicting the long-term response to PDT treatment (Chen et al. 2003; Finlay and Foster 2004; Gross et al. 2003; Stratonnikov and Loschenov 2001; Yu et al. 2005).

Metronomic PDT and the low-fluence-rate PDT studies demonstrated an enhanced apoptosis in tumors along with limited damage in normal brain cells (Bisland et al. 2004; Bogaards et al.

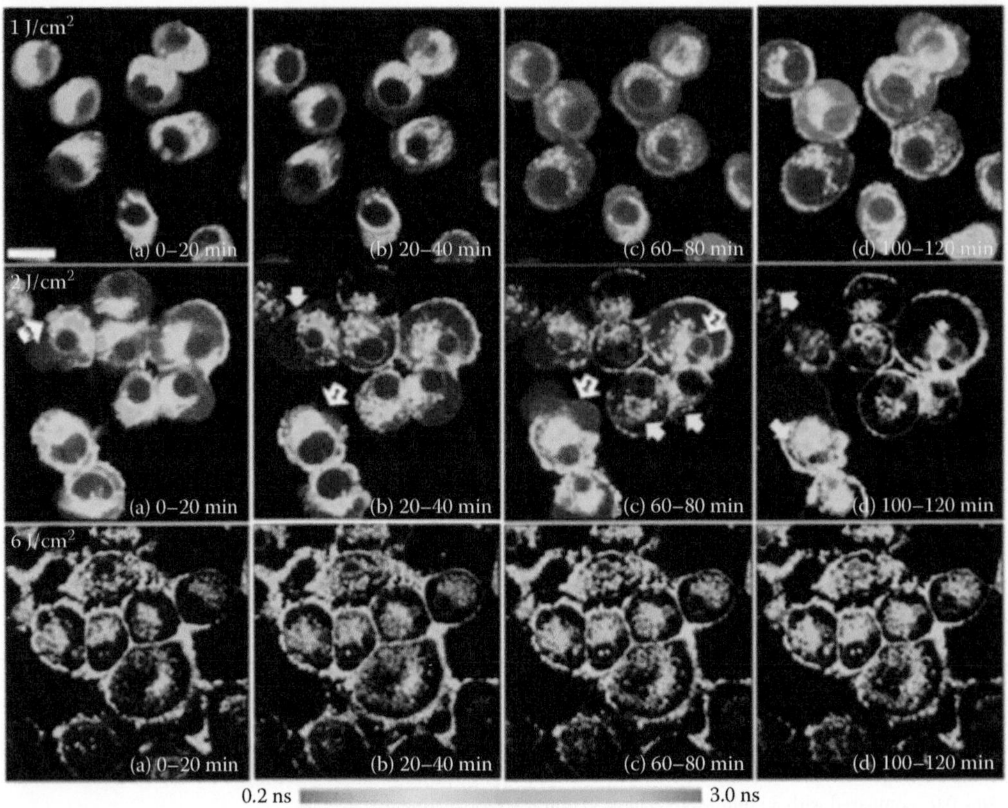

FIGURE 10.4 Time course of NADH fluorescence lifetime changes recorded from H1299 cells treated with 5-ALA and irradiated with the light cell fluence of 1, 2, and 6 J/cm^2. The images were taken at time points of (a) 0–20 min, (b) 20–40 min, (c) 60–80 min, and (d) 100–120 min after 4 h treatment with serum-free medium and subsequent light irradiation. The scale bar is 20 μm. (Adapted from Su, G.-C., et al., *Optics Express*, 19(22), 21145, 2011. With permission.)

2005) and an outstanding long-term tumor control (Henderson et al. 2004, 2006), respectively. In metronomic PDT, slow delivery of photosensitizer and light exposure are continuously applied at low rates in order to selectively destroy the tumor cells via apoptosis over extended periods of time (Bisland et al. 2004; Bogaards et al. 2005). Noninvasive monitoring of cell apoptosis immediately after metronomic PDT would provide a good way to determine the biological consequence and thus the efficiency of cancer treatment.

We have demonstrated that the PDT-induced cell death and the underlying mechanism (i.e., apoptosis versus necrosis) can be directly monitored using both the lifetime and intensity of intracellular NADH autofluorescence (Su et al. 2011). In these studies, we performed 5-aminolevulinic acid (ALA)-PDT in H1299 non-small cell lung carcinoma cell culture as a function of the light fluence. The corresponding dynamics of NADH autofluorescence lifetime exhibited sensitivity to PDT-induced slight cytotoxicity, primary apoptosis, or necrosis (Figure 10.4). Specifically, NADH fluorescence lifetime increased only when the cells died primarily through the apoptotic pathway, which was confirmed by DNA fragmentation analysis, cell morphology, and caspase-3 activation. When cells underwent necrosis or slight cell death (cell viability was approximately 80% at 3 h after PDT and 60%–70% at 4 h after PDT), there was no significant change of NADH fluorescence lifetime within the 2 h observation period after PDT.

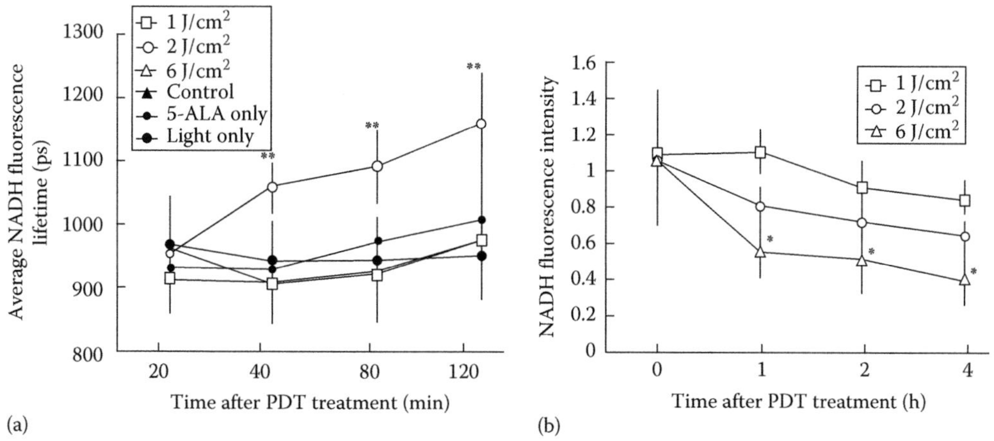

FIGURE 10.5 Changes of the (a) NADH fluorescence lifetime and (b) intensity (average and standard deviation) over time after cell incubation with 5-ALA (4 h) and irradiation with different light fluences. Symbols * and ** indicate a statistically significant difference from control values ($p < 0.05$ and $p < 0.005$, respectively). NADH fluorescence intensity was normalized to cell numbers. (Adapted from Su, G.-C. et al., *Optics Express*, 19(22), 21145, 2011. With permission.)

A decrease in NADH fluorescence intensity was observed for all three outcomes within 4 h after PDT light illumination with cells dying primarily through necrosis exhibiting the greatest decrease, followed by cells undergoing apoptosis and slight cell death (Figure 10.5). As a result, we conclude that the autofluorescence lifetime and intensity of intrinsic NADH is useful for optimizing the treatment efficacy in metronomic PDT. Furthermore, a combined monitoring of NADH autofluorescence lifetime and intensity can be used to differentiate between apoptosis and necrosis.

10.5 Summary

The fluorescence lifetime of intracellular NADH has a potential to become an intrinsic biomarker for cell death, particularly at the early stages of in vivo apoptosis. The observed increase in NADH fluorescence lifetime during several cell death pathways was attributed predominantly to increased protein-protein interaction. The specific protein(s) involved in these pathways, however, remains unknown. Furthermore, the observed increase in the intrinsic NADH fluorescence lifetime may not be attributed to a simple increase of the metabolic demand in mitochondria, like that during stem cell differentiation, because oxygen consumption reduced during cell death. Significant fluorescence intensity loss and no fluorescence lifetime change of NADH are associated with mitochondrial membrane potential dissipation, advanced stage of apoptosis, and necrosis. As a result, the autofluorescence lifetime and intensity of cellular NADH may serve as a natural probe for differentiating between apoptosis and necrosis that is particularly critical for clinical metronomic PDT, in which multiple treatment factors involved and varied both dynamically and interdependently. Effective monitoring of cell apoptosis during metronomic PDT would greatly improve the treatment outcome.

Acknowledgments

I acknowledge useful discussions with Dr. Han-Wen Guo, Mr. Guan-Chin Su, Ms. Jia-Sin Yu, and Professor Yau-Huei Wei.

References

Bisland, S. K., L. Lilge, A. Lin, R. Rusnov, and B. C. Wilson. 2004. Metronomic photodynamic therapy as a new paradigm for photodynamic therapy: Rationale and preclinical evaluation of technical feasibility for treating malignant brain tumors. *Photochemistry and Photobiology* 80 (1): 22–30.

Blankenberg, F. G. 2008. In vivo imaging of apoptosis. *Cancer Biology and Therapy* 7 (10): 1525–1532.

Blankenberg, F. G. 2009. Imaging the molecular signatures of apoptosis and injury with radiolabeled annexin V. *Proceedings of the American Thoracic Society* 6 (5): 469–476.

Bogaards, A., A. Varma, K. Zhang et al. 2005. Fluorescence image-guided brain tumour resection with adjuvant metronomic photodynamic therapy: Pre-clinical model and technology development. *Photochemical and Photobiological Sciences* 4 (5): 438–442.

Brewer, M., U. Utzinger, Y. Li et al. 2002. Fluorescence spectroscopy as a biomarker in a cell culture and in a nonhuman primate model for ovarian cancer chemopreventive agents. *Journal of Biomedical Optics* 7 (1): 20–26.

Buchholz, T. A., D. W. Davis, D. J. McConkey et al. 2003. Chemotherapy-induced apoptosis and Bcl-2 levels correlate with breast cancer response to chemotherapy. *The Cancer Journal* 9 (1): 33–41.

Chen, B., B. W. Pogue, I. A. Goodwin et al. 2003. Blood flow dynamics after photodynamic therapy with verteporfin in the RIF-1 tumor. *Radiation Research* 160 (4): 452–459.

Czarnota, G. J., M. C. Kolios, J. Abraham et al. 1999. Ultrasound imaging of apoptosis: High-resolution non-invasive monitoring of programmed cell death in vitro, in situ and in vivo. *British Journal of Cancer* 81 (3): 520–527.

Dolmans, D. E. J. G. J., D. Fukumura, and R. K. Jain. 2003. Photodynamic therapy for cancer. *Nature Reviews Cancer* 3 (5): 380–387.

Dubray, B., C. Breton, J. Delic et al. 1998. In vitro radiation-induced apoptosis and early response to low-dose radiotherapy in non-Hodgkin's lymphomas. *Radiotherapy and Oncology* 46 (2): 185–191.

Finlay, J. C. and T. H. Foster. 2004. Hemoglobin oxygen saturations in phantoms and in vivo from measurements of steady-state diffuse reflectance at a single, short source-detector separation. *Medical Physics* 31: 1949–1959.

Gross, S., A. Gilead, A. Scherz, M. Neeman, and Y. Salomon. 2003. Monitoring photodynamic therapy of solid tumors online by BOLD-contrast MRI. *Nature Medicine* 9 (10): 1327–1331.

Guo, H.-W., Y.-H. Wei, and H.-W. Wang. 2011. Reduced nicotinamide adenine dinucleotide fluorescence lifetime detected poly (adenosine-5′-diphosphate-ribose) polymerase-1-mediated cell death and therapeutic effect of pyruvate. *Journal of Biomedical Optics* 16 (6): 068001.

Hamstra, D. A., T. L. Chenevert, B. A. Moffat et al. 2005. Evaluation of the functional diffusion map as an early biomarker of time-to-progression and overall survival in high-grade glioma. *Proceedings of the National Academy of Sciences of the United States of America* 102 (46): 16759–16764.

Henderson, B. W., T. M. Busch, and J. W. Snyder. 2006. Fluence rate as a modulator of PDT mechanisms. *Lasers in Surgery and Medicine* 38 (5): 489–493.

Henderson, B. W., S. O. Gollnick, J. W. Snyder et al. 2004. Choice of oxygen-conserving treatment regimen determines the inflammatory response and outcome of photodynamic therapy of tumors. *Cancer Research* 64 (6): 2120–2126.

Hetts, S. W. 1998. To die or not to die. *The Journal of the American Medical Association* 279 (4): 300–307.

Jarvi, M. T., M. J. Niedre, M. S. Patterson, and B. C. Wilson. 2006. Singlet oxygen luminescence dosimetry (SOLD) for photodynamic therapy: Current status, challenges and future prospects. *Photochemistry and Photobiology* 82 (5): 1198–1210.

Kim, R., M. Emi, K. Tanabe, Y. Uchida, and K. Arihiro. 2006. The role of apoptotic or nonapoptotic cell death in determining cellular response to anticancer treatment. *European Journal of Surgical Oncology (EJSO)* 32 (3): 269–277.

Lemar, K. M., M. A. Aon, S. Cortassa, B. O'Rourke, C. T. Müller, and D. Lloyd. 2007. Diallyl disulphide depletes glutathione in *Candida albicans*: Oxidative stress-mediated cell death studied by two-photon microscopy. *Yeast* 24 (8): 695–706.

Levitt, J. M., A. Baldwin, A. Papadakis et al. 2006. Intrinsic fluorescence and redox changes associated with apoptosis of primary human epithelial cells. *Journal of Biomedical Optics* 11 (6): 064012.

Moffat, B. A., T. L. Chenevert, T. S. Lawrence et al. 2005. Functional diffusion map: A noninvasive MRI biomarker for early stratification of clinical brain tumor response. *Proceedings of the National Academy of Sciences of the United States of America* 102 (15): 5524–5529.

Mokrý, M., P. Gal, M. Harakaľová et al. 2007. Experimental study on predicting skin flap necrosis by fluorescence in the FAD and NADH bands during surgery. *Photochemistry and Photobiology* 83 (5): 1193–1196.

Morbidelli, L., M. Monici, N. Marziliano et al. 2005. Simulated hypogravity impairs the angiogenic response of endothelium by up-regulating apoptotic signals. *Biochemical and Biophysical Research Communications* 334 (2): 491–499.

Neves, A. A. and K. M. Brindle. 2006. Assessing responses to cancer therapy using molecular imaging. *Biochimica et Biophysica Acta (BBA)—Reviews on Cancer* 1766 (2): 242–261.

Niedre, M. J., A. J. Secord, M. S. Patterson, and B. C. Wilson. 2003. In vitro tests of the validity of singlet oxygen luminescence measurements as a dose metric in photodynamic therapy. *Cancer Research* 63 (22): 7986–7994.

Petit, P. X., M.-C. Gendron, N. Schrantz et al. 2001. Oxidation of pyridine nucleotides during Fas-and ceramide-induced apoptosis in Jurkat cells: Correlation with changes in mitochondria, glutathione depletion, intracellular acidification and caspase 3 activation. *Biochemical Journal* 353: 357–367.

Schon, E. A. and G. Manfredi. 2003. Neuronal degeneration and mitochondrial dysfunction. *Journal of Clinical Investigation* 111 (3): 303–312.

Stratonnikov, A. A. and V. B. Loschenov. 2001. Evaluation of blood oxygen saturation in vivo from diffuse reflectance spectra. *Journal of Biomedical Optics* 6 (4): 457–467.

Su, G.-C., Y.-H. Wei, and H.-W. Wang. 2011. NADH fluorescence as a photobiological metric in 5-aminolevlinic acid (ALA)-photodynamic therapy. *Optics Express* 19 (22): 21145–21154.

Wang, H.-W., V. Gukassyan, C.-T. Chen et al. 2008a. Differentiation of apoptosis from necrosis by dynamic changes of reduced nicotinamide adenine dinucleotide fluorescence lifetime in live cells. *Journal of Biomedical Optics* 13 (5): 054011.

Wang, H.-W., Y.-H. Wei, and H.-W. Guo. 2009. Reduced nicotinamide adenine dinucleotide (NADH) fluorescence for the detection of cell death. *Anti-Cancer Agents in Medicinal Chemistry* 9 (9): 1012–1017.

Wang, K. K.-H., S. Mitra, and T. H. Foster. 2007. A comprehensive mathematical model of microscopic dose deposition in photodynamic therapy. *Medical Physics* 34: 282–293.

Wang, K. K.-H., S. Mitra, and T. H. Foster. 2008b. Photodynamic dose does not correlate with long-term tumor response to mTHPC-PDT performed at several drug-light intervals. *Medical Physics* 35: 3518–3526.

Wilson, B. C., M. S. Patterson, and L. Lilge. 1997. Implicit and explicit dosimetry in photodynamic therapy: A new paradigm. *Lasers in Medical Science* 12 (3): 182–199.

Yu, G., T. Durduran, C. Zhou et al. 2005. Noninvasive monitoring of murine tumor blood flow during and after photodynamic therapy provides early assessment of therapeutic efficacy. *Clinical Cancer Research* 11 (9): 3543–3552.

Yu, J.-S., H.-W. Guo, C.-H. Wang, Y.-H. Wei, and H.-W. Wang. 2011. Increase of reduced nicotinamide adenine dinucleotide fluorescence lifetime precedes mitochondrial dysfunction in staurosporine-induced apoptosis of HeLa cells. *Journal of Biomedical Optics* 16 (3): 036008.

Zhang, Z., D. Blessington, H. Li et al. 2004. Redox ratio of mitochondria as an indicator for the response of photodynamic therapy. *Journal of Biomedical Optics* 9 (4): 772–778.

11

Monitoring Stem Cell Differentiation in Engineered Tissues

Kyle P. Quinn
Tufts University

Irene Georgakoudi
Tufts University

11.1 Introduction

The ability of stem cells to self-renew and/or differentiate into a variety of specialized cell types underlies their tremendous potential for the treatment and repair from disease or trauma. Noninvasive monitoring of the endogenous fluorescence of these cells can provide a unique insight into the dynamic cellular changes occurring during cell differentiation and formation of specialized tissues. In this chapter, we provide an overview of general approaches to monitor and study stem cell cultures and engineered tissues. We also highlight the particular advantages of noninvasive optical imaging and the specific roles of NAD(P)H and FAD during stem cell differentiation. We then describe current methods for monitoring stem cell cultures and 3D engineered tissues using nonlinear microscopy. Finally, we summarize the current challenges in optical monitoring of these tissues and provide an outlook on the future of this approach in stem cell studies.

11.2 Stem Cell Biology and Tissue Engineering Applications

For therapeutic and regenerative medicine applications, both multipotent and pluripotent stem cells offer a great promise. Pluripotent stem cells can differentiate into cells from any of the three germ layers formed during embryogenesis: the ectoderm, mesoderm, or endoderm. As a result, pluripotent stem cells can potentially give rise to any adult cell type (Cameron et al. 2006). Stem cells with such pluripotency can be harvested from the inner cell mass of the blastocyst and are termed embryonic stem cells (ESCs) (Thomson et al. 1998). Circumventing the controversial use of human embryos, induced pluripotent stem cells can also be derived from reprogrammed human somatic cells, which may prove useful in regenerative medicine applications (Yu et al. 2007).

On the other hand, multipotent stem cells can be harvested from specific sites in adults, most notably bone marrow (Alhadlaq and Mao 2004; Parekkadan and Milwid 2010). Human bone marrow–derived mesenchymal stem cells (hMSCs) are frequently used for tissue engineering approaches in regenerative medicine (Karageorgiou et al. 2006; Kim et al. 2005; Li et al. 2005; Mauney et al. 2007; Parekkadan and Milwid 2010). Human adipose-derived mesenchymal stem cells (hASCs) can also be harvested from liposuction aspirates and have been used for tissue engineering applications (Correia et al. 2012; Dubois et al. 2008; Kang et al. 2009; Mauney et al. 2007).

Collectively, adult stem cells have been used to develop engineered tissue equivalents of bone, cartilage, muscle, tendon, and adipose tissue (Krampera et al. 2006; Pittenger et al. 1999). While multipotent adult stem cells are more limited than ESCs in the range of differentiated cell types they can give rise to, the relative ease of accessibility and their ability to easily differentiate into a range of musculoskeletal cell types make them particularly amenable to tissue engineering applications.

11.2.1 Traditional Assessments of Cell Differentiation and Functional Tissue Development

Allografts and autografts are often required to facilitate tissue regeneration and restoration of function following disease or trauma. However, these transplants can often result in suboptimal outcomes (Coleman 2006; Eppley et al. 1990; Faour et al. 2011; Jackson et al. 1993). Through the use of stem cells, tissue engineering holds a great potential for improving tissue repair and regeneration. Engineered tissues are typically developed by seeding stem cells onto a biomaterial scaffold and promoting differentiation over the course of weeks or months. However, developing a viable and functional 3D engineered tissue remains a challenge in most cases. Accordingly, assessment of tissue characteristics to evaluate and optimize complex culture protocols is necessary to advance the field. Western blots, quantitative polymerase chain reaction (qPCR) techniques, histology, immunohistochemistry (IHC), and electron microscopy have typically been used to evaluate the biochemical and microstructural characteristics of tissue at specific time points (Choi et al. 2011; Kim et al. 2005; Li et al. 2005; Rice et al. 2010). Western blots and qPCR are commonly performed to detect proteins or mRNAs that are specific to the desired differentiation outcome. Histology, IHC, and electron microscopy can be used to evaluate structural organization that is indicative of a specific tissue type. However, in order to enable such visualizations, these 3D samples must be processed and cut into thin sections. The destructive nature of all of these traditional methods does not allow for repeated assessments of tissue development over time. Furthermore, such techniques cannot be used to evaluate dynamically the integration of engineered tissue with the native counterpart following implantation. These limitations are addressed by fluorescence microscopy techniques, which allow for a quantitative, noninvasive approach to assess tissue differentiation status and structural organization within 3D engineered tissues.

11.2.2 Advantages of Fluorescence Measurements for Stem Cell Applications

As described in previous chapters, two-photon excited fluorescence (TPEF) microscopy offers a number of advantages over traditional approaches for imaging 3D tissues. The use of near-infrared light allows for deeper tissue penetration and less out-of-focus photodamage than confocal microscopy (Denk et al. 1990; Helmchen and Denk 2005). Additionally, because two-photon absorption is almost exclusively limited to the focal plane, there is an inherent ability to acquire depth-resolved images without the need for confocal pinhole detection. Through two-photon excitation, the endogenous fluorescence of nicotinamide and flavin adenine dinucleotides (NADH and FAD, respectively), emanating primarily from cell mitochondria, can be imaged (Chance et al. 1964; Eng et al. 1989; Scholz et al. 1969). NADH and FAD are cofactors that are integral to metabolic processes in cells. A similar structure, nicotinamide adenine dinucleotide phosphate (NADPH), is another reducing agent found primarily in the cytosol that has similar autofluorescent properties as NADH. However, the majority of fluorescence emitted from the cell in the blue range is produced by mitochondrial NADH, rather than by cytosolic NADH or NADPH (Blinova et al. 2005, 2008; Eng et al. 1989; Huang et al. 2002). Nonetheless, since the spectral emission profiles of NADH and NADPH are similar, we usually use NAD(P)H when describing the detected fluorescence to indicate that in principle both chromophores could contribute to the signal. It is also worth noting that the relative amount of NADPH with respect to NADH varies among different cell/tissue types.

While morphological changes to cells undergoing differentiation can be identified from the spatial organization of NAD(P)H and FAD fluorescence within cells, a redox ratio based on the FAD and NAD(P)H fluorescence intensity levels (i.e., FAD/[NAD(P)H + FAD]) can be used to probe metabolic changes associated with differentiation (Georgakoudi and Quinn 2012; Quinn et al. 2012; Rice et al. 2007, 2010). In addition, NAD(P)H fluorescence lifetime imaging microscopy (FLIM) (see Chapter 4) can provide an additional source of contrast in order to assess any changes in the relative amount of NAD(P)H and/or FAD that is bound to mitochondrial proteins during the differentiation of stem cells (Guo et al. 2008; König et al. 2011; Stringari et al. 2011, 2012).

The noninvasive nature of nonlinear optical microscopy techniques in acquiring depth-resolved images of tissues makes them particularly amenable to characterizing 3D engineered tissues. By making repeated measurements on the same tissues using these techniques, the statistical power of differentiation studies can be improved while at the same time reducing costs and logistical challenges. Furthermore, unlike traditional quantitative methods to evaluate tissue status, such as qPCR, nonlinear microscopy techniques can be used to localize the spatial patterns of differentiation within developing tissues. In fact, substantial heterogeneity exists within even 2D stem cell cultures undergoing differentiation (Figure 11.1). Yet, because NAD(P)H and FAD are omnipresent in eukaryotic cells and are utilized

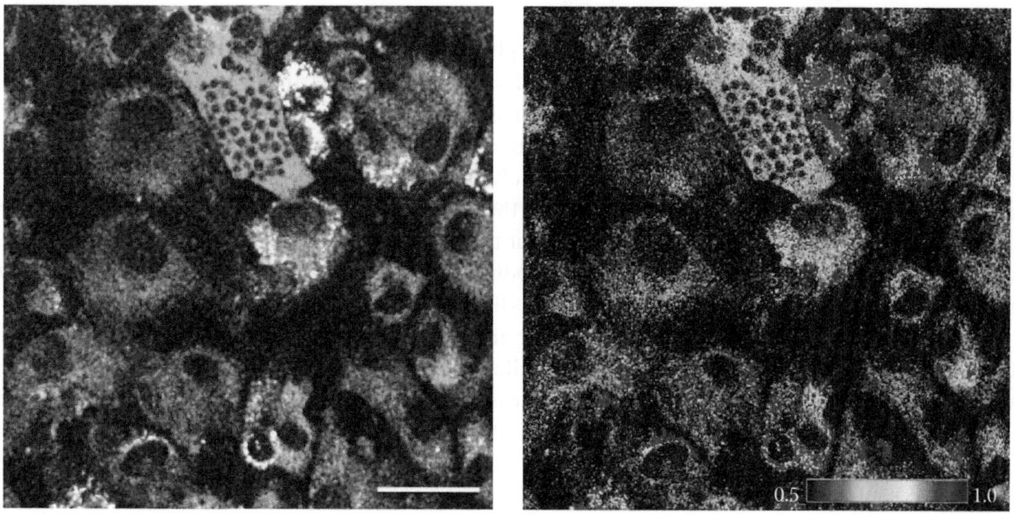

(a) (b)

FIGURE 11.1 Heterogeneity of differentiating stem cell populations. Human mesenchymal stem cells 28 days after induction of adipogenic differentiation exhibit a wide variety of (a) fluorescence intensities and (b) redox ratios. In (a), fluorescence intensity at 755 nm two-photon excitation and 460 nm emission corresponds to green, 860 nm excitation and 525 nm emission corresponds to blue, and 860 nm excitation and 460 nm emission corresponds to red. Scale bar is 50 μm. Color map in (b) shows the redox ratio of FAD/[NAD(P)H + FAD] for the same site.

in numerous cell processes, a detailed understanding of their roles in the metabolism of both stem cells and specialized cell types is necessary for interpreting changes in cofactor ratios during differentiation.

11.3 NAD(P)H and FAD Changes during Cell Differentiation

The electron carriers involved in cell metabolism exist in either oxidized ($NAD(P)^+$, FAD) or reduced (NAD(P)H, $FADH_2$) forms, but only NAD(P)H and FAD are significantly fluorescent (Mayevsky and Chance 1982; Mayevsky and Rogatsky 2007). Free and protein-bound NADH can be found in the cytosol and primary organelles, of which mitochondria have the highest amount of this cofactor. As shown in earlier chapters, the fluorescence of NADH is usually enhanced upon binding to proteins. By comparison, cellular FAD exists primarily in a protein-bound form. In contrast to NADH, however, the FAD autofluorescence can be substantially quenched when bound to proteins such as succinate dehydrogenase (Kunz and Kunz 1985). The majority of flavin fluorescence is thought to be produced by lipoamide dehydrogenase (LipDH)-containing enzyme complexes, such as pyruvate dehydrogenase complex (PDHC) and α-ketoglutarate dehydrogenase complex (Huang et al. 2002; Kunz and Kunz 1985; Saks et al. 1998). These complexes serve as an intermediary electron carrier and are in direct equilibrium with the ratio of NAD^+/NADH in the mitochondria (Georgakoudi and Quinn 2012; Huang et al. 2002). As a result, the relative changes in NAD(P)H and FAD fluorescence intensities can serve as a sensitive indicator of potential metabolic variation during cell differentiation.

11.3.1 Mitochondrial Biogenesis

A hallmark of cell differentiation is an increase in mitochondrial content (Beck and Greenawalt 1976; Duguez et al. 2002; Moyes et al. 1997; Wilson-Fritch et al. 2003). For example, proteomic analysis of hASCs confirmed that while mitochondrial proteins only make up 8% of the total protein content, they represent 18% of the upregulated proteins during adipogenic differentiation (DeLany et al. 2005). Substantial mitochondrial biogenesis is not specific to just adipogenic differentiation—it has been identified during skeletal muscle regeneration (Duguez et al. 2002) and osteogenic differentiation (Chen et al. 2008).

Proliferating stem cells do not rely on oxidative phosphorylation for energy production to nearly the same extent as normal terminally differentiated cells (Vander Heiden et al. 2009). The anabolic requirements of the growth phase leading up to mitosis include a number of glycolysis and tricarboxylic acid (TCA) cycle metabolites. As a result, proliferating cells appear to undergo more glycolysis than the quiescent differentiated cells. Such increase in glycolysis can also serve as a source of ATP, obviating the need for vast mitochondrial networks throughout the cell. However, as cells differentiate and lose their ability to proliferate, there is a reduced demand for these glycolytic metabolites. Under this condition, cells shift toward a more efficient mode of ATP production via oxidative phosphorylation by generating a larger, more organized network of mitochondria (Figure 11.2).

Mitochondrial biogenesis can be detected via monitoring the NAD(P)H and FAD intensity levels over time. Quinn et al. (2012) monitored changes in cellular NAD(P)H and FAD fluorescence in long-term studies of adipogenic differentiation of hACSs. The average NAD(P)H fluorescence intensity increased by approximately twofold over the course of the first 3 weeks of differentiation and remained elevated albeit with substantial variation in intensity between discrete time points. An average 2.7-fold increase in the NAD(P)H fluorescence intensity was detected 6 months after the addition of adipogenic differentiation factors. Unlike NAD(P)H, FAD fluorescence did not substantially increase during the first 6 weeks, but rather between weeks 14 and 26, ultimately resulting in a 3.4-fold increase relative to the beginning of the experiment. It is important to note that the different timing of these increases in NAD(P)H and FAD fluorescence is likely related to changes in the cell redox state observed to occur during differentiation.

In addition to an increase in NAD(P)H and FAD fluorescence, mitochondrial biogenesis may lead to a more reduced state. Using a 3T3-L1 cell line, a 5–8-fold increase in exogenous mitochondrial fluorescence

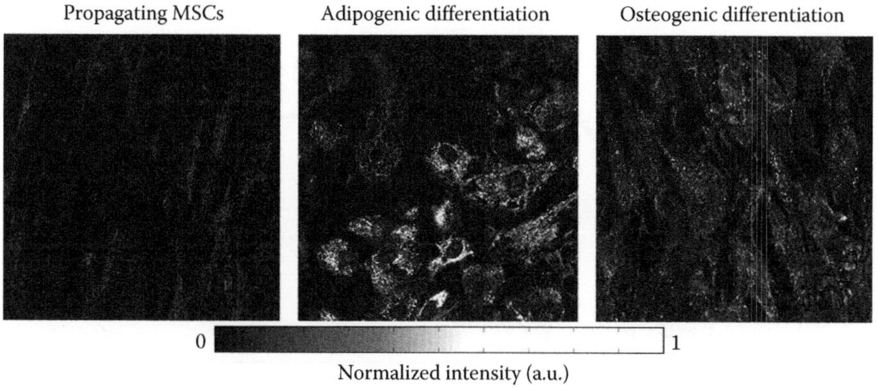

FIGURE 11.2 MitoTracker Orange fluorescence intensity from propagating MSCs and MSCs that have undergone adipogenic and osteogenic differentiation over 2 weeks. An increase in mitochondrial content and organization is evident following differentiation.

staining was identified during adipogenesis in addition to a 20-fold increase in the expression of some mitochondrial proteins (Wilson-Fritch et al. 2003). Despite such increase in putative mitochondrial content, only a twofold increase in the rate of oxygen consumption was observed following differentiation in that study. Furthermore, treatment with carbonyl cyanide *p*-trifluoromethoxy-phenylhydrazone (FCCP), a mitochondrial uncoupler, suggested that those newly differentiated adipocytes were only consuming oxygen at a rate that was approximately 25% of the cell's capability, compared with 60% of the capability of undifferentiated cells. A smaller percentage of the total mitochondria undergoing oxidative phosphorylation during adipogenic differentiation would suggest that mitochondrial biogenesis occurs prior to a substantial increase in oxidative phosphorylation. With an increase in mitochondrial content prior to the associated increase in oxidative phosphorylation, a transient decrease in the NAD^+/NADH ratio would likely occur, which is consistent with the observed reduction of the redox ratio during differentiation (Quinn et al. 2012; Rice et al. 2010).

11.3.2 Adipogenic Differentiation and Fatty Acid Synthesis

Mitochondrial biogenesis may occur during adipogenic differentiation in order to facilitate fatty acid synthesis, rather than an increase in ATP demand. In an experiment conducted by Rice et al., hMSCs or hASCs were induced into adipogenic differentiation through the addition of soluble factors to their culture media and analyzed using TPEF imaging (Rice et al. 2010). Initial studies of adipogenic differentiation in 2D cultures of hMSCs on glass coverslips showed a slight increase in the optical redox ratio of FAD/[NAD(P)H + FAD] during the first 8 days of differentiation, followed by a significant reduction through day 21. These changes coincided with lipid droplet accumulation measured by Oil Red O staining on day 21 and the increased expression of lipoprotein lipase mRNA over the course of the experiment. Expanding upon that work, hASCs that underwent 7 days of differentiation were seeded onto porous silk scaffolds and cultured for 6 months over which optical redox ratios were calculated at 14 different time points (Quinn et al. 2012). Between days 4 and 57 after ASC seeding, the redox ratio in these cells decreased, with a significant reduction between days 8 and 15 ($p = 0.05$), as well as days 30 and 43 ($p = 0.0145$) (Figure 11.3). Of note, significant increases in redox ratio relative to previous time points were measured on days 78 ($p = 0.0002$), 99 ($p < 0.0001$), and 155 ($p = 0.0018$) (Figure 11.3). Lipid droplet numbers and total lipid droplet area, measured through Oil Red O staining, increased between days 15, 30, and 57, which coincided with reduction in the redox ratio. Yet, no additional lipid droplet accumulation was observed between days 57 and 183 when the redox ratio began increasing. This study

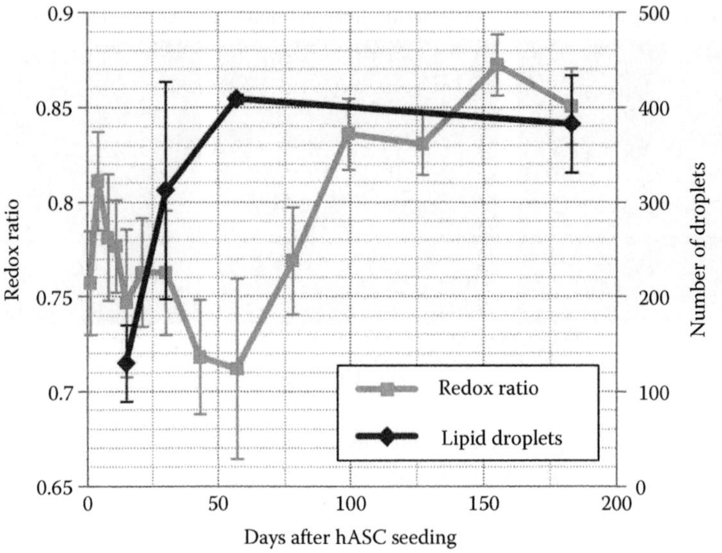

FIGURE 11.3 Decreases in mean optical redox ratio are evident during the first 57 days of adipogenic differentiation in 3D cultures containing hASCs. The decrease in redox ratio corresponded to lipid droplet accumulation. (From Quinn, K.P. et al., *Biomaterials*, 33(21), 5341, 2012.)

demonstrates that the sensitivity of the optical redox ratio to adipogenic differentiation may be related to the storage of lipid droplets.

The role of mitochondria in fatty acid synthesis may explain the decrease in redox ratio during lipid droplet production. One of the primary functions of adipocytes is to store energy in the form of triglycerides, which consist of three fatty acids and glycerol. Fatty acids can be synthesized by first breaking down glucose into pyruvate and shuttling it into the mitochondria (Figure 11.4). Pyruvate is then either converted to oxaloacetate by pyruvate carboxylase or to acetyl-CoA by PDHC, which reduces NAD^+ to NADH. During the first step of the TCA cycle, acetyl-CoA and oxaloacetate are condensed into citrate, which is then shuttled out of the mitochondria and cleaved back into acetyl-CoA and oxaloacetate in the cytosol. This cytosolic acetyl-CoA is used as the carbon supply for fatty acid synthesis. Increased glycolysis and PDHC activity for fatty acid synthesis would produce an increase in mitochondrial NADH; without a proportional increase in ATP demand, the corresponding increase in concentrations of mitochondrial NADH and reduced $FADH_2$ within PDHC will ultimately cause a reduction in the optical redox ratio during lipogenesis (Figure 11.4). Supporting this theory was metabolic flux modeling of differentiating preadipocytes from a 3T3-L1 lineage, which demonstrated that an increase in the PDHC activity coincides with an increase in mitochondrial NADH flux (Si et al. 2007). Additionally, more recent work demonstrates redox ratio sensitivity to the introduction of exogenous fatty acids during adipogenic differentiation, further supporting this hypothesis (Quinn et al. 2013).

11.3.3 Osteogenic Differentiation and Collagen Synthesis

In one of the initial studies to quantify autofluorescence during differentiation, goat mesenchymal stem cells (MSCs) that underwent osteogenic differentiation were found to have an increased ratio of FAD/NAD(P)H relative to the undifferentiated cells (Reyes et al. 2006). While the stage of differentiation was not reported, a higher redox ratio is consistent with a differentiated cell type that relies more on oxidative phosphorylation than glycolysis to produce ATP. More recently, the redox ratio during osteogenic differentiation in both hypoxic and normoxic conditions was monitored in hMSCs, and no significant

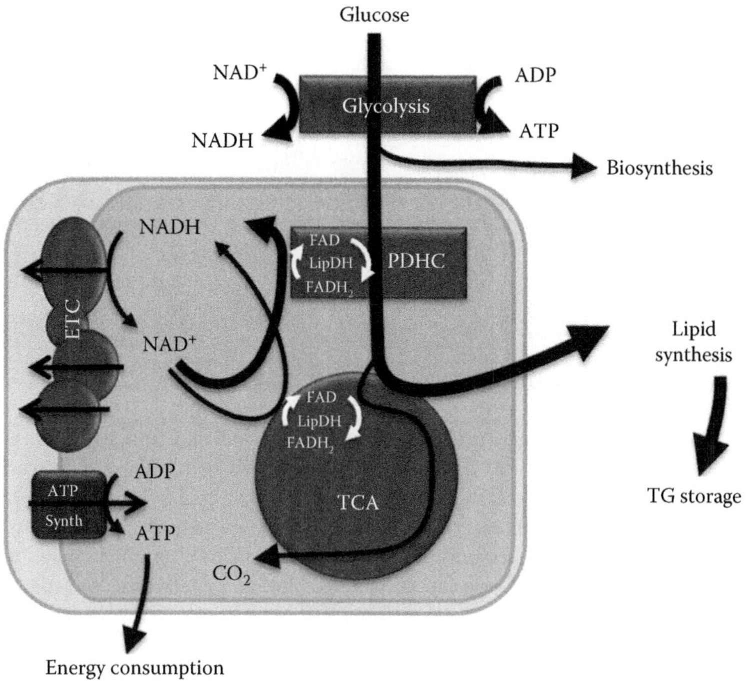

FIGURE 11.4 Adipogenic differentiation causes an increase in flux through lipid synthesis pathways, which can result in an increase in NADH levels.

differences between days 0 and 16 of differentiation were observed (Rice et al. 2010). However, a small transient decrease in redox ratio was observed at day 4 in that study. In contrast to cells undergoing differentiation, a substantial increase in the redox ratio was observed over time for propagating hMSCs. These results suggest that the corresponding autofluorescence is capable of discriminating osteoblastic differentiation from hMSC propagation.

Using new approaches for acquiring and normalizing fluorescence intensity in our lab recently, we identified a significant transient reduction in the redox ratio during osteoblastic differentiation of hMSCs. In hMSCs cultured under hypoxic conditions (5% O_2), the redox ratio decreased during the first 21 days after induction of differentiation and then increased for the remainder of the experiment. This transient decrease in the redox ratio of the osteoblastic cultures produced significantly lower redox values between days 7 and 28 relative to control cultures of propagating hMSCs ($p \leq 0.0001$) (Figure 11.5). In addition to the different imaging time points and analysis methods among these studies, the differences in the magnitude and timing of redox ratio changes could be explained by donor-related variability in stem cell potential toward osteoblastic differentiation.

Using FLIM, an increase in the mean fluorescence lifetime of NAD(P)H was reported on hMSC cultures, undergoing 7, 14, and 21 days of osteogenic differentiation (Guo et al. 2008), suggesting an increase in protein-bound NADH relative to its free counterpart. Using the same culture protocol, it was also demonstrated that intracellular ATP, oxygen consumption, and number of protein subunits of respiratory enzymes increased after 14 days of osteogenic differentiation (Chen et al. 2008). The increase in ATP availability following mitochondrial biogenesis and accumulation of mitochondrial-bound NADH further suggests that the cell redox status is altered toward a more reduced state during the switch from a glycolysis-reliant mode to oxidative phosphorylation.

The underlying mechanism of a decreased redox ratio during osteogenic differentiation may be related to the enhanced collagen synthesis that requires a more reduced state of the cell. An increase in lactate

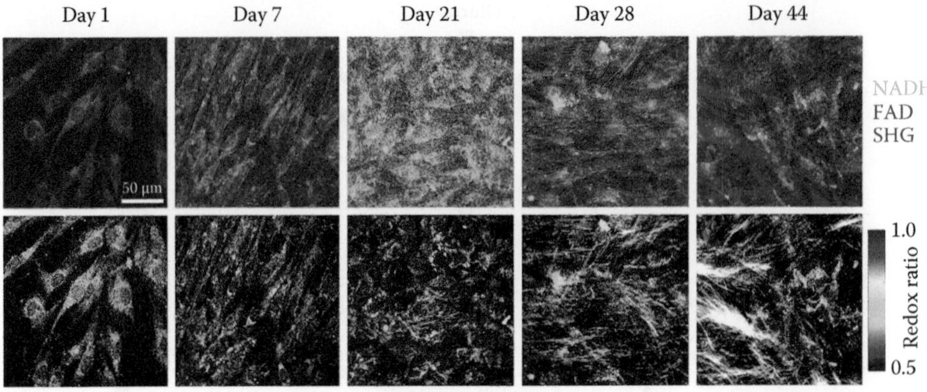

FIGURE 11.5 Upper row shows changes in NAD(P)H and FAD fluorescence and SHG intensity during osteogenic differentiation. An increase in NAD(P)H fluorescence through day 21 produces a decrease in redox ratio. At day 21, collagen deposition becomes clearly visible and continues increasing through day 44, while the redox ratio also begins to increase. Top row shows NAD(P)H fluorescence intensity at 755 nm two-photon excitation and 460 nm emission in green, FAD fluorescence at 860 nm excitation and 525 nm emission in blue, and SHG generation with 800 nm excitation and 400 nm collection in red. Bottom row shows redox ratio. Collagen signal is color-coded with white.

production has been associated with collagen deposition by fibroblasts during wound healing (Trabold et al. 2003) and by hMSCs during osteoblastic differentiation (Park et al. 2010). In a reversible reaction in the cytosol, lactate dehydrogenase converts pyruvate and NADH to lactate and NAD^+ (Figure 11.6). An increase in both NADH and lactate is generally associated with enhanced glycolysis relative to oxidative phosphorylation, which often occurs in hypoxic environments that require anaerobic glycolysis for energy production (Mayevsky and Rogatsky 2007). In normoxic environments, however, aerobic glycolysis also occurs to produce metabolites for anabolism (Lunt and Vander Heiden 2011; Vander Heiden et al. 2009). Collagen deposition and lactate production during wound healing appears relatively insensitive to oxygen levels (Ghani et al. 2003; Trabold et al. 2003); hypoxia is also known to preserve hMSC "stemness" and inhibit osteoblastic differentiation (Yang et al. 2011). Therefore, an increase in lactate concentration during osteoblastic differentiation may be the result of a cell preference toward aerobic glycolysis. In fact, studies of both intact bone preparations and isolated cells have also suggested the potential for aerobic glycolysis in osteoblasts (Felix et al. 1978; Neuman 1976).

A number of mechanisms have been proposed that link the observed increase in aerobic glycolysis and lactate production to enhanced collagen synthesis. Increased lactate concentrations will inhibit the conversion of NADH to NAD^+ by lactate dehydrogenase. Limited NAD^+ availability will reduce the level of ADP-ribosylation in the cell (Figure 11.6). This posttranslational modification is typically associated with DNA repair and apoptosis, but a downregulation of ADP-ribosylation also leads to an increase in collagen mRNA and ultimately collagen synthesis (Trabold et al. 2003). Furthermore, a decreased level of ADP-ribose helps to activate prolyl hydroxylase, which is needed for collagen synthesis (Hussain et al. 1989). Regardless of the presence of oxygen, an increase in lactate concentration will also help to maintain the presence of hypoxia-inducible factor-1 (HIF-1) transcription factors (Lu et al. 2005). It has been demonstrated that increases in HIF-1α transcription factor also amplify prolyl-4-hydroxylase expression, a critical enzyme in collagen synthesis (Takahashi et al. 2000).

The transient decrease in optical redox ratio during osteoblastic differentiation has been directly correlated with collagen deposition. Under hypoxic conditions, a significant increase in the average second harmonic generation (SHG) signal was detected between days 1 and 28 after the addition of osteoblastic induction factors ($p \leq 0.0190$) (Figure 11.5). This indicates a substantially higher collagen deposition at days 21 and 28 relative to the hMSC propagation group ($p < 0.0001$). This period of collagen synthesis was

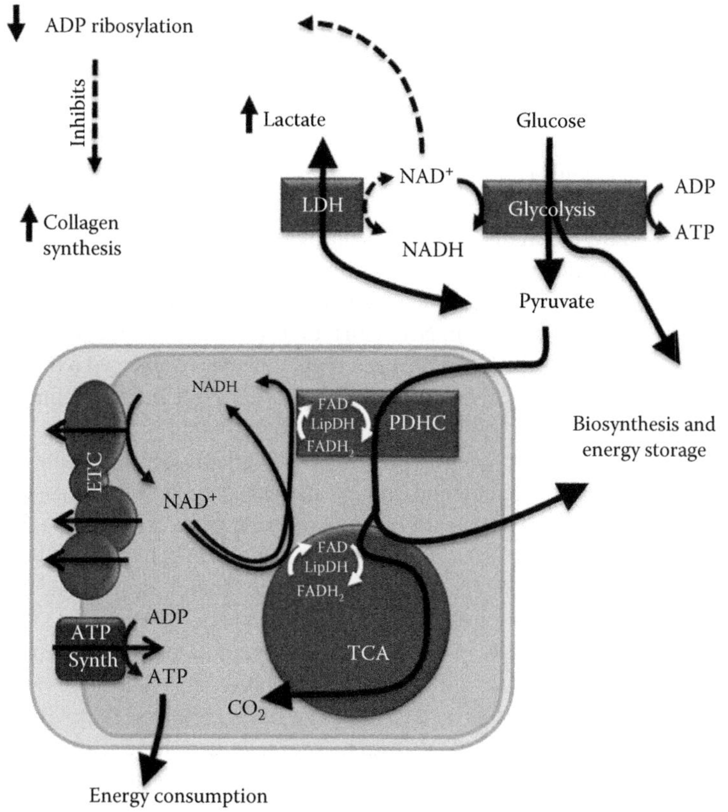

FIGURE 11.6 Schematic representation of the possible link between low redox ratio during osteogenic differentiation and collagen synthesis. Increases in lactate concentrations can inhibit ADP-ribosylation and consequently promote collagen synthesis.

associated with a decrease in the redox ratio between days 7 and 28. Collagen organization into larger bundles, measured by the power exponent of the radially sampled 2D power spectral density, was significantly greater in the osteoblastic group compared to the propagating hMSCs at day 28 ($p < 0.0001$). Substantially, more calcium deposition ($p < 0.0001$) was indeed detected in the osteoblastic cultures (6.80 ± 0.16 μg/cm^2) as compared to the propagation cultures (0.50 ± 0.05 μg/cm^2) at day 44. The redox ratio followed the same pattern with the increase observed at days 28 and 44 (Figure 11.5). An inverse relationship between NAD^+/NADH ratios and the rate of collagen synthesis has also been observed in liver cell cultures in which cirrhosis was induced through carbon tetrachloride (Hernández-Muñoz et al. 1994). Collectively these studies suggest collagen synthesis and cell metabolism are highly interconnected, which makes the optical redox ratio a particularly useful tool for the detection of osteoblastic differentiation.

11.3.4 Differentiation in Developmental Biology

Aside from monitoring the induction of adult stem cell differentiation, autofluorescence imaging has utility in understanding biological processes during development. Embryonic development is highly characterized in *Caenorhabditis elegans*; germ cells of this nematode are particularly well studied, making them a suitable model to understand changes in NAD(P)H and FAD during embryogenesis (Hubbard 2007). Using phasor analysis of fluorescence lifetime images, the spatial transition of germ cells undergoing mitosis, followed by early differentiation, and ultimately meiosis has been investigated

(Stringari et al. 2011). During this early differentiation, FAD fluorescence decreased and the amount of protein-bound NAD(P)H increased relative to free NAD(P)H. Furthermore, a blueshift in the emission spectra of these germ cells was noted in the differentiation transition region of the *C. elegans* relative to the distal undifferentiated mitotic region. These results suggest an increase in NAD(P)H relative to FAD during germ cell differentiation, which is in agreement with findings from adult stem cell differentiation studies described in the previous sections.

Using the same phasor approach, an increase in NADH intensity with slightly longer lifetimes was observed during human ESC differentiation in 2D cultures toward a trophectoderm or neurogenic lineage (Stringari et al. 2012). Differentiation of hESCs into neural stem cells is associated with lower ATP turnover and an increase in glycolysis relative to oxidative phosphorylation (Birket et al. 2011), both of which may lead to the observed increase in NAD(P)H levels (Stringari et al. 2012). However, as neural stem cells continue to a terminally differentiated cell type, a greater reliance on oxidative phosphorylation for ATP production has been reported (Birket et al. 2011) in agreement with findings during adult stem cell differentiation.

Interestingly, an additional fluorophore was identified with a fluorescence lifetime that is longer than that for NAD(P)H, which was assigned to lipid droplet–associated granules (Stringari et al. 2012). These granules could be excited only in the range of 720–760 nm and, therefore, appear to be different from previous reports of lipofuscin granules accumulating in adult stem cells with strong fluorescence at 860 nm excitation (Rice et al. 2010). Stringari et al. hypothesized that these punctate fluorescent objects were the by-product of high concentrations of reactive oxygen species (Stringari et al. 2012).

Large multicellular mouse ESC colonies have also been studied using autofluorescence imaging. These colonies were formed by seeding mouse ESCs on glass coverslips that contained a feeder layer of embryonic fibroblasts (Xu et al. 2011). Over 2–3 days, the ESCs grew into colonies of 200–440 μm in diameter, and a lower redox ratio of FAD/[NAD(P)H + FAD] was detected in the colony core relative to the periphery. Additionally, increased Oct4 expression in the core suggested that the lower redox ratio was associated with increased pluripotency. Although an increase in redox ratio during ESC differentiation contradicts that observed in adult stem cells, the limited resolution (50 μm per pixel) in that study may have masked any transient decrease in redox ratio between the pluripotent core and the more differentiated surface of the colony. Nonetheless, this work demonstrates that ESC colonies can contain a heterogeneous mix of cells with different differentiation potentials, which further highlights the importance of stem cell analysis approaches capable of characterizing the spatial variability in both 3D in vitro cell cultures and embryonic development.

11.4 Monitoring 3D Tissue Cultures

Cell spheroids have frequently been used as a model for studying embryonic development (Desbaillets et al. 2000), and therefore the ability to quantify stem cell differentiation in such a complex 3D environment is critical. Furthermore, new approaches for tissue engineering are increasingly moving toward establishing 3D tissues capable of in vivo implantation. The variable nutrient and oxygen levels within large 3D tissues are likely to produce different effects on cell metabolism and differentiation potential. Nonlinear optical techniques with their noninvasive depth-resolved imaging capacity are essential for assessments of cell status and function within 3D tissues at high resolution. Furthermore, these noninvasive measurements provide a unique opportunity to assess dynamic changes within in vitro disease models. In this section, we provide specific examples of studies in which NAD(P)H and/or FAD fluorescence has been used for understanding cell differentiation within 3D tissues.

11.4.1 Engineered Epithelial Tissues

Organotypic epithelia can be developed by seeding human foreskin keratinocytes onto a collagen gel containing J2 3T3 fibroblasts and raising the tissue to the air–liquid interface (Meyers et al. 1997).

After 10 days of sample preparation, Levitt et al. (2011) observed a high level of differentiation through the depth of epithelial tissue with TPEF imaging. The differentiation status was delineated both through cell morphology and redox state. The morphology of the cells ranged from smaller cell sizes in the proliferating basal layer to larger, flatter cells in the differentiated, cornified layer at the tissue surface. The redox ratio through the depth of the tissue reached a minimum somewhere between the cornified and basal layers, suggesting a transient decrease in redox ratio during cell differentiation. In HPV-transfected epithelia, the absence of distinct differentiated layers was also observed, and depth-dependent fluctuations in redox ratio were attenuated. These findings suggest that the optical redox ratio may prove useful observable for diagnostic purposes of certain disease states, such as precancer, in which cell differentiation patterns are altered.

11.4.2 Stem Cell Spheroids and Embryoid Bodies

The use of FLIM to assess stem cell differentiation status in 3D cell spheroid models has been demonstrated (Konig et al. 2011), where human salivary gland stem cells were isolated and pellets of 200,000 cells were cultured in adipogenic differentiation media. A blueshift of the emission spectra of cellular autofluorescence during differentiation was consistent with an increase in NAD(P)H relative to FAD. Furthermore, mature adipocytes displayed a longer mean lifetime at 750 nm excitation compared to cells in the same culture that did not produce lipid droplets. This shift was primarily attributed to an increase in the relative amount of bound NAD(P)H in adipocytes. Interestingly, the overall intensity of both NAD(P)H and FAD was substantially lower in mature adipocytes relative to less differentiated neighboring cells. The authors attributed this lower fluorescence intensity in adipocytes to a shift toward the nonfluorescent forms of the cofactors $NAD(P)^+$ and $FADH_2$ during differentiation and an increase in oxygen consumption. However, accumulation of strongly autofluorescent lipofuscin or lipid droplet–associated granules may be disproportionately greater in undifferentiated stem cells, which could also explain the observed difference in fluorescence intensity between stem cells and adipocytes within the spheroid (Rice et al. 2010; Stringari et al. 2012).

Embryoid bodies (i.e., spherical aggregates of ESCs) typically span up to 500 μm in diameter and can also be imaged using multiphoton microscopy techniques. Buschke et al. used NAD(P)H fluorescence to investigate the effect of apoptosis on the functional development of the murine embryoid bodies (Buschke et al. 2012). Apoptosis was induced in that study by the addition of staurosporine, an ATP-competitive kinase inhibitor, and the extent of cell death was found to be positively correlated with NAD(P)H intensity. The observed increase in cell death, as estimated by NAD(P)H fluorescence intensity, provided a predictive measure of whether the embryoid body would ultimately give rise to functional cardiomyocytes. The enhanced NAD(P)H level during the early phases of cell death has also been observed in 2D cell cultures (Levitt et al. 2006; Wang et al. 2008) (see also Chapter 10).

11.4.3 Engineered Adipose Tissues

To develop large engineered tissues suitable for implantation, multicellular spheroids are often insufficient in size, and therefore a biomaterial scaffold is typically needed. Adipose tissue was engineered previously by seeding stem cells onto a porous silk scaffold (Kang et al. 2009; Mauney et al. 2007). TPEF imaging was used to quantify the redox state of hASCs undergoing adipogenic differentiation in 3D cultures (Quinn et al. 2012). To accurately quantify cell redox state in these engineered tissues, the intrinsic fluorescence of the silk scaffold had to be characterized for quantitative image analysis. To this end, a 3D segmentation algorithm was developed that automatically classified individual fluorescent objects as either a cell or silk biomaterial using linear discriminant analysis (Quinn et al. 2012). Through these algorithms, 3D projections containing both silk scaffold fluorescence and individual cell redox ratios were produced (Figure 11.7).

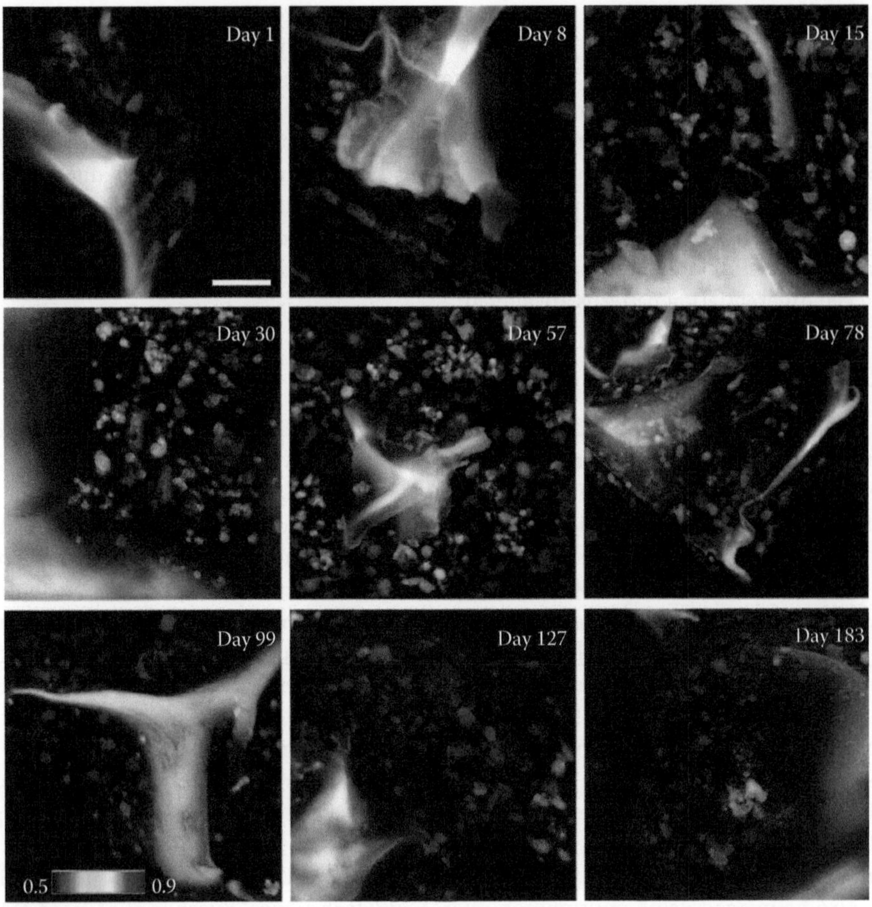

FIGURE 11.7 Axial projections of 3D image volumes containing differentiating hASCs within a fluorescent silk scaffold. Cells and silk were classified based on their different fluorescence characteristics, and cells were color-coded according to their average redox ratio. Scale bar corresponds to 50 μm.

As the field of tissue engineering begins to scale engineered constructs up to a size necessary for human implantation, incorporating endothelial cells into tissue cultures and forming a functional vascular network become increasingly necessary for nutrient and oxygen delivery (Kang et al. 2009). To assess the effect of endothelial cells on the metabolism of developing adipose tissue constructs, comparisons have been made between scaffolds containing hASCs and those containing a coculture of hASCs and human microvascular endothelial cells (hMVECs) (Quinn et al. 2012). These cocultures exhibited a higher average redox ratio ($p = 0.0386$) just 1 day after hASC seeding compared to the hASC monocultures. Differences in cell number and redox ratio with respect to the distance from the scaffold surface were assessed through the image acquisition at three specific depths (0–40, 80–120, and 160–200 μm). Significantly, more cells were detected within the intermediate (80–120 μm) and deep (160–200 μm) zones in the coculture group, and a greater proportion of hASCs resided in the superficial depths of the monoculture group as compared to the coculture cells. Despite significantly larger number of cells residing deeper within scaffolds in the coculture group, no differences between the redox ratios of the two groups were observed at different depths. After TPEF imaging, CD-31 staining of the tissues demonstrated clusters of hMVECs at various depths exceeding 200 μm that appeared to promote hASC viability deeper within the tissues.

11.5 Challenges and Future Outlook

Quantifying the steady-state redox ratio of stem cell cultures, at various stages of differentiation, requires careful calibration and a tight control of instrumentation parameters and biological variables. A major challenge in these autofluorescence studies is to properly isolate FAD and NAD(P)H emission from the potential contributions of other fluorophores. Lipofuscin and lipid droplet–associated granules have been identified as components that can potentially confound results (Rice et al. 2010; Stringari et al. 2012). Fluorescence lifetime measurements may help facilitate better fluorophore discrimination, but such equipment and analysis can be costly and time consuming (see Chapter 4).

Additionally, longitudinal monitoring requires an understanding of how the fluorescence intensity may depend on detector gain and laser power. Aside from potential daily fluctuations in laser power, cell fluorescence intensities can increase greatly over time due to mitochondrial biogenesis, which often requires changes in detector gain and/or laser power to prevent pixel saturation and photodamage. Finally, the local environment of cells must be controlled to ensure consistent temperature, pH, CO_2, and glucose levels between samples and time points.

The majority of studies that quantify changes in redox ratio over extended time periods or between groups of cells report ratios based on fluorescence intensity values. The extent of differentiation within the culture and time course of this process can be identified by the relative magnitude of changes in these optical redox ratios. Relating NAD(P)H and FAD fluorescence to specific intracellular concentrations is particularly valuable for certain applications using engineered tissues, such as drug screening. However, defining a precise relationship may prove challenging if substantial changes in protein binding with FAD or NAD(P)H occur, which will cause changes in the fluorescence emission spectrum and/or quantum yield. Variability in the redox ratio between different stem cell types makes data interpretation rather challenging, but this may also provide additional opportunities to characterize tissues and heterogeneous cell populations.

Variability between stem cell populations of different donors has been reported and presents additional challenges in experimental design and interpretation. Variability of fluorescence intensity among cultures can be attributed to different levels of mitochondrial content among cells. Therefore, the use of a redox ratio instead of individual chromophore intensities may be a more robust metric. In addition, the noninvasive measurements using autofluorescence microscopy can be repeated on the same samples multiple times over extended periods and thereby allow for monitoring relative changes within individual cell cultures with greater statistical power than invasive assessments of different samples over the same time period.

Despite the limitations of cell redox ratio metrics, a number of opportunities exist for nonlinear microscopy approaches to provide valuable insights into dynamic changes during stem cell differentiation and engineered tissue development. Certainly, depth penetration will be a limiting factor for any optical technique used to monitor 3D tissues. Minimally invasive probes that have been used so far to assess brain signaling may be useful for some tissue assessments (Flusberg et al. 2005; Levene et al. 2004). Embedding optical probes into engineered tissue scaffolds may also offer a potential solution, particularly with optical elements derived from biodegradable silk fibroin (Lawrence et al. 2008). The miniaturization of two-photon imaging systems will also be critical for advancing in vivo implants of engineered tissue.

These technological advances, combined with a growing understanding of the biological mechanisms that relate metabolism to cell differentiation and function, should greatly enhance the utility of endogenous fluorescence imaging techniques for assessing the potential of stem cells and engineered tissues both in basic research and human health.

11.6 Conclusions

Techniques based on a redox ratio of FAD and NAD(P)H fluorescence intensities, as well as NAD(P)H fluorescence lifetime, have demonstrated sensitivity to stem cell differentiation. Although a higher redox ratio is often detected in fully differentiated cells compared to proliferating stem cells, a transient decrease in the redox ratio is frequently observed during the process of differentiation. The underlying biological mechanisms driving a change in the optical redox ratio likely differs from the changes in metabolism thought to provide optical sensitivity to other phenomena, such as cancer. Furthermore, different metabolic pathways likely contribute to the transient decrease in redox ratio observed in adipogenic and osteoblastic differentiation of adult stem cells. Fluorescence lifetime measurements can provide information on the protein-bound state of NAD(P)H, which may help in future work to further contextualize the changes observed in the redox ratio. Both lifetime and steady-state techniques uniquely enable the nondestructive mapping of cell metabolism within 3D stem cell cultures.

References

Alhadlaq, A. and J. J. Mao. 2004. Mesenchymal stem cells: Isolation and therapeutics. *Stem Cells and Development* 13 (4): 436–448.

Beck, D. P. and J. W. Greenawalt. 1976. Biogenesis of mitochondrial membranes in *Neurospora crassa* during cellular differentiation: Changes in oxidative phosphorylation and synthesis of mitochondrial phospholipids. *Journal of General Microbiology* 92 (1): 111–119.

Birket, M. J., A. L. Orr, A. A. Gerencser et al. 2011. A reduction in ATP demand and mitochondrial activity with neural differentiation of human embryonic stem cells. *Journal of Cell Science* 124 (3): 348–358.

Blinova, K., S. Carroll, S. Bose et al. 2005. Distribution of mitochondrial NADH fluorescence lifetimes: Steady-state kinetics of matrix NADH interactions. *Biochemistry* 44 (7): 2585–2594.

Blinova, K., R. L. Levine, E. S. Boja et al. 2008. Mitochondrial NADH fluorescence is enhanced by complex I binding. *Biochemistry* 47 (36): 9636–9645.

Buschke, D. G., J. M. Squirrell, J. J. Fong, K. W. Eliceiri, and B. M. Ogle. 2012. Cell death, non-invasively assessed by intrinsic fluorescence intensity of NADH, is a predictive indicator of functional differentiation of embryonic stem cells. *Biology of the Cell* 104 (6): 352–364.

Cameron, C. M., W.-S. Hu, and D. S. Kaufman. 2006. Improved development of human embryonic stem cell-derived embryoid bodies by stirred vessel cultivation. *Biotechnology and Bioengineering* 94 (5): 938–948.

Chance, B., R. W. Estabrook, and A. Ghosh. 1964. Damped sinusoidal oscillations of cytoplasmic reduced pyridine nucleotide in yeast cells. *Proceedings of the National Academy of Sciences of the United States of America* 51 (6): 1244–1251.

Chen, C.-T., Y.-R. V. Shih, T. K. Kuo, O. K. Lee, and Y.-H. Wei. 2008. Coordinated changes of mitochondrial biogenesis and antioxidant enzymes during osteogenic differentiation of human mesenchymal stem cells. *Stem Cells* 26 (4): 960–968.

Choi, J. H., E. Bellas, G. Vunjak-Novakovic, and D. L. Kaplan. 2011. Adipogenic differentiation of human adipose-derived stem cells on 3D silk scaffolds. In *Adipose-Derived Stem Cells*, J. M. Gimble and B. A. Bunnell (eds.). Vol. 702 of Methods in Molecular Biology, J. M. Walker (ed.), pp. 319–330. New York: Humana Press.

Coleman, S. R. 2006. Structural fat grafting: More than a permanent filler. *Plastic and Reconstructive Surgery* 118 (3S): 108S–120S.

Correia, C., S. Bhumiratana, L.-P. Yan et al. 2012. Development of silk-based scaffolds for tissue engineering of bone from human adipose-derived stem cells. *Acta Biomaterialia* 8 (7): 2483–2492.

DeLany, J. P., Z. E. Floyd, S. Zvonic et al. 2005. Proteomic analysis of primary cultures of human adipose-derived stem cells modulation by adipogenesis. *Molecular and Cellular Proteomics* 4 (6): 731–740.

Denk, W., J. H. Strickler, and W. W. Webb. 1990. Two-photon laser scanning fluorescence microscopy. *Science* 248 (4951): 73–76.

Desbaillets, I., U. Ziegler, P. Groscurth, and M. Gassmann. 2000. Embryoid bodies: An in vitro model of mouse embryogenesis. *Experimental Physiology* 85 (6): 645–651.

Dubois, S. G., E. Z. Floyd, S. Zvonic et al. 2008. Isolation of human adipose-derived stem cells from biopsies and liposuction specimens. In *Mesenchymal Stem Cells*, D. J. Prockop, B. A. Bunnell, and D. G. Phinney (eds.). Vol. 449 of Methods in Molecular Biology, J. M. Walker (ed.), pp. 69–79. New York: Springer.

Duguez, S., L. Féasson, C. Denis, and D. Freyssenet. 2002. Mitochondrial biogenesis during skeletal muscle regeneration. *American Journal of Physiology—Endocrinology and Metabolism* 282 (4): E802–E809.

Eng, J., R. M. Lynch, and R. S. Balaban. 1989. Nicotinamide adenine dinucleotide fluorescence spectroscopy and imaging of isolated cardiac myocytes. *Biophysical Journal* 55 (4): 621–630.

Eppley, B. L., P. G. Smith, A. M. Sadove, and J. J. Delfino. 1990. Experimental effects of graft revascularization and consistency on cervicofacial fat transplant survival. *Journal of Oral and Maxillofacial Surgery* 48 (1): 54–62.

Faour, O., R. Dimitriou, C. A. Cousins, and P. V. Giannoudis. 2011. The use of bone graft substitutes in large cancellous voids: Any specific needs? *Injury* 42: S87–S90.

Felix, R., W. F. Neuman, and H. Fleisch. 1978. Aerobic glycolysis in bone: Lactic acid production by rat calvaria cells in culture. *American Journal of Physiology—Cell Physiology* 234 (1): C51–C55.

Flusberg, B. A., J. C. Jung, E. D. Cocker, E. P. Anderson, and M. J. Schnitzer. 2005. In vivo brain imaging using a portable 3.9 gram two-photon fluorescence microendoscope. *Optics Letters* 30 (17): 2272–2274.

Georgakoudi, I. and K. P. Quinn. 2012. Optical imaging using endogenous contrast to assess metabolic state. *Annual Review of Biomedical Engineering* 14: 351–367.

Ghani, Q. P., S. Wagner, and M. Z. Hussain. 2003. Role of ADP-ribosylation in wound repair. The contributions of Thomas K. Hunt, MD. *Wound Repair and Regeneration* 11 (6): 439–444.

Guo, H.-W., C.-T. Chen, Y.-H. Wei et al. 2008. Reduced nicotinamide adenine dinucleotide fluorescence lifetime separates human mesenchymal stem cells from differentiated progenies. *Journal of Biomedical Optics* 13 (5): 050505.

Helmchen, F. and W. Denk. 2005. Deep tissue two-photon microscopy. *Nature Methods* 2 (12): 932–940.

Hernández-Muñoz, R., M. Díaz-Muñoz, and V. C. de Sánchez. 1994. Possible role of cell redox state on collagen metabolism in carbon tetrachloride-induced cirrhosis as evidenced by adenosine administration to rats. *Biochimica et Biophysica Acta (BBA)—General Subjects* 1200 (2): 93–99.

Huang, S., A. A. Heikal, and W. W. Webb. 2002. Two-photon fluorescence spectroscopy and microscopy of NAD(P)H and flavoprotein. *Biophysical Journal* 82 (5): 2811–2825.

Hubbard, A. E. J. 2007. *Caenorhabditis elegans* germ line: A model for stem cell biology. *Developmental Dynamics* 236 (12): 3343–3357.

Hussain, M. Z., Q. P. Ghani, and T. K. Hunt. 1989. Inhibition of prolyl hydroxylase by poly(adp-ribose) and phosphoribosyl-amp. Possible role of adp-ribosylation in intracellular prolyl hydroxylase regulation. *Journal of Biological Chemistry* 264 (14): 7850–7855.

Jackson, D. W., E. S. Grood, J. D. Goldstein et al. 1993. A comparison of patellar tendon autograft and allograft used for anterior cruciate ligament reconstruction in the goat model. *The American Journal of Sports Medicine* 21 (2): 176–185.

Kang, J. H., J. M. Gimble, and D. L. Kaplan. 2009. In vitro 3D model for human vascularized adipose tissue. *Tissue Engineering Part A* 15 (8): 2227–2236.

Karageorgiou, V., M. Tomkins, R. Fajardo et al. 2006. Porous silk fibroin 3-D scaffolds for delivery of bone morphogenetic protein-2 in vitro and in vivo. *Journal of Biomedical Materials Research Part A* 78 (2): 324–334.

Kim, H. J., U.-J. Kim, G. Vunjak-Novakovic, B.-H. Min, and D. L. Kaplan. 2005. Influence of macroporous protein scaffolds on bone tissue engineering from bone marrow stem cells. *Biomaterials* 26 (21): 4442–4452.

König, K., A. Uchugonova, and E. Gorjup. 2011. Multiphoton fluorescence lifetime imaging of 3D-stem cell spheroids during differentiation. *Microscopy Research and Technique* 74 (1): 9–17.

Krampera, M., G. Pizzolo, G. Aprili, and M. Franchini. 2006. Mesenchymal stem cells for bone, cartilage, tendon and skeletal muscle repair. *Bone* 39 (4): 678–683.

Kunz, W. S. and W. Kunz. 1985. Contribution of different enzymes to flavoprotein fluorescence of isolated rat liver mitochondria. *Biochimica et Biophysica Acta (BBA)—General Subjects* 841 (3): 237–246.

Lawrence, B. D., M. Cronin-Golomb, I. Georgakoudi, D. L. Kaplan, and F. G. Omenetto. 2008. Bioactive silk protein biomaterial systems for optical devices. *Biomacromolecules* 9 (4): 1214–1220.

Levene, M. J., D. A. Dombeck, K. A. Kasischke, R. P. Molloy, and W. W. Webb. 2004. In vivo multiphoton microscopy of deep brain tissue. *Journal of Neurophysiology* 91 (4): 1908–1912.

Levitt, J. M., A. Baldwin, A. Papadakis, S. Puri, J. Xylas et al. 2006. Intrinsic fluorescence and redox changes associated with apoptosis of primary human epithelial cells. *Journal of Biomedical Optics* 11 (6): 064012.

Levitt, J. M., M. E. McLaughlin-Drubin, K. Munger, and I. Georgakoudi. 2011. Automated biochemical, morphological, and organizational assessment of precancerous changes from endogenous two-photon fluorescence images. *PLoS ONE* 6 (9): e24765.

Li, W.-J., R. Tuli, C. Okafor et al. 2005. A three-dimensional nanofibrous scaffold for cartilage tissue engineering using human mesenchymal stem cells. *Biomaterials* 26 (6): 599–609.

Lu, H., C. L. Dalgard, A. Mohyeldin, T. McFate, A. S. Tait, and A. Verma. 2005. Reversible inactivation of HIF-1 prolyl hydroxylases allows cell metabolism to control basal HIF-1. *Journal of Biological Chemistry* 280 (51): 41928–41939.

Lunt, S. Y. and M. G. Vander Heiden. 2011. Aerobic glycolysis: Meeting the metabolic requirements of cell proliferation. *Annual Review of Cell and Developmental Biology* 27: 441–464.

Mauney, J. R., T. Nguyen, K. Gillen, C. Kirker-Head, J. M. Gimble, and D. L. Kaplan. 2007. Engineering adipose-like tissue in vitro and in vivo utilizing human bone marrow and adipose-derived mesenchymal stem cells with silk fibroin 3D scaffolds. *Biomaterials* 28 (35): 5280–5290.

Mayevsky, A. and B. Chance. 1982. Intracellular oxidation-reduction state measured in situ by a multichannel fiber-optic surface fluorometer. *Science* 217 (4559): 537–540.

Mayevsky, A. and G. G. Rogatsky. 2007. Mitochondrial function in vivo evaluated by NADH fluorescence: From animal models to human studies. *American Journal of Physiology—Cell Physiology* 292 (2): C615–C640.

Meyers, C., T. J. Mayer, and M. A. Ozbun. 1997. Synthesis of infectious human papillomavirus type 18 in differentiating epithelium transfected with viral DNA. *Journal of Virology* 71 (10): 7381–7386.

Moyes, C. D., O. A. Mathieu-Costello, N. Tsuchiya, C. Filburn, and R. G. Hansford. 1997. Mitochondrial biogenesis during cellular differentiation. *American Journal of Physiology—Cell Physiology* 272 (4): C1345–C1351.

Neuman, W. F. 1976. Aerobic glycolysis in bone in the context of membrane-compartmentalization. *Calcified Tissue Research* 22 (1): 169–178.

Parekkadan, B. and J. M. Milwid. 2010. Mesenchymal stem cells as therapeutics. *Annual Review of Biomedical Engineering* 12: 87–117.

Park, S.-H., E. S. Gil, H. Shi, H. J. Kim, K. Lee, and D. L. Kaplan. 2010. Relationships between degradability of silk scaffolds and osteogenesis. *Biomaterials* 31 (24): 6162–6172.

Pittenger, M. F., A. M. Mackay, S. C. Beck et al. 1999. Multilineage potential of adult human mesenchymal stem cells. *Science* 284 (5411): 143–147.

Quinn, K. P., E. Bellas, N. Fourligas, K. Lee, D. L. Kaplan, and I. Georgakoudi. 2012. Characterization of metabolic changes associated with the functional development of 3D engineered tissues by non-invasive, dynamic measurement of individual cell redox ratios. *Biomaterials* 33 (21): 5341–5348.

Quinn, K. P., G. V. Sridharan, R. S. Hayden et al. 2013. Quantitative metabolic imaging using endogenous fluorescence to detect stem cell differentiation. *Scientific Reports* 3: 3432.

Reyes, J. M. G., S. Fermanian, F. Yang et al. 2006. Metabolic changes in mesenchymal stem cells in osteogenic medium measured by autofluorescence spectroscopy. *Stem Cells* 24 (5): 1213–1217.

Rice, W. L., D. L. Kaplan, and I. Georgakoudi. 2007. Quantitative biomarkers of stem cell differentiation based on intrinsic two-photon excited fluorescence. *Journal of Biomedical Optics* 12 (6): 060504.

Rice, W. L., D. L. Kaplan, and I. Georgakoudi. 2010. Two-photon microscopy for non-invasive, quantitative monitoring of stem cell differentiation. *PLoS One* 5 (4): e10075.

Saks, V. A., V. I. Veksler, A. V. Kuznetsov et al. 1998. Permeabilized cell and skinned fiber techniques in studies of mitochondrial function in vivo. In *Bioenergetics of the Cell: Quantitative Aspects*, V. A. Saks, R. Ventura-Clapier, X. Leverve et al. (eds.). Vol. 25 of Developments in Molecular and Cellular Biochemistry, pp. 81–100. New York: Springer.

Scholz, R., R. G. Thurman, J. R. Williamson, B. Chance, and T. Bücher. 1969. Flavin and pyridine nucleotide oxidation-reduction changes in perfused rat liver I. Anoxia and subcellular localization of fluorescent flavoproteins. *Journal of Biological Chemistry* 244 (9): 2317–2324.

Si, Y., J. Yoon, and K. Lee. 2007. Flux profile and modularity analysis of time-dependent metabolic changes of de novo adipocyte formation. *American Journal of Physiology—Endocrinology and Metabolism* 292 (6): E1637–E1646.

Stringari, C., A. Cinquin, O. Cinquin, M. A. Digman, P. J. Donovan, and E. Gratton. 2011. Phasor approach to fluorescence lifetime microscopy distinguishes different metabolic states of germ cells in a live tissue. *Proceedings of the National Academy of Sciences of the United States of America* 108 (33): 13582–13587.

Stringari, C., R. Sierra, P. J. Donovan, and E. Gratton. 2012. Label-free separation of human embryonic stem cells and their differentiating progenies by phasor fluorescence lifetime microscopy. *Journal of Biomedical Optics* 17 (4): 0460121.

Takahashi, Y., S. Takahashi, Y. Shiga, T. Yoshimi, and T. Miura. 2000. Hypoxic induction of prolyl 4-hydroxylase α(I) in cultured cells. *Journal of Biological Chemistry* 275 (19): 14139–14146.

Thomson, J. A., J. Itskovitz-Eldor, S. S. Shapiro et al. 1998. Embryonic stem cell lines derived from human blastocysts. *Science* 282 (5391): 1145–1147.

Trabold, O., S. Wagner, C. Wicke et al. 2003. Lactate and oxygen constitute a fundamental regulatory mechanism in wound healing. *Wound Repair and Regeneration* 11 (6): 504–509.

Vander Heiden, M. G., L. C. Cantley, and C. B. Thompson. 2009. Understanding the Warburg effect: The metabolic requirements of cell proliferation. *Science* 324 (5930): 1029–1033.

Wang, H.-W., V. Gukassyan, C.-T. Chen et al. 2008. Differentiation of apoptosis from necrosis by dynamic changes of reduced nicotinamide adenine dinucleotide fluorescence lifetime in live cells. *Journal of Biomedical Optics* 13 (5): 054011.

Wilson-Fritch, L., A. Burkart, G. Bell et al. 2003. Mitochondrial biogenesis and remodeling during adipogenesis and in response to the insulin sensitizer rosiglitazone. *Molecular and Cellular Biology* 23 (3): 1085–1094.

Xu, H. N., S. Nioka, B. Chance, and L. Z. Li. 2011. Heterogeneity of mitochondrial redox state in premalignant pancreas in a PTEN null transgenic mouse model. In *Oxygen Transport to Tissue XXXII*, J. C. LaManna, M. A. Puchowicz, K. Xu, D. K. Harrison, and D. F. Bruley (eds.). Vol. 701 of Advances in Experimental Medicine and Biology, I. R. Cohen, A. Lajtha, R. Paoletti, and J. D. Lambris (eds.), pp. 207–213. New York: Springer.

Yang, D.-C., M.-H. Yang, C.-C. Tsai, T.-F. Huang, Y.-H. Chen, and S.-C. Hung. 2011. Hypoxia inhibits osteogenesis in human mesenchymal stem cells through direct regulation of RUNX2 by TWIST. *PLoS One* 6 (9): e23965.

Yu, J., M. A. Vodyanik, K. Smuga-Otto et al. 2007. Induced pluripotent stem cell lines derived from human somatic cells. *Science* 318 (5858): 1917–1920.

IV

Autofluorescence as a Diagnostic Tool in Medicine and Health

12

Autofluorescence-Assisted Examination of Cardiovascular System Physiology and Pathology

Alzbeta Marcek Chorvatova
International Laser Centre

12.1 Introduction

Cardiovascular disease (CVD)—a class of diseases affecting heart and blood vessels—is the leading cause of death worldwide. The World Health Organization estimates there will be about 20 million deaths caused by CVD in 2015, accounting for 30% of all deaths worldwide. Conventional standard techniques for clinical diagnosis and investigation of CVD, such as histological examination, lack spatial resolution, cannot be used for the direct studies in vivo, and require tissue processing, which leads to structural alterations. Recent advances in nonlinear imaging and time- and spectrally resolved spectroscopies provide a new venture to visualize pathological changes related to CVD in vivo or in vitro at high resolution without laborious staining and sample preparation. Label-free imaging of the cardiovascular system (CVS) based on endogenous fluorophores is routinely utilized nowadays to monitor oxidative metabolic state in cardiac and vascular tissues and to conduct structural and morphological analysis of the extracellular matrix (ECM) of the heart and arteries (Table 12.1). Consequently, such autofluorescence (AF)-assisted examination is getting more popular in the study of the CVS physiology and pathology (Table 12.2).

In this chapter, we review the major sources of AF in the CVS and the methods employed for AF imaging and data analysis. We then provide an overview of the AF properties of cardiovascular (CV) cells and tissues and illustrate AF-assisted diagnostics of physiological and pathological modifications in the CVS with a few application cases. In conclusion, we discuss the advantages and limitations of the state-of-the-art use of AF as a diagnostic tool in the CVS physiology and pathology.

12.2 Molecular Origins of AF in the CVS

12.2.1 NAD(P)H and Flavins

Mitochondrial nicotinamide adenine dinucleotide reduced (NADH) and flavin prosthetic groups of respiratory complexes are the most pronounced endogenous fluorophores in CV cells, responsible for the blue and yellow/green band of the AF, respectively (Chance et al. 1979; Eng et al. 1989). In the heart, several flavin groups, including electron-transfer flavoproteins and dehydrogenases, contribute to flavoproteins AF (Chorvat et al. 2005; Eng et al. 1989; Huang et al. 2002; Koke et al. 1980; Romashko et al. 1998) (summarized in Table 12.1).

Both NADH and flavins are components of the mitochondrial oxidative phosphorylation, which supplies the majority of the adenosine triphosphate (ATP)—primary molecular energy source for the contraction of the heart and vessels (Nelson and Cox 2008). The first step in this pathway is binding of the coenzyme NADH to Complex I of the mitochondrial respiratory chain and its dehydrogenation by the flavin mononucleotide prosthetic group (detailed in Chapter 1). As outlined in Chapter 2, binding, oxidation, and reduction of nicotinamide adenine dinucleotides and flavins can be monitored by spectroscopy and microscopy methods, thereby providing information on the metabolic state of the cells.

Fluorescent properties of NADH are also shared by its phosphate form, which, however, has a different function in the cell—NADPH is an important cofactor for several enzymes involved in antioxidant processes put in place to counteract the generation of reactive oxygen species by oxidative respiration and/or oxidative stress (Benderdour et al. 2004). For example, NADPH is an important cofactor for glutathione reductase reactions, facilitating glutathione recycling by converting its oxidized form to reduced glutathione. Oxidative stress can also modulate cellular NADPH content through the release of peroxides and various by-products that have been shown to decrease the activity of several enzymes, such as NADP-isocitrate dehydrogenase. Similar to NADH, fluorescence of NADPH has been used for noninvasive fluorescence probing of the metabolic state in the CVS (Chorvatova et al. 2013a; Table 12.1).

12.2.2 Elastin and Collagen

Collagen and elastin are structural proteins found in the ECM of cardiac and vascular tissues. The main function of collagen is structural support and mechanical reinforcement of the heart and blood vessels (Schenke-Layland et al. 2006, 2007). Fluorescence properties of collagen are highly dependent on the experimental conditions—location in the body, excitation wavelength, and hydration. In the arterial wall, collagen exhibits a broad fluorescence in the blue and green range (360–510 nm) when excited at 337 nm; long fluorescence lifetime of 5.3 ns was retrieved for collagen at 390 nm (Table 12.1) (Maarek et al. 2000a). The noncentrosymmetric structure of fibrillar collagen also allows to image it by second harmonic generation (SHG) microscopy (see Section 12.3.3.2).

Elastin is an intrinsically fluorescing protein, which allows blood vessels and heart to change shape and undergo a substantial deformation over the life span of the organism (Berezin and Achilefu 2010).

TABLE 12.1 Molecular Sources of AF in the CVS

CVS	Fluorophore	Excitation (nm)	Emission (nm)	Lifetime (ns)	Source	Method	References
Cardiac mitochondria	NADH	366	420–480		Pigeon heart mitochondria	Time-sharing fluorometer	Chance et al. (1979)
	NADH	350	459 with 65 nm bandwidth		Pig heart submitochondrial particles	Fluorescence system	Hanley et al. (2002)
	NAD(P)H	335	440	0.4 1.9 5.7	Intact porcine heart mitochondria	Time-resolved fluorescence	Blinova et al. (2005)
	FAD	460 or 436	520–570		Pigeon heart mitochondria	Time-sharing fluorometer	Chance et al. (1979)
	NAD(P)H/FAD	330–570	Offset 20–300		Rat cardiac myocyte	Spectrofluorometer excitation/emission matrix	Chorvat et al. (2004)
Cardiac myocytes	NADH	350	400–520		Rat and guinea pig LV myocytes	Epifluorescence with photon counting	Griffiths et al. (1998)
	NAD(P)H/FAD	750	410–490 and 510–650		Adult dog epicardial myocytes	In vitro 2P spectroscopy and imaging	Huang et al. (2002)
	NAD(P)H	375	397 long pass	0.65 1.98 9.34	Rat LV myocytes	Time-resolved spectroscopy	Cheng et al. (2007)
	NAD(P)H	375	397 long pass	0.62 2.33 13.64	Human myocytes	Time-resolved spectroscopy	Cheng et al. (2007)
	NADH	340	400–550		Permeabilized rat myocytes	Spectrofluorometer	Jepihhina et al. (2011)
	NAD(P)H	375	397 long pass	First component: 0.32 and 1.61 Second component: 0.42 by linear unmixing	Human	Time-resolved spectroscopy by unmixing	Chorvat et al. (2010)
	NAD(P)H	345	397 long pass	First component: 1.94 Second component: 0.55 by linear unmixing	Isolated rat myocytes	Time-resolved spectroscopy by unmixing	Chorvatova et al. (2012b, 2013b)
	FAD	900			Adult dog epicardial myocytes	In vitro 2P spectroscopy	Huang et al. (2002)

(*Continued*)

TABLE 12.1 (*Continued*) Molecular Sources of AF in the CVS

CVS	Fluorophore	Excitation (nm)	Emission (nm)	Lifetime (ns)	Source	Method	References
	FAD	488	Unmixed maxima at 500, 530, 560		Rat cardiac myocytes	Spectrally resolved confocal microscopy by unmixing	Chorvat et al. (2004)
	FAD	438	470 long pass	0.15–0.20 0.46–0.94 1.48–3.67	Rat LV myocytes	Time-resolved spectroscopy	Chorvat and Chorvatova (2006)
	FAD	465	500–600		Permeabilized rat myocytes	Spectrofluorometer	Jepihhina et al. (2011)
	Myosin	450			Chicken cardiac myocytes	SHG	Plotnikov et al. (2006)
	Myosin	415			Sheep fetus cardiac myocytes	SHG	Wallace et al. (2008)
Cultured myocytes	Ceroid/lipofuscin	Blue	Yellow/orange		Cultured rat myocyte	Confocal imaging	Terman and Brunk (1998)
	Lipofuscins				Neonatal rat myocytes		Gao et al. (1994)
Working heart	NADH/NAD+	UV			Langendorff-perfused rat epicardium	Decrease of AF of NADH with increased heartbeat	Ashruf et al. (1995)
	Hb, Mb, cytochrome aa3	400–900			Open-chest calves LV	Multichannel spectrometer, diffuse reflectance spectroscopy	Lindbergh et al. (2010, 2011)
	Hb, Mb	650 and 1050			Isolated pig heart: LV	NIR spectroscopic imaging	Nighswander-Rempel et al. (2005, 2006)
Heart conduction system	Elastin of ECM	760			Heart valves of native tissues	SHG	Schenke-Layland et al. (2007)
	Collagen of ECM	840			Heart valves of native tissues	SHG	Schenke-Layland et al. (2007)
	Collagen and elastin					2P LSCM	Gerson et al. (2012)
His bundle	Collagen	330–380	Above 460			Ratios	Bagdonas et al. (2008)
	Unspecified tissue AF	330–380	Around 460		Healthy human heart specimens		Venius et al. (2011)
Aorta	Collagen	337	360–510	5.3 at 390 Type I: 5.2 Type m: 2.95	Arterial tissue: atherosclerotic lesions vs. commercial samples	Time-resolved laser-induced fluorescence spectroscopy (TRLIFS)	Maarek et al. (2000b); Marcu et al. (2001b)

	Collagen types III and I	340	395		Porcine aorta tissues	Fluorescence spectroscopy	Liu et al. (2010a)
	Elastin	337	360–510	2.3	Arterial tissue: atherosclerotic lesions vs. commercial samples	TR LIFS	Maarek et al. (2000b); Marcu et al. (2001b)
	Elastin	340	445		Porcine aorta tissues	Fluorescence spectroscopy	Liu et al. (2010a)
	Cholesterols	337	360–510	1.3 Free: 1.5 Linoleate: 0.9 Oleate: 1.0	Arterial tissue: atherosclerotic lesions vs. commercial samples	TR LIFS	Maarek et al. (2000b); Marcu et al. (2001b)
	Lipoproteins	337	360–510	LDL: 0.95; VLDL: 0.85	Commercial samples	TR LIFS	Marcu et al. (2001b)
	Tryptophan	300	350		Porcine aorta tissues	Fluorescence spectroscopy	Liu et al. (2010a)
	Unspecified tissue AF		360–600		Rabbit aorta	TR LIFS	Jo et al. (2004)
	Unspecified tissue AF	337			Rabbit intima of atherosclerotic plaques in aorta	TR LIFS	Marcu et al. (2005)
	Unspecified tissue AF	337	390/452		Excised human atherosclerotic aorta	Time-resolved fluorescence spectroscopy	Sun et al. (2011)
	Unspecified tissue AF	337	377/50	0.84	Human atherosclerotic plaques	TR LIFS	Phipps et al. (2011)
	Unspecified tissue AF	460	460/60	0.60 0.54	Human atherosclerotic plaques	TR LIFS	Phipps et al. (2011)
Carotid artery	Elastin	860			Vascular tissue	2P	Beaurepaire et al. (2005)
	Collagen	860			Vascular tissue	SHG	Beaurepaire et al. (2005)
	Unspecified tissue AF	337	360–550		Human carotid plaques	TR LIFS	Jo et al. (2005, 2006); Marcu et al. (2009)
	Unspecified tissue AF	410	650–850		Carotid artery rings	Fluorescence spectroscopy	Pery et al. (2009)
Coronary artery	Unspecified tissue AF	337	360–590, peak at 380	0.9 5.8	Excised human coronary artery	TR LIFS	Marcu et al. (1999)
	Unspecified tissue AF	337	360–510, peak at 390	1.05 5.75	Coronary artery segments postmortem	TR LIFS	Marcu et al. (2001a, 2005)

TABLE 12.2 AF-Assisted Examination of CV Physiology and Pathology

CV Physiopathology	Fluorophore	Excitation (nm)	Emission (nm)	Lifetime (ns)	Source	Findings	References
Allograft rejection	Unspecified tissue AF	442	480–800		Rat	Spectral difference between rejecting and nonrejecting hearts.	Morgan et al. (1999)
	Unspecified tissue AF	337, 440, 486	340–700		Frozen human endomyocardial biopsy	Fluorescence spectroscopy revealed no change in spectra.	Yamani et al. (2000)
	NAD(P)H	375	397 long pass	First component: 0.32 1.61 Second component: 0.42 by linear unmixing	Human isolated LV cardiac myocytes	Time-resolved spectroscopy showed increased lifetime of the first component to 0.44 and 2.18 ns in patients with rejection of transplanted hearts.	Chorvat et al. (2010)
Stem cell heart repair	Collagen and elastin ECM					Native and tissue-engineered heart valve leaflets.	Schenke-Layland et al. (2004)
	Collagen					Stem-cell-treated regenerating heart in rat.	Wallenburg et al. (2011)
Ischemia–reperfusion	NADH				Cardiac tissues	AF increase in ischemia.	Hulsmann et al. (1993)
	NADH	365			Open-chest rabbit model	Myocardial ischemia–reperfusion injury.	Ranji et al. (2006)
	NADH and flavins	340	460		Ventricular myocytes and Langendorff-perfused hearts	Search for protection against damage by metabolic inhibition and reperfusion.	Pasdois et al. (2008); Rodrigo et al. (2002)
LV remodeling in pregnancy	NAD(P)H	375	397 long pass	Resolved lifetime components	Isolated LV rat cardiac myocytes	Metabolic oxidative state change in the presence of lactate in pregnancy.	Bassien-Capsa et al. (2011)
	NAD(P)H	375	397 long pass	Resolved lifetime components	Isolated LV rat cardiac myocytes	Metabolic oxidative state change in the presence of ouabain in pregnancy.	Elzwiei et al. (2013)

Drug evaluation	NADH	350	459 with 65 nm bandwidth		Pig heart submitochondrial particles	Halothane, isoflurane, and sevoflurane reversibly increased AF.	Hanley et al. (2002)
	NADH				Rat ventricular myocytes	Protective effects of dinitrophenol were studied.	Rodrigo et al. (2002)
	NAD(P)H	375	397 long pass	Resolved lifetime components	Rat LV myocytes	Ouabain decreased component 2 (free NADH) amplitude and consequently free/bound component ratio.	Chorvatova et al. (2012b)
	NADH	360	450			Ouabain decreased NADH fluorescence.	Liu et al. (2010b)
	NAD(P)H	375	397 long pass	Resolved lifetime components	Rat LV myocytes	HNE and H_2O_2 reduced amplitude of both resolved components.	Chorvatova et al. (2013a)
Atherosclerosis	Collagen	337	360–510	5.3 at 390 Type I: 5.2 Type m: 2.95	Arterial tissue: atherosclerotic lesions	Breakdown of long-lived collagens.	Maarek et al. (2000b); Marcu et al. (2001b)
	Lipids	337	360–510		Arterial tissue: atherosclerotic lesions	Accumulation of short-lived components of lipids.	Maarek et al. (2000b); Marcu et al. (2001b)
	Unspecified tissue AF	337	370–510	2.4	Human normal aortic samples	Increased lifetime in advanced atherosclerotic lesions to 3.9.	Maarek et al. (2000a)
Vascular damage	AGE	300–450, peak 370	460–600		Patients	AGE fluorescence accumulation in skin.	Hartog et al. (2011); Meerwaldt et al. (2008); Mulder et al. (2006)
	Hb-AGE	308	375			Tested for CVD evaluation.	Vigneshwaran et al. (2005)

Similar to collagen, elastin emits a broad fluorescence with the peak around 410–450 nm, which complicates spectroscopic separation of these proteins (Table 12.1) (Liu et al. 2010a,b; Maarek et al. 2000a,b; Marcu et al. 2001a,b). However, the fluorescence lifetime of elastin is much shorter than that of collagen—measurements in arterial tissues showed the average lifetime of 2.3 ns (Maarek et al. 2000a; Marcu et al. 2001b).

As components of the ECM in the CVS, collagen and elastin are often modified in pathological conditions with corresponding changes in their fluorescence properties (Richards-Kortum and Sevick-Muraca 1996), which makes them convenient biomarkers for a number of disorders.

12.2.3 Contractile Apparatus: Myosin Filaments

An integral part of muscle sarcomeres in cardiac cells' contractile apparatus, myosin filaments can be imaged with SHG microscopy (Plotnikov et al. 2006; Wallace et al. 2008; see also Table 12.1). Plotnikov et al. (2006) found that the SHG signal arises from within the coiled rod region of myosin thick filaments and does not depend upon the functional state of myosin head domains or the actin filaments with which they interact.

12.2.4 Lipids: Lipofuscins, Cholesterol, Lipoproteins, and Advanced Glycation End Products

Lipofuscin is a mixture of fluorescent compounds that accumulate increasingly with age in humans (Heinsen 1981). Stored in lysosomal cell compartments as granules of 1–5 μm in diameter, lipofuscin consists of proteins (30%–58%), primarily malonaldehyde, and lipid-like material highly enriched in metals—possibly oxidation products of polyunsaturated fatty acids (19%–51%) (Jolly et al. 2002). These compounds are highly insoluble and reactive with a characteristic ultraviolet (UV) absorption around 360–420 nm and broad emission spectra of 540–650 nm (Moore 1981).

Accumulation of lipofuscin is a characteristic manifestation of aging and was detected in cardiac myocytes of old humans and rats (Monserrat et al. 1995). In AF-assisted lipofuscinogenesis studies conducted in cultured neonatal rat myocytes, accumulation of lipofuscin was shown to be initiated by a reduction in glutathione levels and enhanced by oxidative reactions (Gao et al. 1994; Marzabadi et al. 1995; Terman and Brunk 1998).

Cholesterol and lipoproteins, the main components of atherosclerotic plaques, are fluorescent at UV excitation. Marcu et al. characterized the spectra of the commercial samples of these compounds (emission of 360–510 nm at 337 nm excitation) and used these data to analyze the composition of atherosclerotic plaques (Maarek et al. 2000a; Marcu et al. 2001b).

Another lipidic source of endogenous fluorescence in tissues is advanced glycation end products (AGEs), the final products of complex chemical reactions between sugars, proteins, lipids, and nucleic acids (Singh et al. 2001), which accumulate during certain conditions, including aging and diabetes. Consequently, AGE fluorescence of the skin is now used as a marker of vascular damage. To study AGE fluorescence, the skin is typically excited between 300 and 450 nm (with peak excitation at 370 nm), and the fluorescence is detected in the 460–600 nm range (Hartog et al. 2011; Meerwaldt et al. 2007).

12.2.5 Tissue and Blood Oxygenation: Hemoglobin, Myoglobin, and Cytochrome c Oxidase

Tissue and blood oxygenation can be monitored spectroscopically by changes of hemoglobin (Hb) and myoglobin (Mb) absorption and emission properties.

A metalloprotein of red blood cells, Hb, transports oxygen from the respiratory organs to the tissues through the bloodstream. Although oxygenated HbO_2 and deoxygenated Hb differ in their absorption spectra, at several wavelengths, both species have the same molar absorption coefficient. An example of such isosbestic point is 585 nm, while at 577 nm illumination, HbO_2 has higher absorption. Subtraction of the fluorescence at 585 nm from that measured at 577 nm, therefore, provides a parameter that correlates with the blood oxygenation at the microcirculatory level (Mayevsky et al. 2002; Nighswander-Rempel et al. 2006) (see also Chapter 6).

In the heart tissue, Mb can enhance intracellular oxygen transport by adding an extra flux of oxygen to the mitochondria; in hypoxia, Mb-facilitated oxygen diffusion increases (Wittenberg and Wittenberg 1985, 1989, 2003). The absorbance of the Mb is very close to that of Hb. In order to evaluate cardiac tissue oxygenation in arrested and beating hearts, Nighswander-Rempel et al. introduced the ratio of oxygenated (Hb + Mb) to total (Hb + Mb) using the near-infrared (NIR) reflectance signals collected at the excitation of 650–1050 nm (Nighswander-Rempel et al. 2006).

Monitoring cytochrome c oxidation status is another approach with the potential for application in clinical CV medicine. Cytochrome c oxidase, also known as cytochrome aa3, is the final electron transporter enzyme in the mitochondrial respiratory chain. As the reduction of the cytochrome aa3 status results in decreased ATP production, monitoring of cytochrome aa3 can therefore be useful in clinical CV medicine (Heineman et al. 1992), although it is noteworthy that cytochrome aa3 is only partially (approximately 10%) reduced in resting cardiac myocytes (Wittenber and Wittenberg 1985). Such monitoring is possible by the diffuse reflectance spectroscopy method, which utilizes differences in the absorbance properties of the reduced and oxidized cytochrome aa3. This approach was used, for example, to assess intramyocardial oxygen transport during open-chest surgery in calves (Lindbergh et al. 2010, 2011). Takahashi et al. also proposed that the balance between the supply of reducing equivalents and that of O_2 at cytochrome c oxidase can be indicated by mitochondrial NAD(P)H (Takahashi et al. 2000).

12.3 AF-Assisted Examination Methods for the CVS

12.3.1 Advantages and Challenges of Optical Imaging

Over the last decades, a number of fluorescence spectroscopy and optical imaging techniques have been proposed for noninvasive observation of endogenous fluorophores (Andersson-Engels et al. 1997; Bachman et al. 2006; Chorvat and Chorvatova 2009). Noninvasive fluorescence imaging has the potential to provide in vivo diagnostic information for many clinical specialties and plays an important role in CV investigations (Ranji et al. 2006). Study of the AF in CVS tissues is challenging because of low intrinsic levels of their fluorescence intensity, overlap of excitation and emission spectra of many endogenous fluorophores, and strong scattering and heterogeneity of the biological tissues. For these reasons, a multimodal approach, combining absorption and emission spectroscopies, fluorescence imaging, and time-resolved spectroscopy, is often necessary. Such multimodal approach can prove to be instrumental for the monitoring of metabolic status in CVS cells and tissues.

12.3.2 Confocal Microscopy

Confocal microscopy with its capacity for optical sectioning and high-resolution imaging (Chapter 3) is one of the common methods for monitoring cellular AF components. Through high-resolution imaging of NADH, Takahashi et al. investigated Mb-assisted oxygen delivery to the mitochondria in isolated cardiomyocytes (Takahashi et al. 2000). Confocal imaging, especially in combination with spectral resolution, is suitable for investigating flavoproteins (Chorvat et al. 2004, 2005; Kuznetsov et al. 1998) (Figure 12.1a). Confocal microscopy of AF was also used for the reconstruction of the 3D shape and estimation of the left ventricular (LV) cardiac cell volume (Figure 12.1b) (Chorvatova et al. 2012a).

Although confocal microscopy has found utility in high-resolution AF imaging of cell cultures, limited penetration depth and high level of photodamage at UV excitation make this approach not suitable for in vivo studies. Such limitations are addressed by nonlinear microscopy techniques, discussed in the following.

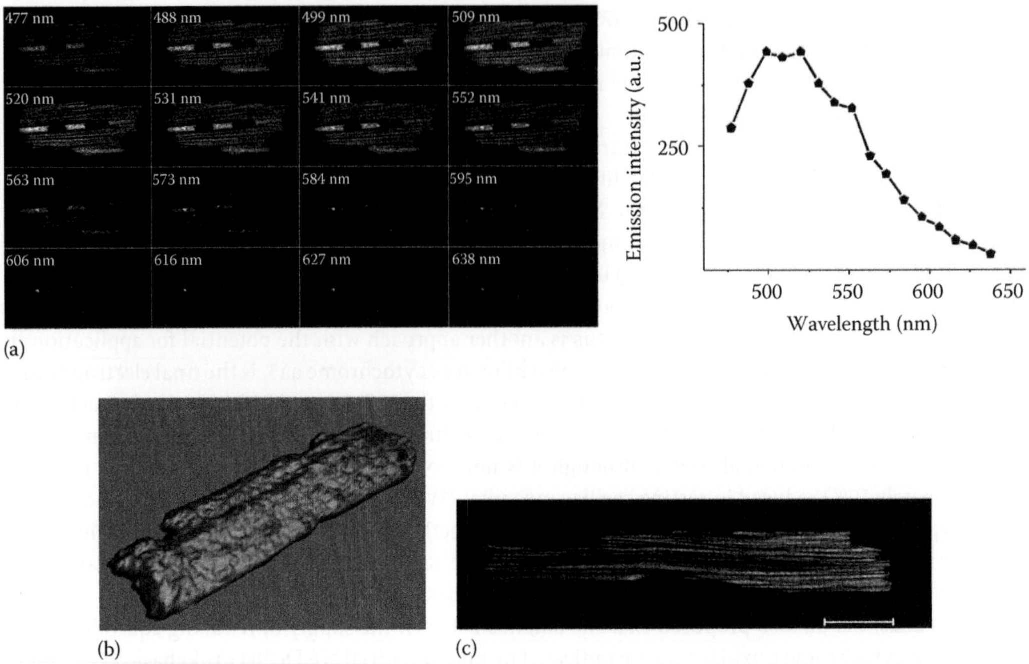

FIGURE 12.1 Confocal microscopy of cardiac cells. (a) Spectrally resolved confocal image of a rat cardiomyocyte flavin fluorescence (recorded with LSM META after excitation at 458 nm Ar/ion, emission at 477–638 nm with 11 nm step). (b) Volume rendering in an unlabeled cardiac myocyte (recorded after excitation at 488 nm Ar/ion, LP 505, 1.2 μm step in the z-stack). (c) 2P microscopy image of NAD(P)H fluorescence in a single cardiac myocyte (excitation 2P at 777 nm with tunable femtosecond oscillator Coherent Chameleon, HFT KP 700/488 dichroic filter, and 450–470 nm spectral range for emission detection).

12.3.3 Nonlinear Microscopy of CV Cells and Tissues

12.3.3.1 Two-Photon Excitation Microscopy

Two-photon excitation (TPE) microscopy offers unique advantages over conventional single-photon confocal microscopy and histological staining. Reduced scattering at longer excitation wavelengths increases the penetration depth of the excitation beam. The quadratic dependence of the excitation efficiency on the excitation intensity leads to the confinement of the excitation to the focal plane and eliminates the need for the confocal pinhole (explained in details in Section 7.3). This results in the increase of the collection efficiency and significant reduction of the photodamage and photobleaching (Williams et al. 1994; Zipfel et al. 2003a,b). Moreover, infrared lasers, used in TPE microscopy, can produce excitation of UV-absorbing endogenous fluorophores without damaging the tissues and thereby allowing for noninvasive time-lapse monitoring of the metabolic state of cells or tissues. These properties are particularly advantageous in monitoring mitochondrial function in the living heart and its cells (see an example of a TPE NAD(P)H fluorescence image in a cardiac myocyte at Figure 12.1c). Two-photon excitation, emission, and absorption cross section of NAD(P)H and flavins in isolated dog cardiomyocytes were thoroughly characterized in Huang et al. (2002).

The potential of TPE microscopy of endogenous fluorophores in CV medicine has been demonstrated by a number of studies, outlined in Table 12.1. For example, monitoring of NAD(P)H, flavin, and elastin fluorescence by TPE microscopy was used in the studies of physiology and pathology in mouse and rat hearts (Huang et al. 2009; Matsumoto-Ida et al. 2006; Ragan et al. 2007; Wallace et al. 2008; Wallenburg et al. 2011), in heart valve tissues (Schenke-Layland et al. 2009), and in aorta

(Lin et al. 2000; Wang et al. 2008; van Zandvoort et al. 2004). The combination of two-photon microscopy with another nonlinear technique, SHG microscopy, provides with a complex approach for the study of pathologies in the CVS.

12.3.3.2 Second Harmonic Generation Microscopy

SHG microscopy uses nonlinear photon scattering caused by the noncentrosymmetric loci to generate an image (Zipfel et al. 2003a,b). In CVS tissues, such loci can be found at myosin filaments in cardiomyocytes (Plotnikov et al. 2006; Wallace et al. 2008) and in the collagen of ECM (Schenke-Layland et al. 2007); both of these filamentous proteins are commonly imaged by SHG microscopy and serve as biomarkers for tissue pathology (Tables 12.1 and Figure 12.2). Thus, this approach was used to examine collagen fiber orientation and structure of connective tissues in muscle (Vanzi et al. 2012), as well as in arterial walls (Schriefl 2013) through monitoring the SHG signal from myosin filaments. Beaurepaire et al. (2005) and Boulesteix et al. (2004) were able to measure sarcomere contraction in unstained cardiomyocytes at 20 nm precision and used this metric to quantify the subcellular physiological response to toxins, drugs, and ionic solutions. SHG microscopy of the elastic fibers and ECM was also used to evaluate the damage of human cardiac tissues caused by cryopreservation (Schenke-Layland et al. 2007).

Multimodal imaging approach, combining two photon-excited fluorescence (TPEF) and SHG microscopies, provides complementary information that allows noninvasive complex characterization of tissues. Wallace et al. (2008) used simultaneous collection of TPEF and SHG signals to determine cardiomyocyte volume and myosin filament content. A multimodal approach also allows a composition analysis of the ECM: elastin and collagen content can be determined by TPEF and SHG microscopies, respectively (Schenke-Layland et al. 2006). In addition to that, a multimodal approach is instrumental in 3D studies of atherosclerotic lesions: fibrous caps, saturated with collagen, can be demarcated with the SHG imaging from the underlying necrotic core and healthy artery, which primarily have the TPEF signal from elastin (Le et al. 2007; Lilledahl et al. 2007).

12.3.4 Time-Resolved Fluorescence Spectroscopy and Microscopy

With all the advantages provided by high-resolution confocal and nonlinear optical imaging methods, separation of some endogenous fluorophores or different states of individual fluorophores (i.e., free vs. bound) is still a challenging task due to the subtle differences between their spectra. As described in Chapter 4, time-resolved fluorescence spectroscopy and/or imaging allows to distinguish between such species by providing a complementary contrast, based on the fluorescence lifetimes of intrinsic emitters.

Time-resolved spectroscopy measurements of NAD(P)H and flavins (Figure 12.3a and b) were shown to be suitable as optical biomarkers of metabolism in the CVS (Chorvat and Chorvatova 2009; Marcu

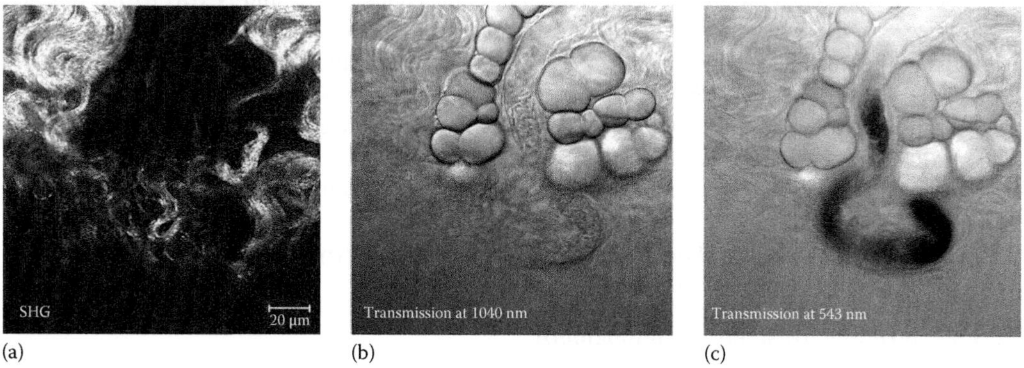

(a) (b) (c)

FIGURE 12.2 SHG imaging of unstained aorta. (a) Unstained aorta imaged by SHG (excitation, t-Pulse 20 Laser, 50 MHz, 1038 nm), (b) NIR transmission at 1038 nm, and (c) visible transmission at 543 nm.

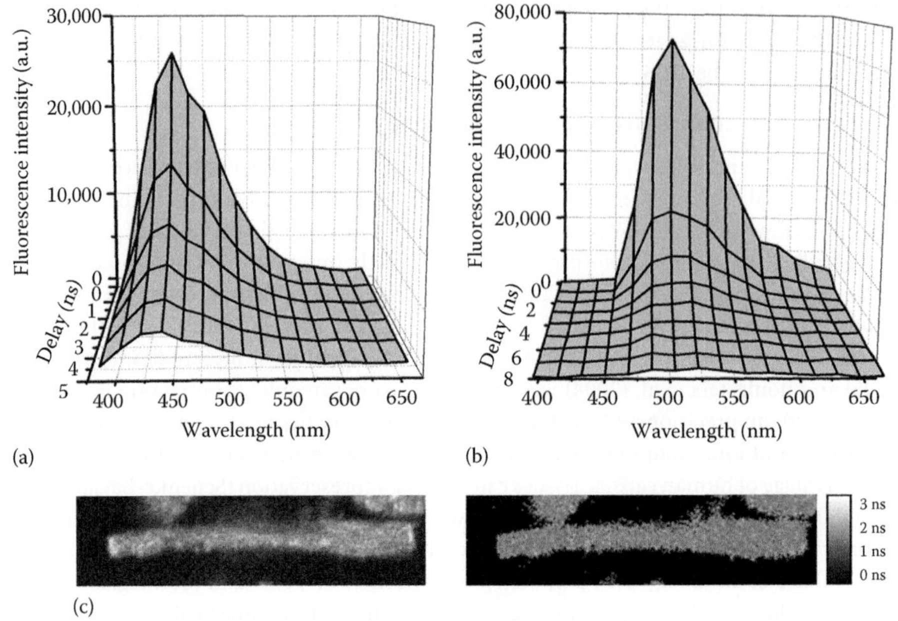

FIGURE 12.3 Time-resolved spectroscopy, and fluorescence lifetime imaging microscopy (FLIM) of cardiac cells. Time-resolved spectroscopy recording of (a) NAD(P)H and (b) flavin fluorescence from isolated rat cardiac myocytes (excitation BDL-375 or BDL-438 nm, emission separated by 397 or 470 nm long-pass filter, detection by 16-channel photomultiplier array PML-16). (c) FLIM image of a cardiac cell (excitation, t-Pulse 20 Laser, 50 MHz, 1038 nm, 2P). Intensity image (left; emission at 530–540 nm) and time-resolved image (right; scale from 0 to 3.0 ns on the right) are illustrated.

2010). Typically, NAD(P)H in cardiac tissue cells exhibits three lifetime pools—a short lifetime component has been shown to originate from free NAD(P)H, and two longer lifetime components are attributed to protein-bound NAD(P)H (Blinova et al. 2005); similar fluorescence lifetime pools were also found in isolated cardiac cells (Cheng et al. 2007; Chorvat et al. 2010; Chorvatova et al. 2013b).

Recently developed spectrally resolved time-correlated single-photon counting (TCSPC) of various endogenously fluorescing molecules is a multidimensional approach to AF imaging with discrimination between endogenous fluorophores and their thorough characterization as of the structure, conformation, microenvironment, and free/bound state (Table 12.1 and Figure 12.3a through c) (Bachmann et al. 2006; Becker 2005; Chorvat and Chorvatova 2006). Endogenous flavins and NAD(P)H fluorescence lifetimes were studied with such approach in rat cardiomyocytes to evaluate the influence of pharmaceuticals for the treatment of heart failure and to monitor cellular metabolism in contracting cardiac myocytes (Chorvat et al. 2008; Chorvatova et al. 2011). In atherosclerotic tissues, spectrotemporal spectroscopy was also proved highly suitable for the detection of atherosclerotic plaques (Marcu et al. 1999, 2001a,b, 2005, 2009). Consequently, this approach facilitates structural discrimination of anatomical features of cellular morphology or changes of metabolic state in various pathological conditions.

12.3.5 Nonmicroscopy Techniques for the Monitoring of Oxidation Status in CVS

12.3.5.1 NIR Spectroscopy of Hemoglobin

NIR spectroscopy (700–2500 nm) is an analytical technique with relatively large tissue penetration depth, which enables real-time noninvasive detection of regional Hb oxygenation using transmission or reflection measurements (reviewed in Ferrari and Quaresima 2012). NIR spectroscopy was used to monitor various heart tissue samples. The initial NIR spectroscopy studies showed the potential of this method to distinguish between myocardium and aorta (Nilsson et al. 1997) and allowed to monitor muscle reserve capacity following tissue oxygen extraction in different pathophysiological conditions (Nighswander-Rempel et al. 2005, 2006).

12.3.5.2 Diffuse Reflectance Spectroscopy of Cytochrome c Oxidase

Optical steady-state diffuse reflectance spectroscopy of chromophores involved in myocardial oxygen transport has been used in a number of in vitro studies to identify myocardial tissue status. In particular, diffuse reflectance spectroscopy of cytochrome aa3 based on the different absorbances of reduced and oxidized cytochrome aa3 species has been instrumental in monitoring intramyocardial oxygen consumption in CVS (Leisey et al. 1994; Lindbergh et al. 2010).

12.3.6 Data Analysis

12.3.6.1 Separation of Individual AF Components

With several algorithms developed to fit multidimensional data, choosing the physically relevant model based on the general knowledge about the investigated system is extremely important to ensure the authenticity of the recovered parameters (Lakowicz 2000; Straume et al. 1991). The analysis of complex multiexponential decays of AF emitters in cells and tissues and separation of individual components still represent a tough scientific challenge that has not yet been fully resolved.

Multiple solutions have been proposed for the analysis of complex, multidimensional, spectrally and/or time-resolved endogenous fluorescence data (reviewed in Chorvat and Chorvatova 2009). Spectral unmixing algorithms for multispectral imaging have made it possible to simultaneously measure and identify multiple fluorescing molecules in biological samples (Dickinson et al. 2001; Zimmermann et al. 2003). Additional resolving power can be obtained by complementing spectral techniques with time-resolved spectroscopy/imaging (Bird et al. 2004; Chorvat et al. 2005; Chorvatova et al. 2013b; Zimmermann et al. 2003).

In the CVS, based on spectral decomposition of spectrally resolved TCSPC AF signals (Chorvat et al. 2007), individual spectral components of NAD(P)H fluorescence recorded by multiwavelength fluorescence lifetime spectroscopy were resolved in isolated cardiomyocytes (Chorvatova et al. 2012b, 2013b; Klaidman et al. 1995). In atherosclerotic tissues, spectrotemporal data recorded in vivo from the aorta were analyzed using the Laguerre deconvolution (Jo et al. 2004, 2005).

12.3.6.2 Assessment of Metabolic Oxidative State

Since the pioneering work of Chance and Thorell (1959), which demonstrated that fluorescence of endogenous fluorophores varies with cellular redox state, a few methods to quantify redox changes have been proposed.

For fluorescence intensity measurements, the NADH/FAD ratio has been commonly used as a quantitative indicator of tissue metabolism, based on the facts that only reduced NAD and oxidized FAD are fluorescent and that differences in partial oxygen pressure can be determined by measuring alterations in the AF of the redox pairs of coenzymes (NAD^+/NADH and FAD^+/$FADH_2$) (Barlow and Chance 1976; Chance et al. 1962). Takahashi et al. evaluated NAD(P)H oxidation by dividing hypoxic NADH image by corresponding anaerobic NADH fluorescence image (Takahashi et al. 2000). An alternative approach is the calculation of the NADH/(NADH + FAD) and Fp/(Fp + NADH) redox ratios, which were proposed as a sensitive index of the steady-state mitochondrial metabolism (Xu et al. 2009).

The ratio of the relative amplitudes of free NADH (short fluorescence lifetime component) and bound NADH (long fluorescence lifetime component), calculated from the multiexponential fluorescence decays, is another popular discrimination method of metabolic state changes (Jameson et al. 1989; Lakowicz et al. 1992). This ratio was proposed to be related to the NADH/NAD^+ ratio (Bird et al. 2005). Supporting evidence to that hypothesis comes from the studies showing that

NADH/NAD^+ redox potential is the driving force of oxidative phosphorylation and its increase leads to a linear rise of maximal respiration rate in isolated heart mitochondria (Mootha et al. 1997; Moreno-Sánchez 1985).

The $NADP^+$/NADPH ratio is normally about 0.005, around 200 times lower than the NAD^+/NADH ratio that ranges from 1 to 10 between the mitochondria and cytosol, respectively, indicating that the reduced form of this coenzyme is dominant (Lin and Guarente 2003; Veech et al. 1969). The $NADP^+$/NADPH ratio is kept very low because NADPH is needed to drive redox reactions as a strong reducing agent. These different ratios are keys to the different metabolic roles of NADH and NADPH.

12.4 AF in the CVS

12.4.1 Cardiac Tissue

12.4.1.1 Cardiac Mitochondria

Cardiac mitochondria are crucial for the heart oxidative metabolic status, as they contain respiratory chain responsible for the ATP production in cardiac cells. One of the earliest characterizations of cardiac cells, AF components was performed by Chance et al. (1979). They measured the 366 nm excited emission spectra from oxidized and reduced isolated pigeon heart mitochondria and demonstrated peaks around 460 and 510 nm, attributed to NADH and flavins, respectively. The AF excitation/emission matrix of a population of isolated rat cardiomyocyte mitochondria, conducted later by Chorvat et al., revealed comparable peaks to those recorded in cardiac myocytes (Chorvat et al. 2004). Hanley et al. (2002) measured the effect of volatile anesthetics on NADH fluorescence in pig heart submitochondrial particles and demonstrated reversible increase in the fluorescence intensity with drugs such as halothane, isoflurane, and/or sevoflurane.

Blinova et al. complemented the intensity-based studies by conducting time-resolved characterization of NADH in intact porcine heart mitochondria (Blinova et al. 2005). Three fluorescence lifetime pools were discovered in that study: the shorter lifetime component of 0.4 ns was attributed to free NADH (63% of the total NADH content), while the longer lifetime pools, 1.8 ns (30%) and 5.7 ns (7%), were consistent with the bound NADH.

12.4.1.2 Cardiac Myocytes

Cardiac myocytes underlie the heart contraction. A thorough characterization of NAD(P)H and flavin fluorescence at different metabolic states in living cardiac myocytes was conducted by TPE microscopy (Huang et al. 2002) and time-resolved fluorescence spectroscopy methods (Chorvat and Chorvatova 2006; Chorvat et al. 2005, 2010; Chorvatova et al. 2012b, 2013b). Takahashi et al. mapped mitochondrial oxygen supply by NAD(P)H fluorescence microscopy and demonstrated its radial Mb-assisted distribution in isolated myocytes (Takahashi et al. 2000). Metabolic oxidative state was reproducibly determined directly in living left ventricular cardiac cells using time-resolved fluorescence spectroscopy by monitoring their naturally occurring endogenous NAD(P)H (Chorvat et al. 2010; Chorvatova et al. 2012b, 2013a,b) or flavin fluorescence (Chorvat and Chorvatova 2006).

AF microscopy and spectroscopy methods were also used to study the energy metabolism of cardiomyocytes at contraction. NADH levels have been shown not to change in contracting single rat myocytes (Griffiths et al. 1997, 1998; White and Wittenberg 1993). Similar results were obtained with time-resolved spectroscopy of flavins (Chorvat et al. 2008) or NAD(P)H (Chorvatova et al. 2011) fluorescence—no differences were observed at the peak of contraction and at rest. Cultured neonatal rat myocardial cells (Nakagami et al. 2008) were employed to evaluate lipofuscin accumulation (Gao et al. 1994; Marzabadi et al. 1995; Terman and Brunk 1998). These studies demonstrated an increase in the accumulation of lipofuscins in cardiac myocytes during oxidative stress (Gao et al. 1994). They also suggested that neonatal myocytes in culture are suitable model for the evaluation of oxygen radical–induced myocardial damage (Marzabadi et al. 1995).

12.4.1.3 Whole Heart

AF microscopy and spectroscopy techniques are commonly used for noninvasive studies of the physiology and pathology of the heart. The feasibility of this approach can be illustrated, for example, with the studies of Ashruf et al. who demonstrated that an increase in the Langendorff-perfused heart is accompanied by a decrease in NADH fluorescence (Ashruf et al. 1995). Wallenburg et al. used TPEF and SHG microscopies for the label-free noninvasive study of healthy, infarcted, and stem-cell-treated regenerating rat hearts (Wallenburg et al. 2011). This study demonstrated that while healthy tissues are mostly composed of densely packed myocytes and very little collagen, infarcted regions are characterized by sparse myocytes and high collagen content.

In CV structures, such as heart valves, the interstitial cells are surrounded by primary ECM components, which perform many essential functions, including mechanical support and physical strength (Schenke-Layland et al. 2006). A thorough evaluation of elastic and collagenous fiber content in the ECM is crucial for the complete understanding of healthy and diseased valve tissues. For that purpose, TPEF microscopy imaging is conducted at two excitation wavelengths: 760 nm for elastin and 840 nm for collagen (Schenke-Layland et al. 2004, 2006). Collagen and elastin were also visualized in leaflets and conduits of cryopreserved cardiac valve allografts to evaluate the ECM integrity (Gerson et al. 2012). Schenke-Layland used confocal microscopy to assess the ECM content in CV progenitor cell line (Schenke-Layland et al. 2011).

Heart conduction system, or the His bundle, is a specific muscular tissue where a heartbeat originates and initiates the depolarization of the ventricle. AF detected in this region showed the most distinct differences from the surrounding tissues at 330 and 380 nm excitation indicative of different collagen and elastin content in these regions (Venius et al. 2011). For the purpose of this investigation, the authors detected the fluorescence only at 460 nm for both excitation wavelengths and used the ratio between the 330 nm excited fluorescence and 380 nm excited fluorescence as the measure for ECM characterization. Comparably, spectral differences between the His bundle and the connective tissues were displayed by comparing the ratios of fluorescence intensities at above 460 nm under the preferred excitation of elastin and collagen (366 nm) (Bagdonas et al. 2008).

12.4.2 Vascular Tissue

12.4.2.1 Aorta

Noninvasive microscopy and spectroscopy of the aortic wall have proven to be a useful technique to investigate atherosclerosis. AF features of atherosclerotic plaques could be employed intravascularly through TPE microscopy to characterize plaque composition without the use of contrast agents. A recent study of Phipps et al. demonstrated that clinically relevant features, including ratios of lipid vs. collagen vs. elastin, can be evaluated with a fluorescence lifetime imaging of endogenous fluorophores (Phipps et al. 2011). Consequently, time-resolved microscopy images of human atherosclerotic plaques can serve to derive parameters that discriminate between luminal areas that are elastin-rich, elastin and macrophage-rich, collagen-rich, and/or lipid-rich only using intrinsic fluorescence decay dynamics. Maarek et al. thoroughly characterized spectral and temporal features for graded levels of atherosclerosis in human aorta wall using time-resolved fluorescence emission spectra; the results were interpreted in terms of morphological and compositional changes that accompany each stage of atherosclerotic development (Maarek et al. 2000a,b). Simultaneous time- and wavelength-resolved fluorescence spectroscopy was also tested by Sun et al. (2011) for the dynamic characterization of atherosclerotic tissue ex vivo and arterial vessels in vivo. New analytical deconvolution method for fluorescence lifetime imaging microscopy was used to evaluate changes during atherosclerosis, as demonstrated using animal model (Jo et al. 2004). Allen et al. applied an alternative approach for the study of lipid-rich plaque—spectroscopic photoacoustic imaging in the 700–1400 nm range (Allen et al. 2012).

12.4.2.2 Carotid Artery

Correlations between rheological and optical properties of carotid artery rings before and after cryopreservation were studied by fluorescence spectroscopy of AF components (Choserot et al. 2005; Pery et al. 2009). Freshly excised carotid plaques from patients were studied by time-resolved fluorescence spectroscopy in the aim to detect vulnerable (e.g., rupture-prone) plaques (Jo et al. 2005, 2006; Marcu et al. 2009). In these studies, the artery samples were excited at 337 nm and the fluorescence was detected in the range of 360–550 nm. This approach demonstrated a potential for rapid clinical investigation of atherosclerotic plaques and detection of those at risk.

12.4.2.3 Coronary Artery

In the coronary artery, time- and spectrally resolved spectroscopy was used to evaluate differences between normal and atherosclerotic regions (Marcu et al. 1999, 2005). This technique served to differentiate lipid rich from fibrous lesions and distinguish between plaques and normal arterial wall (Marcu 2010).

12.5 AF-Assisted Examination of CVS Physiology and Pathology

12.5.1 AF as a Noninvasive Marker of Cardiovascular Disease

Numerous diseases of the heart, including hypertension and diabetes, are often linked to alterations in mitochondrial energy metabolism associated with the mitochondrial dysfunction (Dhalla et al. 1993). Chronic alterations of fuel metabolism and oxidative stress status are factors that could impair the capacity of the mitochondria to fulfill their crucial role in energy production (Lesnefsky et al. 2001) and thereby contribute to the activation of pathways governing cell death and/or disease (Borutaite and Brown 2003; Cortassa et al. 2003). This creates room for numerous applications of noninvasive AF measurements in the identification and study of CV physiology and pathology.

12.5.1.1 Detection of Ischemia and Myocardial Infarction

The heart is a pump converting chemical energy into mechanical work by oxidizing carbon fuels supplied by blood flow. Such oxidative metabolism is primarily the function of the mitochondria through the process of oxidative phosphorylation. Because of the high oxidative metabolism, heart cells have a high oxidative capacity, demonstrated by their ultrastructure: 25%–35% of total cardiomyocyte volume is occupied by the mitochondria (Figure 12.1c). When deprived of oxygen, cardiac cells can maintain ATP levels by glycolytic ATP production and can then revert smoothly to original conditions upon reperfusion (Das and Harris 1990). However, if blood flow is restricted, as in myocardial infarct, oxygen deprivation is accompanied with a decrease in cytoplasmic pH to 5.5–6 and accumulation of lactate in cells (Dennis et al. 1991; Smith et al. 1996; Vuorinen et al. 1995). This condition, known as ischemia, can damage cardiac cells irreversibly. Paradoxically, the major damage to ischemic cells comes from the reintroduction of oxygen. During such reperfusion, the cells typically undergo further contraction (hypercontracture) and membrane damage, followed by cell death (Piper et al. 2004; Schluter et al. 1996). It is widely acknowledged that ischemia and reperfusion lead to mitochondrial and cellular damage in cardiac myocytes (Lesnefsky et al. 2001; Piper et al. 1993, 1994).

During hypoxia or ischemia, the supply of O_2 to the respiratory chain fails, leading to the blocking of the Krebs cycle, accumulation of cytoplasmic NADH at a rapid (minute-based) scale (Hulsmann et al. 1993), and thus increase in the NAD(P)H fluorescence and decrease in the percent of flavoprotein oxidation (Pasdoris et al. 2008). Measurement of NADH and flavin fluorescence was therefore proposed as a tool to study cardiac ischemia (Chance 1976; Ranji et al. 2007; Schaffer et al. 1978) and develop protection against damage by metabolic inhibition and reperfusion (Pasdois et al. 2008; Rodrigo et al. 2002). Ranji et al. used this approach to study an open-chest rabbit model of myocardial ischemia–reperfusion injury (Ranji et al. 2006). Fluorescence of both coenzymes was also used to monitor ROS production in cardiac mitochondria during hypoxia and reoxygenation (Korge et al. 2008).

12.5.1.2 Cardiac Transplantation

Cardiac rejection upon transplantation involves several processes, including activation and proliferation of T-lymphocyte subsets (leading to lymphocyte infiltration and immune destruction of the graft tissue), as well as graft vessel injury and thrombosis (Hoffman 2005). Although molecular mechanisms of the rejection remain unclear, alloantigen-dependent and alloantigen-independent factors are known as contributors (Vassalli et al. 2003). Ischemia–reperfusion injury is the most influential alloantigen-independent factor (Gaudin et al. 1994). Some observations suggest that cardiac cells undergo rapid modifications in their oxidative state and oxidative metabolism with the progression of cardiac rejection as a result of ischemic hypoxia (Tanaka et al. 2004a,b). Evaluation of the oxidative metabolism can therefore serve as an early indication of the rejection of transplanted hearts. AF spectroscopy was attempted for the examination of transplanted tissues. A strong correlation between the changes in AF spectra and the rejection grade was found in rat heart allograft model (Morgan et al. 1999). In case of human tissues, however, the results of fluorescence spectroscopy studies were not straightforward possibly due to the use of frozen fractions (Yamani et al. 2000). Better results were obtained with time-resolved AF spectroscopy of freshly isolated cardiac cells from endomyocardial biopsy (Cheng et al. 2009; Chorvat et al. 2010). In this study, a reduction of the average fluorescence lifetime and increased NAD(P)H fluorescence intensity was observed from the samples of the hearts with mild rejection. These results show the potential of AF spectroscopy for the early identification of transplant rejection.

Application of pluripotent stem cells for the regeneration of the cardiac tissue lost after myocardial infarction is currently investigated as an approach alternative to heart transplantation (He et al. 2009; Nussbaum et al. 2007). TPE and SHG microscopies of myocytes and collagen were used to evaluate modifications in stem-cell-treated regenerating heart in rat (Wallenburg et al. 2011). A similar approach was used to evaluate ECM composition in tissue-engineered heart valve leaflets (Schenke-Layland et al. 2004).

12.5.1.3 Remodeling of the Left Ventricular Myocytes

Remodeling of the left ventricular myocytes in pregnancy includes adaptations in metabolism, as well as hemodynamic and cardiac parameters, such as cardiomyocyte dimensions (Bassien-Capsa et al. 2006, 2011). To study the metabolic oxidative state modifications of the cardiomyocytes in pregnancy, Bassien-Capsa et al. (2011) used spectrally resolved lifetime detection of NAD(P)H fluorescence to demonstrate modifications in metabolic oxidative state, namely, in the presence of lactate, in pregnancy. The same approach was applied to monitor the metabolic state of cardiac cells following the application of ouabain, the Na^+ pump inhibitor, which rises in pregnancy—this compound, also used as a pharmaceutical drug, was found to decrease the NAD(P)H fluorescence in rat cardiac cells (Chorvatova et al. 2012b; Elzwiei et al. 2013; Liu et al. 2010a,b).

12.5.1.4 Pharmacological Studies

AF offers great potential in the investigation of cellular responses to treatment by pharmaceutical drugs. In addition to the studies of the ouabain effect on metabolism, described earlier, noninvasive AF evaluation was employed by Hanley et al., to measure the reaction of NADH to volatile anesthetics such as halothane, isoflurane, and sevoflurane (Hanley et al. 2002), and Rodrigo et al., who studied protective effects of 9,10-dinitrophenol on the cellular damage induced by metabolic inhibition and reperfusion in freshly isolated rat ventricular myocytes (Rodrigo et al. 2002).

12.5.2 AF in the Examination of Vascular Damage

12.5.2.1 Diagnostics of Atherosclerotic Plaques and Lesions

Early detection and treatment of rupture-prone vulnerable atherosclerotic plaques are critical to reducing patient mortality associated with CVD. Depending on the anatomical location of the plaque, the clinical consequence can be myocardial infarction, stroke, or limb ischemia. Composition of the plaque is one of the important predictive factors for the plaque stability. Laser-induced fluorescence spectroscopy was shown to play an important role in the detection of human atherosclerosis (Kittrell et al. 1985; van de Poll et al. 1999) and composition of atherosclerotic plaques (Marcu et al. 2001a,b). Measurement of endogenous fluorescence has displayed the potential for diagnosing atherosclerotic plaques and thus for clinical applications of fluorescence spectroscopy to the detection of atherosclerotic lesions (Maarek et al. 2000a,b; Marcu et al. 2001a,b; Richards Kortum and Sevick-Muraca 1996). The combination of reflectance, fluorescence, and Raman spectroscopy provides detailed biochemical information about the tissue and can detect vulnerable plaque features (Šćepanović et al. 2011). Nonlinear optical microscopy techniques were found to be well suited for the study of arterial cells, atherosclerosis, and restenosis (Le et al. 2007; van Zandvoort et al. 2004; Wang et al. 2008). Time-resolved microscopy was used for the characterization and diagnostics of atherosclerotic plaques (Marcu 2010) and their composition (Marcu et al. 2001a,b); Maarek et al. tested time-resolved microscopy for the diagnosis of atherosclerotic lesions, and they were able to discriminate such advanced injuries based on increased tissue fluorescence lifetime in this condition (Maarek et al. 2000a,b). The capability of imaging significant components of an arterial wall and distinctive stages of atherosclerosis in a label-free manner suggests the potential application of multimodal nonlinear optical microscopy to monitor the onset and progression of arterial diseases.

12.5.2.2 Skin AF of AGE as a Marker of Vascular Damage

Skin AF associated with AGE is used as a noninvasive marker of vascular damage in patients with type 2 (Lutgers et al. 2006; Meerwaldt et al. 2007) or type 1 diabetes (Araszkiewicz et al. 2011; Samborski et al. 2011). Similar fluorescence pattern was observed also for patients with CVD, including acute myocardial infarction (Mulder et al. 2006) or chronic heart failure (Hartog et al. 2011). AGEs also accumulate in formerly preeclamptic women (Coffeng et al. 2011), and its measurement also has clinical relevance for vascular surgery (Meerwaldt et al. 2008). In these experiments, the AF reader illuminates skin surface with an excitation light source between 300 and 420 nm and emission between 430 and 450 nm (for details, see reviews Hartog et al. 2011; Meerwaldt et al. 2008). In addition, AF of Hb-AGE, which possesses a characteristic AF at 308/345, is also tested for the evaluation of CVD (Vigneshwaran et al. 2005).

12.6 Conclusions

In regard to developments in the last decade, AF-assisted examination is more and more applied in the study of the CVS physiology and pathology. Thanks to the possibility to examine unstained samples, such studies proved to have great potential for clinical use. An overview of the main endogenous fluorophores in the CVS given in this chapter provides a new insight into various applications of AF-assisted examination in the monitoring and diagnosis of physiological and pathological modifications of CV cells and tissues.

Acknowledgments

Support from Integrated Initiative of European Laser Infrastructures LASERLAB-EUROPE III (grant agreement no. 284464, EC's Seventh Framework Programme) and the research grant agency of the Ministry of Education, Science, Research and Sport of the Slovak Republic VEGA no. 1/0296/11 and APVV-0242-11 is acknowledged. This publication was also supported by the Centre of Excellence for design, preparation, and diagnostics of nanostructures in electronics and photonics, NanoNet 2, ITMS

26240120018, funded by the Research and Development Operational Programme from the ERDF. The author would like to thank Dusan Chorvat Jr., Anton Mateasik, Jana Kirchnerova, and Martin Uherek from the International Laser Centre, Bratislava, for help with the figures.

References

Allen, T. J., A. Hall, A. P. Dhillon, J. S. Owen, and P. C. Beard. 2012. Spectroscopic photoacoustic imaging of lipid-rich plaques in the human aorta in the 740 to 1400 nm wavelength range. *Journal of Biomedical Optics* 17 (6): 061209.

Andersson-Engels, S., C. Klinteberg, K. Svanberg, and S. Svanberg. 1997. In vivo fluorescence imaging for tissue diagnostics. *Physics in Medicine and Biology* 42 (5): 815–824.

Araszkiewicz, A., D. Naskret, P. Niedzwiecki, P. Samborski, B. Wierusz-Wysocka, and D. Zozulinska-Ziolkiewicz. 2011. Increased accumulation of skin advanced glycation end products is associated with microvascular complications in type 1 diabetes. *Diabetes Technology and Therapeutics* 13 (8): 837–842.

Ashruf, J. F., J. M. Coremans, H. A. Bruining, and C. Ince. 1995. Increase of cardiac work is associated with decrease of mitochondrial NADH. *American Journal of Physiology—Heart and Circulatory Physiology* 269 (3): H856–H862.

Bachmann, L., D. M. Zezell, A. da Costa Ribeiro, L. Gomes, and A. Siuiti Ito. 2006. Fluorescence spectroscopy of biological tissues—A review. *Applied Spectroscopy Reviews* 41 (6): 575–590.

Bagdonas, S., E. Zurauskas, G. Streckyte, and R. Rotomskis. 2008. Spectroscopic studies of the human heart conduction system ex vivo: Implication for optical visualization. *Journal of Photochemistry and Photobiology B: Biology* 92 (2): 128–134.

Barlow, C. H. and B. Chance. 1976. Ischemic areas in perfused rat hearts: Measurement by NADH fluorescence photography. *Science* 193 (4256): 909–910.

Bassien-Capsa, V., F. M. Elzwiei, S. Aneba, J.-C. Fouron, B. Comte, and A. Chorvatova. 2011. Metabolic remodelling of cardiac myocytes during pregnancy: The role of mineralocorticoids. *Canadian Journal of Cardiology* 27 (6): 834–842.

Bassien-Capsa, V., J.-C. Fouron, B. Comte, and A. Chorvatova. 2006. Structural, functional and metabolic remodeling of rat left ventricular myocytes in normal and in sodium-supplemented pregnancy. *Cardiovascular Research* 69 (2): 423–431.

Beaurepaire, E., T. Boulesteix, A.-M. Pena et al. 2005. Multiphoton microscopy using intrinsic signals for pharmacological studies in unstained cardiac and vascular tissue. *Proceedings of SPIE* 5699: 67–74.

Becker, W. 2005. *Advanced Time-Correlated Single Photon Counting Techniques*. New York: Springer.

Benderdour, M., G. Charron, B. Comte et al. 2004. Decreased cardiac mitochondrial $NADP^+$-isocitrate dehydrogenase activity and expression: A marker of oxidative stress in hypertrophy development. *American Journal of Physiology—Heart and Circulatory Physiology* 287 (5): H2122–H2131.

Berezin, M. Y. and S. Achilefu. 2010. Fluorescence lifetime measurements and biological imaging. *Chemical Reviews* 110 (5): 2641–2684.

Bird, D. K., K. W. Eliceiri, C.-H. Fan, and J. G. White. 2004. Simultaneous two-photon spectral and lifetime fluorescence microscopy. *Applied Optics* 43 (27): 5173–5182.

Bird, D. K., L. Yan, K. M. Vrotsos et al. 2005. Metabolic mapping of MCF10A human breast cells via multiphoton fluorescence lifetime imaging of the coenzyme NADH. *Cancer Research* 65 (19): 8766–8773.

Blinova, K., S. Carroll, S. Bose et al. 2005. Distribution of mitochondrial NADH fluorescence lifetimes: Steady-state kinetics of matrix NADH interactions. *Biochemistry* 44 (7): 2585–2594.

Borutaite, V. and G. C. Brown. 2003. Mitochondria in apoptosis of ischemic heart. *FEBS Letters* 541 (1): 1–5.

Boulesteix, T., E. Beaurepaire, M.-P. Sauviat, and M.-C. Schanne-Klein. 2004. Second-harmonic microscopy of unstained living cardiac myocytes: Measurements of sarcomere length with 20-nm accuracy. *Optics Letters* 29 (17): 2031–2033.

Chance, B. 1976. Pyridine nucleotide as an indicator of the oxygen requirements for energy-linked functions of mitochondria. *Circulation Research* 38 (5 Suppl. 1): I31–I38.

Chance, B., P. Cohen, F. Jobsis, and B. Schoener. 1962. Intracellular oxidation-reduction states in vivo the microfluorometry of pyridine nucleotide gives a continuous measurement of the oxidation state. *Science* 137 (3529): 499–508.

Chance, B., B. Schoener, R. Oshino, F. Itshak, and Y. Nakase. 1979. Oxidation-reduction ratio studies of mitochondria in freeze-trapped samples. NADH and flavoprotein fluorescence signals. *Journal of Biological Chemistry* 254 (11): 4764–4771.

Chance, B. and B. Thorell. 1959. Fluorescence measurements of mitochondrial pyridine nucleotide in aerobiosis and anaerobiosis. *Nature* 184: 931–934.

Cheng, Y., D. Chorvat Jr., N. Poirier, J. Miro, N. Dahdah, and A. Chorvatova. 2007. Spectrally and time-resolved study of NAD(P)H autofluorescence in cardiac myocytes from human biopsies. *Proceedings of SPIE* 6771: 677104.

Cheng, Y., A. Mateasik, N. Poirier et al. 2009. Analysis of NAD(P)H fluorescence components in cardiac myocytes from human biopsies: A new tool to improve diagnostics of rejection of transplanted patients. *Proceedings of SPIE* 7183: 71830K.

Chorvat Jr., D., S. Abdulla, F. Elzwiei, A. Mateasik, and A. Chorvatova. 2008. Screening of cardiomyocyte fluorescence during cell contraction by multi-dimensional TCSPC. *Proceedings of SPIE* 6860: 686029.

Chorvat Jr., D., V. Bassien-Capsa, M. Cagalinec et al. 2004. Mitochondrial autofluorescence induced by visible light in single rat cardiac myocytes studied by spectrally resolved confocal microscopy. *Laser Physics* 14 (2): 220–230.

Chorvat Jr., D. and A. Chorvatova. 2006. Spectrally resolved time-correlated single photon counting: A novel approach for characterization of endogenous fluorescence in isolated cardiac myocytes. *European Biophysics Journal* 36 (1): 73–83.

Chorvat Jr., D. and A. Chorvatova. 2009. Multi-wavelength fluorescence lifetime spectroscopy: A new approach to the study of endogenous fluorescence in living cells and tissues. *Laser Physics Letters* 6 (3): 175–193.

Chorvat Jr., D., J. Kirchnerova, M. Cagalinec, J. Smolka, A. Mateasik, and A. Chorvatova. 2005. Spectral unmixing of flavin autofluorescence components in cardiac myocytes. *Biophysical Journal* 89 (6): L55–L57.

Chorvat Jr., D., A. Mateasik, Y. Cheng et al. 2010. Rejection of transplanted hearts in patients evaluated by the component analysis of multi-wavelength NAD(P)H fluorescence lifetime spectroscopy. *Journal of Biophotonics* 3 (10–11): 646–652.

Chorvat Jr., D., A. Mateasik, J. Kirchnerova, and A. Chorvatova. 2007. Application of spectral unmixing in multi-wavelength time-resolved spectroscopy. *Proceedings of SPIE* 6771: 677105.

Chorvatova, A., S. Aneba, A. Mateasik, D. Chorvat Jr., and B. Comte. 2013a. Time-resolved fluorescence spectroscopy investigation of the effect of 4-hydroxynonenal on endogenous NAD(P)H in living cardiac myocytes. *Journal of Biomedical Optics* 18 (6): 067009.

Chorvatova, A., M. Cagalinec, A. Mateasik, and D. Chorvat Jr. 2012a. Estimation of single cell volume from 3D confocal images using automatic data processing. *Proceedings of SPIE* 8427: 842704.

Chorvatova, A., F. Elzwiei, A. Mateasik, and D. Chorvat Jr. 2012b. Effect of ouabain on metabolic oxidative state in living cardiomyocytes evaluated by time-resolved spectroscopy of endogenous NAD(P)H fluorescence. *Journal of Biomedical Optics* 17 (10): 1015051.

Chorvatova, A., A. Mateasik, and D. Chorvat Jr. 2011. Laser-induced photobleaching of NAD(P)H fluorescence components in cardiac cells resolved by linear unmixing of TCSPC signals. *Proceedings of SPIE* 7903: 790326.

Chorvatova, A., A. Mateasik, and D. Chorvat Jr. 2013b. Spectral decomposition of NAD(P)H fluorescence components recorded by multi-wavelength fluorescence lifetime spectroscopy in living cardiac cells. *Laser Physics Letters* 10 (7): 125703.

Choserot, C., E. Péry, J.-C. Goebel, D. Dumas, J. Didelon, J.-F. Stoltz, and W. C. P. M. Blondel. 2005. Experimental comparison between autofluorescence spectra of constrained fresh and cryopreserved arteries. *Clinical Hemorheology and Microcirculation* 33 (3): 235–242.

Coffeng, S. M., J. Blaauw, E. T. D. Souwer et al. 2011. Skin autofluorescence as marker of tissue advanced glycation end-products accumulation in formerly preeclamptic women. *Hypertension in Pregnancy* 30 (2): 231–242.

Cortassa, S., M. A. Aon, E. Marbán, R. L. Winslow, and B. O'Rourke. 2003. An integrated model of cardiac mitochondrial energy metabolism and calcium dynamics. *Biophysical Journal* 84 (4): 2734–2755.

Das, A. M. and D. A. Harris. 1990. Regulation of the mitochondrial ATP synthase in intact rat cardiomyocytes. *Biochemical Journal* 266: 355–361.

Dennis, S. C., W. Gevers, and L. H. Opie. 1991. Protons in ischemia: Where do they come from; where do they go to? *Journal of Molecular and Cellular Cardiology* 23 (9): 1077–1086.

Dhalla, N. S., N. Afzal, R. E. Beamish, B. Naimark, N. Takeda, and M. Nagano. 1993. Pathophysiology of cardiac dysfunction in congestive heart failure. *The Canadian Journal of Cardiology* 9 (10): 873–887.

Dickinson, M. E., G. Bearman, S. Tille, R. Lansford, and S. E. Fraser. 2001. Multi-spectral imaging and linear unmixing add a whole new dimension to laser scanning fluorescence microscopy. *Biotechniques* 31 (6): 1272–1279.

Elzwiei, F., V. Bassien-Capsa, J. St-Louis, and A. Chorvatova. 2013. Regulation of the sodium pump during cardiomyocyte adaptation to pregnancy. *Experimental Physiology* 98 (1): 183–192.

Eng, J., R. M. Lynch, and R. S. Balaban. 1989. Nicotinamide adenine dinucleotide fluorescence spectroscopy and imaging of isolated cardiac myocytes. *Biophysical Journal* 55 (4): 621–630.

Ferrari, M. and V. Quaresima. 2012. Review: Near infrared brain and muscle oximetry: From the discovery to current applications. *Journal of Near Infrared Spectroscopy* 20: 1.

Gao, G., K. Öllinger, and U. T. Brunk. 1994. Influence of intracellular glutathione concentration on lipofuscin accumulation in cultured neonatal rat cardiac myocytes. *Free Radical Biology and Medicine* 16 (2): 187–194.

Gaudin, P. B., B. K. Rayburn, G. M. Hutchins et al. 1994. Peritransplant injury to the myocardium associated with the development of accelerated arteriosclerosis in heart transplant recipients. *The American Journal of Surgical Pathology* 18 (4): 338–346.

Gerson, C. J., R. C. Elkins, S. Goldstein, and A. E. Heacox. 2012. Structural integrity of collagen and elastin in SynerGraft® decellularized–cryopreserved human heart valves. *Cryobiology* 64 (1): 33–42.

Griffiths, E. J., H. Lin, and M. S. Suleiman. 1998. NADH fluorescence in isolated guinea-pig and rat cardiomyocytes exposed to low or high stimulation rates and effect of metabolic inhibition with cyanide. *Biochemical Pharmacology* 56 (2): 173–179.

Griffiths, E. J., S.-K. Wei, M. C. P. Haigney, C. J. Ocampo, M. D. Stern, and H. S. Silverman. 1997. Inhibition of mitochondrial calcium efflux by clonazepam in intact single rat cardiomyocytes and effects on NADH production. *Cell Calcium* 21 (4): 321–329.

Hanley, P. J., J. Ray, U. Brandt, and J. Daut. 2002. Halothane, isoflurane and sevoflurane inhibit NADH:ubiquinone oxidoreductase (complex I) of cardiac mitochondria. *The Journal of Physiology* 544 (3): 687–693.

Hartog, J. W. L., S. Willemsen, D. J. van Veldhuisen et al. 2011. Effects of alagebrium, an advanced glycation endproduct breaker, on exercise tolerance and cardiac function in patients with chronic heart failure. *European Journal of Heart Failure* 13 (8): 899–908.

He, Q., P. T. Trindade, M. Stumm et al. 2009. Fate of undifferentiated mouse embryonic stem cells within the rat heart: Role of myocardial infarction and immune suppression. *Journal of Cellular and Molecular Medicine* 13 (1): 188–201.

Heineman, F. W., V. V. Kupriyanov, R. Marshall, T. A. Fralix, and R. S. Balaban. 1992. Myocardial oxygenation in the isolated working rabbit heart as a function of work. *American Journal of Physiology—Heart and Circulatory Physiology* 262 (1): H255–H267.

Heinsen, H. 1981. Regional differences in the distribution of lipofuscin in Purkinje cell perikarya. *Anatomy and Embryology* 161 (4): 453–464.
Hoffman, F. M. 2005. Outcomes and complications after heart transplantation: A review. *Journal of Cardiovascular Nursing* 20 (5S): S31–S42.
Huang, H., C. MacGillivray, H.-S. Kwon et al. 2009. Three-dimensional cardiac architecture determined by two-photon microtomy. *Journal of Biomedical Optics* 14 (4): 044029.
Huang, S., A. A. Heikal, and W. W. Webb. 2002. Two-photon fluorescence spectroscopy and microscopy of NAD (P) H and flavoprotein. *Biophysical Journal* 82 (5): 2811–2825.
Hulsmann, W. C., J. F. Ashruf, H. A. Bruining, and C. Ince. 1993. Imminent ischemia in normal and hypertrophic Langendorff rat hearts; effects of fatty acids and superoxide dismutase monitored by NADH surface fluorescence. *Biochimica et Biophysica Acta (BBA)—Molecular Basis of Disease* 1181 (3): 273–278.
Jameson, D. M., V. Thomas, and D.-M. Zhou. 1989. Time-resolved fluorescence studies on NADH bound to mitochondrial malate dehydrogenase. *Biochimica et Biophysica Acta (BBA)—Protein Structure and Molecular Enzymology* 994 (2): 187–190.
Jepihhina, N., N. Beraud, M. Sepp, R. Birkedal, and M. Vendelin. 2011. Permeabilized rat cardiomyocyte response demonstrates intracellular origin of diffusion obstacles. *Biophysical Journal* 101 (9): 2112–2121.
Jo, J. A., Q. Fang, T. Papaioannou, and L. Marcu. 2004. Novel ultra-fast deconvolution method for fluorescence lifetime imaging microscopy based on the Laguerre expansion technique. In *Proceedings of 26th Annual International Conference of Engineering in Medicine and Biology Society*, San Francisco, CA, pp. 1271–1274.
Jo, J. A., Q. Fang, T. Papaioannou et al. 2005. Application of the Laguerre deconvolution method for time-resolved fluorescence spectroscopy to the characterization of atherosclerotic plaques. In *Proceedings of 27th Annual International Conference of Engineering in Medicine and Biology Society*, Santa Barbara, CA, pp. 6559–6562.
Jo, J. A., Q. Fang, T. Papaioannou et al. 2006. Diagnosis of vulnerable atherosclerotic plaques by time-resolved fluorescence spectroscopy and ultrasound imaging. In Vol. 1 of the *Proceedings of 28th Annual International Conference of Engineering in Medicine and Biology Society*, New York, pp. 2663–2666.
Jolly, R. D., D. N. Palmer, and R. R. Dalefield. 2002. The analytical approach to the nature of lipofuscin (age pigment). *Archives of Gerontology and Geriatrics* 34 (3): 205–217.
Kittrell, C., R. L. Willett, C. de Los Santos-Pacheo et al. 1985. Diagnosis of fibrous arterial atherosclerosis using fluorescence. *Applied Optics* 24 (15): 2280–2281.
Klaidman, L. K., A. C. Leung, and J. D. Adams. 1995. High-performance liquid chromatography analysis of oxidized and reduced pyridine dinucleotides in specific brain regions. *Analytical Biochemistry* 228 (2): 312–317.
Koke, J. R., W. Wylie, and M. Wills. 1980. Sensitivity of flavoprotein fluorescence to oxidative state in single isolated heart cells. *Cytobios* 32 (127–128): 139–145.
Korge, P., P. Ping, and J. N. Weiss. 2008. Reactive oxygen species production in energized cardiac mitochondria during hypoxia/reoxygenation modulation by nitric oxide. *Circulation Research* 103 (8): 873–880.
Kuznetsov, A. V., O. Mayboroda, D. Kunz, K. Winkler, W. Schubert, and W. S. Kunz. 1998. Functional imaging of mitochondria in saponin-permeabilized mice muscle fibers. *The Journal of Cell Biology* 140 (5): 1091–1099.
Lakowicz, J. R. 2000. On spectral relaxation in proteins. *Photochemistry and Photobiology* 72 (4): 421–437.
Lakowicz, J. R., H. Szmacinski, K. Nowaczyk, and M. L. Johnson. 1992. Fluorescence lifetime imaging of free and protein-bound NADH. *Proceedings of the National Academy of Sciences of the United States of America* 89 (4): 1271–1275.

Le, T. T., I. M. Langohr, M. J. Locker, M. Sturek, and J.-X. Cheng. 2007. Label-free molecular imaging of atherosclerotic lesions using multimodal nonlinear optical microscopy. *Journal of Biomedical Optics* 12 (5): 054007.

Leisey, J. R., D. A. Scott, L. W. Grotyohann, and R. C. Scaduto Jr. 1994. Quantitation of myoglobin saturation in the perfused heart using myoglobin as an optical inner filter. *American Journal of Physiology—Heart and Circulatory Physiology* 267 (2): H645–H653.

Lesnefsky, E. J., S. Moghaddas, B. Tandler, J. Kerner, and C. L. Hoppel. 2001. Mitochondrial dysfunction in cardiac disease: Ischemia–reperfusion, aging, and heart failure. *Journal of Molecular and Cellular Cardiology* 33 (6): 1065–1089.

Lilledahl, M. B., O. A. Haugen, C. de Lange Davies, and L. Othar Svaasand. 2007. Characterization of vulnerable plaques by multiphoton microscopy. *Journal of Biomedical Optics* 12 (4): 044005.

Lin, S.-J. and L. Guarente. 2003. Nicotinamide adenine dinucleotide, a metabolic regulator of transcription, longevity and disease. *Current Opinion in Cell Biology* 15 (2): 241–246.

Lin, X., F. Sun, H. Ma, J. Zhao, L. Jin, and D. Y. Chen. 2000. Two-photon fluorescence imaging of rat aorta. *Proceedings of SPIE* 4224: 7–12.

Lindbergh, T., E. Häggblad, H. Ahn, E. G. Salerud, M. Larsson, and T. Strömberg. 2011. Improved model for myocardial diffuse reflectance spectra by including mitochondrial cytochrome aa3, methemoglobin, and inhomogenously distributed RBC. *Journal of Biophotonics* 4 (4): 268–276.

Lindbergh, T., M. Larsson, Z. Szabó, H. Casimir-Ahn, and T. Strömberg. 2010. Intramyocardial oxygen transport by quantitative diffuse reflectance spectroscopy in calves. *Journal of Biomedical Optics* 15 (2): 027009.

Liu, C. H., W. B. Wang, V. Kartazaev, H. Savag, and R. R. Alfano. 2010a. Changes of collagen, elastin and tryptophan contents in laser welded porcine aorta tissues studied using fluorescence spectroscopy. *Proceedings of SPIE* 7561: 756115.

Liu, T., D. A. Brown, and B. O'Rourke. 2010b. Role of mitochondrial dysfunction in cardiac glycoside toxicity. *Journal of Molecular and Cellular Cardiology* 49 (5): 728–736.

Lutgers, H. L., R. Graaff, T. P. Links et al. 2006. Skin autofluorescence as a noninvasive marker of vascular damage in patients with type 2 diabetes. *Diabetes Care* 29 (12): 2654–2659.

Maarek, J.-M. I., L. Marcu, M. C. Fishbein, and W. S. Grundfest. 2000b. Time-resolved fluorescence of human aortic wall: Use for improved identification of atherosclerotic lesions. *Lasers in Surgery and Medicine* 27 (3): 241–254.

Maarek, J.-M. I., L. Marcu, W. J. Snyder, and W. S. Grundfest. 2000a. Time-resolved fluorescence spectra of arterial fluorescent compounds: Reconstruction with the Laguerre expansion technique. *Photochemistry and Photobiology* 71 (2): 178–187.

Marcu, L. 2010. Fluorescence lifetime in cardiovascular diagnostics. *Journal of Biomedical Optics* 15 (1): 011106.

Marcu, L., Q. Fang, J. A. Jo et al. 2005. In vivo detection of macrophages in a rabbit atherosclerotic model by time-resolved laser-induced fluorescence spectroscopy. *Atherosclerosis* 181 (2): 295–303.

Marcu, L., M. C. Fishbein, J.-M. I. Maarek, and W. S. Grundfest. 2001a. Discrimination of human coronary artery atherosclerotic lipid-rich lesions by time-resolved laser-induced fluorescence spectroscopy. *Arteriosclerosis, Thrombosis, and Vascular Biology* 21 (7): 1244–1250.

Marcu, L., W. S. Grundfest, and J.-M. I. Maarek. 2001b. Arterial fluorescent components involved in atherosclerotic plaque instability: Differentiation by time-resolved fluorescence spectroscopy. *Proceedings of SPIE* 4244: 428–433.

Marcu, L., J. A. Jo, Q. Fang et al. 2009. Detection of rupture-prone atherosclerotic plaques by time-resolved laser-induced fluorescence spectroscopy. *Atherosclerosis* 204 (1): 156–164.

Marcu, L., J.-M. I. Maarek, M. C. Fishbein, and W. S. Grundfest. 1999. Time-resolved laser-induced fluorescence of normal and atherosclerotic coronary artery. *Proceedings of SPIE* 3600: 182–191.

Marzabadi, M. R., C. Jones, and J. Rydström. 1995. Indenoindole depresses lipofuscin formation in cultured neonatal rat myocardial cells. *Mechanisms of Ageing and Development* 80 (3): 189–197.

Matsumoto-Ida, M., M. Akao, T. Takeda, M. Kato, and T. Kita. 2006. Real-time 2-photon imaging of mitochondrial function in perfused rat hearts subjected to ischemia/reperfusion. *Circulation* 114 (14): 1497–1503.

Mayevsky, A., E. Ornstein, S. Meilin, N. Razon, and G. E. Ouaknine. 2002. The evaluation of brain CBF and mitochondrial function by a fiber optic tissue spectroscope in neurosurgical patients. In *Intracranial Pressure and Brain Biochemical Monitoring*, M. Czosnyka, J. D. Pickard, P. Kirkpatrick et al. (eds.). Vol. 81 of Acta Neurochirurgica Supplements, H.-J. Steiger (ed.), pp. 367–371. Vienna, Austria: Springer.

Meerwaldt, R., H. L. Lutgers, T. P. Links et al. 2007. Skin autofluorescence is a strong predictor of cardiac mortality in diabetes. *Diabetes Care* 30 (1): 107–112.

Meerwaldt, R., M. G. van der Vaart, G. M. van Dam, R. A. Tio, J.-L. Hillebrands, A. J. Smit, and C. J. Zeebregts. 2008. Clinical relevance of advanced glycation endproducts for vascular surgery. *European Journal of Vascular and Endovascular Surgery* 36 (2): 125–131.

Monserrat, A. J., S. H. Benavides, A. Berra, S. Fariña, S. C. Vicario, and E. A. Porta. 1995. Lectin histochemistry of lipofuscin and certain ceroid pigments. *Histochemistry and Cell Biology* 103 (6): 435–445.

Moore, R. Y. 1981. Fluorescence histochemical methods. In *Neuroanatomical Tract-Tracing Methods*, L. Heimer and M. J. Robards (eds.), pp. 457–458. New York: Plenum Press.

Mootha, V. K., A. E. Arai, and R. S. Balaban. 1997. Maximum oxidative phosphorylation capacity of the mammalian heart. *American Journal of Physiology—Heart and Circulatory Physiology* 272 (2): H769–H775.

Moreno-Sánchez, R. 1985. Regulation of oxidative phosphorylation in mitochondria by external free Ca^{2+} concentrations. *Journal of Biological Chemistry* 260 (7): 4028–4034.

Morgan, D. C., J. E. Wilson, C. E. MacAulay et al. 1999. New method for detection of heart allograft rejection validation of sensitivity and reliability in a rat heterotopic allograft model. *Circulation* 100 (11): 1236–1241.

Mulder, D. J., T. Van De Water, H. L. Lutgers et al. 2006. Skin autofluorescence, a novel marker for glycemic and oxidative stress-derived advanced glycation endproducts: An overview of current clinical studies, evidence, and limitations. *Diabetes Technology and Therapeutics* 8 (5): 523–535.

Nakagami, T., H. Tanaka, P. Dai et al. 2008. Generation of reentrant arrhythmias by dominant-negative inhibition of connexin43 in rat cultured myocyte monolayers. *Cardiovascular Research* 79 (1): 70–79.

Nelson, D. L. and M. M. Cox. 2008. *Lehninger Principles of Biochemistry.* New York: W. H. Freeman & Co.

Nighswander-Rempel, S. P., V. V. Kupriyanov, and R. A. Shaw. 2005. Relative contributions of hemoglobin and myoglobin to near-infrared spectroscopic images of cardiac tissue. *Applied Spectroscopy* 59 (2): 190–193.

Nighswander-Rempel, S. P., V. V. Kupriyanov, and R. A. Shaw. 2006. Regional cardiac tissue oxygenation as a function of blood flow and pO_2: A near-infrared spectroscopic imaging study. *Journal of Biomedical Optics* 11 (5): 054004.

Nilsson, A. M. K., D. Heinrich, J. Olajos, and S. Andersson-Engels. 1997. Near infrared diffuse reflection and laser-induced fluorescence spectroscopy for myocardial tissue characterisation. *Spectrochimica Acta Part A: Molecular and Biomolecular Spectroscopy* 53 (11): 1901–1912.

Nussbaum, J., E. Minami, M. A. Laflamme et al. 2007. Transplantation of undifferentiated murine embryonic stem cells in the heart: Teratoma formation and immune response. *The FASEB Journal* 21 (7): 1345–1357.

Pasdois, P., B. Beauvoit, L. Tariosse, B. Vinassa, S. Bonoron-Adèle, and P. Dos Santos. 2008. Effect of diazoxide on flavoprotein oxidation and reactive oxygen species generation during ischemia-reperfusion: A study on Langendorff-perfused rat hearts using optic fibers. *American Journal of Physiology—Heart and Circulatory Physiology* 294 (5): H2088–H2097.

Pery, E., W. C. P. M. Blondel, J. Didelon, A. Leroux, and F. Guillemin. 2009. Simultaneous characterization of optical and rheological properties of carotid arteries via bimodal spectroscopy: Experimental and simulation results. *IEEE Transactions on Biomedical Engineering* 56 (5): 1267–1276.

Phipps, J., Y. Sun, R. Saroufeem, N. Hatami, M. C. Fishbein, and L. Marcu. 2011. Fluorescence lifetime imaging for the characterization of the biochemical composition of atherosclerotic plaques. *Journal of Biomedical Optics* 16 (9): 096018.

Piper, H. M., Y. Abdallah, and C. Schäfer. 2004. The first minutes of reperfusion: A window of opportunity for cardioprotection. *Cardiovascular Research* 61 (3): 365–371.

Piper, H. M., T. Noll, and B. Siegmund. 1994. Mitochondrial function in the oxygen depleted and reoxygenated myocardial cell. *Cardiovascular Research* 28 (1): 1–15.

Piper, H. M., B. Siegmund, Y. V. Ladilov, and K.-D. Schlüter. 1993. Calcium and sodium control in hypoxic-reoxygenated cardiomyocytes. *Basic Research in Cardiology* 88 (5): 471–482.

Plotnikov, S. V., A. C. Millard, P. J. Campagnola, and W. A. Mohler. 2006. Characterization of the myosin-based source for second-harmonic generation from muscle sarcomeres. *Biophysical Journal* 90 (2): 693–703.

Ragan, T., J. D. Sylvan, K. H. Kim et al. 2007. High-resolution whole organ imaging using two-photon tissue cytometry. *Journal of Biomedical Optics* 12 (1): 014015.

Ranji, M., S. Kanemoto, M. Matsubara et al. 2006. Fluorescence spectroscopy and imaging of myocardial apoptosis. *Journal of Biomedical Optics* 11 (6): 064036.

Ranji, M., M. Matsubara, M. A. Grosso et al. 2007. Fluorescence spectroscopy to assess apoptosis in myocardium. In *Biomedical Optics (BiOS) 2007*, Munich, Germany, pp. 64380J–64380J. Bellingham, WA: International Society for Optics and Photonics.

Richards-Kortum, R. and E. Sevick-Muraca. 1996. Quantitative optical spectroscopy for tissue diagnosis. *Annual Review of Physical Chemistry* 47 (1): 555–606.

Rodrigo, G. C., C. L. Lawrence, and N. B. Standen. 2002. Dinitrophenol pretreatment of rat ventricular myocytes protects against damage by metabolic inhibition and reperfusion. *Journal of Molecular and Cellular Cardiology* 34 (5): 555–569.

Romashko, D. N., E. Marban, and B. O'Rourke. 1998. Subcellular metabolic transients and mitochondrial redox waves in heart cells. *Proceedings of the National Academy of Sciences of the United States of America* 95 (4): 1618–1623.

Samborski, P., D. Naskręt, A. Araszkiewicz, P. Niedźwiecki, D. Zozulińska-Ziółkiewicz, and B. Wierusz-Wysocka. 2011. Assessment of skin autofluorescence as a marker of advanced glycation end product accumulation in type 1 diabetes. *Polskie Archiwum Medycyny Wewnętrznej* 121 (3): 67–72.

Šćepanović, O. R., M. Fitzmaurice, A. Miller et al. 2011. Multimodal spectroscopy detects features of vulnerable atherosclerotic plaque. *Journal of Biomedical Optics* 16 (1): 011009.

Schaffer, S. W., B. Safer, C. Ford, J. Illingworth, and J. R. Williamson. 1978. Respiratory acidosis and its reversibility in perfused rat heart: Regulation of citric acid cycle activity. *American Journal of Physiology—Heart and Circulatory Physiology* 234 (1): H40–H51.

Schenke-Layland, K., A. Nsair, B. Van Handel et al. 2011. Recapitulation of the embryonic cardiovascular progenitor cell niche. *Biomaterials* 32 (11): 2748–2756.

Schenke-Layland, K., I. Riemann, O. Damour, U. A. Stock, and K. König. 2006. Two-photon microscopes and in vivo multiphoton tomographs—Powerful diagnostic tools for tissue engineering and drug delivery. *Advanced Drug Delivery Reviews* 58 (7): 878–896.

Schenke-Layland, K., I. Riemann, F. Opitz, K. König, K. J. Halbhuber, and U. A. Stock. 2004. Comparative study of cellular and extracellular matrix composition of native and tissue engineered heart valves. *Matrix Biology* 23 (2): 113–125.

Schenke-Layland, K., U. A. Stock, A. Nsair et al. 2009. Cardiomyopathy is associated with structural remodelling of heart valve extracellular matrix. *European Heart Journal* 30 (18): 2254–2265.

Schenke-Layland, K., J. Xie, S. Heydarkhan-Hagvall et al. 2007. Optimized preservation of extracellular matrix in cardiac tissues: Implications for long-term graft durability. *The Annals of Thoracic Surgery* 83 (5): 1641–1650.

Schlüter, K. D., G. Jakob, M. Ruiz-Meana, D. Garcia-Dorado, and H. M. Piper. 1996. Protection of reoxygenated cardiomyocytes against osmotic fragility by nitric oxide donors. *The American Journal of Physiology* 271 (2 Pt 2): H428–H434.

Schriefl, A. J. 2013. *Quantification of Collagen Fiber Morphologies in Human Arterial Walls.* Graz, Austria: Verlag des Technischen Universitat Graz.

Singh, R., A. Barden, T. Mori, and L. Beilin. 2001. Advanced glycation end-products: A review. *Diabetologia* 44 (2): 129–146.

Smith, D. R., D. Stone, and V. M. Darley-Usmar. 1996. Stimulation of mitochondrial oxygen consumption in isolated cardiomyocytes after hypoxia-reoxygenation. *Free Radical Research* 24 (3): 159–166.

Straume, M., S. G. Frasier-Cadoret, and M. L. Johnson. 1991. Least-squares analysis of fluorescence data. In Vol. 2 of *Topics in Fluorescence Spectroscopy*, J. R. Lakowicz (ed.), pp. 177–240. New York: Springer.

Sun, Y., Y. Sun, D. Stephens et al. 2011. Dynamic tissue analysis using time-and wavelength-resolved fluorescence spectroscopy for atherosclerosis diagnosis. *Optics Express* 19 (5): 3890–3891.

Takahashi, E., H. Endoh, and K. Doi. 2000. Visualization of myoglobin-facilitated mitochondrial O_2 delivery in a single isolated cardiomyocyte. *Biophysical Journal* 78 (6): 3252–3259.

Tanaka, M., G. K. Mokhtari, R. D. Terry et al. 2004a. Overexpression of human copper/zinc superoxide dismutase (SOD1) suppresses ischemia–reperfusion injury and subsequent development of graft coronary artery disease in murine cardiac grafts. *Circulation* 110 (11 Suppl. 1): II-200–II-206.

Tanaka, M., S. Nakae, R. D. Terry et al. 2004b. Cardiomyocyte-specific Bcl-2 overexpression attenuates ischemia-reperfusion injury, immune response during acute rejection, and graft coronary artery disease. *Blood* 104 (12): 3789–3796.

Terman, A. and U. T. Brunk. 1998. On the degradability and exocytosis of ceroid/lipofuscin in cultured rat cardiac myocytes. *Mechanisms of Ageing and Development* 100 (2): 145–156.

van de Poll, S. W. E., R. R. Dasari, and J. R. Kramer. 1999. The role of laser-induced fluorescence spectroscopy in the detection of human atherosclerosis. *Current Science* 77: 934–941.

van Zandvoort, M., W. Engels, K. Douma et al. 2004. Two-photon microscopy for imaging of the (atherosclerotic) vascular wall: A proof of concept study. *Journal of Vascular Research* 41 (1): 54–63.

Vanzi, F., L. Sacconi, R. Cicchi, and F. S. Pavone. 2012. Protein conformation and molecular order probed by second-harmonic-generation microscopy. *Journal of Biomedical Optics* 17 (6): 060901.

Vassalli, G., A. Gallino, M. Weis et al. 2003. Alloimmunity and nonimmunologic risk factors in cardiac allograft vasculopathy. *European Heart Journal* 24 (13): 1180–1188.

Veech, R. L., L. V. Eggleston, and H. A. Krebs. 1969. The redox state of free nicotinamide-adenine dinucleotide phosphate in the cytoplasm of rat liver. *Biochemical Journal* 115: 609–619.

Venius, J., S. Bagdonas, E. Žurauskas, and R. Rotomskis. 2011. Visualization of human heart conduction system by means of fluorescence spectroscopy. *Journal of Biomedical Optics* 16 (10): 107001.

Vigneshwaran, N., G. Bijukumar, N. Karmakar, S. Anand, and A. Misra. 2005. Autofluorescence characterization of advanced glycation end products of hemoglobin. *Spectrochimica Acta Part A: Molecular and Biomolecular Spectroscopy* 61 (1): 163–170.

Vuorinen, K., K. Ylitalo, K. Peuhkurinen, P. Raatikainen, A. Ala-Rämi, and I. E. Hassinen. 1995. Mechanisms of ischemic preconditioning in rat myocardium roles of adenosine, cellular energy state, and mitochondrial F_1F_o-ATPase. *Circulation* 91 (11): 2810–2818.

Wallace, S. J., J. L. Morrison, K. J. Botting, and T. W. Kee. 2008. Second-harmonic generation and two-photon-excited autofluorescence microscopy of cardiomyocytes: Quantification of cell volume and myosin filaments. *Journal of Biomedical Optics* 13 (6): 064018.

Wallenburg, M. A., J. Wu, R.-K. Li, and I. A. Vitkin. 2011. Two-photon microscopy of healthy, infarcted and stem-cell treated regenerating heart. *Journal of Biophotonics* 4 (5): 297–304.

Wang, H.-W., V. Simianu, M. J. Locker, M. Sturek, and J.-X. Cheng. 2008. Imaging arterial cells, atherosclerosis, and restenosis by multimodal nonlinear optical microscopy. *Proceedings of SPIE* 6860: 68600W.

White, R. L. and B. A. Wittenberg. 1993. NADH fluorescence of isolated ventricular myocytes: Effects of pacing, myoglobin, and oxygen supply. *Biophysical Journal* 65 (1): 196–204.

Williams, R. M., D. W. Piston, and W. W. Webb. 1994. Two-photon molecular excitation provides intrinsic 3-dimensional resolution for laser-based microscopy and microphotochemistry. *The FASEB Journal* 8 (11): 804–813.

Wittenberg, B. A. and J. B. Wittenberg. 1985. Oxygen pressure gradients in isolated cardiac myocytes. *Journal of Biological Chemistry* 260 (11): 6548–6554.

Wittenberg, B. A. and J. B. Wittenberg. 1989. Transport of oxygen in muscle. *Annual Review of Physiology* 51 (1): 857–878.

Wittenberg, J. B. and B. A. Wittenberg. 2003. Myoglobin function reassessed. *Journal of Experimental Biology* 206 (12): 2011–2020.

Xu, H. N., B. Wu, S. Nioka, B. Chance, and L. Z. Li. 2009. Calibration of redox scanning for tissue samples. *Proceedings of SPIE BiOS* 7174: 71742F.

Yamani, M. H., S. W. E. van de Poll, N. B. Ratliff et al. 2000. Fluorescence spectroscopy of endomyocardial tissue post–human heart transplantation: Does it correlate with histopathology? *The Journal of Heart and Lung Transplantation* 19 (11): 1077–1080.

Zimmermann, T., J. Rietdorf, and R. Pepperkok. 2003. Spectral imaging and its applications in live cell microscopy. *FEBS Letters* 546 (1): 87–92.

Zipfel, W. R., R. M. Williams, R. Christie, A. Y. Nikitin, B. T. Hyman, and W. W. Webb. 2003a. Live tissue intrinsic emission microscopy using multiphoton-excited native fluorescence and second harmonic generation. *Proceedings of the National Academy of Sciences of the United States of America* 100 (12): 7075–7080.

Zipfel, W. R., R. M. Williams, and W. W. Webb. 2003b. Nonlinear magic: Multiphoton microscopy in the biosciences. *Nature Biotechnology* 21 (11): 1369–1377.

13

Autofluorescence Perspective of Cancer Diagnostics

Lin Z. Li
University of Pennsylvania

Nannan Sun
University of Pennsylvania and Huazhong University of Science and Technology

13.1 Introduction

Various histological and genetic biomarkers are currently utilized for cancer diagnostics and treatment; however, conventional biopsy and histology methods are invasive and time consuming. Autofluorescence spectroscopy and imaging techniques provide an alternative approach with major advantages of minimum- or noninvasiveness and the capacity to obtain real-time information about tumor heterogeneity and metastasis.

While other intrinsic fluorophores, such as collagen, elastin, keratin, lipofuscin, and porphyrins, may contribute to autofluorescence in the visible range (see Chapter 2), NAD(P)H and oxidized flavoproteins (Fps) are the major intracellular biomolecules that contribute to autofluorescence in the range of 400–490 and 490–580 nm, respectively (Li et al. 2010; Palmer et al. 2003; Ramanujam 2000).

In this chapter, we review the potential of NAD(P)H and Fp autofluorescence, as well as reduction-oxidation (redox) ratios (Fp/NAD(P)H, Fp/(Fp + NAD(P)H), NAD(P)H/Fp, etc.), as natural biomarkers for distinguishing between normal tissues and benign or malignant tumors and for the assessment of cancer aggressiveness. Lastly, we discuss various technical and biological issues toward the future development of cellular autofluorescence as reliable, noninvasive, and fast diagnostic tool.

13.1.1 Clinical Needs in Cancer Diagnosis

In clinical cancer management, noninvasive biomarkers are primarily sought for (1) distinguishing normal, precancerous, and cancerous tissues and (2) assessment of cancer aggressiveness or prognosis. Early diagnosis of precancer or cancer can result in much improved prognosis for patients. Cancer aggressiveness factors, particularly, state of metastasis, are among other important factors that can affect the prognosis and determine the treatment plan. Occurrence of metastases is one of the major indicators of poor prognosis: 90% of cancer patients succumb to cancer due to the spread of the disease beyond the initial organ or tissue. Evaluation of the metastatic potential of a primary tumor or the metastatic risk is extremely important for effective therapy. Thus, less aggressive treatment can be implemented for less or nonmetastatic tumors in order to minimize the side effects associated with overtreatment. Quite often, cancer proliferation rate or tumor growth rate is used as an indicator of cancer aggressiveness. However, whether the growth rate is an appropriate tumor aggressiveness marker in the clinic may depend on the cancer types. For example, fast-growing brain tumors, metastatic or not, usually have poor prognosis because of the detrimental effects of large tumors on the central nervous system. For body sites other than the brain, it is known that some benign or low-grade neoplasia can grow faster than high-grade or metastatic tumors. Cancer aggressiveness can be determined also by tumor response to treatment. Although biomarkers for such response are of much clinical interest, they are beyond the scope of this chapter. Throughout this chapter, increased aggressiveness will refer to higher tumor grade, higher metastatic potential/risk, and poor prognosis unless otherwise indicated.

13.1.2 Role of NAD and Flavin Groups in Cancer Metabolism

Cellular metabolism has become an increasingly active area of cancer research in recent decades, triggered by the recognition of enhanced aerobic glycolysis in cancer transformation, known as the Warburg effect (Garber 2004; Zhivotovsky and Orrenius 2009). Although the significance of the Warburg effect and the details of related metabolic pathway aberrations are not fully understood, it is widely recognized that the shift from oxidative phosphorylation to aerobic glycolysis provides biosynthetic building blocks (e.g., NADPH) needed for tumor redox balance and growth (Heiden et al. 2009). In recent years, other cancer-related abnormalities in intermediary metabolism were discovered, including the elevation of pyruvate kinase M2 levels (Christofk et al. 2008; Heiden et al. 2010) and inactivation of mitochondrial metabolic enzymes fumarate hydratase, succinate dehydrogenase, and isocitrate dehydrogenase, which were identified as tumor suppressors (Thompson 2009). These and other studies have highlighted the important role of metabolism for cancer cell transformation, growth, apoptosis, and identified mitochondrial metabolic pathways as a potential target of cancer therapy.

The activities of the aforementioned mitochondrial metabolic enzymes, as well as lactate dehydrogenase, are mediated by the redox potentials maintained by nicotinamide adenine dinucleotides (NADs)—both oxidized form NAD^+ and reduced form NADH—and flavin adenine dinucleotides (FADs). Redox state plays an important mediating role for many biological activities (Figure 13.1) including cellular metabolism, growth, survival, apoptosis, cell cycle, free radical generation, signaling, and gene expression (Li 2012). As discussed in Chapter 1, the free-energy carrier NADH mediates a number of redox reactions in cellular energy metabolism. By controlling glycolysis in cytosol and the TCA cycle in mitochondria, the tissue redox potential NAD^+/NADH is linked to the phosphorylation potential [ATP]/([ADP][Pi]) in living tissues and represents a key parameter for the metabolic control of normal and diseased phenotypes (Veech 2006). In addition, NAD^+/NADH is a key component in cellular redox homeostasis through the coupling to NADP(H) by transhydrogenase activity and, therefore, can indirectly affect the redox couples of glutathione and thioredoxin systems (Lemasters and Nieminen 2001). These redox couples and related redox-sensitive enzymes may affect the major signaling pathways, including p53, PI3K, and MAPK, which play an important role in cancer growth, survival, motility, and metabolism (Adler et al. 1999; Olovnikov et al. 2009). NAD can mediate many cellular activities including signaling, cell growth, differentiation, survival, apoptosis, Ca^{2+} release, modification of transcription factors, and generation of reactive oxygen species (ROS) (Banerjee 2007; Orrenius et al. 2007; Ying 2008; Ziegler 2005). The mitochondrial NAD^+-coupled redox state, for example, is involved in the generation of ROS by complexes I and III in the electron transport chain (ETC). A mutation in mitochondrial DNA encoding a subunit of complex I was shown to result in excessive ROS, thus driving the development of metastasis in animal models (Ishikawa et al. 2008). Recently, the NAD metabolome was proposed as the key determinant of cancer cell biology (Chiarugi et al. 2012).

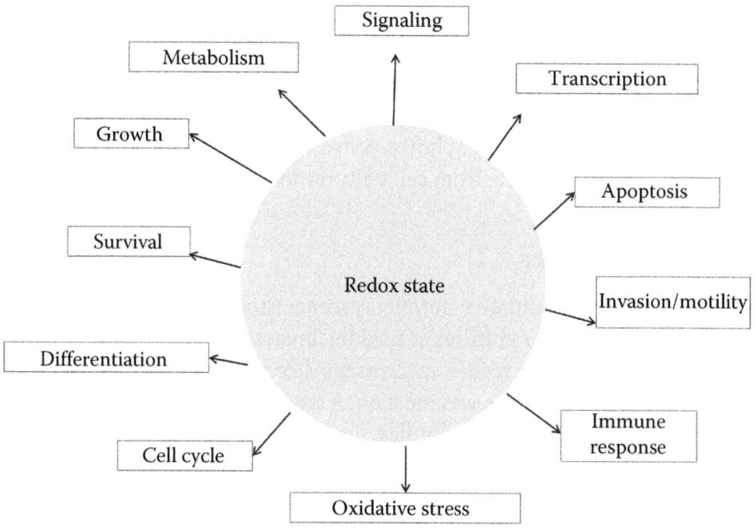

FIGURE 13.1 The central biological roles of redox state. (Reproduced from Li, L.Z., *J. Bioenerg. Biomembr.*, 44(6), 645, 2012. With permission.)

In addition to NAD, another group of redox-important molecules, flavin nucleotides, play an important role in various biological processes including metabolism and signaling events (Becker et al. 2011; Lehninger et al. 1993; Senda et al. 2009; Taylor et al. 2001). Oxidized flavins, flavin mononucleotide (FMN) and FAD, can be found in cells as coenzymes or prosthetic groups for various mitochondrial Fps, such as complex I, complex II, α-ketoglutarate dehydrogenase, pyruvate dehydrogenase, electron-transferring Fps, and glycerol-3-phosphate dehydrogenase (Lehninger et al. 1993). Flavin prosthetic groups also induce redox-dependent conformational and functional changes in Fps, which are important for protein transcription, signaling pathways, and environmental adaptation (Becker et al. 2011; Senda et al. 2009; Taylor et al. 2001). In contrast, the reduced form ($FADH_2$) is a free-energy carrier in mitochondrial ETC. FAD-coupled redox potential FAD/$FADH_2$ regulates key reactions in the TCA cycle, oxidative phosphorylation, and fatty acid metabolism (Lehninger et al. 1993).

In summary, the intrinsic fluorescence signals of NADH and oxidized Fps FAD and FMN may provide useful information for understanding biology, pathology, and diagnostic biomarkers for various diseases including cancer.

13.2 Autofluorescence Studies in Cancer

The first report on tumor autofluorescence (which was later assigned predominantly to porphyrins) was published by Policard in 1924 (Arens et al. 2004). A few decades later, Duysens and Amesz were the first to report a detailed study of NAD(P)H fluorescence in intact cells (Duysens and Amesz 1957). Beginning in the 1950s, Chance and collaborators have systematically studied mitochondrial NAD(P)H and flavins by various absorption/fluorescence spectroscopic and imaging methods and demonstrated that the redox ratio (Fp/NAD(P)H, Fp/(Fp + NAD(P)H), or NAD(P)H/Fp) can serve as a sensitive indicator for cellular metabolism both in vitro and in vivo and a surrogate biomarker for NAD^+/NADH (Chance 1991; Chance and Baltscheffsky 1958; Chance and Jobsis 1959; Chance et al. 1962, 1979; Hassinen and Chance 1968; Ozawa et al. 1992). It has been established by Chance and coworkers that metabolically more active mitochondria (State 3) exhibit higher Fp/NAD(P)H ratio or more oxidized state compared to inactive state such as State 4 (see detailed discussion of the metabolic states in Section 6.2).

Chance et al. have also shown that the NAD(P)H fluorescence of tumor cells of ascites—peritoneal cavity fluid—was higher than that of normal cells (Chance and Thorell 1959). The perspective of autofluorescence for cancer diagnosis was demonstrated by Alfano et al. (1984) through a significant difference in the fluorescence spectrum between normal and malignant tissues, which was attributed to variations in the local environment of intrinsic fluorophores. Since then, there have been many autofluorescence studies on various cancer models raging from cell cultures to animals and human subjects.

13.2.1 In Vitro Cell Studies

Table 13.1 is a summary of representative autofluorescence studies on cancer cells over the last two decades. Suspensions and monolayer cultures of bladder, breast, cervix, esophagus, skin, and lung cancer cell lines were studied with fluorescence spectroscopy, confocal, or two-photon imaging methods. Emission in the NAD(P)H spectral range was most often studied, while Fp fluorescence and redox ratios were investigated in ~70% and 40% of these studies, respectively.

The majority of these studies consistently report increased NAD(P)H and decreased or unchanged Fp levels in cancer cells with only a few exceptions (Palmer et al. 2003; Pitts et al. 2001), which can be explained by unknown technical or biological model issues (Villette et al. 2006). Collectively, these results indicate that cancer cells can be recognized by an increased NAD(P)H/Fp ratio compared to normal cells.

The more aggressive/metastatic cancer cells have less NAD(P)H and, correspondingly, lower NAD(P)H/Fp (or higher Fp/NAD(P)H) compared to less aggressive/metastatic cancer cells. For example, Glassman et al. reported that HTB22 (a breast cancer cell line from adenocarcinoma metastatic site) has higher Fp/NAD(P)H ratio than HTB126 (a nonmetastatic clinical carcinosarcoma cell line) (Glassman et al. 1994). Similar results were obtained by Pradhan et al., who compared various metastatic and nonmetastatic cancer cell lines from melanoma, sarcoma, and lung cancer (Pradhan et al. 1995). In this study, the metastatic cell lines exhibited a decreased NAD(P)H fluorescence intensity and lifetime as compared to nonmetastatic cells. Using fluorescence spectroscopy measurements on suspensions of two bladder epithelial cancer cell lines, UM-UC-6 (wild-type p53) and a more drug-resistant T24 (mutated p53), Kirkpatrick et al. have shown a decrease in the NAD(P)H signal with an increased Fp/(Fp + NAD(P)H) redox ratio in T24 cell line compared to UM-UC-6 (Kirkpatrick et al. 2005).

Ostrander et al. distinguished breast cancer lines with different estrogen receptor (ER) status, namely, MDA-MB-231, MDA-MB-435, MDA-MB-468, BT-474, T47D, and MCF-7, using confocal imaging of NAD(P)H, Fp, and the redox ratio NAD(P)H/Fp (Ostrander et al. 2010). It was found that ER+ lines (BT-474, T47D, and MCF-7) have higher NAD(P)H/Fp ratios than the ER– lines (MDA-MB-231, MDA-MB-435, and MDA-MB-468). Put together, the data from Ostrander et al. and other published reports on the invasive potentials and metastatic potential of these cell lines (Freund et al. 2003; Kirschmann et al. 2000, 2002) show that the Fp/NAD(P)H redox ratios exhibit a good correlation with the degree of aggressiveness (Figure 13.2). It should be noted that cancer cell metabolism and presumably the exact invasiveness order depend on the cell culture conditions (e.g., availability of nutrients glucose, glutamine). Therefore, further studies are needed in order to investigate the correlation between cancer cell redox status and invasiveness under controlled microenvironment, especially with simultaneous measurement of redox ratios and invasive potential.

Several studies also showed that cancer transformations lead to the increase of free NAD(P)H and/or decrease of bound NAD(P)H in cells (Heikal 2010; Li et al. 2009a; Villette et al. 2006; Yu and Heikal 2009). The increase of free/bound NAD(P)H ratio was shown to cause the oxidative stress in $HepG_2$ liver cancer cells under Cd treatment, correlating with the increase of FAD/NAD(P)H (Yang et al. 2008). Nevertheless, the FAD/NAD(P)H seems to decrease, while free/bound NAD(P)H ratio increases in cancer cells compared to normal cells under resting conditions (Heikal 2010; Yu and Heikal 2009). These results suggest different relationships between NAD(P)H free/bound state and the redox status in cancer cells at the basal level and under treatment.

TABLE 13.1 Summary of Representative Autofluorescence Studies of Cancer Cells

Cancer/Precancer	Condition	Meas. Type[a]	NAD(P)H-Like	Fp Like	Redox Ratio	Others	References
Bladder epithelial cancer cell lines: T24 (p53 mutant) more drug resistant than UM-UC-6 (p53 wt)	Cell suspension	Spectra	UM-UC-6 > T24	→[a]	FP/(FP + NAD(P)H) ↑[a] in T24	—	Kirkpatrick et al. (2005)
Breast cell lines HTB125 (normal), HTB126 (carcinosarcoma), and HTB22 (adenocarcinoma pleural effusion)	Cells centrifuged and packed into cuvette	Spectra	↑, HTB22 > HTB126> HTB125, life time ↑, less blue shift than normal (bound NAD(P)H ↓)	↓,[a] HTB126 <HTB22 <HTB125	450/525nm ↑ HTB126 > HTB22 > HTB125	—	Glassman et al. (1994)
Breast epithelial cells (T47D, MDA231, MCF10, and MCF-10 ras transfected)	Cell suspension and monolayers	Spectra and two-photon imaging (2-P)	→ In cell suspension, but 2P of monolayer shows NAD(P)H ↓	→ In cell suspension	—	Tryptophan ↓ in cell suspension	Palmer et al. (2003)
Breast cancer Hs578T (HTB126) and normal Hs578Bst (HTB125) cells	Cell monolayers	2-P	↑, lifetime ↓, free/bound NAD(P)H ↑	↓	—	—	Yu and Heikal (2009); Heikal (2010)
Breast tumor cell lines (MDA-231, MDA-435, MDA-468, BT-20, BT-474, MDA-361, MCF-7, T47D, and ZR-75-1) and benign MCF-10A and normal HMEC cells	Cell monolayers	Confocal imaging	↑	↓	NAD(P)H/FP ↑ MDA231 ~ MDA435 < MDA468 < MCF7 ~ BT474	—	Ostrander et al. (2010)
Breast tumor cell lines (BT474, MDA-MB-231, MCF7, and SKBr3) and normal MCF-10A cells	Cell monolayers	Confocal imaging	↑	↓	NAD(P)H/FP ↑	—	Walsh et al. (2012)
Bronchial epithelial cells (BEAS-2B vs. transformed BEAS-2BNNK)	Suspension and monolayer	Spectra	Invisible in cell suspension; monolayer ↓	Invisible in cell suspension; monolayer ↓	—	Tryptophan → (cell suspension)	Pitts et al. (2001)
Cervical cancerous SiHa and normal ECT1 cells	Cell monolayer	2-P	NAD(P)H ↑, free/bound NAD(P)H ↑	—	Free NAD(P)H/tryptophan ↑, bound NAD(P)H/tryptophan →	—	Li et al. (2009a)

(Continued)

TABLE 13.1 (*Continued*) Summary of Representative Autofluorescence Studies of Cancer Cells

Cancer/Precancer	Condition	Meas. Type[a]	NAD(P)H-Like	Fp Like	Redox Ratio	Others	References
Esophageal squamous cell carcinoma (OE21) and adenocarcinoma in Barrett's esophagus (OE33) and normal human cells (NC)	Cell suspensions, cell monolayers	Spectra and imaging[b]	Total ↑, free ↑, bound →, free/bound ratio ↑ (OE33 > OE21 > NC)	—	—	—	Villette et al. (2006)
Rat fibroblast and H-ras-transfected fibroblast	Cell suspensions	Spectra	↑	—	—	—	Shirogane et al. (2010)
Normal human foreskin keratinocytes (HFKs) and papilloma virus immortalized HFK cells (HKc/DR)	Cell suspensions	Spectra	↑	↓	Fp/(Fp + NAD(P)H) ↓	Tryptophan ↑	Mujat et al. (2008)
Metastatic, nonmetastatic, and nontumorigenic cell lines from different species including melanoma, sarcoma, and lung	Cell suspension in PBS	Spectra	Nonmetastatic > metastatic > normal; lifetime metastatic < nonmetastatic/normal indicating free NAD(P)H ↑	—	—	Tryptophan ↑Nonmetastatic ≥ metastatic > normal; tryptophan/NAD(P)H ratio metastatic cells ≥ nonmetastatic	Pradhan et al. (1995)

Note: Only studies with a focus on NAD(P)H and/or Fp were included.

[a] Meas., measurement; →, about same; ↑, increase; ↓, decrease compared to the normal cells unless specified otherwise. Same for the rest of Tables 13.1 through 13.3.

[b] Imaging—one-photon fluorescence imaging. Same for the rest of Tables 13.1 through 13.3 unless specified otherwise.

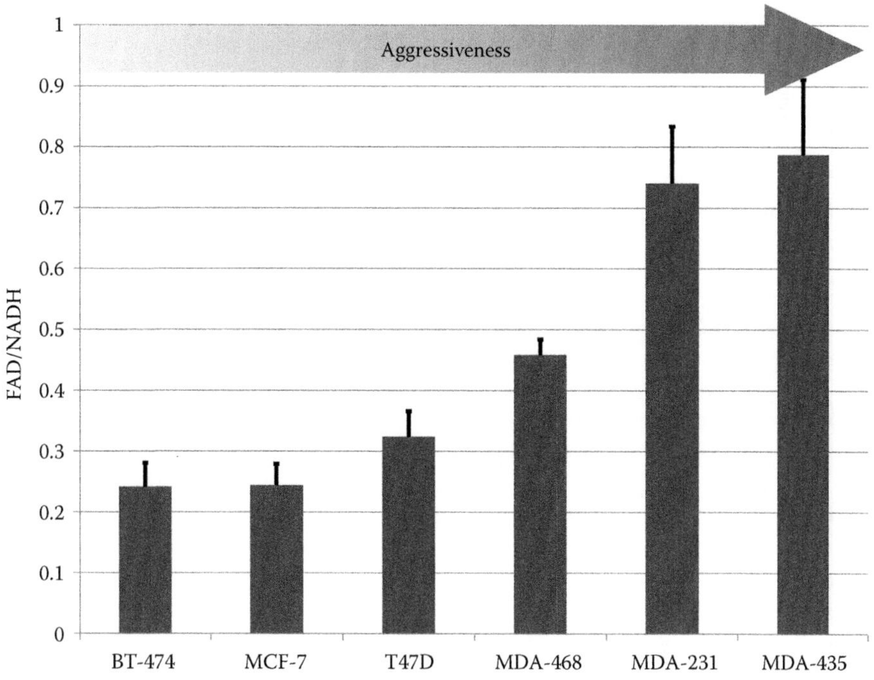

FIGURE 13.2 Redox ratio Fp/NADH versus cancer aggressiveness falling in the rank order MDA-MB-435 ≥ MDA-MB-231 > MDA-MB-468 > T47D > MCF-7 ≥ BT-474. (Redox ratio data taken from Ostrander, J.H. et al., *Cancer Res.*, 70(11), 4759, 2010.)

13.2.2 Animal Models

In vivo and ex vivo autofluorescence studies of various precancers and cancers in animal models are summarized in Table 13.2. Similar to the in vitro cell models, majority of these studies focused on NAD(P)H signals rather than the Fp and redox ratio quantification. Changing patterns of the NAD(P)H and Fp fluorescence signals were less consistent among these studies compared to the cell culture models. This lack of consistency is probably due to more technical differences (e.g., selection of excitation/emission wavelength, spectra vs. imaging) and biological challenges posed by tissue studies, such as tissue and autofluorescence heterogeneity, penetration depth, tissue preparations, and other physiological variables.

The most consistent results showing lower NAD(P)H signals in tumors were generated by two-photon imaging studies, which allow better control of the probing depth in tissues and exclude the signals from the submucosal or subepithelium compartments (Skala et al. 2005, 2007; Zheng et al. 2011). For example, Zheng et al. studied oral precancer in hamster cheek pouch models by time-resolved two-photon fluorescence spectroscopy imaging to measure the total NAD(P)H and the free/bound NAD(P)H ratio (Zheng et al. 2011). Both precancerous and cancerous tissues exhibited lower NAD(P)H signal than normal tissues, whereas the higher-grade tumors revealed lower NAD(P)H autofluorescence intensity than the lower-grade tumors (Figure 13.3a). These observations were attributed to the thickening of precancerous epithelium and abnormal metabolism in precancer. As the precancer progresses to thicker epithelium, less excitation light reaches the stratum basale due to enhanced scattering and absorption. An opposite trend was shown for free/bound NAD(P)H ratio, which increased with the precancer grades (Figure 13.3b).

In order to discover tissue features that may serve as biomarkers for tumor aggressiveness, many studies utilized cryogenic NAD(P)H/Fp fluorescence imaging, also known as the Chance redox scanner (Gu et al. 2002; Li et al. 2009b; Quistorff et al. 1985). The scanner allows 3D tissue redox state

TABLE 13.2 Summary of Representative Autofluorescence Studies in Animal Models of Cancers and Precancers

Cancer/Precancer	Animal	Condition	Meas. Type	NAD(P)H Like	Fp Like	Redox Ratio	Others	References
Breast cancer	Mouse xenograft (HT1080)	Ex vivo (frozen tissue)	Redox scanner	↑ 14× Normal	↑ 3× Normal	NAD(P)H/ (NAD(P)H + Fp) ↑	Cathepsin B ↑ 7×, ICG ↑ (830 nm) 14× normal	Zhou et al. (2003)
Breast carcinoma in situ	Mouse transgenic model	Ex vivo (fixed tissue section)	2-P	↑, Lifetime ↑	↑, Lifetime ↑	—	—	Conklin et al. (2009)
Breast cancer	Mouse xenograft	Ex vivo	Redox scanner	Tumor core ↓ with increased invasive potential	Tumor core ↑	Tumor core ↑ with increased invasive potential	Tumor heterogeneity	Xu et al. (2010)
Oral neoplasia	Hamster buccal pouch model (DMBA-induced carcinogenesis)	Ex vivo (excised frozen tissue unfrozen for observation, 1 mm thick)	Spectra	↑ (Ex 330 nm) as the normal tissue progressed to dysplasia and tumor	—	—	380 nm ↓	Chen et al. (1998)
Oral neoplasia	Hamster buccal pouch model	Ex vivo (same as given earlier, <5 min, thick 0.8–1.2 mm)	Spectra	↓ (Ex 360) as the normal tissue progressed to dysplasia and tumor	—	—	Ex 320/Em 470 ↑, Ex 20/Em 385 ↓, Ex 360/Em 640 ↑	Wang et al. (1999a)
Oral neoplasia	Hamster cheek pouch model	In vivo (surface)	Spectra (EEM)	↑ (Ex 360–370/Em 450) most diagnostic (neoplasia sensitivity and specificity >90%)	↓ (Ex 460–470/ Em 530)	—	↑ (Porphyrin, Ex 410/ Em 630)	Coghlan et al. (2000)
Oral neoplasia	Hamster cheek pouch model	In vivo (surface)	Spectra	Ex 330/Em 460 ↑, normal < dysplasia < squamous cell carcinoma	—	—	380 nm ↓, normal > Dysplasia > squamous cell carcinoma; I380/ I460 as an indicator	Tsai et al. (2001)
Oral precancer and cancer	Hamster cheek pouch model	Ex vivo	2-P	↓ Normal > preca. > ca.	—	—	Keratin ↓	Skala et al. (2005)

Oral precancer	Hamster cheek pouch model	In vivo	2-P	Bound ↓, bound lifetime ↓ (normal > preca.)	Bound ↓, bound lifetime ↑ (normal < preca.)	Fp/NAD(P)H →	Heterogeneity ↑, lifetime intracellular variabilities of the ratio, bound Fp, and bound NAD(P)H ↑ (normal < preca.)	Skala et al. (2007)
Oral precancer	Hamster cheek pouch model	In vivo	2-P	↓, Normal > low > mid > high grade; free/bound ratio ↑, normal < low < mid < high grade	—	—	—	Zheng et al. (2011)
Oral precancer and cancer	Hamster buccal pouch model	Ex vivo	Spectra (Ex 330)	↑; Lifetime dysplasia < carcinoma in situ < carcinoma < normal	—	—	380/collagen↓, 635/porphyrin ↑	Farwell et al. (2010)
Melanoma	Mouse xenograft (K1735P, 3–5 mm in ear)	In vivo	Spectra and imaging	↑ (Ex 365/430–440)	Ex 430/Em 400–600 ↑	—	↑ 630 nm (porphyrin) and 670 nm (food related)	Sterenborg et al. (1995)
Melanoma	Mouse xenografts (A375P, A375M, A375P10, A375P5, C8161)	Ex vivo (frozen tissue)	Redox scanner	Tumor core ↓ with increased invasive potential	Tumor core ↑	Tumor core ↑ with increased invasive potential	Tumor heterogeneity more in metastatic tumors	Li et al. (2009c)
Melanoma	Mouse xenografts	Ex vivo	Redox scanner	Average ↑ compared to muscle; tumor core ↓	Tumor core ↑	Tumor core ↑ correlate with metastatic potential	Tumor heterogeneity more in metastatic tumors	Li et al. (2007)
Skin cancer	Mouse (nude)	In vivo	A confocal reflectance/fluorescence tomography	↑	—	—	—	Peng et al. (2012)
Pancreatic acinar tumor	Rat (graft)	Ex vivo	Spectra	↓	—	—	—	Keller et al. (1996)
Pancreas precancer	PTEN null transgenic mouse model	Ex vivo (frozen tissue)	Redox scanner	↓	↓	Fp/(Fp + NAD(P)H) ↑	Redox ratio heterogeneity ↑ in precancer	Xu et al. (2011, 2013b)
Squamous cell carcinoma	Mouse	In vivo (window chamber)	Multifunctional microscope	—	—	Fp/NAD(P)H vary with time (cycling) and location, weakly associated with hemoglobin saturation	—	Skala et al. (2010)

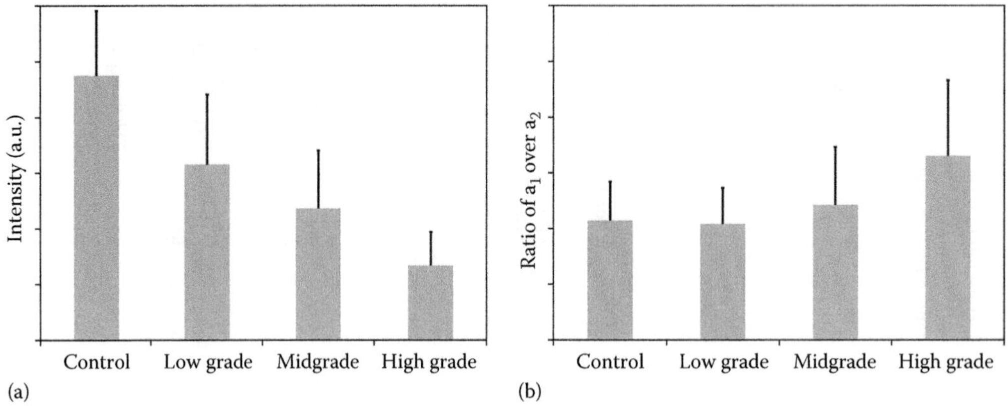

FIGURE 13.3 NAD(P)H intensity (a) and free-to-bound NAD(P)H ratio (b) versus the grade of oral precancer in check pouch. (Reproduced from Zheng, W. et al., *Proc. SPIE*, 7890, 789001, 2011. With permission.) Control, normal (N = 20); low-grade precancer, mild dysplasia (N = 30); midgrade precancer, moderate dysplasia (N = 7); high-grade cancer, severe dysplasia and carcinoma in situ (N = 7).

imaging at submillimeter resolution (down to 50 μm) by tissue snap freezing, sectioning, and raster scanning at low temperature of liquid nitrogen. The criterion of tumor aggressiveness is based on the invasive potential or metastatic potential of the cancer rather than on tumor growth rate. For example, the melanoma xenograft C8161 is more metastatic/invasive and faster growing than the relatively indolent A375P, while the invasive breast cancer xenograft MDA-MB-231 has lower growth rate than the less-metastatic MCF-7. Redox scanning studies revealed both oxidized and reduced regions in metastatic tumors and that more metastatic/invasive xenografts had more oxidized and heterogeneous redox state than less-metastatic/invasive xenografts regardless of the growth rates (Li et al. 2007, 2009c; Xu et al. 2010) (Figure 13.4). A statistically significant correlation (Figure 13.5) was demonstrated between the redox status in oxidized regions and the invasive potentials in five melanoma (Li et al. 2009c) and three breast cancer xenografts (manuscript in preparation) with different levels of aggressiveness. Note that these results are consistent with the correlation of Fp redox ratio (Fp/[Fp + NAD(P)H]) with cancer invasiveness in cell culture (Figure 13.2). Similar correlation was also demonstrated for HCT116 colon cancer xenografts with different p53 status. In these studies, the p53-null HCT116 exhibited more intratumor heterogeneity and more oxidized redox core regions than the p53-wild-type HCT116 tumors (Xu et al. 2013a). The importance of the redox state in cancer transformation was further established in a transgenic model of premalignant pancreas with the pancreatic specific deletion of PTEN gene encoding phosphatase and tensin homolog (Xu et al. 2011, 2013b). PTEN is a key inhibitory regulator of the PI3K/Akt pathway, which plays an important role in pancreatic cancer development and drug resistance, and therefore is a potential target for chemotherapy (Asano et al. 2004; Yan et al. 2006). In this transgenic model of pancreas, the PTEN deletion infrequently resulted in cancer transformation and a shift of the redox state to a more oxidized and heterogeneous as compared to normal pancreas. Collectively, these studies provided the initial evidences that support tissue redox state as an important mediator for tumor progression.

The correlation of redox state with the aggressiveness of tumors or premalignancies appeared even more pronounced when the heterogeneity of redox state in tissues was taken into account. The whole-tumor averaging of redox state in five melanomas produced a less significant correlation with their invasive potentials compared to the redox states in the oxidized tumor cores (Li et al. 2009c). In the aforementioned study of HCT116 mouse xenografts, the global averaging over the whole tumor did not produce any significant difference in NAD(P)H, Fp, and redox ratios between the p53 null and wild type. This finding appeared to be consistent with a previous PET study of the same models that

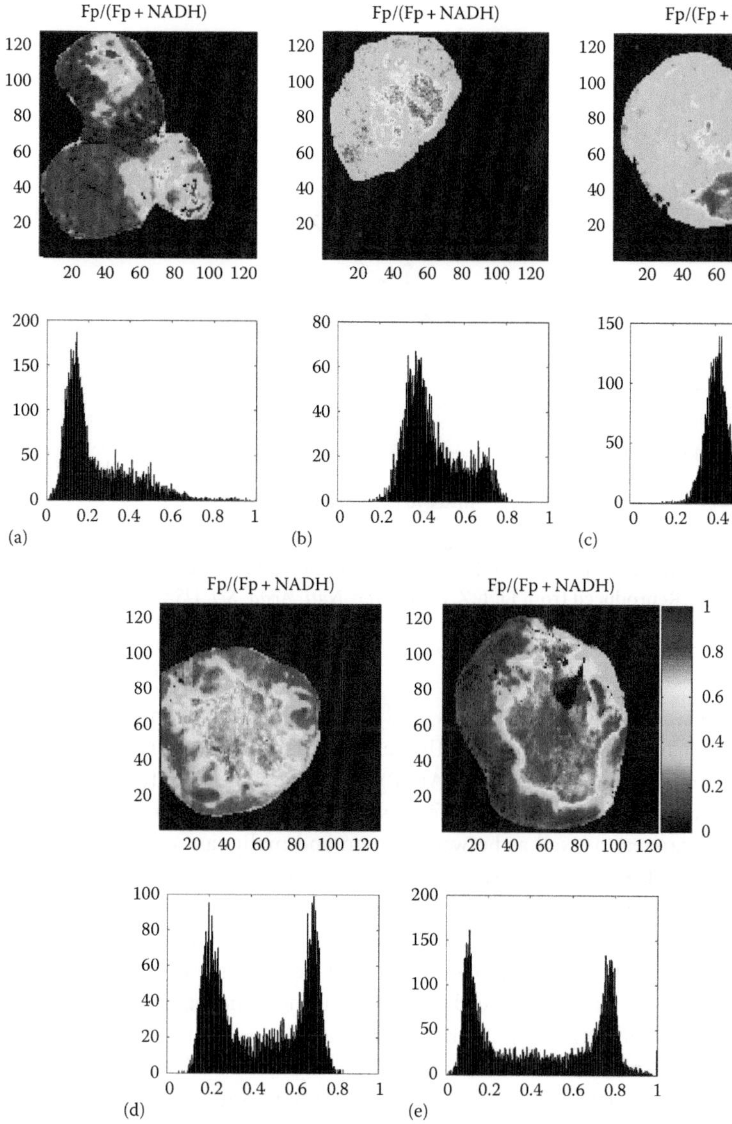

FIGURE 13.4 Images of the redox ratios Fp/(Fp + NAD(P)H) for five melanoma xenografts with increasing aggressiveness from (a) to (e). (Reproduced from Li, L.Z. et al., *Proc. Natl. Acad. Sci. USA*, 106(16), 6608, 2009c. With permission.)

showed no metabolic differences (Wang et al. 2007). However, consideration of the in-plane heterogeneity and tissue signal variation across the tissue thickness and imaging depth led to the identification of significant differences in redox state.

13.2.3 Human Patients

Tissue autofluorescence has been tested as diagnostic tool for a number of cancer conditions, including breast, cervical, lung, oral, and colon cancers (Table 13.3). These trials demonstrated high specificity and sensitivity of tissue autofluorescence for the differentiation between cancerous, precancerous, benign, and normal tissues. For example, the 325 nm excited autofluorescence from clinical biopsy

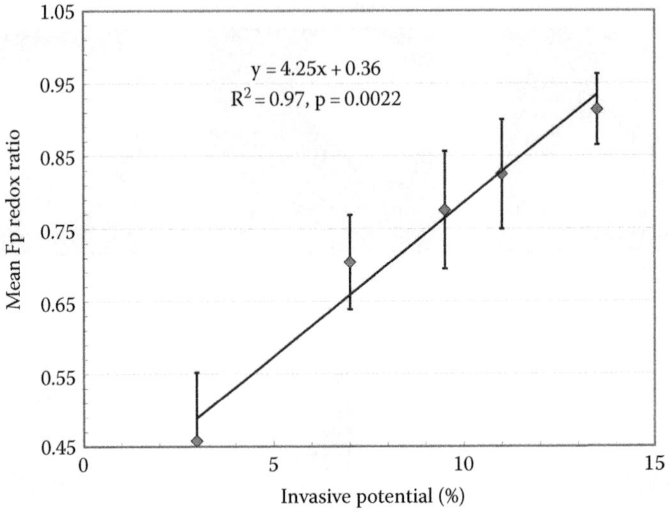

FIGURE 13.5 The redox ratios Fp/(Fp + NAD(P)H) in oxidized tumor cores correlate with the invasive potential of five melanoma cell lines. (Reproduced from Li, L.Z. et al., *Proc. Natl. Acad. Sci. USA*, 106(16), 6608, 2009c. With permission.)

samples, isolated from breast cancer patients, was used to distinguish between normal tissues, benign, and malignant tumors (Chowdary et al. 2009). The relative quantities of intrinsic fluorophores were determined using spectral deconvolution. The NAD(P)H versus collagen content plot yielded a sensitivity and specificity that was greater than 93% for discriminating between normal and pathological conditions (benign and cancer), as well as between benign and malignant conditions. The malignant tissues had higher NAD(P)H content, followed by normal tissues and then benign tissues. In addition, the bound/free NAD(P)H ratio increased from normal to pathological conditions with relatively higher values measured from the benign breast lesions than from the malignant tissues.

Delgado et al. used the 365 nm excited autofluorescence of freshly resected mammary tissues, where the collagen contribution was minimized, to differentiate normal and pathological tissues (Delgado et al. 2003). The predominant NAD(P)H autofluorescence at 465 nm was found to decrease in the following order: dysplasia > breast lesion > normal. It was also observed that the NAD(P)H signal was reduced by 10%–20% during the 2 min data acquisition and the decay time constants were different for cancer lesion, dysplasia, and normal tissues. By comparison, Xu et al. used the Chance redox scanner to image the NAD(P)H and Fp in breast biopsy specimens frozen within 15 min after the tissue removal from body (Xu et al. 2013c,d). Although the difference in intrinsic NAD(P)H signal was invariant with the tissue type, up to 10-fold higher Fp autofluorescence and significantly higher redox ratio Fp/(Fp + NAD(P)H) were observed in cancerous tissues than in the corresponding periphery normal tissues.

Among the studies reviewed and summarized in Table 13.3, the NAD(P)H signal was consistently lower mainly in oral cancers, while the Fp signals decreased in all cancer types except for three studies. Two of them showed increase (Pantalone et al. 2007; Xu et al. 2013c,d), while the third one showed little change in Fp signals (Izuishi et al. 1999).

It is important to realize that autofluorescence may originate from multiple tissues or tissue compartments in clinical samples. For example, at endoscopic observation of autofluorescence in colon, cross talk between mucosal and submucosal fluorescence should be taken into consideration (Imaizumi et al. 2012). It is also known that subepithelial stroma may have stronger fluorescence than epithelia tissue (Pavlova et al. 2008; Ramanujam et al. 2001). While the thickness of epithelium varies (tens to hundreds of micrometer), the penetration depth of the 370 nm excitation light may reach as far as 400 μm

TABLE 13.3 Summary of Representative Human Autofluorescence Studies of Cancers and Precancers

Cancer/Precancer	Condition	Meas. Type	NAD(P)H Like	Fp Like	Redox Ratio	Others	References
Breast carcinoma and skin cancers (squamous cell carcinoma [SCC] and basal cell carcinoma [BCC])	Ex vivo (surgically resected ~1–1.5 h after tissue removal)	Selected Ex and Em band	Breast cancer: both dysplasia and cancer > normal tissue; Skin cancer: BCC ↓, SCC mixed	—	—	NAD(P)H ↓ with time within minutes	Delgado et al. (2003)
Breast cancer (infiltrating ductal carcinoma [IDC], fibroadenoma—benign, normal)	Ex vivo (biopsy <0.5 h tissue removal)	Spectra (Ex 325)	Bound + free NAD(P)H level: ↑ IDC > normal > benign; NAD(P)H bound/free ratio: benign > IDC > normal	—	—	Em 460: normal > IDC > benign; Em 390 ↑ vs. normal	Chowdary et al. (2009)
Breast cancer (tumor vs. surrounding normal tissue)	Ex vivo (biopsy samples frozen <15 min tissue removal)	Redox scanner	→	↑	Fp/(Fp + NAD(P)H) ↑	Tissue heterogeneity ↑	Xu et al. (2013c,d)
Colon cancer	In vivo	Imaging	—	—	—	Em 400–700: adenoma > hyperplastic (benign) ≥ polyps normal	Wang et al. (1999b)
Colon	Ex vivo (frozen sections), in vitro (HPLC)	HPLC, fluor. microscopy	—	→	—	—	Izuishi et al. (1999)
Colon	Ex vivo (slice section)	Imaging	↑ In adenoma	—	—	Ex 365/Ex 405 ↑, refl 550/fluo—405ex ↑	Imaizumi et al. (2011)
Colon	Fresh biopsy (0.3–0.9 mm thick, <1 h)	2-P imaging	—	—	NAD(P)H/Fp normal < precancer < cancer	Nuclear cytoplasmic ratio ↑	Zhuo et al. (2011)
Colon	Ex vivo (in 4°C PBS, sliced specimen at 3°C for imaging, <2 h after resection)	Imaging	→	↓(Ex 405)	—	Ex 365/Ex 405 ↑	Imaizumi et al. (2012)

(*Continued*)

TABLE 13.3 (*Continued*) Summary of Representative Human Autofluorescence Studies of Cancers and Precancers

Cancer/Precancer	Condition	Meas. Type	NAD(P)H Like	Fp Like	Redox Ratio	Others	References
Colon	Ex vivo (biopsy or resected tissue, <0.5 h, moistured with saline); spectral stable at least till 1 h	Spectra	↓	↓ Proportionally as NAD(P)H	? (Fp/NAD(P)H might ↑)	—	Luo et al. (2012)
Esophageal dysplasia	In vivo	Endoscopy	↓ (Ex 410)	—	—	—	Panjehpour et al. (1996)
Esophageal squamous cell cancer, adenocarcinoma of the esophagus, and adenocarcinoma of the stomach	In vivo	Endoscopy	—	↓	—	630 nm ↓	Mayinger et al. (2001)
Laryngeal precancer and cancer	In vivo	Endoscopy	—	Green light ↓	—	—	Malzahn et al. (2002)
Laryngeal	Ex vivo resected tissue and frozen	Endoscopy and fluorescence microscopy	—	Green light ↓	—	Collagen ↓	Palasz et al. (2003)
Laryngeal	In vivo	Autofluorescence endoscopy	→	515 nm channel decreased by 70%	—	—	Arens et al. (2006)
Laryngeal	In vivo	Microlaryngoscopy	—	Green light ↓	—	—	Mehlmann et al. (1999)
Lung bronchial	In vivo	Spectra	↓ (Ex 405)	↓	—	—	Hung et al. (1991)
Lung bronchial	In vivo	Imaging	—	↓	—	Red light ↓	Wagnieres et al. (2002)
Lung adenocarcinoma (LAC) and lung squamous cell carcinoma (SCC)	Ex vivo (frozen $3 \times 3 \times 1$ mm³ from tissue bank, 10 min in room temp.)	Imaging	—	—	SCC: cellular Fp/NAD(P)H ↓ (data limited)	Collagen ↓	Wang et al. (2010)
Lymph node biopsies from patients with adenopathy	Ex vivo	Spectra and imaging	↑ Connective tissue compared to lymphocytes	—	—	—	Rigacci et al. (2000)

Metastatic lymph nodes of colorectal and gastric tumors	Ex vivo	Spectra and imaging	—	Green ↑	—	Red autofluor. ↑, collagen contribution ↓	Pantalone et al. (2007)
Nasopharyngeal carcinoma	Ex vivo	Spectra	Ex 340/Em 455 ↓	—	—	NAD(P)H/collagen ↑, 340/380 (collagen) ↓	Lin et al. (2012)
Oral	In vivo	Spectra	—	↓	—	—	Betz et al. (2002)
Oral	In vivo	Spectra	—	—	—	Red/blue light ratio ↑	Gillenwater et al. (1998)
Oral	Ex vivo (excised frozen tissue unfrozen for observation, 1 mm thick)	Spectra	↓ (Ex 330)	—	—	380 nm ↓	Chen et al. (1998)
Oral	Ex vivo (biopsy)	Spectra	↓	↓	—	—	Majumder et al. (1998)
Oral	In vivo	Spectra	↓	↓	—	Tryptophan →	Pauli et al. (2007)
Oral	In vivo	Spectra	—	—	—	Ex 400/Ex 350 ↓ (Em 472)	Heintzelman et al. (2000b)
Oral	In vivo	Imaging	—	Green light ↓	—	—	Betz et al. (1999)
Oral	In vivo	Spectra (Ex 320)	↑	—	—	Collagen ↑	Haris et al. (2009)
Ovarian	In vitro	Spectra (Ex 325)	↑	—	—	Collagen ↓	Kamath et al. (2009)
Ovarian	Ex vivo	Spectra (measurement depth 50–120 μm)	↓	↓	—	Tryptophan ↑	George et al. (2012)
Ovarian	Ex vivo	Spectra	Em 400–600 ↓	—	—	—	Renkoski et al. (2012)
Ovarian	Ex vivo	2-P	—	—	Fp/(Fp + NAD(P)H) ↑ from normal to low risk, normal to high risk, to cancer	Collagen structure change	Kirkpatrick et al. (2007)

(Continued)

TABLE 13.3 (*Continued*) Summary of Representative Human Autofluorescence Studies of Cancers and Precancers

Cancer/Precancer	Condition	Meas. Type	NAD(P)H Like	Fp Like	Redox Ratio	Others	References
Skin: basal cell carcinomas (BCCs)	In vivo and ex vivo	2-P	Lifetime tend to ↑	Lifetime tend to ↑	—	—	Patalay et al. (2011)
Skin	In vivo	Spectra	↓	—	—	—	Stender and Wulf (2001)
Skin	In vivo	Spectra	Ex 385/Em 450–700, BCC ↓, SCC ↑	Ex 405/Em 450–700, keratoacanthoma ↑	—	Ex 405 BCC advanced stage higher than initial stage	Borisova et al. (2009)
Various cancer (stomach, esophagus, tongue, mandible, and bladder)	Ex vivo	Spectra	—	560 nm peak	—	Peaks at 630 and 690 nm ↑	Yang et al. (1987)
Oral and oropharyngeal	In vivo	Spectra	↓	↓	—	—	Dhingra et al. (1996)
Cervical lesions and Barrett's esophagus	In vivo	Fluoresence and reflectance spectra	↑	—	—	Collagen ↓	Georgakoudi et al. (2002)

(Imaizumi et al. 2012). Therefore, the decreased Fp intensity in cancer-transformed tissues may be partially explained by the increased thickness of epithelium in addition to the enhanced blood absorption and scattering. It is important to consider these factors while designing wide-field, single-photon fluorescence microscopy measurements on tissues. By comparison, confocal or two-photon fluorescence microscopy with optical sectioning capability provides a better alternative for tissue imaging.

It is worth mentioning that the redox ratio values in the studies reviewed were controversial. While some demonstrated elevated Fp/(Fp + NAD(P)H) or reduced NAD(P)H/Fp in cancerous tissues compared to normal or low-risk tissues (Kirkpatrick et al. 2007; Xu et al. 2013c,d), others reported increase of the NAD(P)H/Fp redox ratio from normal tissues to precancer and/or cancer (Wang et al. 2010; Zhuo et al. 2011). Pavlova et al. reported decrease of Fp/(Fp + NAD(P)H) with poorly differentiated carcinoma (Pavlova et al. 2008), which is commonly considered as an indicator of increased tumor aggressiveness.

13.3 Discussion on Some Biological and Technical Issues

Some inconsistencies among autofluorescence studies on animal and human specimens indicate the technical and biological complexities that exist in the field. Here, we briefly discuss some issues and approaches to fix them.

13.3.1 Ratiometric Approach

Autofluorescence of NAD(P)H and Fp may vary greatly among normal cells, either normal or cancer (Heintzelman et al. 2000a), presumably due to mitochondria density variation. Ratiometric approach, commonly used in research to reduce experimental errors and artifacts, has the advantage of being insensitive to cell/mitochondria densities. Chance first employed the ratio Fp/NAD(P)H to represent mitochondrial metabolic state and showed that the ratio is insensitive to hemodynamic factors, such as blood volume (Chance et al. 1979). Later, Fp/(Fp + NAD(P)H) and NAD(P)H/(Fp + NAD(P)H) ratios were introduced and have been commonly used as sensitive indices of the metabolic state of mitochondria. Other less popular but instrumental ratios include tryptophan/NAD(P)H (Pradhan et al. 1995), free/bound NAD(P)H (Villette et al. 2006), and the signal ratios at different excitation or emission wavelengths (Huang et al. 2002; Zheng et al. 2001). Ratiometric approach may also help avoiding some artifacts due to, for example, thickening of epithelium in neoplasia and increased blood volume that lower NAD(P)H and Fp fluorescence in cancerous/precancerous epithelium (Galeotti et al. 1970; Zheng et al. 2011).

13.3.2 Meaning of Redox Ratio

It is regarded that "the ratio of NADH to FAD is a measurement of the balance between glycolysis (seen in tumor cells) and oxidative phosphorylation (seen in untransformed cells)" and the redox ratio NAD(P)H/FAD is "the ratio of glycolysis to oxidative metabolism" (Walsh et al. 2012). However, since both mitochondrial and cytosolic NAD(P)H can contribute to the cellular autofluorescence, the exact weighing of the signal from two compartments may be cell specific. As a result, whether an increase of NAD(P)H and NAD(P)H/Fp ratio may indicate a shift of mitochondrial respiration to aerobic glycolysis remains an open question. In addition, the correlation of the redox ratio Fp/NAD(P)H or NAD(P)H/Fp with metabolic activity remains controversial and would benefit from further studies. Importantly, what holds true for normal cells may not be applicable to tumor cells.

13.3.3 Tumor Heterogeneity and Bioenergetics

The association of tissue heterogeneity (in structure, function, metabolism, and genetics) with malignancy is well established. The more aggressive tumors seem to exhibit an increased heterogeneity such that a metastatic tumor would reveal a tumor core with high signals in the Fp channel and more oxidized redox state Fp/(Fp + NAD(P)H) (Li 2012) with the underlying biological mechanism under investigation. The redox signal in this oxidized region seems to predict the aggressiveness of tumors better than the global averages (Li et al. 2009c; Xu et al. 2013a). Thus, it is important to take into account the tissue heterogeneity when trying to identify biomarkers for cancer aggressiveness and understand the tumor progression to metastasis.

The inconsistency among various studies may be attributed in part to the variations of cell type, size, and density as well as the nuclei-to-cytoplasmic ratio. In order to elucidate the bioenergetics of cancer tissues using autofluorescence, it would be useful to account for those variations toward a meaningful comparison with in vitro cell culture studies.

13.3.4 Temporal Changes and Other Issues

Another issue that is often overlooked is autofluorescence fluctuation with time. For example, significant changes in the NAD(P)H autofluorescence of pulmonary melanoma metastases were observed over the course of 25 days (Hua et al. 2012). Similarly, Dewhirst and coworkers demonstrated the dynamic nature of in vivo blood hemodynamics and metabolic signals (FAD and NAD(P)H) using a window chamber tumor model in mice (Skala et al. 2010).

The metabolic state or intermediary metabolism of cancer cells is also dependent on microenvironment and nutrition supply. Based on our own experience, it is rather hard to compare results between cell, animal, and human studies because of different cancer cell microenvironment parameters (e.g., pH, pO_2, oxygen, and nutritional supply), which are key for meaningful interpretation of those intrinsic metabolic biomarkers.

Contribution of other fluorophores (e.g., tryptophan, collagen, elastin, and lipofuscin) to tissue autofluorescence should also be considered in both experimental design and data interpretation (see Chapter 2 for the review of common intrinsic fluorophores). Lipofuscin is particularly a concern for autofluorescence studies of metabolic activities due to its broad spectra in visible range and the potential spectral overlap with that of Fp as shown in neurons and stem cells. Although its presence in tumors has been reported (Almeida et al. 2013; Sehdev et al. 2001; Shields et al. 1976; Son et al. 2004; Watson et al. 2013), the exact nature and role of lipofuscin in tumor progression remains far from being understood.

13.4 Summary

There is a growing interest in developing autofluorescence-based diagnostic tools for cancer. Many studies on cells, animal models, and human specimens have demonstrated that autofluorescence, dominated by the NAD(P)H and Fp emission as metabolic biomarkers, may provide useful information for differentiating among normal, precancerous, and cancerous tissues. However, several technical and biological challenges (hence opportunities) remain toward achieving consistent results. For example, researchers may devote some attention to intratumor heterogeneity, temporal evolution, microenvironment, nutritional status, and other physiological and experimental factors in order to minimize artifacts, have meaningful data interpretation, and enhance confidence level in autofluorescence-based diagnostics of cancer, especially at early stages.

Acknowledgments

The authors would like to thank the general guidance and support from the late Dr. Britton Chance, Dr. Shoko Nioka, Dr. Jerry D. Glickson, and the editors of this book. Thanks also to the support of the National Institutes of Health (NIH) grant R01CA155348 (Lin Z. Li.).

References

Adler, V., Z. M. Yin, K. D. Tew, and Z. Ronai. 1999. Role of redox potential and reactive oxygen species in stress signaling. *Oncogene* 18 (45): 6104–6111.

Alfano, R., D. Tata, J. Cordero et al. 1984. Laser induced fluorescence spectroscopy from native cancerous and normal tissue. *IEEE Journal of Quantum Electronics* 20 (12): 1507–1511.

Almeida, A., S. Kaliki, and C. L. Shields. 2013. Autofluorescence of intraocular tumours. *Current Opinion in Ophthalmology* 24 (3): 222–232.

Arens, C., T. Dreyer, H. Glanz, and K. Malzahn. 2004. Indirect autofluorescence laryngoscopy in the diagnosis of laryngeal cancer and its precursor lesions. *European Archives of Oto-Rhino-Laryngology and Head and Neck* 261 (2): 71–76.

Arens, C., D. Reussner, H. Neubacher, J. Woenckhaus, and H. Glanz. 2006. Spectrometric measurement in laryngeal cancer. *European Archives of Oto-Rhino-Laryngology and Head and Neck* 263 (11): 1001–1007.

Asano, T., Y. Yao, J. Zhu et al. 2004. The PI 3-kinase/Akt signaling pathway is activated due to aberrant Pten expression and targets transcription factors NF-κB and c-Myc in pancreatic cancer cells. *Oncogene* 23 (53): 8571–8580.

Banerjee, R., ed. 2007. *Redox Biochemistry*. Hoboken, NJ: John Wiley & Sons.

Becker, D. F., W. Zhu, and M. A. Moxley. 2011. Flavin redox switching of protein functions. *Antioxidants and Redox Signaling* 14 (6): 1079–1091.

Betz, C. S., M. Mehlmann, K. Rick et al. 1999. Autofluorescence imaging and spectroscopy of normal and malignant mucosa in patients with head and neck cancer. *Lasers in Surgery and Medicine* 25 (4): 323–334.

Betz, C. S., H. Stepp, P. Janda et al. 2002. A comparative study of normal inspection, autofluorescence and 5-ALA-induced PPIX fluorescence for oral cancer diagnosis. *International Journal of Cancer* 97 (2): 245–252.

Borisova, E., D. Dogandjiiska, I. Bliznakova et al. 2009. Multispectral autofluorescence diagnosis of non-melanoma cutaneous tumors. *Proceedings of SPIE* 7368: 736823.

Chance, B. 1991. Optical method. *Annual Review of Biophysics and Biophysical Chemistry* 20: 1–28.

Chance, B. and H. Baltscheffsky. 1958. Respiratory enzymes in oxidative phosphorylation VII. Binding of intramitochondrial reduced pyridine nucleotide. *Journal of Biological Chemistry* 233 (3): 736–739.

Chance, B., P. Cohen, F. Jobsis, and B. Schoener. 1962. Intracellular oxidation-reduction states in vivo. *Science* 137 (3529): 499–508.

Chance, B. and F. Jobsis. 1959. Changes in fluorescence in a frog sartorius muscle following a twitch. *Nature* 184 (4681): 195–196.

Chance, B., B. Schoener, R. Oshino, F. Itshak, and Y. Nakase. 1979. Oxidation-reduction ratio studies of mitochondria in freeze-trapped samples. NADH and flavoprotein fluorescence signals. *Journal of Biological Chemistry* 254 (11): 4764–4771.

Chance, B. and B. Thorell. 1959. Localization and kinetics of reduced pyridine nucleotide in living cells by microfluorometry. *Journal of Biological Chemistry* 234 (11): 3044–3050.

Chen, C. T., H. K. Chiang, S. N. Chow et al. 1998. Autofluorescence in normal and malignant human oral tissues and in DMBA-induced hamster buccal pouch carcinogenesis. *Journal of Oral Pathology and Medicine* 27 (10): 470–474.

Chiarugi, A., C. Dolle, R. Felici, and M. Ziegler. 2012. The NAD metabolome—A key determinant of cancer cell biology. *Nature Reviews Cancer* 12 (11): 741–752.

Chowdary, M. V., K. K. Mahato, K. K. Kumar et al. 2009. Autofluorescence of breast tissues: Evaluation of discriminating algorithms for diagnosis of normal, benign, and malignant conditions. *Photomedicine and Laser Surgery* 27 (2): 241–252.

Christofk, H. R., M. G. Vander Heiden, M. H. Harris et al. 2008. The M2 splice isoform of pyruvate kinase is important for cancer metabolism and tumour growth. *Nature* 452 (7184): 230–233.

Coghlan, L., U. Utzinger, R. Drezek et al. 2000. Optimal fluorescence excitation wavelengths for detection of squamous intra-epithelial neoplasia: Results from an animal model. *Optics Express* 7 (12): 436–446.

Conklin, M. W., P. P. Provenzano, K. W. Eliceiri, R. Sullivan, and P. J. Keely. 2009. Fluorescence lifetime imaging of endogenous fluorophores in histopathology sections reveals differences between normal and tumor epithelium in carcinoma in situ of the breast. *Cell Biochemistry and Biophysics* 53 (3): 145–157.

Delgado, J. A., L. Anasagasti, I. Quesada, J. C. Cruz, and A. Y. Joan. 2003. Ex-vivo autofluorescence measurements of human tissues. *AIP Conference Proceedings* 682: 30–37.

Dhingra, J. K., D. F. Perrault, Jr., K. McMillan et al. 1996. Early diagnosis of upper aerodigestive tract cancer by autofluorescence. *Archives of Otolaryngology—Head & Neck Surgery* 122 (11): 1181–1186.

Duysens, L. N. and J. Amesz. 1957. Fluorescence spectrophotometry of reduced phosphopyridine nucleotide in intact cells in the near-ultraviolet and visible region. *Biochimica et Biophysica Acta* 24 (1): 19–26.

Farwell, D. G., J. D. Meier, J. Park et al. 2010. Time-resolved fluorescence spectroscopy as a diagnostic technique of oral carcinoma: Validation in the hamster buccal pouch model. *Archives of Otolaryngology—Head & Neck Surgery* 136 (2): 126–133.

Freund, A., C. Chauveau, J. P. Brouillet et al. 2003. IL-8 expression and its possible relationship with estrogen-receptor-negative status of breast cancer cells. *Oncogene* 22 (2): 256–265.

Galeotti, T., G. D. van Rossum, D. H. Mayer, and B. Chance. 1970. On the fluorescence of NAD(P)H in whole-cell preparations of tumours and normal tissues. *European Journal of Biochemistry* 17 (3): 485–496.

Garber, K. 2004. Energy boost: The Warburg effect returns in a new theory of cancer. *Journal of the National Cancer Institute* 96 (24): 1805–1806.

Georgakoudi, I., B. C. Jacobson, M. G. Muller et al. 2002. NAD(P)H and collagen as in vivo quantitative fluorescent biomarkers of epithelial precancerous changes. *Cancer Research* 62 (3): 682–687.

George, R., M. Michaelides, M. A. Brewer, and U. Utzinger. 2012. Parallel factor analysis of ovarian autofluorescence as a cancer diagnostic. *Lasers in Surgery and Medicine* 44 (4): 282–295.

Gillenwater, A., R. Jacob, R. Ganeshappa et al. 1998. Noninvasive diagnosis of oral neoplasia based on fluorescence spectroscopy and native tissue autofluorescence. *Archives of Otolaryngology—Head & Neck Surgery* 124 (11): 1251–1258.

Glassman, W. S., M. Steinberg, and R. R. Alfano. 1994. Time resolved and steady state fluorescence spectroscopy from normal and malignant cultured human breast cell lines. *Lasers in the Life Sciences* 6: 91–98.

Gu, Y., Z. Qian, J. Chen et al. 2002. High-resolution three-dimensional scanning optical image system for intrinsic and extrinsic contrast agents in tissue. *Review of Scientific Instruments* 73 (1): 172–178.

Haris, P. S., A. Balan, R. S. Jayasree, and A. K. Gupta. 2009. Autofluorescence spectroscopy for the in vivo evaluation of oral submucous fibrosis. *Photomedicine and Laser Surgery* 27 (5): 757–761.

Hassinen, I. and B. Chance. 1968. Oxidation-reduction properties of the mitochondrial flavoprotein chain. *Biochemical and Biophysical Research Communication* 31 (6): 895–900.

Heiden, M. G. V., L. C. Cantley, and C. B. Thompson. 2009. Understanding the Warburg effect: The metabolic requirements of cell proliferation. *Science* 324 (5930): 1029–1033.

Heiden, M. G. V., J. W. Locasale, K. D. Swanson et al. 2010. Evidence for an alternative glycolytic pathway in rapidly proliferating cells. *Science* 329 (5998): 1492–1499.

Heikal, A. A. 2010. Intracellular coenzymes as natural biomarkers for metabolic activities and mitochondrial anomalies. *Biomarkers in Medicine* 4 (2): 241–263.

Heintzelman, D. L., R. Lotan, and R. R. Richards-Kortum. 2000a. Characterization of the autofluorescence of polymorphonuclear leukocytes, mononuclear leukocytes and cervical epithelial cancer cells for improved spectroscopic discrimination of inflammation from dysplasia. *Photochemistry and Photobiology* 71 (3): 327–332.

Heintzelman, D. L., U. Utzinger, H. Fuchs et al. 2000b. Optimal excitation wavelengths for in vivo detection of oral neoplasia using fluorescence spectroscopy. *Photochemistry and Photobiology* 72 (1): 103–113.

Hua, D. Z., S. H. Qi, H. Li, Z. H. Zhang, and L. Fu. 2012. Monitoring the process of pulmonary melanoma metastasis using large area and label-free nonlinear optical microscopy. *Journal of Biomedical Optics* 17 (6): 066002.

Huang, S., A. A. Heikal, and W. W. Webb. 2002. Two-photon fluorescence spectroscopy and microscopy of NAD(P)H and flavoprotein. *Biophysical Journal* 82 (5): 2811–2825.

Hung, J., S. Lam, J. C. Leriche, and B. Palcic. 1991. Autofluorescence of normal and malignant bronchial tissue. *Lasers in Surgery and Medicine* 11 (2): 99–105.

Imaizumi, K., Y. Harada, N. Wakabayashi et al. 2011. Autofluorescence ratio imaging of human colonic adenomas. *Proceedings of SPIE* 7902: 79020E.

Imaizumi, K., Y. Harada, N. Wakabayashi et al. 2012. Dual-wavelength excitation of mucosal autofluorescence for precise detection of diminutive colonic adenomas. *Gastrointestinal Endoscopy* 75 (1): 110–117.

Ishikawa, K., K. Takenaga, M. Akimoto et al. 2008. ROS-generating mitochondrial DNA mutations can regulate tumor cell metastasis. *Science* 320 (5876): 661–664.

Izuishi, K., H. Tajiri, T. Fujii et al. 1999. The histological basis of detection of adenoma and cancer in the colon by autofluorescence endoscopic imaging. *Endoscopy* 31 (7): 511–516.

Kamath, S. D., R. A. Bhat, S. Ray, and K. K. Mahato. 2009. Autofluorescence of normal, benign, and malignant ovarian tissues: A pilot study. *Photomedicine and Lasers Surgery* 27 (2): 325–335.

Keller, P., M. Sowinska, V. Tassetti et al. 1996. Photodynamic imaging of a rat pancreatic cancer with pheophorbide a. *Photochemistry and Photobiology* 63 (6): 860–867.

Kirkpatrick, N. D., M. A. Brewer, and U. Utzinger. 2007. Endogenous optical biomarkers of ovarian cancer evaluated with multiphoton microscopy. *Cancer Epidemiology Biomarkers and Prevention* 16 (10): 2048–2057.

Kirkpatrick, N. D., C. Zou, M. A. Brewer et al. 2005. Endogenous fluorescence spectroscopy of cell suspensions for chemopreventive drug monitoring. *Photochemistry and Photobiology* 81 (1): 125–134.

Kirschmann, D. A., R. A. Lininger, L. M. G. Gardner et al. 2000. Down-regulation of HP1Hsα expression is associated with the metastatic phenotype in breast cancer. *Cancer Research* 60 (13): 3359–3363.

Kirschmann, D. A., E. A. Seftor, S. F. T. Fong et al. 2002. A molecular role for lysyl oxidase in breast cancer invasion. *Cancer Research* 62 (15): 4478–4483.

Lehninger, A. L., D. L. Nelson, and M. M. Cox. 1993. *Principles of Biochemistry*, 2nd edn. New York: Worth Publishers.

Lemasters, J. J. and A.-L. Nieminen, eds. 2001. *Mitochondria in Pathogenesis*. New York: Kluwer Academic.

Li, C., C. Pitsillides, J. M. Runnels, D. Côté, and C. P. Lin. 2010. Multiphoton microscopy of live tissues with ultraviolet autofluorescence. *IEEE Journal of Selected Topics in Quantum Electronics* 16 (3): 516–523.

Li, D., W. Zheng, and J. Y. Qu. 2009a. Two-photon autofluorescence microscopy of multicolor excitation. *Optics Letters* 34 (2): 202–204.

Li, L. Z. 2012. Imaging mitochondrial redox potential and its possible link to tumor metastatic potential. *Journal of Bioenergetics and Biomembranes* 44 (6): 645–653.

Li, L. Z., H. N. Xu, and M. Ranji et al. 2009b. Mitochondrial redox imaging for cancer diagnostic and therapeutic studies. *Journal of Innovative Optical Health Sciences* 2: 325–341.

Li, L. Z., R. Zhou, H. N. Xu et al. 2009c. Quantitative magnetic resonance and optical imaging biomarkers of melanoma metastatic potential. *Proceedings of the National Academy of Sciences of the United States of America* 106 (16): 6608–6613.

Li, L. Z., R. Zhou, T. Zhong et al. 2007. Predicting melanoma metastatic potential by optical and magnetic resonance imaging. In *Oxygen Transport to Tissue XXVIII*, D. J. Maguire, D. F. Bruley, and D. K. Harrison (eds.). Vol. 599 of Advances in Experimental Medicine and Biology, I. R. Cohen, A. Lajtha, R. Paoletti, and J. D. Lambris (eds.), pp. 67–78. New York: Springer

Lin, L. S., F. W. Yang, and S. S. Xie. 2012. Extracting autofluorescence spectral features for diagnosis of nasopharyngeal carcinoma. *Laser Physics* 22 (9): 1431–1434.

Luo, X. J., B. Zhang, J. G. Li, X. A. Luo, and L. F. Yang. 2012. Autofluorescence spectroscopy for evaluating dysplasia in colorectal tissues. *Zeitschrift Fur Medizinische Physik* 22 (1): 40–47.

Majumder, S. K., A. Uppal, and P. K. Gupta. 1998. Autofluorescence spectroscopy of oral mucosa. *Proceedings of SPIE* 3252: 158–168.

Malzahn, K., T. Dreyer, H. Glanz, and C. Arens. 2002. Autofluorescence endoscopy in the diagnosis of early laryngeal cancer and its precursor lesions. *Laryngoscope* 112 (3): 488–493.

Mayinger, B., P. Horner, M. Jordan et al. 2001. Endoscopic fluorescence spectroscopy in the upper GI tract for the detection of GI cancer: Initial experience. *The American Journal of Gastroenterology* 96 (9): 2616–2621.

Mehlmann, M., C. S. Betz, H. Stepp et al. 1999. Fluorescence staining of laryngeal neoplasms after topical application of 5-aminolevulinic acid: Preliminary results. *Lasers in Surgery and Medicine* 25 (5): 414–420.

Mujat, C., C. Greiner, A. Baldwin et al. 2008. Endogenous optical biomarkers of normal and human papillomavirus immortalized epithelial cells. *International Journal of Cancer* 122 (2): 363–371.

Olovnikov, I. A., J. E. Kravchenko, and P. M. Chumakov. 2009. Homeostatic functions of the p53 tumor suppressor: Regulation of energy metabolism and antioxidant defense. *Seminars in Cancer Biology* 19 (1): 32–41.

Orrenius, S., V. Gogvadze, and B. Zhivotovsky. 2007. Mitochondrial oxidative stress: Implications for cell death. *Annual Review of Pharmacology and Toxicology* 47: 143–183.

Ostrander, J. H., C. M. McMahon, S. Lem et al. 2010. Optical redox ratio differentiates breast cancer cell lines based on estrogen receptor status. *Cancer Research* 70 (11): 4759–4766.

Ozawa, K., B. Chance, A. Tanaka et al. 1992. Linear correlation between acetoacetate beta-hydroxybutyrate in arterial blood and oxidized flavoprotein reduced pyridine-nucleotide in freeze-trapped human liver-tissue. *Biochimica et Biophysica Acta* 1138 (4): 350–352.

Palasz, Z., A. Grobelny, E. Pawlik et al. 2003. Investigation of normal and malignant laryngeal tissue by autofluorescence imaging technique. *Auris Nasus Larynx* 30 (4): 385–389.

Palmer, G. M., P. J. Keely, T. M. Breslin, and N. Ramanujam. 2003. Autofluorescence spectroscopy of normal and malignant human breast cell lines. *Photochemistry and Photobiology* 78 (5): 462–469.

Panjehpour, M., B. F. Overholt, T. Vo-Dinh et al. 1996. Endoscopic fluorescence detection of high-grade dysplasia in Barrett's esophagus. *Gastroenterology* 111 (1): 93–101.

Pantalone, D., F. Andreoli, F. Fusi et al. 2007. Multispectral imaging autofluorescence microscopy in colonic and gastric cancer metastatic lymph nodes. *Clinical Gastroenterology and Hepatology* 5 (2): 230–236.

Patalay, R., C. Talbot, Y. Alexandrov et al. 2011. Non-invasive imaging of skin cancer with fluorescence lifetime imaging using two photon tomography. *Proceedings of SPIE* 8087: 808718.

Pauli, R., C. Betz, M. Havel et al. 2007. Multiple fluorophore-analysis (MFA) for qualitative tissue diagnosis in the oral cavity. *Proceedings of SPIE* 6628: 66280D.

Pavlova, I., M. Williams, A. El-Naggar, R. Richards-Kortum, and A. Gillenwater. 2008. Understanding the biological basis of autofluorescence imaging for oral cancer detection: High-resolution fluorescence microscopy in viable tissue. *Clinical Cancer Research* 14 (8): 2396–2404.

Peng, T., H. Xie, Y. Ding et al. 2012. CRAFT: Multimodality confocal skin imaging for early cancer diagnosis. *Journal of Biophotonics* 5 (5–6): 469–476.

Pitts, J. D., R. D. Sloboda, K. H. Dragnev, E. Dmitrovsky, and M. A. Mycek. 2001. Autofluorescence characteristics of immortalized and carcinogen-transformed human bronchial epithelial cells. *Journal of Biomedical Optics* 6 (1): 31–40.

Pradhan, A., P. Pal, G. Durocher et al. 1995. Steady state and time-resolved fluorescence properties of metastatic and non-metastatic malignant cells from different species. *Journal of Photochemistry and Photobiology B* 31 (3): 101–112.

Quistorff, B., J. C. Haselgrove, and B. Chance. 1985. High spatial resolution readout of 3-D metabolic organ structure: An automated, low-temperature redox ratio-scanning instrument. *Analytical Biochemistry* 148 (2): 389–400.

Ramanujam, N. 2000. Fluorescence spectroscopy of neoplastic and non-neoplastic tissues. *Neoplasia* 2 (1–2): 89–117.

Ramanujam, N., R. Richards-Kortum, S. Thomsen et al. 2001. Low temperature fluorescence imaging of freeze-trapped human cervical tissues. *Optics Express* 8 (6): 335–343.

Renkoski, T. E., K. D. Hatch, and U. Utzinger. 2012. Wide-field spectral imaging of human ovary autofluorescence and oncologic diagnosis via previously collected probe data. *Journal of Biomedical Optics* 17 (3): 036003.

Rigacci, L., R. Alterini, P. A. Bernabei et al. 2000. Multispectral imaging autofluorescence microscopy for the analysis of lymph-node tissues. *Photochemistry and Photobiology* 71 (6): 737–742.

Sehdev, A. E. S., A. W. Levi, T. Nadasdy et al. 2001. The pigmented "black" neuroendocrine tumor of the pancreas—A question of origin. *Cancer* 92 (7): 1984–1991.

Senda, T., M. Senda, S. Kimura, and T. Ishida. 2009. Redox control of protein conformation in flavoproteins. *Antioxidants & Redox Signaling* 11 (7): 1741–1766.

Shields, J. A., M. M. Rodrigues, L. K. Sarin, W. S. Tasman, and W. H. Annesley. 1976. Lipofuscin pigment over benign and malignant choroidal tumors. *Transactions—Section on Ophthalmology. American Academy of Ophthalmology and Otolaryngology* 81 (5): 871–881.

Shirogane, R., H. Sasabe, N. Sato, and L. M. Li. 2010. Discrimination of normal and cancer cells from autofluorescence spectra by laser excitation. *Molecular Crystals and Liquid Crystals* 519: 163–168.

Skala, M. C., A. Fontanella, L. Lan, J. A. Izatt, and M. W. Dewhirst. 2010. Longitudinal optical imaging of tumor metabolism and hemodynamics. *Journal of Biomedical Optics* 15 (1): 011112.

Skala, M. C., K. M. Riching, A. Gendron-Fitzpatrick et al. 2007. In vivo multiphoton microscopy of NADH and FAD redox states, fluorescence lifetimes, and cellular morphology in precancerous epithelia. *Proceedings of the National Academy of Sciences of the United States of America* 104 (49): 19494–19499.

Skala, M. C., J. M. Squirrell, K. M. Vrotsos et al. 2005. Multiphoton microscopy of endogenous fluorescence differentiates normal, precancerous, and cancerous squamous epithelial tissues. *Cancer Research* 65 (4): 1180–1186.

Son, H. J., Y. J. Jeong, J. H. Kim, and M. J. Chung. 2004. Phyllodes tumor of the seminal vesicle: Case report and literature review. *Pathology International* 54 (12): 924–929.

Stender, R. N. I. and H. C. Wulf. 2001. Can autofluorescence demarcate basal cell carcinoma from normal skin? A comparison with protoporphyrin IX fluorescence. *Acta Dermato-Venereologica* 81 (4): 246–249.

Sterenborg, H. J., S. Thomsen, S. L. Jacques, and M. Motamedi. 1995. In vivo autofluorescence of an unpigmented melanoma in mice. Correlation of spectroscopic properties to microscopic structure. *Melanoma Research* 5 (4): 211–216.

Taylor, B. L., A. Rebbapragada, and M. S. Johnson. 2001. The FAD-PAS domain as a sensor for behavioral responses in *Escherichia coli*. *Antioxidants & Redox Signaling* 3 (5): 867–879.

Thompson, C. B. 2009. Metabolic enzymes as oncogenes or tumor suppressors. *The New England Journal of Medicine* 360 (8): 813–815.

Tsai, T. M., C. T. Chen, C. Y. Wang, C. P. Chiang, and Y. L. Ku. 2001. Identification of oral carcinogenesis using autofluorescence spectroscopy: An in vivo study. *Proceedings of SPIE* 4597: 115–120.

Veech, R. L. 2006. The determination of the redox states and phosphorylation potential in living tissues and their relationship to metabolic control of disease phenotypes. *Biochemistry and Molecular Biology Education* 34 (3): 168–179.

Villette, S., S. Pigaglio-Deshayes, C. Vever-Bizet, P. Validire, and G. Bourg-Heckly. 2006. Ultraviolet-induced autofluorescence characterization of normal and tumoral esophageal epithelium cells with quantitation of NAD(P)H. *Photochemical and Photobiological Sciences* 5 (5): 483–492.

Wagnieres, G., M. Zellweger, P. Grosjean et al. 2002. In vivo fluorescence spectroscopy to optimize the detection of early bronchial carcinoma by autofluorescence imaging. *Proceedings of SPIE* 4615: 1–12.

Walsh, A., R. S. Cook, B. Rexer, C. L. Arteaga, and M. C. Skala. 2012. Optical imaging of metabolism in HER2 overexpressing breast cancer cells. *Biomedical Optics Express* 3 (1): 75–85.

Wang, C. C., F. C. Li, R. J. Wu et al. 2010. Differentiation of normal and cancerous lung tissues by multiphoton imaging. *Journal of Biomedical Optics* 14 (4): 044034.

Wang, C. Y., C. T. Chen, C. P. Chiang et al. 1999a. A probability-based multivariate statistical algorithm for autofluorescence spectroscopic identification of oral carcinogenesis. *Photochemistry and Photobiology* 69 (4): 471–477.

Wang, S., A. Mintz, K. Mochizuki et al. 2007. Multimodality optical imaging and 18F-FDG uptake in wild-type p53-containing and p53-null human colon tumor xenografts. *Cancer Biology & Therapy* 6 (10): 1649–1653.

Wang, T. D., J. M. Crawford, M. S. Feld et al. 1999b. In vivo identification of colonic dysplasia using fluorescence endoscopic imaging. *Gastrointestinal Endoscopy* 49 (4): 447–455.

Watson, J. M., S. L. Marion, P. F. Rice et al. 2013. Two-photon excited fluorescence imaging of endogenous contrast in a mouse model of ovarian cancer. *Lasers in Surgery and Medicine* 45 (3): 155–166.

Xu, H. N., M. Feng, L. Moon et al. 2013a. Redox imaging of the p53-dependent mitochondrial redox state in colon cancer ex vivo. *Journal of Innovative Optical Health Sciences* 6 (3): 1350016.

Xu, H. N., S. Nioka, B. Chance, and L. Z. Li. 2011. Heterogeneity of mitochondrial redox state in premalignant pancreas in a PTEN null transgenic mouse model. In *Oxygen Transport to Tissue XXXII*, J. C. LaManna, M. A. Puchowicz, K. Xu, D. K. Harrison, and D. F. Bruley (eds.). Vol. 701 of Advances in Experimental Medicine and Biology, I. R. Cohen, A. Lajtha, R. Paoletti, and J. D. Lambris (eds.), pp. 207–213. New York: Springer.

Xu, H. N., S. Nioka, J. D. Glickson, B. Chance, and L. Z. Li. 2010. Quantitative mitochondrial redox imaging of breast cancer metastatic potential. *Journal of Biomedical Optics* 15 (3): 036010.

Xu, H. N., S. Nioka, and L. Z. Li. 2013b. Imaging heterogeneity in the mitochondrial redox state of premalignant pancreas in the pancreas-specific PTEN-null transgenic mouse model. *Biomarker Research* 1: 6.

Xu, H. N., J. Tchou, B. Chance, and L. Z. Li. 2013c. Imaging the redox states of human breast cancer core biopsies. In *Oxygen Transport to Tissue XXXIV*, W. Welch, F. Palm, D. F. Bruley, and D. K. Harrison (eds.). Vol. 765 of Advances in Experimental Medicine and Biology, I. R. Cohen, A. Lajtha, R. Paoletti, and J. D. Lambris (eds.), pp. 343–349. New York: Springer.

Xu, H. N., J. Tchou, and L. Z. Li. 2013d. Redox imaging of human breast cancer core biopsies: A preliminary investigation. *Academic Radiology* 20 (6): 764–768.

Yan, X., H. Shen, H. Jiang et al. 2006. External Qi of Yan Xin Qigong differentially regulates the Akt and extracellular signal-regulated kinase pathways and is cytotoxic to cancer cells but not to normal cells. *The International Journal of Biochemistry & Cell Biology* 38 (12): 2102–2113.

Yang, M. S., D. Li, T. Lin et al. 2008. Increase in intracellular free/bound NAD(P)H as a cause of Cd-induced oxidative stress in the HepG(2) cells. *Toxicology* 247 (1): 6–10.

Yang, Y. L., Y. M. Ye, F. M. Li, Y. F. Li, and P. Z. Ma. 1987. Characteristic autofluorescence for cancer diagnosis and its origin. *Lasers in Surgery and Medicine* 7 (6): 528–532.

Ying, W. 2008. NAD^+/NADH and $NADP^+$/NADPH in cellular functions and cell death: Regulation and biological consequences. *Antioxidants & Redox Signaling* 10 (2): 179–206.

Yu, Q. and A. A. Heikal. 2009. Two-photon autofluorescence dynamics imaging reveals sensitivity of intracellular NADH concentration and conformation to cell physiology at the single-cell level. *Journal of Photochemistry and Photobiology B* 95 (1): 46–57.

Zheng, W., D. Li, Y. Zeng, and J. A. Y. Qu. 2011. In vivo multiphoton fluorescence microscopy of epithelial precancer. *Proceedings of SPIE* 7890: 789001.

Zheng, W., M. Olivo, and S. K. Chee. 2001. Diagnosis of oral tumours using light-induced autofluorescence spectra at multiple excitation wavelengths. *Proceedings of SPIE* 4597: 133–138.

Zhivotovsky, B. and S. Orrenius. 2009. The Warburg effect returns to the cancer stage. *Seminars in Cancer Biology* 19 (1): 1–3.

Zhou, L. L., D. Blessington, Z. H. Zhang et al. 2003. NIR imaging the delivery of Cathepsin B probe to breast tumors. *Proceedings of SPIE* 4955: 505–512.

Zhuo, S., J. Yan, G. Chen et al. 2011. Label-free monitoring of colonic cancer progression using multiphoton microscopy. *Biomedical Optics Express* 2 (3): 615–619.

Ziegler, M. 2005. A vital link between energy and signal transduction. Regulatory functions of NAD(P). *FEBS Journal* 272 (18): 4561–4564.

14

Dynamic Imaging of Intracellular Coenzyme Autofluorescence in Intact Pancreatic Islets

Alan K. Lam
University of Toronto
and
University Health Network

Jonathan V. Rocheleau
University of Toronto
and
University Health Network

14.1 Introduction

The pancreas is a mixed gland in the body containing both exocrine and endocrine tissues (Mastracci and Sussel 2012). As an exocrine tissue, the pancreas produces various enzymes that participate in the digestion of food. The smaller endocrine tissue consists of pancreatic islets, also known as islets of Langerhans, which are dispersed throughout the pancreas (Figure 14.1a).

The islets of Langerhans *in vivo* are highly vascularized microorgans structured for the effective sensing and control of glucose homeostasis (Sankar et al. 2011). Individual islets vary in size from roughly 100 to 200 μm in diameter and contain 10^2–10^3 cells (Rocheleau and Piston 2008), of which approximately 85% are the β-cells, responsible for the production of insulin in response to blood glucose (Unger 1981) (Figure 14.1b).

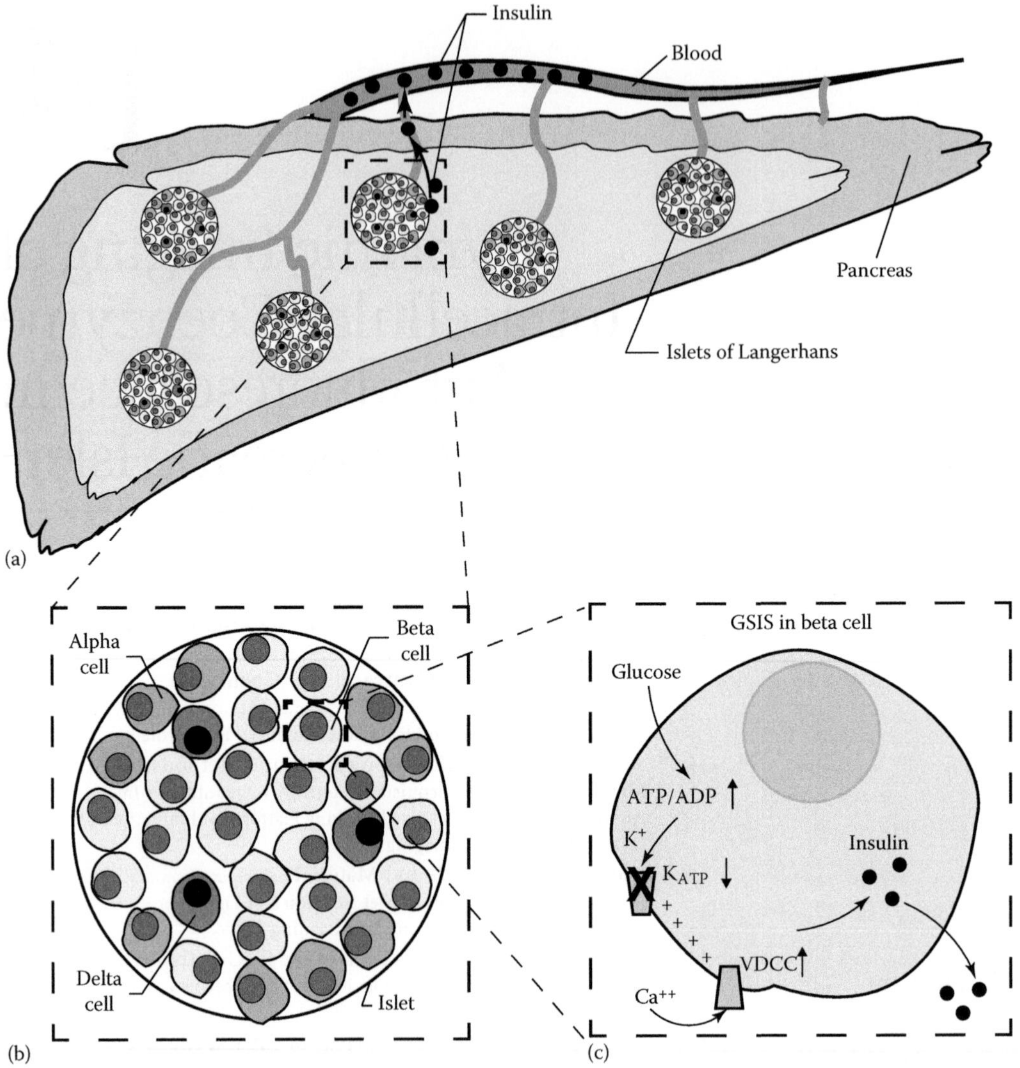

FIGURE 14.1 Pancreatic islets are found in the pancreas and mainly contain insulin-secreting β-cells. (a) Pancreatic islets or the islets of Langerhans *in vivo* are highly vascularized microorgans that are dispersed throughout the pancreas. (b) Pancreatic islet consists of three main cell types of which ~85% are insulin-secreting β-cells. (c) A scheme of GSIS in a pancreatic islet β-cell.

This property of β-cells makes the study of their biology and metabolism critically important to diabetes research. In particular, mitochondria, the major organelles responsible for metabolism in a cell, play a key mechanistic role in many cellular processes, and mitochondrial dysfunction is widely linked to a number of diseases including diabetes (Heikal 2010). To understand the biological perturbations caused by hyperglycemia, one must first appreciate the metabolism of glucose leading to insulin secretion.

14.1.1 Glucose-Stimulated Insulin Secretion

Pancreatic islet β-cells metabolically sense nutrients to maintain blood glucose homeostasis through the regulated multiple steps of glucose-stimulated insulin secretion (GSIS) (Rocheleau et al. 2002).

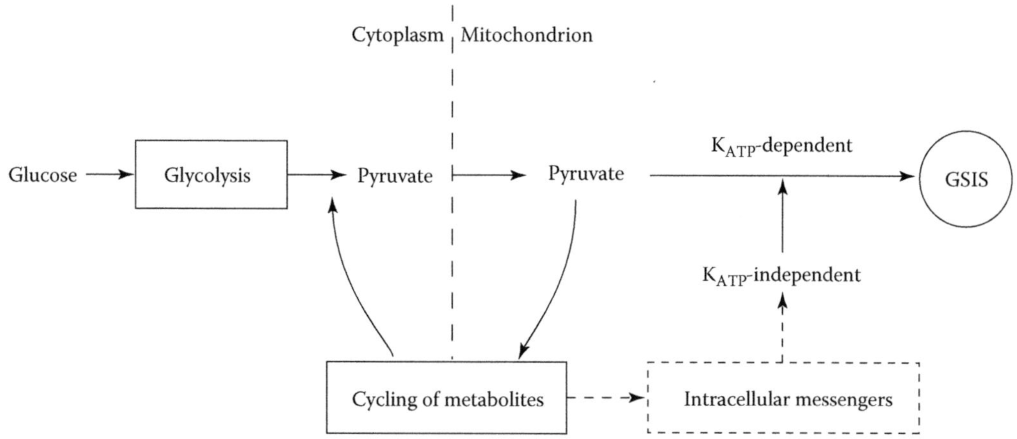

FIGURE 14.2 K_{ATP}-dependent and K_{ATP}-independent pathways of GSIS. Pyruvate generated from glycolysis enters β-cell mitochondrial metabolism to activate both K_{ATP}-dependent (triggering) and K_{ATP}-independent (modulating) pathways of GSIS. The K_{ATP}-dependent process triggers GSIS. K_{ATP}-independent pathway is postulated to modulate GSIS via the production of intracellular messengers through the cycling of pyruvate-associated metabolites between the cytoplasm and mitochondria.

As glucose equilibrates across the plasma membrane, it gets metabolized, leading to a cascade of reactions: increase of the [NADH]:[NAD^+] redox ratio and the cellular [ATP]:[ADP] ratio, closure of the ATP-sensitive potassium channels (K_{ATP}), plasma membrane depolarization, opening of the voltage-dependent calcium channels, influx of intracellular Ca^{2+} ions, and stimulation of insulin exocytosis (Figure 14.1c).

Collectively termed K_{ATP}-dependent GSIS, these steps have long been demonstrated as the major mechanism to link glucose metabolism and insulin secretion in pancreatic islet β-cells (Newgard and McGarry 1995). Yet, it is becoming increasingly clear that GSIS is further potentiated or modulated by mechanisms independent of K_{ATP} channel activity (K_{ATP} independent) including metabolic cycling of pyruvate and generation of intracellular messengers. The postulated secondary messengers include reduced nicotinamide adenine dinucleotide and its phosphate (NAD(P)H), guanosine triphosphate (GTP), and ATP (Jitrapakdee et al. 2010). The impact of K_{ATP}-independent mechanisms on GSIS is particularly amplified under physiological conditions where concomitant nutrients, such as fatty acids and amino acids, exist. These observations have led to new emphasis on understanding the metabolic routes affecting K_{ATP}-independent stimulation and how lipids and other nutrients may contribute to the dysfunction of GSIS and diabetes (Figure 14.2).

14.1.2 Excess Nutrients and Glucolipotoxicity

High capacity of pancreatic islet β-cells for nutrient sensing necessitates limited protection from nutrient-induced toxicity (Nolan and Prentki 2008). Chronic exposure to a mixed supply of excess glucose and lipids has been linked to pancreatic islet β-cell dysfunction and type II diabetes in a process termed glucolipotoxicity (Poitout and Robertson 2002, 2008; Poitout et al. 2010). The interplay between glucose and lipid metabolism remains debated and less clear although a number of mechanisms underlying glucolipotoxicity have been proposed. Therefore, the capacity to dynamically track metabolic events holds promise to elucidate mechanisms contributing to glucolipotoxicity. In this context, the goals of this chapter are the following:

- To establish the connection between intracellular coenzymes and islet metabolism
- To introduce the techniques of confocal ratiometric imaging of mitochondrial flavins and two-photon (2P) excitation microscopy of NAD(P)H
- To demonstrate the utility of intrinsically fluorescent flavins and NAD(P)H as natural biomarkers for islet metabolism via case studies

14.2 Metabolic Linkages of Intracellular Coenzyme Autofluorescence

It is well established that cellular redox state can be monitored by quantitative imaging of NAD(P)H and mitochondrial flavin autofluorescence (Bennett et al. 1996; Piston and Knobel 1999; Rocheleau et al. 2004). To utilize these intrinsic markers in biological applications, one must appreciate their linkages to metabolism.

14.2.1 Reduced Pyridine Nucleotides

The redox state of pancreatic islet β-cells correlates with the metabolic flux in response to extracellular nutrients. Glucose stimulates a rise in the [NADH]:[NAD^+] ratio through glycolysis and TCA cycle metabolism, leading to an increase of the [ATP]:[ADP] ratio and triggering of the K_{ATP}-dependent insulin secretion. A rise in β-cell [NADPH]:[$NADP^+$] ratio can also be stimulated through the cycling of multiple TCA pathway intermediates between the cytoplasm and mitochondria, ultimately impacting K_{ATP}-independent mechanisms of insulin secretion (Jitrapakdee et al. 2010). These changes in β-cell redox state can be assayed using 2P microscopy of NAD(P)H autofluorescence (Bennett et al. 1996; Piston 1999; see also discussion of 2P autofluorescence microscopy advantages and applications in Chapters 4 and 5). Both NADH and NADPH are spectrally indistinguishable and are fluorescent only in their reduced state (Chance and Thorell 1959), thus providing a direct measure of overall glucose-stimulated β-cell metabolism (Piston and Knobel 1999).

14.2.2 Mitochondrial Flavoproteins

Glucose and fatty acid metabolism interact in the mitochondria at both the TCA cycle and the electron transport chain (ETC) (Figure 14.3). Pyruvate generated from glycolysis enters mitochondrial metabolism through pyruvate dehydrogenase (PDH)–dependent production of acetyl CoA leading to a rise of the mitochondrial NADH and ATP levels and—eventually—to the stimulation of K_{ATP}-dependent GSIS. Pyruvate also enters the TCA cycle through pyruvate carboxylase–dependent production of oxaloacetate to ultimately support cycling of the TCA pathway intermediates and K_{ATP}-independent activities. This cycling also stimulates the production of cytoplasmic NADPH. Therefore, NADH and NADPH are likely to reflect the K_{ATP}-dependent and K_{ATP}-independent activities, respectively.

Unfortunately, 2P excitation microscopy of NAD(P)H autofluorescence cannot spectrally differentiate NADH and NADPH. To address this issue, our laboratory previously developed confocal ratiometric imaging of the mitochondrial flavoprotein, lipoamide dehydrogenase (LipDH). LipDH is a part of the multienzyme complex of PDH and has a redox state that is in direct equilibrium with the mitochondrial pool of NADH, as depicted by the following redox reaction:

$$\text{LipDH}(\text{FADH}_2) + \text{NAD}^+ \leftrightarrow \text{LipDH}(\text{FAD}) + \text{NADH} + \text{H}^+ \qquad (14.1)$$

A comparison of LipDH and NAD(P)H autofluorescence therefore provides a means to distinguish the production of NADH, NADPH, or both. For instance, a change in NAD(P)H without a similar change observed in LipDH suggests that the change is likely contributed by NADPH rather than NADH (Rocheleau et al. 2004).

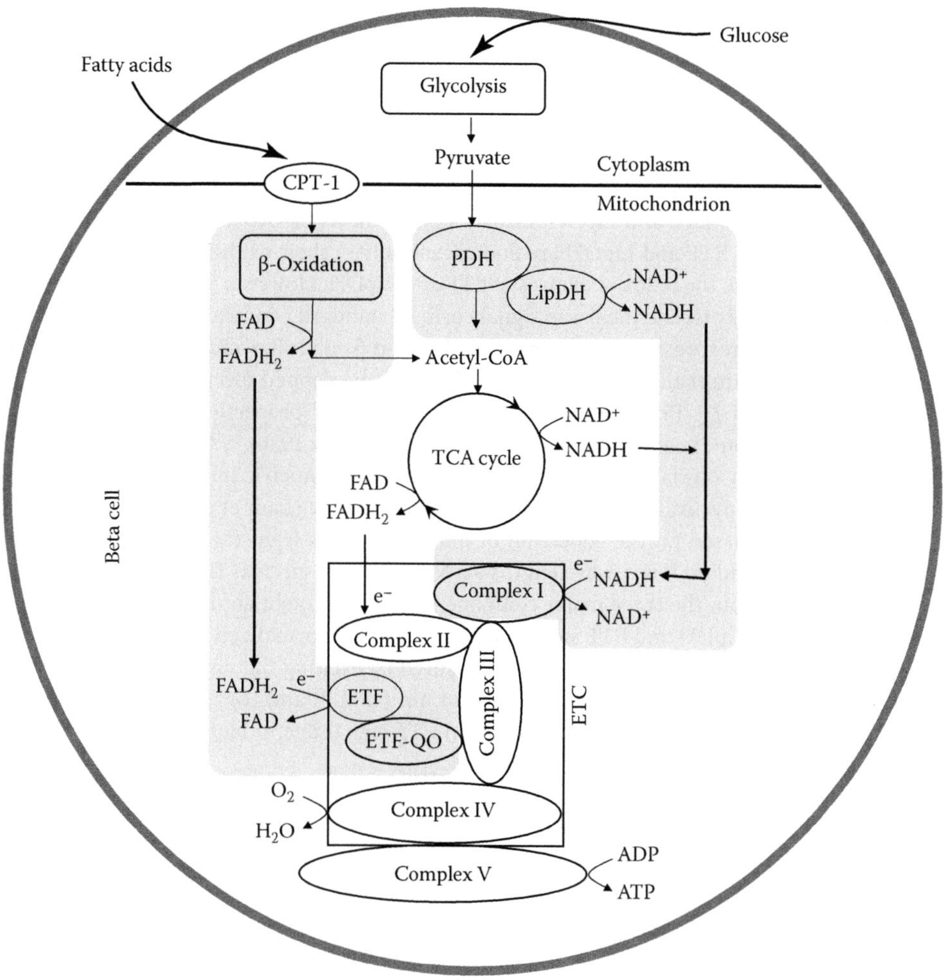

FIGURE 14.3 Flavins involved in β-cell metabolism and GSIS. LipDH participates in pyruvate metabolism via PDH. In contrast, fatty acids enter mitochondria via CPT-1 transporter, and through β-oxidation enter the TCA cycle as acetyl CoA. NADH donates high-energy electrons to Complex I of the ETC, and $FADH_2$ derived from β-oxidation donates high-energy electrons to Complex III via ETF and ETF-QO. The left- and the right-shaded areas emphasize the metabolic activities associated with ETF and LipDH, respectively.

In contrast to glucose, fatty acids are actively transported into mitochondria via the carnitine palmitoyltransferase I (CPT-1) and undergo mitochondrial β-oxidation with repeated cycles of dehydrogenation, hydration, oxidation, and thiolysis. The final product, acetyl CoA, enters the TCA cycle, producing NADH and $FADH_2$, which supply electrons to Complex I and Complex II of the ETC, respectively. In contrast, $FADH_2$ generated by mitochondrial β-oxidation supplies electrons to Complex III via the electron transfer flavoprotein (ETF) and ETF-coenzyme Q (CoQ) oxidoreductase (ETF-QO) (Lam et al. 2012). ETF is a soluble heterodimeric protein found in all kingdoms of life that contains a single equivalent of flavin adenine nucleotide (FAD) (Toogood et al. 2007). It functions as an electron shuttle that accepts electrons from up to nine primary FAD-containing acyl-CoA dehydrogenases involved in mitochondrial fatty acid and amino acid catabolism (Roberts et al. 1996; Watmough and Frerman 2010), as represented by the following redox reaction:

$$\text{Acyl-CoA}(\text{FADH}_2) + \text{ETF}(\text{FAD}) \leftrightarrow \text{Acyl-CoA}(\text{FAD}) + \text{ETF}(\text{FADH}_2) \tag{14.2}$$

Therefore, the redox steady state of ETF is determined by the opposing rates of reduction by ETF-linked metabolism (fatty acid β-oxidation and amino acid catabolism) and oxidization by ETF-QO activity.

It is well established that flavin autofluorescence consists of three major spectral components. The first component is from a number of unbound free flavins that are essentially non-redox responsive. The next two components, ETF and LipDH, display significant redox-dependent changes in intensity. In contrast to NAD(P)H, ETF and LipDH are fluorescent only in their oxidized state. Our goal was to develop methods to isolate the redox responses of ETF and LipDH. However, it was clear that we would never be able to completely isolate the flavin signals using a standard confocal microscope. Figure 14.4 shows the spectra of overexpressed recombinant ETF (α and β subunits) and LipDH in HeLa cells by transient transfection (Lam et al. 2012). The plot shows highly overlapped spectra, however, with LipDH redshifted compared to ETF. Previous studies used the redshifted properties of LipDH to isolate its fluorescence intensity from the other flavin autofluorescence (Kunz 1986, 1988; Kunz and Kunz 1985). Building off these studies, our laboratory established confocal ratiometric imaging of LipDH and ETF to dynamically monitor pyruvate and fat metabolism, respectively (Lam et al. 2012; Rocheleau et al. 2004). The approach is based on the collection of multiple images from the same cells using excitation (458 and 488 nm) and emission wavelengths across the flavin spectra. These images each contain varying contributions from the three major components of flavoprotein autofluorescence. The images with the most dynamic LipDH and ETF signals are subsequently normalized to the images containing the least responsive signal. In particular, ETF is measured by exciting the samples at 458 nm and collecting the images in two spectral bands: (a) 462–525 nm (458BP) and (b) 525–650 nm (458LP). The 458BP image contains the most dynamic response in ETF signal due to blue shoulder of the emitted

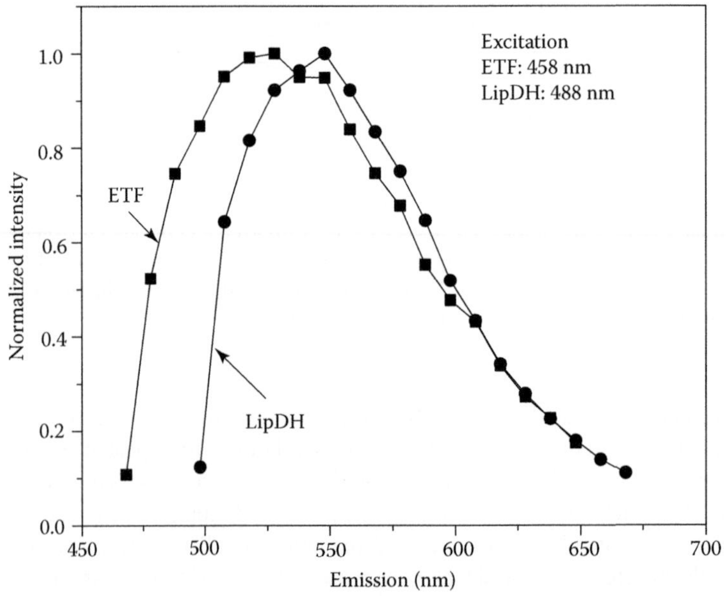

FIGURE 14.4 Fluorescence spectra of mitochondrial flavins. Emission spectra of ETF (ETF-α and ETF-β) and LipDH overexpressed in HeLa cells using confocal excitation of 458 and 488 nm, respectively. (From Lam, A.K., Silva, P.N., Altamentova, S.M., and Rocheleau, J.V., Quantitative imaging of electron transfer flavoprotein autofluorescence reveals the dynamics of lipid partitioning in living pancreatic islets, *Integr. Biol.*, 4(8), 838–846, 2012, Figure 2a. Reproduced by permission of the Royal Society of Chemistry.)

TABLE 14.1 Confocal Ratiometric Imaging versus Previous Studies on ETF and LipDH

	Fluorescence Spectra	
Study	Excitation (nm)	Emission (nm)
Confocal ratiometric techniques:		
ETF (458LP:458BP ratio)	458	(1) 462–525 (458BP)
(Lam et al. 2012)		(2) 525–650 (458LP)
LipDH (458LP:488LP ratio)	458	(1) 525–650 (458LP)
(Rocheleau et al. 2004)	488	(2) 505–650 (488LP)
Scientific literature:		
Spectral properties of fluorescent flavoproteins of isolated rat liver mitochondria (Kunz 1986)	ETF: 435 peak LipDH: 480 peak	ETF: 485 peak LipDH: 545 peak
Acyl coenzyme A dehydrogenases and electron-transferring flavoprotein from beef heart mitochondria (Hall et al. 1976)	ETF: 380 peak	ETF: 495 peak

fluorescence, while the 458LP image contains a significant signal from other redshifted and nonredox responsive flavins. The ratio of these two images (458LP:458BP) is used to quantify the ETF signal. Changes in LipDH are evaluated by the 458LP:488LP ratio, where 458LP is recorded at 458 nm excitation and collection in the range of 525–650 nm; the 488LP denotes the intensity measured in the 505–650 nm band at the excitation with 488 nm. These ratios do not isolate the absolute LipDH and ETF signals, but rather reveal the relative changes in redox state that can be used to monitor dynamics. The confocal ratiometric imaging techniques developed in our laboratory relative to previous scientific literature are shown in Table 14.1.

Two observations can be made from Table 14.1. First, the ETF emission spectra measured using our overexpression model is redshifted relative to previous studies mainly due to the differences in experimental setup and the nonredox responsive flavins. Second, the ETF and LipDH ratios are measured in dimensionless quantities as opposed to arbitrary units used for imaging intensity.

To this end, the utility of these confocal ratiometric techniques will be demonstrated later in this chapter with two case studies on mitochondrial metabolism in living pancreatic islets. To facilitate dynamic imaging in intact tissue, the microscopy setup and the design of a microfluidic device as the imaging platform need to be addressed.

14.3 Autofluorescence Microscopy Setup and Methods

A typical microscopy setup used in ratiometric measurements is based on a laser scanning confocal microscope for the ETF and LipDH signal collection, equipped with a multiphoton laser and a non-descanned detector for recording the NAD(P)H autofluorescence, a microfluidic device to immobilize islets while changing the solution and nutrients, and an incubator to maintain physiological conditions.

14.3.1 Microscopy Setup and Acquisition Settings

Ratiometric images were acquired with a Zeiss LSM 710 laser scanning microscope equipped with a tunable multiphoton titanium–sapphire laser (Chameleon, Coherent) (Figure 14.5). Autofluorescence images were sequentially collected using excitation wavelengths of 705 nm in 2P microscopy and 458/488 nm in confocal microscopy at a scan speed of 12.61 μs/pixel and resolution of 0.27 μm/pixel. Two-photon NAD(P)H emission was collected using the non-descanned dual gallium arsenide phosphide (GaAsP) BiG detector (Carl Zeiss) equipped with a custom 380–550 nm band-pass heat filter

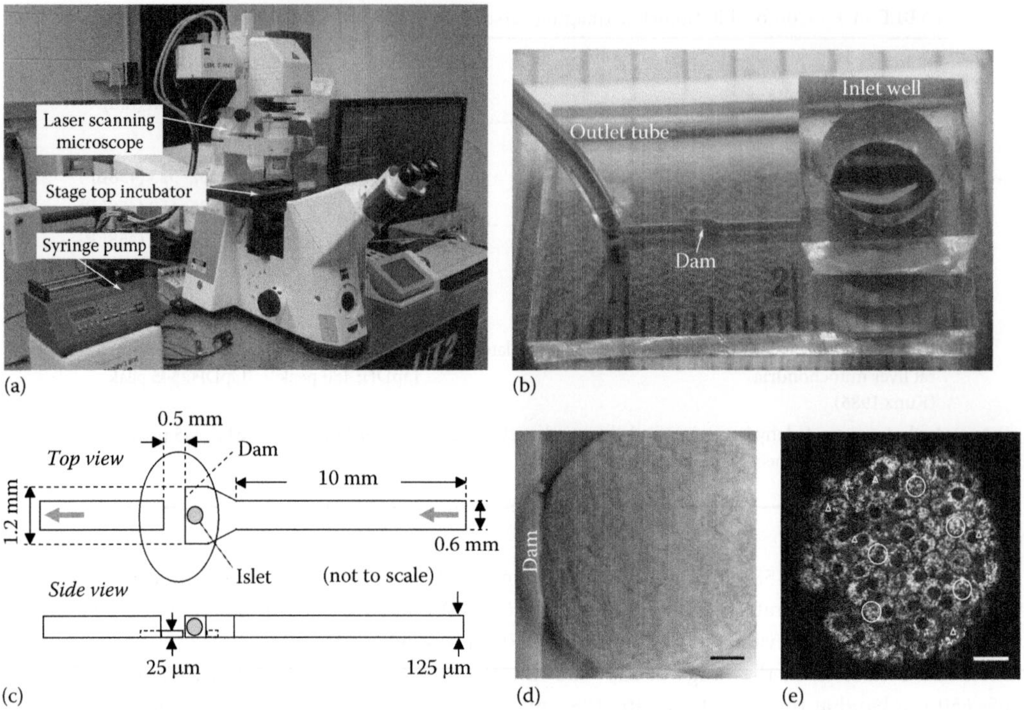

FIGURE 14.5 Autofluorescence microscopy setup. (a) A typical setup includes a laser scanning microscope with both confocal and 2P excitation, an onstage microscope incubator, and a syringe pump. (b) A custom-designed microfluidic device filled with a red dye solution showing a reservoir well at the inlet, a dam in the channel, and a tube at the outlet. The ruler below the device shows 1 cm. (c) A scheme of the two-layer device. (d) A *DIC* image of an islet being held stationary against the dam (*left edge*). (e) NAD(P)H autofluorescence image of an islet stimulated by 20 mM glucose using 2P 705 nm excitation. The five outlined circles show regions away from dark nuclei and non-redox-responsive lipofuscin deposits as marked by the five triangles. The scale bar represents 20 μm.

(Chroma). The 458 and 488 nm excited flavin signal was collected with the descanned multichannel internal detector.

Collecting fluorescence images with a consistently high level of spatial and temporal resolution is essential for a meaningful quantitative analysis of the images. An objective lens with the highest numerical aperture (NA) and ultimate level of chromatic and spherical aberrations correction should be considered. For the case studies described later, a Plan-Apochromat 63 × 1.4 NA oil immersion lens (Carl Zeiss) was used.

To prevent observable photodamage to the sample during repetitive imaging, the 2P laser power was attenuated to ~3.5 mW at the sample (Piston and Knobel 1999; Rocheleau et al. 2004). Low output power was also used for the multiline argon laser (458, 488, 514 nm) in confocal imaging, while the detectors gain was set at near maximum voltage. We further limit the exposure of the living tissue to the confocal laser by bringing the sample into focus using 2P excitation and using the argon laser only during data collection. To establish the standardization of the microscopy setup, laser power levels and detector gain voltages used in imaging were consistently maintained.

14.3.2 Quantitative Image Analysis

As intracellular coenzyme autofluorescence is dominated by its mitochondrial origin (Rocheleau et al. 2004), quantitative analysis of the fluorescence images should be started from defining the bright redox-responsive region of interest (ROI, circled in Figure 14.5e). This means that dark nuclei and non-redox responsive lipofuscin deposits (outlined triangles in Figure 14.5e) need to be avoided. Typically, 30–50 ROIs per islet were selected using *ImageJ* software (National Institutes of Health) to constitute a meaningful sample size for statistical analysis by a commercially available software program, such as OriginPro version 8.5 (OriginLab). The mean intensity per unit area extracted from the selected ROIs can be averaged to represent the mean intensity per unit area of the entire islet in each imaging mode.

14.3.3 Microfluidic-Based Imaging Platform

Pancreatic islets sitting at the bottom of a dish are not sufficiently stationary during changes in solution to allow dynamic imaging with high resolution. To achieve subcellular spatial and sub-minute temporal resolution, we used a custom-designed microfluidic device in combination with fluorescence microcopy (Rocheleau and Piston 2008). Shown in Figure 14.5 is a two-layer microfluidic design with a dam region that is used to limit islets movement in flowing solution (Figure 14.5b and c). More specifically, this device uses a reverse-nozzle-shaped channel where pancreatic islets are immobilized by the channel dam created between the 125 μm upper layer and the 25 μm lower layer (Figure 14.5c). Islets are held stationary in the device and pressed against the glass coverslip even during quick exchange of media.

The microfluidic devices were fabricated using elastomer polydimethylsiloxane (PDMS) (Dow Corning) as documented previously (Duffy et al. 1998; Rocheleau et al. 2004). The cured mold is permanently bonded to a 24 mm × 50 mm No. 1 glass coverslip (VWR) via oxygen plasma treatment (Harrick Scientific) to seal the channel while providing an optical window for fluorescence microscopy imaging. Tygon tubes (Cole-Parmer) are then inserted directly into the inlet and outlet ports of the device to allow loading of islets via gravity-driven flow and media flow (Rocheleau and Piston 2008). The inlet tube can subsequently be replaced by a PDMS well placed over the inlet as a reservoir for the imaging buffer (Figure 14.5b). Solution flow is then maintained by a syringe pump (Braintree Scientific) connected to the outlet tube (Figure 14.5a and b). To maintain physiological pH and temperature, the device loaded with islets was mounted inside a stage top microscope incubator (OkoLab) with controlled heated air (37°C) and CO_2 (5%) mix infusion. HEPES-buffered imaging media was used in short-term experiments to maintain physiological pH. Microfluidic devices should be treated as disposables, lasting up to 3 days of operation (Rocheleau and Piston 2008).

To illustrate the capability of this microfluidic-based platform, Figures 14.5d and e show the differential interference contrast (DIC) and 2P-microscopy-acquired NAD(P)H autofluorescence images of an intact islet stimulated by 20 mM glucose, respectively. It can be seen from Figure 14.5e that the fluorescence image displays a high level of subcellular resolution with discrimination of bright regions (circled) that are sufficiently separated from dark nuclei and non-redox responsive lipofuscin deposits (outlined triangles). This high level of spatial resolution is necessary to achieve sufficient dynamic range during quantitative image analysis.

14.3.4 Limitations of Confocal Ratiometric Imaging

Confocal ratiometric imaging of the mitochondrial flavoproteins ETF and LipDH has several limitations. First, collecting two sets of images instead of one can add to the complexity of analysis and results in a dimensionless ratio that is not readily intuitive. To place these dimensionless ratios into context, one can index them to a preestablished dynamic range. For instance, the ETF ratio can be placed into perspective by indexing to the pharmacological treatments that maximize (3 mM sodium cyanide + 20 mM glucose) and minimize (10 μM rotenone + 2 μM trifluoromethoxy-phenyl hydrazone [FCCP]) cellular redox state. More specifically, islets are routinely imaged 3 min after treatment with a solution containing 10 μM rotenone and 2 μM FCCP to establish the maximum oxidation of ETF (defined as zero) and 5 min after treatment with 3 mM sodium cyanide and 20 mM glucose to establish maximum reduction (defined as one), respectively. We find that subsequently plotting our ETF values relative to these maximum and minimum values provides a more intuitive sense of redox state.

Second, the images cannot be grossly measured. Instead, ROIs away from dark nuclei and lipofuscin deposits should be chosen in order to measure as much of the redox-responsive portion as possible.

Finally, the diminishing fluorescence of mitochondrial flavoproteins upon their reduction makes it more difficult to quantify the recorded signal. Despite these limitations, confocal ratiometric imaging of ETF and LipDH used in conjunction with 2P microscopy of NAD(P)H can yield insightful mechanistic information on mitochondrial metabolism in living pancreatic islets as illustrated by the two case studies described in Sections 14.5 and 14.6.

14.4 Islet-Related Methods and Materials

In this section, we focus on the specific methods and materials used for isolating, culturing, and treating pancreatic islets and applied for the case studies later.

14.4.1 Reagents and Preparation

Pharmacological treatments of etomoxir (product no. E1905), carbonyl cyanide *p*-FCCP (product no. C2920), rotenone (product no. R8875), sodium cyanide (product no. 205222), palmitic acid (product no. P9767), and RPMI 1640 medium (product no. R1383) were purchased from Sigma-Aldrich. Bovine serum albumin (BSA), fatty acid–free BSA, and polyethylenimine (PEI) were purchased from Invitrogen, Bioshop, and Polysciences, respectively. ETF-α (Open Biosystems catalog no. MMM1013-62815) plasmid was obtained from Thermo Scientific, while ETF-β (ATCC no. 10699891) and LipDH (ATCC no. MGC-5874) plasmids were purchased from Cedarlane Labs.

To prepare the stock solutions of the pharmacological treatments, rotenone and FCCP were dissolved in DMSO while sodium cyanide and etomoxir were dissolved in H_2O. The working concentrations of rotenone, sodium cyanide, FCCP, and etomoxir were finally diluted in the imaging buffer to 10 μM, 3 mM, 2 μM, and 50 μM, respectively. To prepare palmitate, palmitic acid was dissolved in 0.1 M NaOH to palmitate at 70°C. Fatty acid–free BSA was dissolved in ddH_2O by shaking at 4°C. Palmitate was then conjugated to the fatty acid–free BSA in ddH_2O using a ratio of 1 mM palmitate to 1% BSA at 60°C. The working concentrations of the BSA-conjugated palmitate solution were finally diluted to 0.4 and 0.04 mM in either islet culture media or imaging buffer as required.

14.4.2 Islet Isolation and Tissue Culture

Pancreatic islets were isolated from 8- to 12-week-old C57BL6 male mice by using collagenase digestion (Scharp et al. 1973; Stefan et al. 1987). Islets were subsequently cultured in full RPMI 1640 medium supplemented with 11 mM glucose, 10% FBS, and 5 U/mL penicillin–streptomycin at 37°C under 5% CO_2 for about 24 h before the first use. To maintain consistent tissue health, only islets cultured up to 3 days after the isolation procedures were used in experiments. The protocol was approved by the Animal Care Committee of the University Health Network (Toronto, Ontario, Canada) in compliance with the policies and guidelines set forth by the Canadian Council on Animal Care (Animal Use Protocol No. 1531).

14.4.3 Imaging Buffer and Islet Loading

Prior to loading to a microfluidic device for microscopy imaging, islets were typically equilibrated in imaging buffer (125 mM NaCl, 5.7 mM KCl, 2.5 mM $CaCl_2$, 1.2 mM $MgCl_2$, 10 mM HEPES, 2 mM glucose, and 0.1% BSA, pH 7.4) for about 1 h with or without the required concentration of palmitate (fatty acids) to achieve a consistent basal activity level. BSA and HEPES were used to reduce sticking of the tissue to the channel's walls and maintain the physiological pH level, respectively. In each experiment, 5–10 islets were loaded into the microfluidic channel through the inlet tubing via gravity-driven flow as described earlier. After the islet loading, the inlet tube was replaced by a PDMS well over the inlet to hold the imaging buffer with or without the stimulation treatments. The loaded device was kept at a constant temperature of 37°C throughout the experiment using the stage top incubator. Flow rate was typically maintained at 200 μL per hour using a syringe pump connected to the outlet tube.

14.5 Case Study I: Characterization of the ETC Activity

The ETC plays an important role in mitochondrial metabolism enabling tight coupling of oxidative phosphorylation to the generation of ATP. In β-cells, glucose-induced rise in ATP is necessary for the exocytosis of insulin. In this study, we characterize the role of the ETC in setting the redox state of ETF and LipDH. Confocal ratiometric imaging and 2P microscopy were used on intact islets that are stimulated with pharmacological treatments targeting the ETC activity.

14.5.1 Redox-Associated Changes in Flavin Autofluorescence

To examine how redox changes in the ETC correlate with the mitochondrial flavin autofluorescence, we applied rotenone and sodium cyanide, which are inhibitors of Complexes I and IV of the ETC, respectively (Figure 14.6a). Upon stimulation with rotenone, the 2P NAD(P)H autofluorescence intensity of the islets turned brighter compared to 2 mM glucose control, while the flavin image intensities (488LP, 458BP, and 458LP), recorded with a confocal microscope, became dimmer. These visual responses are consistent with the rotenone's reducing capacity—only reduced pyridine nucleotides are fluorescent, while reduced mitochondrial flavins are not. Comparison of the paired images also shows dark nuclear regions (non-mitochondrial) and bright lipofuscin deposits that are non-redox responsive. Subsequent addition of sodium cyanide induced relatively little change in the 2P NAD(P)H image intensity. In contrast, the confocal flavin image intensities decreased significantly with sodium cyanide, indicative

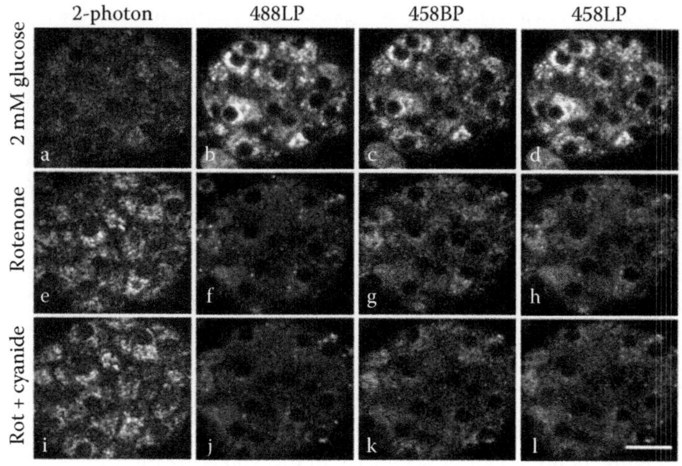

(a)

FIGURE 14.6 Relative changes in islet autofluorescence due to ETC activity. (a) Representative autofluorescence images using 2P 705 nm excitation with a 380–550 nm emission filter (*a*, *e*, and *i*), 488 nm excitation with a long-pass 505 nm emission filter (*b*, *f*, and *j*), 458 nm excitation with a band-pass 462–525 nm emission filter (*c*, *g*, and *k*), and 458 nm excitation with a long-pass 525 nm emission filter (*d*, *h*, and *l*). The scale bar represents 20 μm.

(Continued)

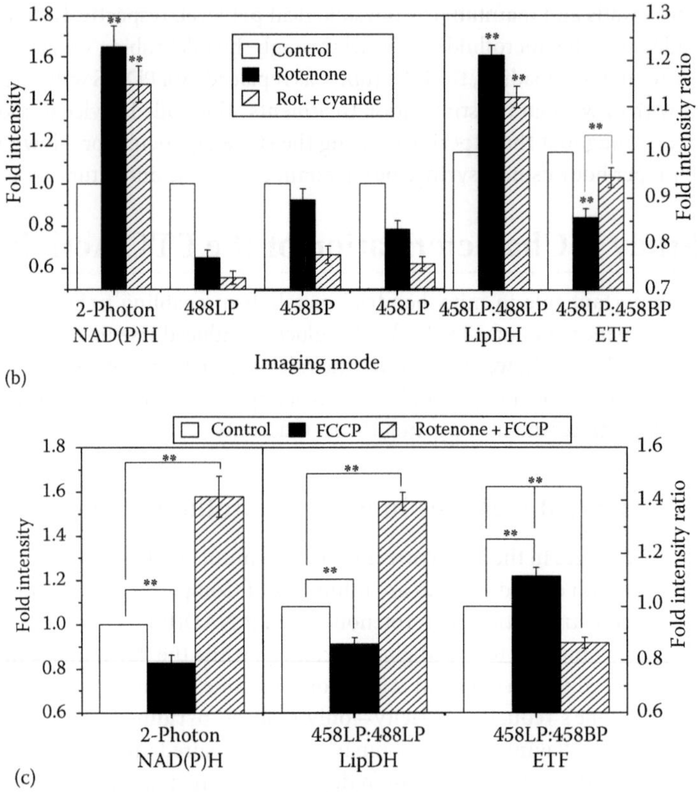

FIGURE 14.6 (*Continued*) Relative changes in islet autofluorescence due to ETC activity. (b) Stimulation with 10 μM rotenone (*black bars*) followed by 3 mM sodium cyanide (*hatched bars*). (c) Stimulations with 2 μM FCCP (*black bars*) and 10 μM rotenone + 2 μM FCCP (*hatched bars*). The fold intensity (*left vertical axis*) and fold intensity ratio (*right vertical axis*) are reported relative to 2 mM glucose (*open bars*). The data are reported as the means ± standard error of the mean ($n = 15$ *islets from two different mice* and $n = 11$ *islets from two different mice* were used for Figures 14.6a through c, respectively). Statistical significance relative to 2 mM glucose (*open bars*) or as indicated was represented by the symbol ** ($P < 0.01$) using one-way analysis of variance (ANOVA) followed by Tukey post hoc test. (From Lam, A.K., Silva, P.N., Altamentova, S.M., and Rocheleau, J.V., Quantitative imaging of electron transfer flavoprotein autofluorescence reveals the dynamics of lipid partitioning in living pancreatic islets, *Integr. Biol.*, 4(8), 838–846, 2012, Figures 4 and 5. Reproduced by permission of the Royal Society of Chemistry.)

of further reduction of flavins downstream of Complex I. Taken together, one can appreciate that the visual responses of flavin autofluorescence are consistent with reduction of the ETC.

14.5.2 Supply Side of the ETC

Rotenone and sodium cyanide block the supply of electrons to the ETC at and downstream of Complex I, respectively. From Figure 14.6b, it can be seen that 10 μM rotenone induced a significant increase in NAD(P)H intensity compared to 2 mM glucose (Figure 14.6b, *black* vs. *open bar*) and decreased the intensity of each of the flavin imaging modes to varying degrees (Figure 14.6a). The LipDH (458LP:488LP) ratio was also increased by rotenone, consistent with accumulation of mitochondrial NADH due to the inhibition of Complex I (Figure 14.6b, *black* vs. *open bar*). In contrast, the ETF (458LP:458BP) ratio was decreased, suggesting oxidation of ETF by increased ETC activity downstream of Complex I

(Figure 14.6b, *black* vs. *open bar*). Taken together, these data show that inhibition of Complex I induces independent responses from the ETF and LipDH imaging ratios.

When 3 mM of sodium cyanide was added subsequently to block Complex IV of the ETC, the NAD(P)H intensity and LipDH (458LP:488LP) ratio were relatively unchanged as compared to islets treated with rotenone alone (Figure 14.6b, *hatched* vs. *black bar*). These results are indicative of no further reduction of mitochondrial NADH (and—hence—of LipDH), with inhibition of Complex IV when Complex I is already inhibited by rotenone (Figure 14.6b, *hatched* vs. *black bar*). In contrast, the ETF (458LP:458BP) ratio was significantly higher than with rotenone alone. These data are consistent with a reduction of ETF by decreased ETC activity with closure of Complex IV. Taken together, one can appreciate that the intensity ratios of LipDH and ETF correctly reflect the supply of electrons to Complex I and downstream of Complex I of the ETC, respectively.

14.5.3 Demand Side of the ETC

Having looked at the supply side of the ETC, we turned our attention to the demand side and examined the response of the ETC to increased demand via carbonyl cyanide *p*-FCCP treatment. FCCP treatment increases electron outflow at Complex IV by dissipating the proton gradient across the inner mitochondrial membrane to effectively decouple Complex IV from oxidative phosphorylation.

The steady-state redox potential of ETF depends on both the rate of reduction by fatty acid metabolism and the rate of oxidation by the ETC downstream of Complex I. However, the ability of the ETC to oxidize ETF may also depend on the availability of Complex IV that can become saturated by electrons from Complex I (NADH). In other words, the flow of electrons to Complex IV may be preoccupied by electrons from Complex I leaving ETF in a reduced state. To determine whether Complex I dominates electron outflow at Complex IV, we applied the pharmacological treatments of (1) 2 μM FCCP and (2) 2 μM FCCP + 10 μM rotenone to stimulate living pancreatic islets (Figure 14.6c).

We focused on the oxidative responses up- and downstream of Complex I by also limiting fuel supply. After FCCP treatment, the 2P NAD(P)H intensity and LipDH redox state decreased significantly relative to the 2 mM glucose control while the ETF redox state increased significantly (Figure 14.6c, *black* vs. *open bars*). These data show that relatively more electrons are drawn from Complex I than ETF-QO to fulfill the FCCP-induced increase in the ETC activity, causing the redox state of LipDH to drop while increasing ETF redox state. Addition of rotenone subsequently increased 2P NAD(P)H intensity and LipDH redox state relative to the 2 mM glucose control (Figure 14.6c, *hatched* vs. *open bars*). These data show inhibition of Complex I induces accumulation of mitochondrial NADH resulting in a rise in both LipDH redox state and 2P NAD(P)H intensity. In contrast, the ETF redox state decreased significantly with rotenone treatment relative to the FCCP and 2 mM glucose control treatments (Figure 14.6c, *hatched* vs. *black and open bars*). These data suggest that as Complex I is blocked and no longer dominates the supply of electrons to the ETC, more electrons are drawn from ETF-QO to decrease ETF redox state. Taken together, Complex I has a higher priority than ETF-QO to supply electrons to the ETC consistent with glucose metabolism being the predominant energy supply in pancreatic islet β-cells.

14.6 Case Study II: Dynamic Interplay of Fat and Glucose Metabolism

As mentioned earlier, chronic exposure of pancreatic islets to an excess supply of fats and sugars in the blood from overeating and lack of physical exercise results in glucolipotoxicity and can lead to type II diabetes. Therefore, studying the role of fatty acids in glucolipotoxicity and pancreatic islet β-cell dysfunction has therapeutic relevance (Tahrani et al. 2011). It was hypothesized that the interplay of fatty acid and glucose metabolism plays a critical role in contributing to glucolipotoxicity due to shared molecular pathways particularly in mitochondrial metabolism (Figure 14.3). To examine this interplay, we dynamically imaged ETF and NAD(P)H autofluorescence in islets stimulated with these two nutrients in solution.

14.6.1 Response Time of Intrinsic Fat Metabolism

The oxidation of fat is an iterative process involving large molecules suggesting that ETF responses may be slow due to the metabolism of fatty acid reserves within mitochondria. To measure ETF dynamics in response to diminished fatty acid oxidation, we applied etomoxir as a well-established inhibitor of CPT-1-dependent transport of fatty acids into the mitochondria. From Figure 14.7a, it can be seen that 0.04 mM palmitate induced a significant rise in the ETF redox index compared to the 2 mM

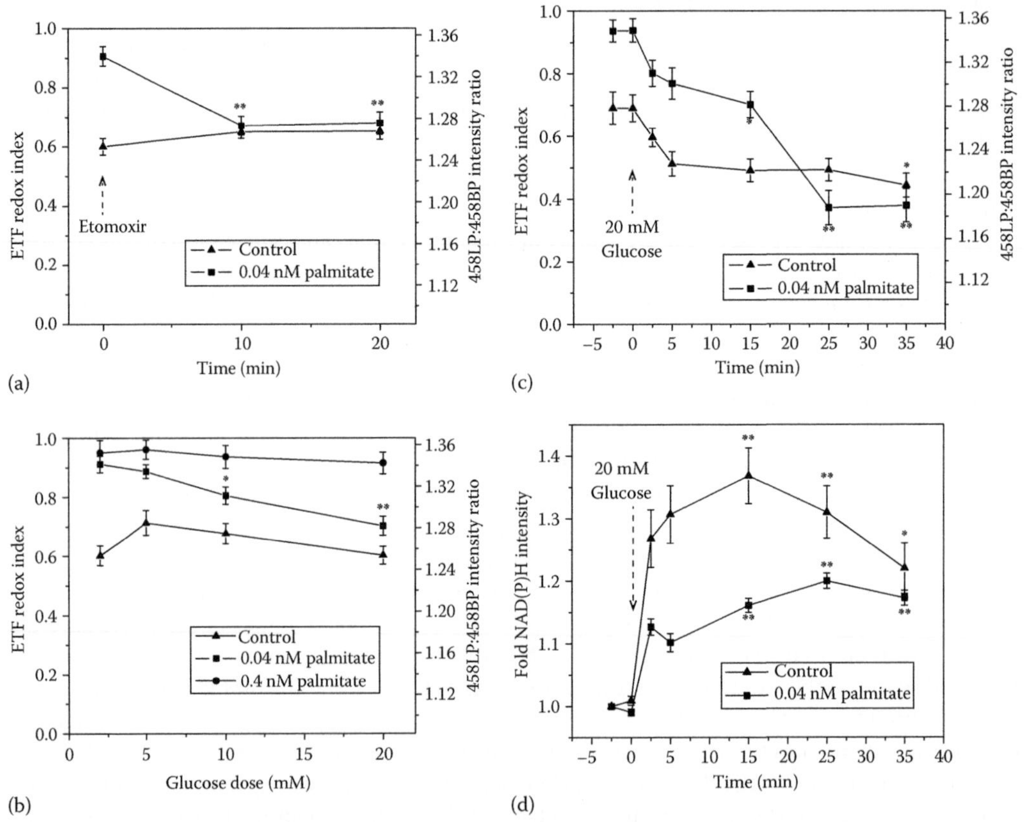

FIGURE 14.7 Interplay of fat and glucose metabolism. (a) Stimulation with 50 μM etomoxir ($n = 17$ *islets from three different mice, symbol* ■; $n = 9$ *islets from two different mice, symbol* ▲). (b) glucose dose response of ETF. The images were collected after ~15 min of stimulation with each of the indicated glucose doses ($n = 18$ *islets from three different mice, symbol* ▲; $n = 16$ *islets from three different mice, symbol* ■; $n = 10$ *islets from two different mice, symbol* ●). (c) Temporal ETF response to a glucose bolus. (d) Temporal NAD(P)H response to a glucose bolus ($n = 18$ *islets from three different mice, symbol* ■; $n = 10$ *islets from three different mice, symbol* ▲ for (c) and (d). The data are reported as mean ± standard error of the mean. Statistical significance relative to the starting point was represented by the symbols * ($P < 0.05$) and ** ($P < 0.01$) using one-way ANOVA followed by Tukey post hoc test. (From Lam, A.K., Silva, P.N., Altamentova, S.M., and Rocheleau, J.V., Quantitative imaging of electron transfer flavoprotein autofluorescence reveals the dynamics of lipid partitioning in living pancreatic islets, *Integr. Biol.*, 4(8), 838–846, 2012, Figures 6 and 7. Reproduced by permission of the Royal Society of Chemistry.)

glucose control. Subsequent addition of 50 μM etomoxir induced a significant decrease in ETF redox state within 10 min. These data show that after turning off the transport of fatty acids into mitochondria, ETF is oxidized within 10 min, suggesting the intrinsic mitochondrial reservoir of fatty acids is diminished within this time.

14.6.2 Glucose-Mediated Fat Metabolism

Pancreatic islet β-cells also become damaged during chronic fatty acid stimulation in the presence of excessive glucose (Prentki and Nolan 2006; Ruderman and Prentki 2004). To examine glucose and fatty acid metabolism further, we measured the ETF redox state at increasing glucose doses (Figure 14.7b). Recall that ETF redox state is measured by indexing the ETF intensity ratio to the predetermined dynamic range (from zero to one). Pancreatic islets were first equilibrated in varying amounts of palmitate (0, 0.04, and 0.4 mM) and 2 mM glucose for 1 h prior to the experiment. The islets were then imaged in the same concentration of palmitate with increasing glucose doses (2–20 mM). The ETF redox index was measured after 15 min of the indicated glucose stimulation to allow sufficient oxidation of the remaining reservoir of fat in the mitochondria. The glucose-induced ETF redox index in the absence of palmitate remained relatively unchanged or flat at ~0.6 consistent with no significant effect of glucose in the absence of fatty acid stimulation. At 0.04 mM palmitate and 2 mM glucose, the ETF redox index at 2 mM glucose was ~0.9 and progressively dropped with increasing amounts of glucose. These results indicate that palmitate metabolism induces a drop in ETF redox state and that fatty acid oxidation is decreased at glucose concentrations above 10 mM. In the presence of tenfold higher 0.4 mM palmitate, the ETF responded with an extremely high redox index of ~0.95 and stayed relatively unchanged: 0.4 mM palmitate oversaturated mitochondrial fatty acid oxidation and continued to supply electrons to ETF even in the presence of >10 mM glucose. Taken together, the results indicate a dominant role of glucose metabolism over fatty acid oxidation in pancreatic islet β-cells when presented with a mixed nutrient condition associated with glucolipotoxicity. More specifically, pancreatic islet β-cells maintain a dynamic glucose response by switching between anabolism and catabolism of fatty acids in a process termed lipid partitioning (Prentki et al. 2002).

14.6.3 Temporal Courses of Fat and Glucose Metabolism

To examine the temporal dynamics associated with glucose-mediated lipid partitioning, we simultaneously imaged the ETF and NAD(P)H responses to a bolus of glucose in the absence and presence of 0.04 mM palmitate (Figure 14.7c and d). These data again show that islets in low glucose treated with palmitate exhibit a significantly higher ETF redox index (Figure 14.7c). In the absence of palmitate, glucose stimulated a straight decrease in ETF redox state within 5 min. In contrast, a two-tier decline in ETF redox was observed in the presence of palmitate. The first drop in ETF occurred at a similar time and to a similar magnitude to the drop observed in the absence of palmitate, suggesting dependence on ETC activity rather than fatty acid oxidation. The second reduction in the ETF redox index occurred at about 20 min consistent with our previous data on glucose-induced inhibition of the mitochondrial fatty acid oxidation (Figure 14.7b). Palmitate also significantly shaped the dynamics of the glucose-stimulated NAD(P)H response (Figure 14.7d). Palmitate-treated islets exhibited a lower and less robust NAD(P)H response relative to glucose-only islets reflecting a net decline in the production of NADH and NADPH during early time points. Notably, the ETF and NAD(P)H responses (Figures 14.7c and d, respectively) are similar at later time points for both palmitate-treated and glucose-only islets suggesting that both eventually turn off fatty acid oxidation and metabolize glucose similarly. Therefore, the unique differences between these sets of islets would have been missed if one were to use end point measurements rather than temporal measurements.

Overall, the case studies presented demonstrate the utility of confocal ratiometric imaging of ETF and LipDH in yielding mechanistic insights into mitochondrial metabolism. Together with 2P microscopy of NAD(P)H, the dynamic imaging of intracellular coenzyme autofluorescence is proving to be a natural test system to effectively monitor the metabolic activities in intact pancreatic islets.

14.7 Summary and Future Directions

Diabetes researchers would like to probe into the dynamics of pancreatic islet metabolism as a means to dissect the molecular mechanisms behind the regulation of insulin secretion and glucose homeostasis. Conventional biochemical assays and extrinsic labels present formidable hurdles to this goal due to their lack of biological context and potential molecular interference, respectively. Advances in fluorescence microscopy and microfluidic technologies create capabilities to bridge the gap. In this chapter, we introduced dynamic imaging of intracellular coenzyme autofluorescence as an alternative way to probe into the metabolic activities in intact pancreatic islets with direct relevance to diabetes. More specifically, we presented two representative techniques, namely, 2P excitation microscopy of NAD(P)H and confocal ratiometric imaging of ETF and LipDH. We also demonstrated the capacity of these autofluorescence techniques through the two case studies on mitochondrial metabolism in the context of the overnutrition and glucolipotoxicity postulated to drive type II diabetes. Despite the limitations of these techniques, dynamic imaging of intracellular coenzyme autofluorescence provides insightful information on the metabolic events as they occur and as naturally as possible in intact pancreatic islets.

Looking to the future, we anticipate that dynamic imaging of intracellular coenzyme autofluorescence will show utility in dissecting nutrient-induced pancreatic islet β-cell dysfunction. Potential therapeutic strategies targeting fat-induced effects on β-cell dysfunction warrant future investigation through the ETF imaging platform. Furthermore, advances in fluorescence microscopy, microfluidic technologies, and technique development will extend autofluorescence imaging applications to an ever-wider range of tissues and diseases.

Acknowledgments

These studies were generously supported by a Natural Sciences and Engineering Research Council of Canada (NSERC) Discovery Research Grant (RGPIN 371705) to Jonathan V. Rocheleau and stipend support from an NSERC CREATE MATCH (Microfluidic Applications and Training in Cardiovascular Health) Program scholarship to Alan K. Lam. The microscope used in these studies was purchased through a Canada Foundation for Innovation (CFI) Leaders Opportunity Fund (18301) to Jonathan V. Rocheleau.

References

Bennett, B. D., T. L. Jetton, G. Ying, M. A. Magnuson, and D. W. Piston. 1996. Quantitative subcellular imaging of glucose metabolism within intact pancreatic islets. *Journal of Biological Chemistry* 271 (7):3647–3651.

Chance, B. and B. Thorell. 1959. Localization and kinetics of reduced pyridine nucleotide in living cells by microfluorometry. *Journal of Biological Chemistry* 234:3044–3050.

Duffy, D. C., J. C. McDonald, O. J. Schueller, and G. M. Whitesides. 1998. Rapid prototyping of microfluidic systems in poly(dimethylsiloxane). *Analytical Chemistry* 70 (23):4974–4984.

Hall, C. L., L. Heijkenskjold, T. Bartfai, L. Ernster, and H. Kamin. 1976. Acyl coenzyme A dehydrogenases and electron-transferring flavoprotein from beef hart mitochondria. *Archives of Biochemistry and Biophysics* 177 (2):402–414.

Heikal, A. A. 2010. Intracellular coenzymes as natural biomarkers for metabolic activities and mitochondrial anomalies. *Biomarkers in Medicine* 4 (2):241–263.

Jitrapakdee, S., A. Wutthisathapornchai, J. C. Wallace, and M. J. MacDonald. 2010. Regulation of insulin secretion: Role of mitochondrial signalling. *Diabetologia* 53 (6):1019–1032.

Kunz, W. S. 1986. Spectral properties of fluorescent flavoproteins of isolated rat liver mitochondria. *FEBS Letters* 195 (1–2):92–96.

Kunz, W. S. 1988. Evaluation of electron-transfer flavoprotein and alpha-lipoamide dehydrogenase redox states by two-channel fluorimetry and its application to the investigation of beta-oxidation. *Biochimica et Biophysica Acta (BBA)—Bioenergetics* 932 (1):8–16.

Kunz, W. S. and W. Kunz. 1985. Contribution of different enzymes to flavoprotein fluorescence of isolated rat liver mitochondria. *Biochimica et Biophysica Acta (BBA)—General Subjects* 841 (3):237–246.

Lam, A. K., P. N. Silva, S. M. Altamentova, and J. V. Rocheleau. 2012. Quantitative imaging of electron transfer flavoprotein autofluorescence reveals the dynamics of lipid partitioning in living pancreatic islets. *Integrative Biology* 4 (8):838–846.

Mastracci, T. L. and L. Sussel. 2012. The endocrine pancreas: Insights into development, differentiation and diabetes. *Wiley Interdisciplinary Reviews: Developmental Biology* 1 (5):609–628.

Newgard, C. B. and J. D. McGarry. 1995. Metabolic coupling factors in pancreatic beta-cell signal transduction. *Annual Review of Biochemistry* 64 (1):689–719.

Nolan, C. J. and M. Prentki. 2008. The islet beta-cell: Fuel responsive and vulnerable. *Trends in Endocrinology and Metabolism* 19 (8):285–291.

Piston, D. W. 1999. Imaging living cells and tissues by two-photon excitation microscopy. *Trends in Cell Biology* 9 (2):66–69.

Piston, D. W. and S. M. Knobel. 1999. Real-time analysis of glucose metabolism by microscopy. *Trends in Endocrinology and Metabolism* 10 (10):413–417.

Poitout, V., J. Amyot, M. Semache et al. 2010. Glucolipotoxicity of the pancreatic beta cell. *Biochimica et Biophysica Acta (BBA)—Molecular and Cell Biology of Lipids* 1801 (3):289–298.

Poitout, V. and R. P. Robertson. 2002. Minireview: Secondary beta-cell failure in type 2 diabetes—A convergence of glucotoxicity and lipotoxicity. *Endocrinology* 143 (2):339–342.

Poitout, V. and R. P. Robertson. 2008. Glucolipotoxicity: Fuel excess and beta-cell dysfunction. *Endocrine Reviews* 29 (3):351–366.

Prentki, M., E. Joly, W. El-Assaad, and R. Roduit. 2002. Malonyl-CoA signaling, lipid partitioning, and glucolipotoxicity: Role in beta-cell adaptation and failure in the etiology of diabetes. *Diabetes* 51 (Suppl. 3):S405–S413.

Prentki, M. and C. J. Nolan. 2006. Islet beta cell failure in type 2 diabetes. *Journal of Clinical Investigation* 116 (7):1802–1812.

Roberts, D. L., F. E. Frerman, and J. J. Kim. 1996. Three-dimensional structure of human electron transfer flavoprotein to 2.1-A resolution. *Proceedings of the National Academy of Sciences of the United States of America* 93 (25):14355–14360.

Rocheleau, J. V., W. S. Head, W. E. Nicholson, A. C. Powers, and D. W. Piston. 2002. Pancreatic islet beta-cells transiently metabolize pyruvate. *Journal of Biological Chemistry* 277 (34):30914–30920.

Rocheleau, J. V., W. S. Head, and D. W. Piston. 2004. Quantitative NAD(P)H/flavoprotein autofluorescence imaging reveals metabolic mechanisms of pancreatic islet pyruvate response. *Journal of Biological Chemistry* 279 (30):31780–31787.

Rocheleau, J. V. and D. W. Piston. 2008. Combining microfluidics and quantitative fluorescence microscopy to examine pancreatic islet molecular physiology. In *Methods in Cell Biology*, Vol. 89, J. J. Correia and H. W. Detrich (eds.), pp. 71–92. Oxford, U.K.: Elsevier.

Ruderman, N. and M. Prentki. 2004. AMP kinase and malonyl-CoA: Targets for therapy of the metabolic syndrome. *Nature Reviews Drug Discovery* 3 (4):340–351.

Sankar, K. S., B. J. Green, A. R. Crocker et al. 2011. Culturing pancreatic islets in microfluidic flow enhances morphology of the associated endothelial cells. *PLoS One* 6 (9):e24904.

Scharp, D. W., C. B. Kemp, M. J. Knight, W. F. Ballinger, and P. E. Lacy. 1973. The use of ficoll in the preparation of viable islets of Langerhans from the rat pancreas. *Transplantation* 16 (6):686–689.

Stefan, Y., P. Meda, M. Neufeld, and L. Orci. 1987. Stimulation of insulin secretion reveals heterogeneity of pancreatic B cells in vivo. *Journal of Clinical Investigation* 80 (1):175–183.

Tahrani, A. A., C. J. Bailey, S. Del Prato, and A. H. Barnett. 2011. Management of type 2 diabetes: New and future developments in treatment. *Lancet* 378 (9786):182–197.
Toogood, H. S., D. Leys, and N. S. Scrutton. 2007. Dynamics driving function: New insights from electron transferring flavoproteins and partner complexes. *FEBS Journal* 274 (21):5481–5504.
Unger, R. H. 1981. The milieu interieur and the islets of Langerhans. *Diabetologia* 20 (1):1–11.
Watmough, N. J. and F. E. Frerman. 2010. The electron transfer flavoprotein: Ubiquinone oxidoreductases. *Biochimica et Biophysica Acta (BBA)—Bioenergetics* 1797 (12):1910–1916.

15

Autofluorescence Diagnostics of Ophthalmic Diseases

Dietrich Schweitzer
University of Jena

15.1 Introduction

First pathological alterations occur in metabolism and they are reversible with both early detection and proper treatment. For that reason, noninvasive methods for detecting functional changes are particularly important, especially before any morphological alterations become detectable.

There are two aspects for in vivo investigation of the metabolism in tissue: measuring the extracellular metabolism and studying the intracellular metabolism. At the present time, research for determination of healthy metabolic conditions and of their pathological changes is directed on the extracellular metabolism in ophthalmology.

The extracellular metabolism in the retina, especially the supply by and the consumption of oxygen, can be characterized by measuring parameters of the microcirculation. One important parameter is blood flow rate (Q) in retinal vessels, which can be calculated from vessel diameter (d) and velocity (v) of blood according to the following:

$$Q = \frac{\pi \cdot d^2}{4} \cdot v = \frac{V}{t} \tag{15.1}$$

This equation describes which volume V of blood penetrates through a vessel per unit of time t. To investigate the supply of oxygen in the retinal tissue, the oxygen saturation (OS) must also be measured.

The oxygen saturation is the relative concentration (c_{Hbo_2}) of oxyhemoglobin in relation to all components of hemoglobin in the blood, for example, reduced hemoglobin c_{Hb}, c_{HbA1c}:

$$OS = \frac{c_{\mathrm{Hbo}_2}}{c_{\mathrm{Hbo}_2} + c_{\mathrm{Hb}} + c_{\mathrm{HbA1c}}} \quad (15.2)$$

Dividing the fundus in quadrants with the optic disk in the center, each quadrant is supplied by a pair of an arteriole and a venule. The product of blood flow and arterial oxygen saturation (OS_a) describes the supply by oxygen:

$$\text{Oxygen supply} = Q \cdot OS_a \cdot c_{total} \cdot \eta_o = \frac{V}{t} \cdot \frac{c_{\mathrm{Hbo}_2}}{c_{\mathrm{Hbo}_2} + c_{\mathrm{Hb}} + c_{\mathrm{HbA1c}}} \cdot c_{total} \cdot \eta_o \quad (15.3)$$

where

c_{total} is the total concentration of hemoglobin in the blood

η_o is a transport coefficient, which describes how many Mol oxygen are transported by 1 Mol of hemoglobin

Measuring both arterial and venous oxygen saturation, the consumption of oxygen in the considered tissue can be calculated:

$$\text{Oxygen consumption} = Q \cdot \left(OS_a - OS_v\right) \cdot c_{total} \cdot \eta_o \quad (15.4)$$

Assuming a blood flow $Q = 21$ µL/s, an arterial oxygen saturation $OS_a = 95\%$, a total hemoglobin concentration/4 $c_{total} = 8.9$ mMol/l, and a transport effectiveness η_o, which means that 1 Mol O_2 (32 g) is transported by 1/Mol hemoglobin (≈16,100 g), then the tissue is supplied by 0.358 ng/s. If the arteriovenous difference $OS_{a-v} = 35\%$, then the consumption of oxygen is about 0.132 ng/s in the supplied retinal tissue (Schweitzer et al. 1995).

Such measurements are done by fundus reflectometry using the extinction spectra of specific chromophores and of wavelength-dependent scattering of light in the tissue. In this chapter, we discuss the methods of metabolism mapping in the eye, data analysis, and interpretation of the results obtained by this approach.

15.2 Characterization of Endogenous Fluorophores in the Eye

In contrast to isolated cells, several endogenous fluorophores contribute to the detectable fluorescence in biological tissue, especially at the living eye ground, which has a layered anatomical structure. To investigate the influence of single fluorophores of the detected fluorescence, substance-specific properties can be investigated. As discussed throughout this book, fluorophores can be discriminated from each other by the excitation and emission spectra and by the decay of fluorescence intensity after pulsed excitation.

When investigating the spectra, it is important also to take into consideration the transmission of the ocular media (Figure 15.1). Although the excitation maximum of reduced nicotinamide adenine dinucleotide (NADH) in phosphate buffered saline (PBS) solution (pH 7.4) is about 350 nm with a corresponding emission maxima at 460 nm, it can also be excited at 446 nm (within the ocular transmission window) with a detectable signal at 530 nm (Schweitzer et al. 2007b).

The excitation spectrum of flavine adenine dinucleotide (FAD) has two maxima, one at about 360 nm and another at 450 nm. When excited at 350 nm, two fluorescence maxima are detectable at 440 and 525 nm. However, under 446 nm excitation, FAD emission peaks at 530 nm. As a result, the short-wavelength maximum of FAD excitation should be taken into account, when exciting NADH at 350 nm (Figure 15.2).

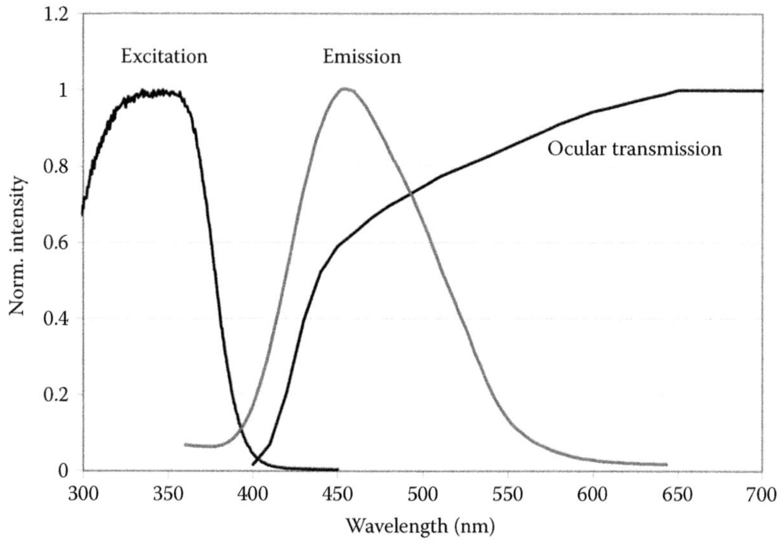

FIGURE 15.1 Comparative plot of the excitation (maximum at 350 nm) and emission (maximum at 460 nm) spectra of NADH in PBS (150 mM, pH 7.4), and the transmission of the ocular media. (From van de Kraats, J. and van Norren, D., *J. Opt. Soc. Am. A*, 24(7), 1842, 2007.)

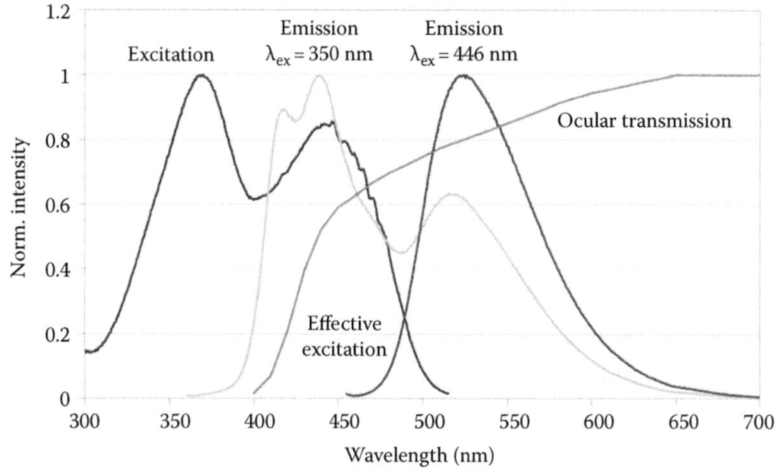

FIGURE 15.2 Excitation and emission spectra of FAD (100 μM in water). (From Schweitzer, D., Quantifying fundus autofluorescence, in: *Fundus Autofluorescence*, Lois, N. and Forrester, J.V., eds., Lippincott Williams & Wilkins, Philadelphia, PA, 2009, pp. 78–95. With permission.)

Although NADH and FAD are mainly the most abundant intrinsic fluorophores throughout the human body, at the fundus, the dominating fluorophore is lipofuscin. The fluorescence spectrum of a 20° fundus field of the macular region is shown in Figure 15.3. The shape of the fundus fluorescence spectrum is dominated by A2E (*N*-retinylidene-*N*-retinylethanolamine), the component VIII in the chloroform extract of retinal lipofuscin (Eldred and Katz 1988). The fundus fluorescence spectrum was corrected for the fluorescence of the crystalline lens, for ocular transmission, and for absorption in the macular pigment xanthophyll. Some local maxima are apparent in the fluorescence emission and are

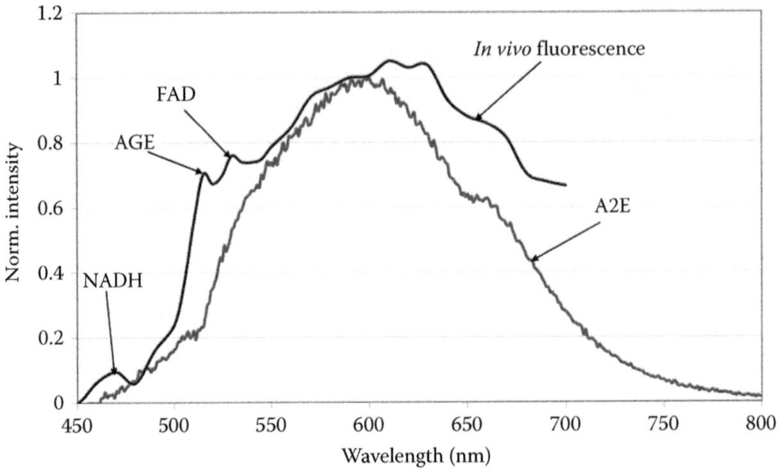

FIGURE 15.3 Comparison of the in vivo measured fundus fluorescence spectrum with the emission spectrum of *N*-retinylidene-*N*-retinylethanolamine (A2E) (1.28 μM in ethanol).

assigned to several weakly emitting fluorophores (e.g., FAD, NADH, advanced glycation end products [AGEs], collagen, and elastin). These fluorophores are not separable in fundus images with a high lateral resolution by measurements of emission spectra. In contrast, fluorescence lifetime measurements permit the discrimination of these fluorophores, especially when these ensemble fluorophores undergo structural changes after provocation tests. For interpretation of fluorescence lifetime measurements, it is necessary to know the specific lifetimes of endogenous fluorophores that may contribute to the detected autofluorescence.

The published lifetime data of fluorophores differ considerably in the literature. Variations in the data can be caused by different measuring techniques, different solvents, or different model functions for the approximation of the decay of fluorescence intensity. Table 15.1 is a compilation of lifetimes of endogenous fluorophores (Meyer 2012).

15.3 Limitations for FLIM Application in Ophthalmology

In contrast to fluorescence lifetime measurement on microscopic samples, there are several serious limitations for fluorescence lifetime imaging microscopy (FLIM) application in ophthalmology. Most limitations result from in vivo measurements of the living human eye under both diseased and healthy conditions. First, the eye must not be damaged during the lifetime measurements even in the worst-case scenario (ANSI 2000). As a consequence, the power of the excitation light should be as low as possible. This requirement is optimal for the application of time-correlated single-photon counting technique (TCSPC) for measuring fluorescence decay in the time domain (discussed in details in Chapter 4). In TCSPC, only one photon (1P) should be detectable in a series of 10–100 excitation pulses (Becker 2005). As a consequence, the accumulation of a sufficient number of photons in each time channel takes a long acquisition time during which the movement of the living eyes cannot be controlled. As a result, the eye movements must be compensated for during data acquisition using, for example, online image registration. Furthermore, the fluorescence of different anatomical layers is likely to contribute to the detected fluorescence decay. There is also a difference in the appearance time of the fluorescence from both the crystalline lens and the layered fundus in the eye. Furthermore, a selective excitation of single fluorophores in the ultraviolet (UV)-spectral region is not possible. The transmission of the ocular media permits spectral measurements only for wavelengths above 400 nm (Geeraets and Berry 1968). The crystalline lens acts as a UV filter with an absorption edge at 400 nm. Thus, the fundus is protected

TABLE 15.1 Fitting Parameters for Fluorescence Lifetime Measurements of Endogenous Fluorophores under Both 1P and 2P Excitations

Fluorophore	Lifetime/Amplitude (%)		Excitation (nm)	Emission (nm)	References
NADH (free)	$\tau = 0.30–0.40$ ns	α = n/a	350 (1P)/700–800 (2P)	450	Berezin and Achilefu (2010)
	$\tau_1 = 0.39$ ns $\tau_2 = 3.65$ ns	$\alpha_1 = 73$ $\alpha_2 = 27$	340 (1P)	470	Schweitzer et al. (2007b)
	$\tau = 0.40–0.50$ ns	α = n/a	375 (1P)	n/a	Schneckenburger et al. (2004)
	$\tau = 0.35–0.45$ ns	α = n/a	760 (2P)	n/a	Bird et al. (2005)
	$\tau_1 = 0.15$ ns (*free extended*) $\tau_2 = 0.60$ ns (*free folded or enzyme bound*)	$\alpha_1 = 44$ $\alpha_2 = 32$	740 (2P)	450–460 (emission filter 350–550)	Vishwasrao et al. (2005)
	$\tau = 0.30$ ns	α = n/a	780 (2P)		Skala et al. (2007)
	$\tau = 0.45$ ns	α = n/a	n/a	n/a	Niesner et al. (2004)
	$\tau = 0.33$ ns	α = n/a	780 (2P)	460	Skala (2007)
	$\tau = 0.40$ ns	$\alpha = 73$	760 (2P)	500–550 (emission filter)	Peters et al. (2011)
	$\tau = 0.40$ ns	$\alpha = 78$	375 (1P)	442	Katikaa et al. (2006)
	$\tau = 0.30$ ns	α = n/a	300–340	460	Koenig (2008)
NADH (bound)	$\tau = 2.30–3.00$ ns	α = n/a	350 (1P)/700–800 (2P)	450	Berezin and Achilefu (2010)
	$\tau = 1.70–3.1$ ns	α = n/a	760 (2P)	n/a	Bird et al. (2005)
	$\tau_3 = 2.20$ ns $\tau_4 = 6.04$ ns	$\alpha_3 = 20$ $\alpha_4 = 4$	740 (2P)	450–460 (emission filter 350–550)	Vishwasrao et al. (2005)
	$\tau = 2.40$ ns	α = n/a	780 (2P)		Skala et al. (2007)
	$\tau = 2.98$ ns	α = n/a	n/a	n/a	Niesner et al. (2004)
	$\tau = 2.20$ ns	α = n/a	780 (2P)	460	Skala (2007)
	$\tau = 2.40$ ns	$\alpha = 27$	760 (2P)	500–550 (emission filter)	Peters et al. (2011)
	$\tau = 2.70$ ns	$\alpha = 18$	375 (1P)	442	Katikaa et al. (2006)
	$\tau = 2.00–2.30$ ns	α = n/a	300–340	440	Koenig (2008)
FAD (free)	$\tau = 2.30–2.90$ ns	α = n/a	n/a	n/a	Berezin and Achilefu (2010)
	$\tau_1 = 2.57$ ns $\tau_2 = 4.42$ ns (*in a pH 7.4 buffer*)	$\alpha_1 = 71$ $\alpha_2 = 29$			Heikal (2010)
	$\tau_1 = 0.33$ ns $\tau_2 = 2.81$ ns	$\alpha_1 = 18$ $\alpha_2 = 82$	450 (1P)	530	Schweitzer et al. (2007a)
	$\tau = 2.30–2.75$ ns	α = n/a	315 (1P)	584 (20°C) 557 (5°C)	Tanaka et al. (1989)
	$\tau_1 = 2.05$ ns	α = n/a	890 (2P)	n/a	Skala (2007)
	$\tau_1 = 2.00$ ns	$\alpha = 15$	760 (2P)	550–700 (emission filter)	Peters et al. (2011)

(*Continued*)

TABLE 15.1 (*Continued*) Fitting Parameters for Fluorescence Lifetime Measurements of Endogenous Fluorophores under Both 1P and 2P Excitations

Fluorophore	Lifetime/Amplitude (%)		Excitation (nm)	Emission (nm)	References
FAD (*bound*)	τ = <0.1 ns	α = n/a	n/a	n/a	Berezin and Achilefu (2010)
	τ_1 = 0.27 ns τ_2 = 2.17 ns τ_3 = 5.30 ns (*bound to LipDH*)	α_1 = 34 α_2 = 26 α_2 = 39		530	Heikal (2010)
	τ_1 = 0.025 ns τ_2 = 0.045 ns (*Dimer*)	α_1 = n/a α_2 = n/a	315 (1P)	584 (20°C)	Tanaka et al. (1989)
	τ_3 = 0.17–0.23 ns (*monomer*)	α_2 = n/a		557 (5°C)	
	τ = 0.10 ns	α = n/a	890 (2P)	n/a	Skala (2007)
	τ = 0.20 ns	α = 85	760 (2P)	550–700 (emission filter)	Peters et al. (2011)
	τ_1 = 0.39 ns τ_2 = 2.20 ns	α_1 = n/a α_2 = n/a	440 (1P)	600	Berezin and Achilefu (2010)
Lipofuscin	(Age = 5–29 years) τ_1 = 0.22 ns τ_2 = 0.66 ns τ_3 = 1.63 ns τ_4 = 6.12 ns	 α_1 = 70 α_2 = 20 α_3 = 8 α_4 = 1	364 (1P)	590–600	Cubeddu et al. (1990)
	(Age = 30–49 years) τ_1 = 0.21 ns τ_2 = 0.61 ns τ_3 = 1.68 ns τ_4 = 6.65 ns	 α_1 = 68 α_2 = 23 α_3 = 8 α_4 = 1	364 (1P)	590–600	
	(Age = >50 years) τ_1 = 0.21 ns τ_2 = 0.73 ns τ_3 = 2.04 ns τ_4 = 6.60 ns	 α_1 = 75 α_2 = 21 α_3 = 4 α_4 = 0.5	364 (1P)	590–600	
	τ_1 = 0.39 ns τ_2 = 2.24 ns	α_1 = 48 α_2 = 52	446 (1P)	510–700	Schweitzer et al. (2007b)
A2E	τ_1 = 0.33 ns τ_2 = 1.90 ns τ_3 = 6.60 ns	α_1 = 35 α_2 = 38 α_3 = 27	413 (1P)	n/a	Cubeddu et al. (1999)
	τ_1 = 0.17 ns τ_2 = 1.12 ns	α_1 = 98 α_2 = 2	446 (1P)	510–700	Schweitzer et al. (2007b)
	τ = 0.012 ns	α = n/a	458 (1P)	560	Haralampus-Grynaviski et al. (2003)
ATR	τ_1 = 0.03 ps τ_2 = 0.37 ps τ_3 = 33–34 ps	α_1 = n/a α_2 = n/a α_2 = n/a	800 (2P)	$\tau_1 \rightarrow$ 430 $\tau_2 \rightarrow$ 440 $\tau_3 \rightarrow$ 560	Takeuchi and Tahara (1997)
Melanin	τ_1 = 0.07 ns τ_2 = 0.60 ns	α_1 = 98 α_2 = 2	760 (2P)	550–700 (emission filter)	Peters et al. (2011)
	τ_1 = 0.28 ns τ_2 = 2.40 ns	α_1 = 70 α_2 = 30	446 (1P)	510–700	Schweitzer et al. (2007b)

(*Continued*)

TABLE 15.1 (*Continued*) Fitting Parameters for Fluorescence Lifetime Measurements of Endogenous Fluorophores under Both 1P and 2P Excitations

Fluorophore	Lifetime/Amplitude (%)			Excitation (nm)	Emission (nm)	References
	$\tau_1 = 0.10$ ns		$\alpha_1 =$ n/a	(2P)	440, 520, 575	Koenig (2008)
	$\tau_2 = 1.90$ ns		$\alpha_2 =$ n/a			
	$\tau_3 = 8.00$ ns		$\alpha_3 =$ n/a			
	$\tau_1 = 0.14$ ns		$\alpha_1 = 93$	760, 800 (2P)	550	Dimitrow et al. (2009)
	$\tau_2 = 1.08$ ns		$\alpha_2 = 7$			
	(Hydrogen peroxide solvent)			470 (1P)	540	Colbert and Heikal (2005)
	$\tau_1 = 0.05$ ns		$\alpha_1 = 21$			
	$\tau_2 = 0.79$ ns		$\alpha_2 = 34$			
	$\tau_3 = 3.26$ ns		$\alpha_3 = 45$			
	(DMSO solvent)			470 (1P)	540	
	$\tau_1 = 0.04$ ns		$\alpha_1 = 48$			
	$\tau_2 = 0.62$ ns		$\alpha_2 = 25$			
	$\tau_3 = 3.24$ ns		$\alpha_3 = 26$			
	(PBS solvent)			470 (1P)	540	
	$\tau_1 = 0.03$ ns		$\alpha_1 = 52$			
	$\tau_2 = 0.61$ ns		$\alpha_2 = 20$			
	$\tau_3 = 3.26$ ns		$\alpha_3 = 24$			
	(Donor = 5–29 years)			364 (1P)	450, 560	Cubeddu et al. (1990)
	$\tau_1 = 0.21$ ns		$\alpha_1 = 45$			
	$\tau_2 = 0.75$ ns		$\alpha_2 = 29$			
	$\tau_3 = 2.15$ ns		$\alpha_3 = 17$			
	$\tau_4 = 7.13$ ns		$\alpha_4 = 9$			
	(Donor = 30–49 years)			364 (1P)	450, 560	
	$\tau_1 = 0.15$ ns		$\alpha_1 = 57$			
	$\tau_2 = 0.56$ ns		$\alpha_2 = 28$			
	$\tau_3 = 2.25$ ns		$\alpha_3 = 11$			
	$\tau_4 = 7.99$ ns		$\alpha_4 = 4$			
	(Donor = >50 years)			364 (1P)	450, 560	
	$\tau_1 = 0.21$ ns		$\alpha_1 = 66$			
	$\tau_2 = 0.78$ ns		$\alpha_2 = 25$			
	$\tau_3 = 2.46$ ns		$\alpha_3 = 7$			
	$\tau_4 = 8.26$ ns		$\alpha_4 = 2$			
Collagen	$\tau_1 = 0.30$ ps		$\alpha_1 =$ n/a	300–340	420–460	Koenig (2008)
	$\tau_2 = 2.00$ ps		$\alpha_2 =$ n/a			
	(Powder)					Koenig and Schneckenburger (1994)
	$\tau_1 = 2.70$ ps		$\alpha_1 =$ n/a			
	$\tau_2 = 8.90$ ps		$\alpha_2 =$ n/a			
	Collagen 1	$\tau_1 = 0.67$ ps	$\alpha_1 = 68$	446 (1P)	510–700	Schweitzer et al. (2007b)
		$\tau_2 = 4.04$ ps	$\alpha_2 = 32$			
	Collagen 2	$\tau_1 = 0.47$ ps	$\alpha_1 = 64$			
		$\tau_2 = 3.15$ ps	$\alpha_2 = 36$			
	Collagen 3	$\tau_1 = 0.35$ ps	$\alpha_1 = 69$			
		$\tau_2 = 2.80$ ps	$\alpha_2 = 31$			
	Collagen 4	$\tau_1 = 0.74$ ps	$\alpha_1 = 70$			
		$\tau_2 = 3.67$ ps	$\alpha_2 = 30$			
Elastin	$\tau_1 = 0.30$ ps		$\alpha_1 =$ n/a	300–340	420–460	Koenig (2008)
	$\tau_2 = 2.00$ ps		$\alpha_2 =$ n/a			

(*Continued*)

TABLE 15.1 (*Continued*) Fitting Parameters for Fluorescence Lifetime Measurements of Endogenous Fluorophores under Both 1P and 2P Excitations

Fluorophore	Lifetime/Amplitude (%)		Excitation (nm)	Emission (nm)	References
	$\tau_1=0.38$ ns $\tau_2=3.59$ ns	$\alpha_1=72$ $\alpha_2=28$	446 (1P)	510–700	Schweitzer et al. (2007b)
	(Powder) $\tau_1=2.00$ ps $\tau_2=6.70$ ps	 $\alpha_1=$ n/a $\alpha_2=$ n/a			Koenig and Schenckenburger (1994)
Tryptophan	(Aqueous solution) $\tau_1=0.38$ ($pH=2.55$)–0.6 ($pH=6.8$) ns $\tau_2=2.33$ ($pH=2.55$)–3.16 ($pH=6.8$) ns	 $\alpha_1=6$–11 $\alpha_2=90$–94			Gudgin et al. (1981)
	(HCl–water solution) $\tau_1=0.13$ ($pH=1.21$)–0.52 ($pH=3.5$) ns $\tau_2=0.34$ ($pH=0.84$)–3.2 ($pH=3.5$) ns	 $\alpha_1=6$–35 $\alpha_2=65$–100			
	(Sodium tetraborate buffer solution) $\tau_1=0.5$ ($pH=10.5$)–0.6 ($pH=7.0$) ns $\tau_2=3.1$ ($pH=10.5$)–3.2 ($pH=7.0$) ns $\tau_3=9$ ns	$\alpha_1=2$–6 $\alpha_2=3$ (pH$=10.5$)–95 ($pH=7.0$) $\alpha_2=1$ ($pH=7.0$)–96 ($pH=10.5$)			
α-Crystallin	(Buffer solution pH 7.4) $\tau_1=0.50$ ns $\tau_2=2.70$–3.50 ns		295	360	Borkman et al. (1993)
	(Guanidine hydrochloride solution pH 7.4) $\tau_1=0.50$ ns $\tau_2=2.70$ ns		295	360	
3-HKG	Purified 3-HKG: $\tau=31\pm4$ ps 3-HKG in homogenate lens: $\tau=44\pm5$ ps 3-HKG in intact human lens: $\tau=61\pm5$ ps	$\alpha=$ n/a	n/a	n/a	Dillon and Atheron (1998)
	$\tau_1=0.62$ ns $\tau_2=7.70$ ns	$\alpha_1=94$ $\alpha_2=6$	365	n/a	Thiagarajan et al. (2002)
3-HK	$\tau_1=$<0.50 ns	$\alpha=$ n/a			Thiagarajan et al. (2002)
XA8OG (xanthurenic acid 8-O-β-D-glucoside)	$\tau_1=1.10$ ns $\tau_2=12.3$ ns	$\alpha_1=23$ $\alpha_2=77$	338	440	Thiagarajan et al. (2002)

(*Continued*)

TABLE 15.1 (*Continued*) Fitting Parameters for Fluorescence Lifetime Measurements of Endogenous Fluorophores under Both 1P and 2P Excitations

Fluorophore	Lifetime/Amplitude (%)		Excitation (nm)	Emission (nm)	References
Pentosidine (AGE)	$\tau_{Methanol} = 3.9$ ns $\tau_{DMSO} = 3.5$ ns $\tau_{Water} = 4.4$ ns $\tau_{alkaline\ water} = 3.9$ ns	$\alpha = n/a$	325–335	375–385	Schneckenburger et al. (2004)
Argpyrimidine (AGE)	Water solution → $\tau_1 = 0.86$ ns; $\tau_2 = 16.6$ ns Acidic solution → $\tau = 0.86$ ns Methanol → $\tau = 7.3$ ns DMSO, acetonitrile → $\tau_1 = 6.6$ ns; $\tau_2 = 12.8$ ns	$\alpha = n/a$	335	400	
AGE	$\tau_1 = 0.86$ ns $\tau_2 = 4.17$ ns	$\alpha_1 = 62$ $\alpha_2 = 28$	446	510–700	Schweitzer et al. (2007b)

against damages that can be created by energy-rich UV radiation. Unfortunately, the absorption of UV light during the life span results in the formation of cataract, which decreases the ocular transmission with age. FLIM measurements are hardly achievable in the eyes with pronounced cataract.

15.4 Laser Scanning Ophthalmoscope for Time-Resolved Autofluorescence

Based on calculations of the permissible exposition (Schweitzer et al. 2000), measurements of pure substances (Schweitzer et al. 2007b), and isolated anatomical structures of porcine eyes (Schweitzer et al. 2007a), a laser scanning ophthalmoscope was developed for measurements of time-resolved autofluorescence of the human eye in vivo. Figure 15.4 shows the setup used in these studies.

The optomechanical basis for these types of measurements is a modified laser scanner ophthalmoscope (HRA II, Heidelberg Engineering). Fiber-coupled pulse laser diodes (HLD 440 or HLD 470, Lasos, Jena, Germany; Becker & Hickl, Berlin, Germany) excite the fluorescence in a 30° fundus field. These diodes emit pulses (FWHM = 70 ps, 80 MHz) at 448 or 468 nm. The fiber end (3 μm) is imaged at the fundus during the scanning process. The average excitation power in the cornea plane is about 120 μW. Thus, the applied exposure is about 1% of the maximal permissible exposure. The fluorescence light is confocally detected via a 100 μm fiber. After blocking the reflected excitation light by a long-pass razor filter at 488 nm (Laser 2000, Wesseling, Germany), the fluorescence emission is separated by a dichroic filter in a short-wavelength channel (ch 1, 490–560 nm) and in a long-wavelength channel (ch 2, 560–700 nm). In each of these spectral channels, single photons are detected by a multichannel photomultiplier (MCP-PMT, HAM-R 3809U-50, Hamamatsu, Herrsching, Germany). A TCSPC board SPC-150 (Becker & Hickl) accumulates the detected photons in 1024 time channels. Thus, a time resolution of 12.5 ps is achieved. The SPC-150 board works in first-in-first-out (FIFO) mode that has direct memory access. Thus, continuous measurements are possible and no interruption is required for data saving. Simultaneously with the excitation beam, the fundus is scanned by an infrared laser (820 nm). Detected by a frame grabber, the contrast-rich infrared reflection fundus image is used for online image registration of the fluorescence in both spectral channels. A security system switches off the lasers if the scanner is not moving. The online image registration system ensures that each photon is accumulated in both the right time and spectral channel for each pixel.

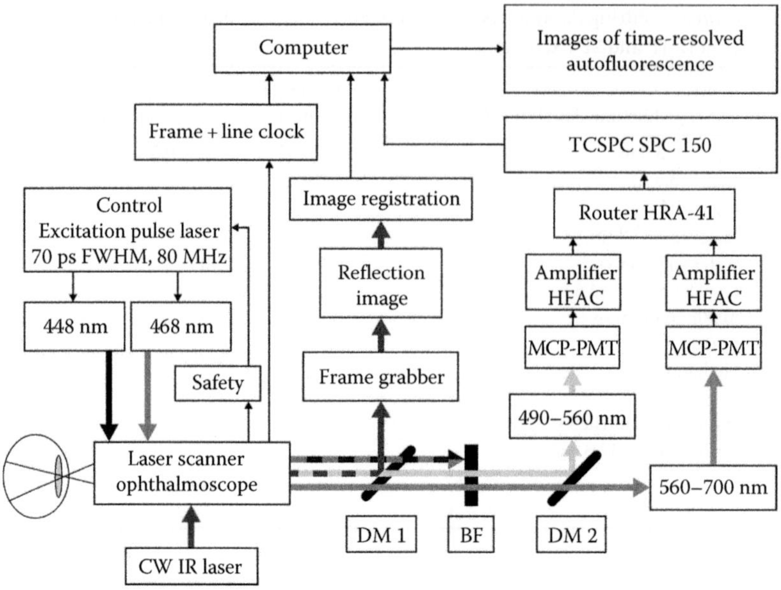

FIGURE 15.4 Schematic diagram of the lifetime laser scanning ophthalmoscope.

15.5 Evaluation of FLIM Measurements

Following data acquisition, the measured fluorescence images in two spectral channels are used for further evaluation. Depending on the appearance time after the excitation pulse, each photon is accumulated in the corresponding time channel. Thus, the photons in all time channels form a histogram for each pixel in the FLIM image. Assuming this process as an exponential decay, a sum of exponential functions can be used as model to fit the experimental histogram in each pixel. The intensity of fluorescence in each pixel is determined by the sum of photons in all time channels (i.e., the integrated fluorescence decay).

The fluorescence decay is mostly assumed as an exponential decay, which can be approximated by a sum of exponential functions (Equation 15.5):

$$\frac{I(t)}{I_o} = \sum_{i=1}^{p} \alpha_i \cdot e^{-t/\tau_i} + b \tag{15.5}$$

where

α_i is the preexponential factor (or amplitude) of *i*th exponent
τ_i is the lifetime of exponent of the *i*th component
b is the background (or baseline)
p is the number of exponential decay components

As the excitation is not a Dirac pulse and also the detector has specific time-dependent response, the measured decay of fluorescence is the convolution of instrumental response function (IRF) with the actual fluorescence deactivation process. The IRF can be measured under the same experimental conditions using 10 µMol Eosin Y added to 5 M of potassium iodide. Another possibility is the detection of the excitation pulse, reflected at a surface mirror. In this case, the blocking filter must be replaced by a neutral density filter with a high optical density to avoid damaging the detectors.

In addition to the residual, the criterion for the fitting process is the minimization of χ_r^2 (Equation 15.6), which is defined as follows:

$$\chi_r^2 = \frac{1}{n-q} \cdot \sum_{j=1}^{n} \frac{\left[N(t_j) - N_c(t_j)\right]^2}{N(t_j)} \tag{15.6}$$

where

$N(t_j)$ is the measured number of photons in the *j*th time channel

$N_c(t_j)$ is the number of expected photons, which are calculated by the convolution of the IRF and the model function

n is the number of time channels

q is the number of free parameters (α_i, τ_i, *b*)

If the detection of photons is a Poisson process, the mean square root error between detected photons and calculated photons is equal to the square root of the detected events:

$$\text{Noise} = \sqrt{\left[N(t_j) - N_c(t_j)\right]^2} = \sqrt{N(t_j)} \tag{15.7}$$

Thus, the ratio in the sum of Equation 15.6 is 1 for each time channel and the sum is *n*, which implies that the limiting value of χ_r^2 is 1. The algorithm is independent of the degree of exponential function, but the calculation time increases with the number of exponents.

For evaluation of lifetime measurements, in addition to the single amplitude and lifetime, other parameters such as mean lifetime (τ_{mean}) and relative contribution (Q_i) of each component to the detected signal are also useful quantities in FLIM analysis. Here, the mean lifetime is defined as follows:

$$\tau_m = \frac{\sum_{i=1}^{3} a_i \cdot \tau_i}{\sum_{i=1}^{3} a_i} \tag{15.8}$$

The relative contribution Q_i of the *i*th species corresponds to the area under the *i*th decay component in the fluorescence decay curve. This value is calculated according to the following:

$$Q_i = \frac{\alpha_i \cdot \tau_i}{\sum_{i=1}^{p} \alpha_i \cdot \tau_i} \tag{15.9}$$

After the fitting calculation is completed, a 2D data matrix is determined for each basic parameter τ_i and α_i, as well as for the derived parameter τ_{mean} and Q_i.

15.5.1 Relation between Components in the Fitting Model and Anatomical Structure

After approximation of the fluorescence decay by the model function, using, for example, SPCImage (Becker & Hickl), 2D or quasi-3D functional images can be presented from the experimental data for visual inspection. Especially for images of amplitudes, there are two different possibilities for presentations. In FLIM images of absolute values of amplitudes, regions of high or low fluorescence intensities become apparent. Unfortunately, the heterogeneity in these images may also reflect inhomogeneous illumination. However, images of the relative amplitudes are independent of the laser illumination. As a result, these relative-amplitude images can be used to highlight the dominant amplitude (or species), which can then be compared with the anatomical structure of the fundus. In addition, the layer of autofluorescence origin associated with a specific amplitude can be determined.

The lifetimes and amplitudes of single layers of porcine eye were measured by two-photon (2P) excitation *in toto*. As shown in Table 15.2, the shortest lifetime ($\tau_1 = 70$ ps) was detected from retinal pigment epithelium (RPE) in both spectral channels. The lifetime τ_1 of the other layers is about 400 ps in the short-wavelength channel and about 200 ps in long-wavelength channel. In vivo measurements are optimally fitted by triple-exponential model function.

Considering cross sections in images of amplitudes or lifetime of human eyes, a certain relation exists between RPE and the component with the shortest lifetime, between neuronal retina and the component with the mean lifetime, while both connective tissue and fluorescence of the crystalline lens are the structures with the longest lifetime.

The profile of the relative amplitude α_1 in Figure 15.5 is in accordance with the anatomical distribution of melanin in RPE where it goes up to 97% in the macula. It is consistent with the highest optical density of melanin in the macula. Furthermore, only the receptor axons form the neuronal retina in this region and no nerve fiber layer exists in the macula. Thus, the relative amplitude α_2 decreases to 2.5% and the relative amplitude α_3 is only 0.5%.

The lifetime τ_1 in Figure 15.6 is about 50 ps in the macula and about 80 ps outside this region. Additionally, the lifetime τ_2 is about 350–400 ps in all regions outside the optic disk. These lifetimes are on the same order of magnitude of the fluorescence lifetimes in RPE and neuronal retina determined from porcine samples.

As no RPE exists in the optic disk, the triple-exponential model fits the fluorescence of connective tissue. Besides the fluorescence of connective tissue, τ_3 is also influenced by the long fluorescence decay of the crystalline lens. It can be demonstrated by two facts: First, the eye lens fluoresces dominantly in the short-wavelength range. Hence, τ_3 is longer in the short-wavelength channel than in the long-wavelength ones. Second, τ_3 is considerably shorter in eyes with artificial intraocular lenses (IOLs) than in eyes with crystalline lenses.

For comparison with the profile of amplitudes in Figure 15.5, the anatomical structure of the human fundus is demonstrated in an image of optical coherence tomography (OCT) in Figure 15.7.

The thickness of neuronal retina is reduced in the macula. As a consequence, the relative amplitude of α_1, corresponding to RPE, is maximal. No RPE exists in the optic disk.

15.5.2 Changes in Fundus Layers

15.5.2.1 Statistical Discrimination of Healthy Subjects and Patients

Because of the relationship between exponents in the model function and anatomical layers, it is likely that the same correlation may exist between the exponents of model function and the state of the eye (i.e., healthy subjects versus diseased patients).

For such diagnostic procedure, regions of interest (ROIs) of the same size and same position are selected in FLIM images of both healthy subjects and patients that are suffering from the same disease. Histograms of lifetimes (τ_1, τ_2, and τ_3), amplitudes (α_1, α_2, and α_3), and relative signal contributions (Q_1, Q_2, and Q_3) will be calculated for both spectral channels.

The application of the Holm–Bonferroni method (Holm 1979) is more important than the statistical comparison of the histograms of healthy subjects and patients using, for example, *t*-test. The test compares the shape of histograms of different groups and takes into account that a histogram of lifetimes is the result of contributions of different substances. For that reason, the effective range is divided into intervals. Each interval contains the number of pixels, having the same lifetime or amplitude or relative signal contribution in the ROI in a healthy subject or a patient. Table 15.3 shows the frequency in intervals of lifetime τ_2 in the short-wavelength channel for diabetic patients having no signs of diabetic

TABLE 15.2 Fluorescence Lifetimes of Fundus Layers of Porcine Eye, Measured In Toto by 2P Excitation at 760 nm

Channel (nm)		NFL	GCL	GCL-c	GCL-m	INL	ONL	PRIS	RPE
500–550	α_1 (%)	72.5 ± 5.7	74.1 ± 6.9	80.0 ± 3.4	66.6 ± 3.7	73.1 ± 4.1	68.6 ± 3.4	76.2 ± 3.3	98.4 ± 0.9
	τ_1 (ns)	0.39 ± 0.05	0.39 ± 0.06	0.36 ± 0.04	0.43 ± 0.05	0.40 ± 0.06	0.39 ± 0.07	0.43 ± 0.06	0.07 ± 0.01
	α_2 (%)	27.5 ± 5.7	25.9 ± 6.9	20.0 ± 3.4	33.4 ± 3.7	26.9 ± 4.1	31.4 ± 3.4	23.8 ± 3.3	1.6 ± 0.9
	τ_2 (ns)	2.35 ± 0.20	2.36 ± 0.25	2.22 ± 0.24	2.48 ± 0.20	2.36 ± 0.24	2.64 ± 0.26	2.33 ± 0.27	0.61 ± 0.23
550–700	α_1 (%)	83.4 ± 2.4	85.3 ± 3.0	86.7 ± 2.2	81.9 ± 2.2	85.7 ± 2.1	88.6 ± 2.1	91.6 ± 2.5	98.3 ± 1.2
	τ_1 (ns)	0.23 ± 0.03	0.21 ± 0.04	0.20 ± 0.04	0.23 ± 0.04	0.20 ± 0.04	0.17 ± 0.03	0.17 ± 0.03	0.07 ± 0.01
	α_2 (%)	16.6 ± 2.4	14.7 ± 3.0	13.3 ± 2.2	18.1 ± 2.2	14.3 ± 2.1	11.4 ± 2.1	8.4 ± 2.5	1.7 ± 1.2
	τ_2 (ns)	1.98 ± 0.20	1.93 ± 0.25	1.78 ± 0.19	2.13 ± 0.18	1.98 ± 0.22	2.07 ± 0.24	1.81 ± 0.19	0.54 ± 0.22

Source: Adapted from Peters, S. et al., *Proc. SPIE*, 8086, 808604, 2011. With permission.

Note: NFL, nerve fiber layer; GCL, ganglion cell layer; INL, inner nuclear layer; ONL, outer nuclear layer; PRIS, photoreceptor inner segment; RPE, retinal pigment epithelium.

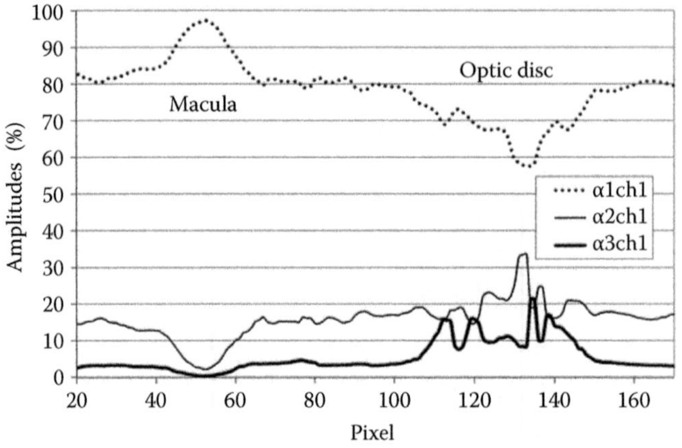

FIGURE 15.5 Relative amplitudes α_1, α_2, and α_3 along a cross section in an image of time-resolved autofluorescence in short-wavelength channel.

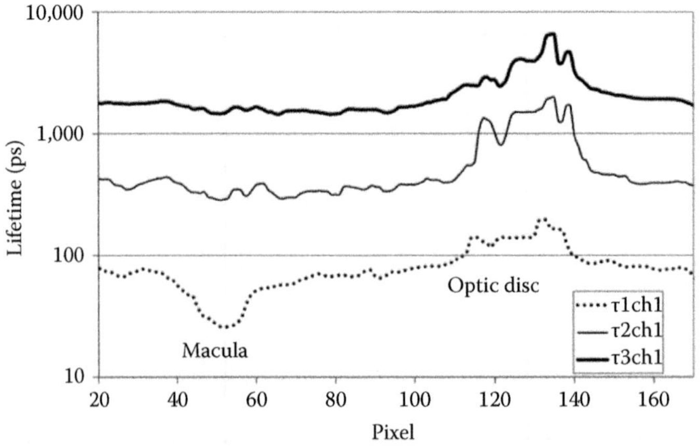

FIGURE 15.6 Lifetimes τ_1, τ_2, and τ_3 along a cross section in an image of time-resolved autofluorescence in short-wavelength channel.

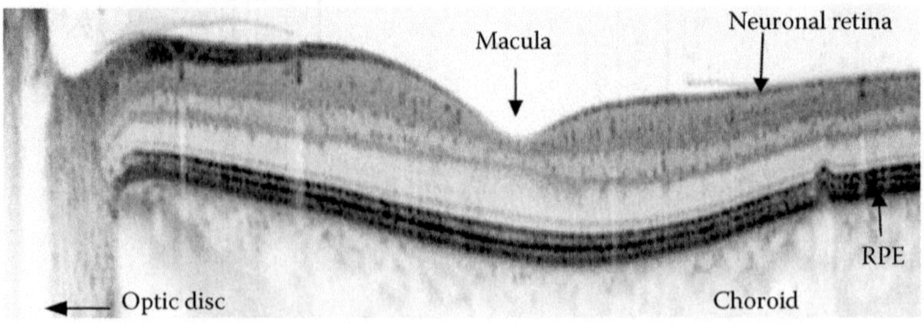

FIGURE 15.7 Anatomy of a living human fundus, determined by OCT.

TABLE 15.3 Frequency of Selected Pixels in τ_2 Intervals in ps of Channel 1 for Diabetic Patients (D003, etc.)

Tau 2	360	370	380	390	400	410	420	430	440	450	460	470
D003	99	127	139	219	196	254	251	316	278	295	311	303
D004	82	165	282	377	481	522	615	642	613	565	402	443
D005	18	31	111	213	427	758	987	1081	914	741	532	445
D006	59	143	291	595	766	792	861	744	658	524	467	405
D008	0	29	73	185	279	515	535	597	558	467	521	515
D009	10	14	18	50	92	124	138	211	181	223	210	237
D010	0	4	21	35	110	213	294	382	489	573	650	589
D011	103	140	268	358	393	547	613	620	674	512	520	418
D014	23	70	119	231	294	506	595	658	715	693	648	510

Note: The ROI was 71 × 101 pixels.

TABLE 15.4 Interval Sizes and Ranges of FLIM Parameters for Holm–Bonferroni Test

Parameter	Interval Size	Range
Amplitude α_1 (%)	0.5	60–100
Amplitude α_2 (%)	0.5	5–35
Amplitude α_3 (%)	0.5	0.5–7
Lifetime τ_1 (ps)	5	30–150
Lifetime τ_2 (ps)	10	340–540
Lifetime τ_3 (ps)	50	1500–4000
Rel. contribution Q_1 (%)	2	0–50
Rel. contribution Q_2 (%)	2	20–50
Rel. contribution Q_3 (%)	2	20–70

retinopathy. To be independent of the size of ROI, the number of pixels in each interval can also be given as percent of all pixels in the ROI.

In the next step, the distribution of parameters (e.g., lifetime) in each interval for healthy subjects and patients suffering from a certain disease will be compared by Wilcoxon rank-sum test. Significant different distributions exists, if the error probability p is equal or lower than the significance level (e.g., 0.05) divided by the number n of intervals. To find the best discrimination between ROIs, the interval size (and the number of intervals) has to be optimized. Table 15.4 shows interval sizes and ranges of parameters, found by experiences. Other interval sizes and ranges of parameters might be better suited for discrimination of special pathological alterations.

In cooperation with the Department of Experimental Ophthalmology at the University of Jena, Germany, a program, "FLIMVis," was developed at the Institute of Biomedical Technique and Informatics at the Technical University Ilmenau, Ilmenau, Germany (Klemm 2012). This program contains several statistical features that are specialized for evaluating FLIM images of the eye. Performing the Holm–Bonferroni test, histograms of FLIM parameters of both patients and healthy subjects are used as basic information. All subjects are grouped according to special conditions. For example, subjects wearing crystalline eye lens and are older than 40 years will represent one group. In addition, we may restrict the subjects in this group to be suffering from an early stage of a disease. Such a classification is necessary because FLIM parameters are likely to depend on age or the state of crystalline lens (Deutsch 2012). The number of pixels (or the percent) in each interval of a fitting parameter (e.g., lifetimes, amplitudes, or relative signal contribution) is compared by the Wilcoxon rank-sum test. As a result, intervals of FLIM parameters with significant differences between the groups are given in decreasing order.

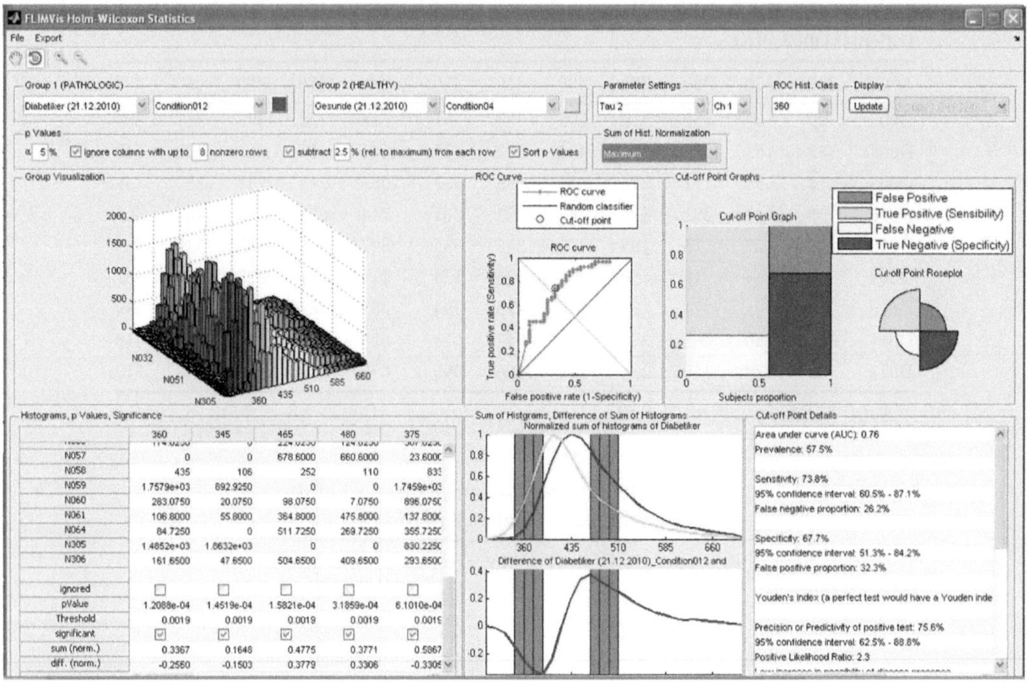

FIGURE 15.8 FLIMVis menu for visualization and statistical evaluation of FLIM measurements.

These intervals are drawn in histograms of the groups under investigation. Furthermore, the receiver operating characteristic (ROC) curve along with sensitivity and specificity is given and the threshold for classification of an individual data set in the corresponding group. These are statistical data that characterize a diagnostic method. Figure 15.8 shows the FLIMVis menu for statistical evaluation of FLIM measurements.

15.5.2.2 Visualization of FLIM Data

Statistical analysis of FLIM images characterizes global alterations while the local alterations are visualized in parameter-specific FLIM images. As can be seen from the screenshot in Figure 15.9, FLIMVis can also be used to visualize amplitudes, lifetimes, or relative contribution-based images of both spectral channels for each subject in 2D or quasi 3D. An ROI can be selected in each image. Furthermore, vertical or horizontal intersections are possible through FLIM images. Individual or group histograms are drawn and characterized by statistical parameter models: the maximum of a distribution value, mean, median, standard deviation, kurtosis, and skewness. In this way, the dependence of median of lifetime, for example, on age, can be investigated. Arithmetic operations can be performed between FLIM parameters, and the images, graphics, or primary date can be exported in different formats for further processing or presentation.

15.6 Clinical Applications

15.6.1 FLIM in Early Stages of Diabetic Retinopathy

Diabetic retinopathy is a long-term outcome of diabetes mellitus and it can lead to blindness if left untreated. As a result, it is important to find metabolic alterations at the fundus as early as possible for personalized treatment and for avoiding pathological degradation of vision. Figure 15.10 shows a fundus

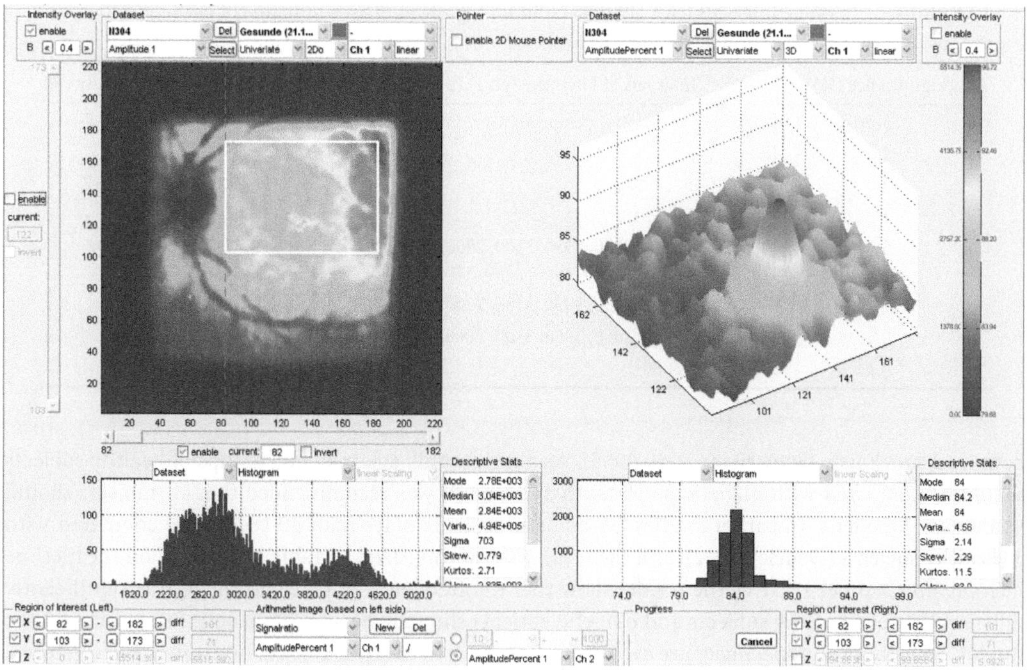

FIGURE 15.9 Presentation of FLIM images by FLIMVis.

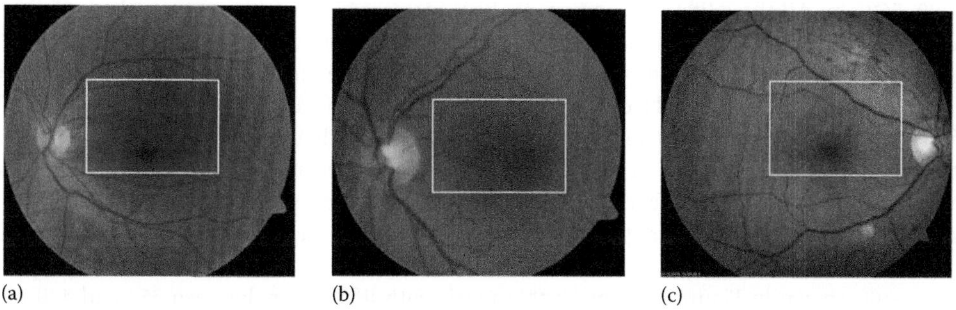

FIGURE 15.10 Fundus images of different stages of diabetic retinopathy in a healthy subject (a), no diabetic retinopathy (b), and heavy diabetic retinopathy (c). The ROI is marked (white box).

image of a healthy subject in comparison with that of diabetic patients at different stages of diabetic retinopathy. No signs of diabetic retinopathy are visible at this stage (named RD 0) in the fundus images as shown in Figure 15.10b. The light scattering in the crystalline lens leads to the loss of contrast in the retinal vessel structure as compared with the eye wearing an artificial IOL. Micro aneurysms, exudates, or bleedings appear in advanced stages of diabetic retinopathy as shown in Figure 15.10c. New retinal vessels are formed in proliferative diabetic retinopathy.

As diabetes mellitus is a systemic disease, the detection of metabolic alteration was assumed at the fundus using FLIM. In earlier studies, age-dependent alternation in metabolism was found using FLIM-based diagnoses (Quick 2009). Furthermore, the TCSPC principle is so sensitive that the influence of fluorescence of the crystalline lens is verifiable on fundus measurements, despite the use of the confocal technique. The limitation of the confocal technique for FLIM measurements at the eye has been

TABLE 15.5 Lifetime Intervals (ch 1, 490–560 nm) for Significant Discrimination of Diabetic Patients in Stage RD 0 with Healthy Subjects

	Significance (%)	Intervals of Lifetimes (ch 1) (ps)	Sensitivity (%)	Specificity (%)
τ_1	5	45, 70	64.1	73.7
	1	45		
τ_2	5	470, 480, 460, 490	76.9	73.7
	1	470		
τ_3	5	3000, 2900, 3050, 2850, 3950, 3100, 2800, 4000, 2950, 3750, 3900, 3850, 3800, 2750, 3700, 3150	89.7	60.5
	1	3000, 2900, 3050, 2850, 3950, 3100, 2800, 4000, 2950		
	0.5	3000, 2900, 3050, 2850, 3950, 3100, 2800, 4000		
	0.1	3000, 2900, 3050		

described previously (Schweitzer et al. 2005). As a result, only age-matched groups of healthy subjects should be compared with diabetic patients. In addition, it was recommended that all subjects should wear crystalline lens. In earlier studies, 39 diabetic patients older than 40 years were compared with 38 healthy subjects (Deutsch 2012), and the results are summarized in Table 15.5. Based on the lifetime intervals in Channel 1 (ch 1), the results show that the frequency of pixels in the ROI is significantly different between healthy subjects and diabetic patients showing no sign of diabetic retinopathy at the fundus. Sensitivity and specificity are calculated according to the frequency of pixels in the first interval in a line. The frequency of pixels in each interval permits the discrimination of the groups under investigation independently from other intervals at the stated level of significance.

Diabetic patients are separable from healthy subjects by lifetimes (τ_1, τ_2, and τ_3) in the spectral channel 490–560 nm. All three lifetime components have increased in diabetes. Our results indicate that τ_3 images provide the best discrimination of diabetic patients down to a significance level of 0.1%. This might be the result of accumulation of AGEs in the lens and at the fundus. The relative amplitude α_3 is low, and its modal value is shifted from 2.5% in healthy subjects to 4% in diabetic retinopathy. In addition, the estimated error in τ_3 is relatively high. For that reason, changes in τ_2 are of more interest for further studies. Both sensitivity and specificity of τ_2 are determined at a high level. The amplitude α_2 is unchanged. That means changes between substances, but not changes in thickness, are responsible for this alteration. The difference in frequencies of lifetime τ_2 (ch 1) in diabetic patients (RD 0) and healthy subjects is shown in Figure 15.11.

The results shown in Figure 15.11 mean that pixels with lifetimes τ_2 between 350 and 420 ps are missing in the ROI in diabetic patients (stage RD 0). On the other hand, additional pixels with lifetimes longer than about 420 ps appear in RD 0. Thus, changes between different fluorophores can be derived from this diagram. This change is attributed to a shift of free NADH ($\tau = 380$ ps) population to protein-bound NADH with a longer fluorescence lifetimes, depending on the protein. As shown in Figure 15.12, the number of yellow pixels ($340 < \tau_2 < 380$ ps) decreases and the number of blue pixels ($460 < \tau_2 < 520$ ps) increases with increasing severity of diabetic retinopathy.

The discrimination between diabetic patients (RD 0) and healthy subjects is also possible using the measured autofluorescence lifetimes in the long-wavelength channel. Table 15.6 provides a summary of lifetime intervals for significant discrimination of both groups.

In contrast to the lifetime intervals in the short-wavelength channel ch 1, no high values of sensitivity and specificity were simultaneously determined in ch 2. Interestingly, the lifetime τ_2 in ch 2 permits only the detection of early metabolic changes in diabetic retinopathy at the 5% significance level.

This investigation demonstrates the detection of metabolic alterations at the human fundus, when no characteristic morphological alterations of diabetic retinopathy are visible. Such early detection using natural coenzymes in the eye is key for successful diagnosis and treatment before irreversible degradation in vision.

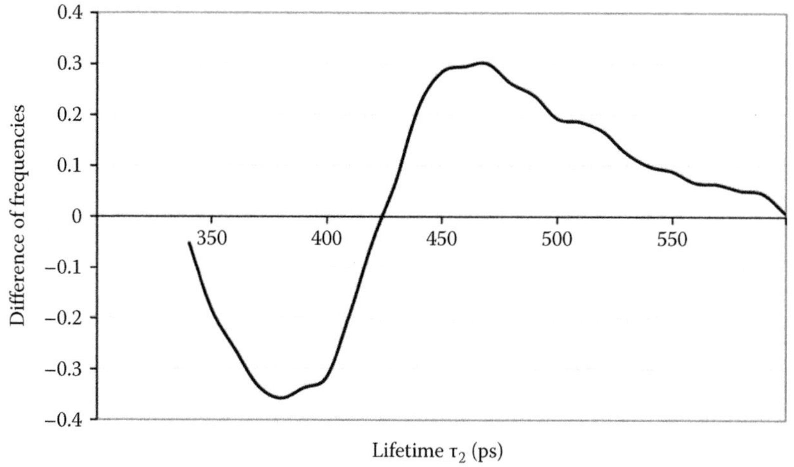

FIGURE 15.11 Difference in frequencies of lifetime τ_2 in diabetic patients and healthy subjects. No signs of diabetic retinopathy were detectable at the fundus in stage RD 0. All probands (both healthy subjects and patients) were older than 40 years.

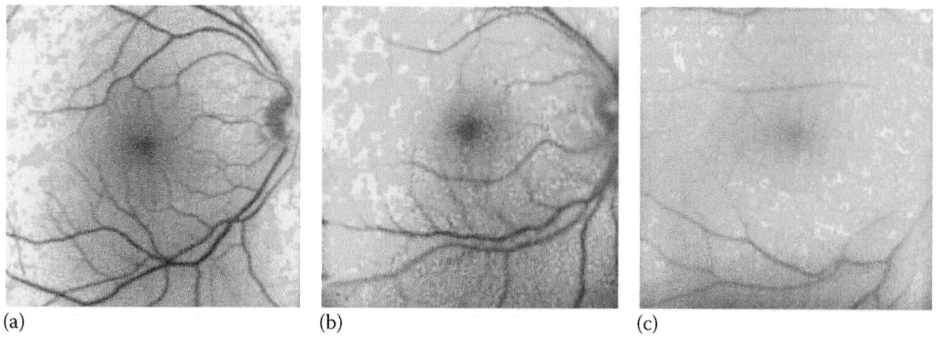

FIGURE 15.12 Changes in lifetime τ_2 in ch 1 with increasing severity of diabetic retinopathy. Film images of healthy fundus (a), RD 0 no diabetic retinopathy (b), and RD 1 mild diabetic retinopathy (c). Key: yellow, $340 < \tau_2 < 380$ ps; blue, $460 < \tau_2 < 520$ ps; green, $300 < \tau_2 < 700$ ps.

TABLE 15.6 Lifetime Intervals (ch 2, 560–700 nm) for Significant Discrimination of Diabetic Patients in Stage RD 0 with Healthy Subjects

	Significance (%)	Intervals of Lifetimes (ch 2) (ps)	Sensitivity (%)	Specificity (%)
τ_1	5	80	82.1	60.5
	1	80		
	0.5	80		
τ_2	5	400, 370	56.4	78.7
τ_3	5	2100, 2000, 2550, 2600, 2750, 2500, 2150, 2650, 2700, 2800, 2850, 2900, 2200	94.9	50
	1	2100, 2000, 2550, 2600		
	0.5	2100, 2000		

15.6.2 FLIM in Age-Related Macular Degeneration

The Holm–Bonferroni test was performed within the frameworks of the age-related eye disease study (AREDS) on 11 patients, suffering from nonexudative age-related macular degeneration (AMD) stage 2 and higher (AREDS 2001). The patients were older than 40 years and have been wearing crystalline lenses. As a control, 38 healthy subjects were chosen for the study similar criteria. No significant change in lifetimes of the short-wavelength channel ch 1 were found for the late stages of AMD and healthy subjects. High values were found for sensitivity and specificity of amplitudes α_1 and α_2 in ch 1, as shown in Table 15.7.

The distribution of relative amplitude α_1 of ch 1 in AMD is shifted to lower values than in healthy subjects (Figure 15.13). In compensation, the relative amplitude α_2 is considerably increased. The modal value of α_3 stays unchanged between both groups (as α_3 is unchanged, no values were found for significant discrimination). As the relative contribution α_1 in both spectral channels corresponds to melanin, its change is attributed to a thinning of the RPE layer. In the long-wavelength channel ch 2, different, highly significant pixel frequencies are achievable in lifetime intervals for τ_1 and τ_2 as shown in Table 15.8.

The fluorescence in the long-wavelength channel (ch 2) is dominated by lipofuscin with emission maxima at about 600 nm and higher. Figure 15.14 shows an elongation of τ_1 in ch 2, which is the result of an increased contribution of lipofuscin in which the lifetime is longer than the lifetime of melanin.

TABLE 15.7 Intervals of Amplitudes (ch 1, 490–560 nm) for Significant Discrimination of Late AMD with Healthy Subjects

	Significance (%)	Intervals of Amplitudes (%)	Sensitivity (%)	Specificity (%)
α_1	5	78, 76.5, 77.5, 85, 85.5, 77, 76, 84.5, 78.5, 79, 86, 86.5, 79.5	72.7	94.7
	1	78, 76.5, 77.5, 85, 85.5, 77, 76		
	0.5	78		
α_2	5	18.5, 19, 18, 19.5, 17.5, 16.5, 11, 17, 12, 11.5, 10, 12.5, 9.5, 16, 13, 9	90.9	97.4
	1	18.5, 19, 18, 19.5, 17.5, 16.5, 11, 17, 12, 11.5, 10, 12.5, 9.5, 16		
	0.5	18.5, 19, 18, 19.5, 17.5, 16.5, 11, 17, 12, 11.5, 10, 12.5, 9.5		
	0.1	18.5, 19, 18, 19.5, 17.5, 16.5, 11, 17, 12, 11.5		

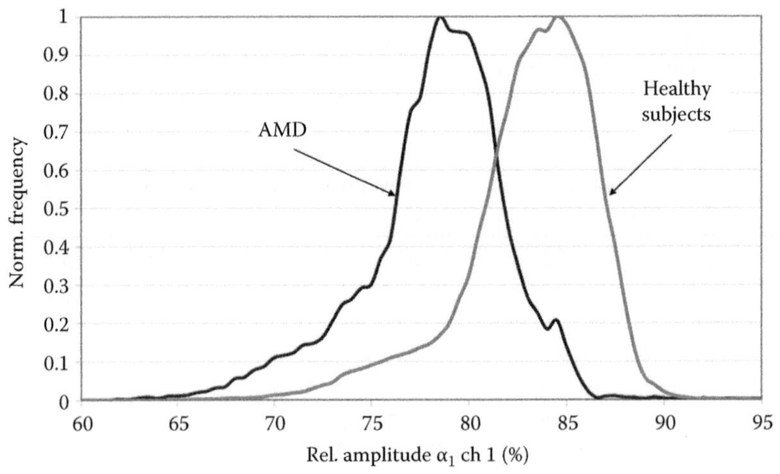

FIGURE 15.13 Normalized histograms of relative amplitude α_1 in late AMD in comparison with healthy subjects.

TABLE 15.8 Intervals of Lifetimes (ch 2, 560–700 nm) for Significant Discrimination of Late AMD with Healthy Subjects

	Significance (%)	Intervals of Lifetimes (ps)	Sensitivity (%)	Specificity (%)
τ_1	5	60, 65, 95, 110, 70, 100, 105, 115	90.9	71.1
	1	60, 65, 95, 110, 70		
	0.5	60		
τ_2	5	380, 370, 390, 360, 400, 480, 470, 490, 460, 500	81.8	97.4
	1	380, 370, 390, 360, 400		
	0.5	380, 370, 390, 360		
	0.1	380, 370		

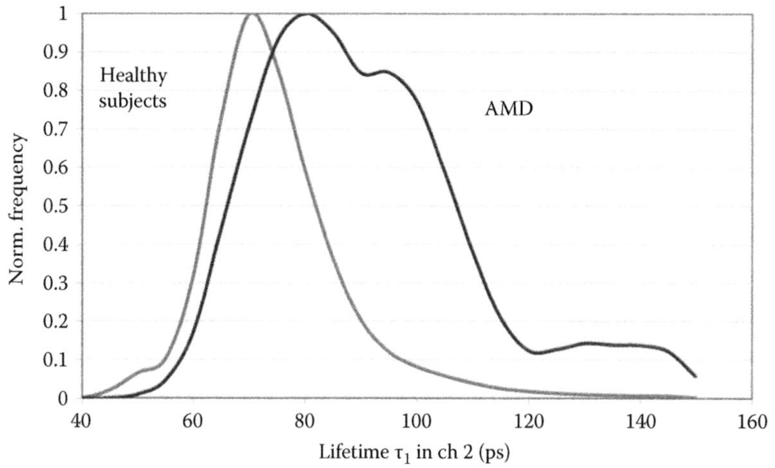

FIGURE 15.14 Histogram of lifetime τ_1 in ch 2 for diabetic patients and age-matched healthy subjects.

This change is clearly demonstrated in the difference histogram of AMD patients and healthy subjects in Figure 15.15.

Intervals for significant discrimination between AMD patients and healthy subjects were found in histograms of lifetime τ_2 in ch 2 (Table 15.8). The reason for this excellent discrimination is the shift to longer lifetime τ_2 in ch 2 in AMD as demonstrated in Figure 15.16.

The difference of τ_2-histograms in ch 2 of AMD patients and healthy subjects (Figure 15.17) shows that the number of pixels with τ_2 at about 390 ps is reduced in AMD, but more pixels have lifetimes τ_2 at about 490 ps. This alteration in lifetime is comparable with the shift of τ_2 in early stage of diabetic retinopathy, but it is the result of different substance-specific contributions. In early diabetic retinopathy, the shift of τ_2 with significantly different lifetime was detected in the spectral range 490–560 nm with negligible changes in the long-wavelength range. In AMD, the spectral dependence of lifetime τ_2 changes followed a reverse trend. No significant discrimination was found between AMD and healthy subjects in the short-wavelength channel, but the distributions of τ_2 were significantly different in the long-wavelength channel.

The amplitudes α_1, α_2, and α_3 in the long-wavelength channel allow for excellent discrimination (i.e., diagnosis) of AMD patients from healthy subjects as shown in Table 15.9.

As in ch 1, the relative amplitude α_1 is also reduced in AMD in comparison with healthy subjects (Figure 15.18). The modal value of the α_2 histogram in ch 2 is increased from 21.5% in healthy subjects to 28% in late AMD in Figure 15.19. Also, the modal value of α_3 is increased from 3% in healthy subjects up to 5.5% in late AMD in Figure 15.20.

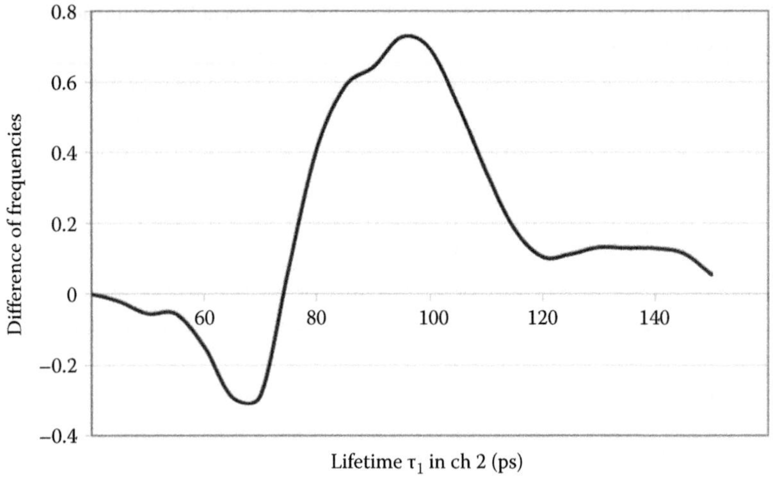

FIGURE 15.15 Difference of lifetime τ_1 histograms (AMD–healthy subjects) in channel 560–700 nm. Pixels with τ_1 of about 70 ps are missing in the selected ROI, and pixels with τ_1 of about 100 ps appear additionally.

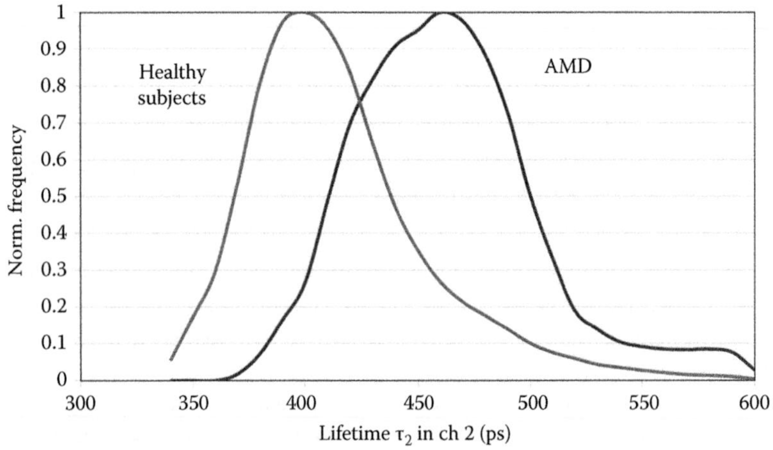

FIGURE 15.16 Elongation of τ_2 in ch 2 for AMD patients in comparison with healthy subjects.

The reduction of the amplitude α_1 in ch 2 is attributed to the thinning of RPE in AMD in the same way as α_1 is also reduced in ch 1. The increase of τ_1 only in ch 2 suggests an accumulation of lipofuscin in RPE.

The increased relative amplitude α_2 in ch 2 is clearly demonstrated in Figure 15.19.

As the fluorescence spectrum of the crystalline lens influences predominantly the τ_3 values in the short-wavelength range, the changes of α_3 and of τ_3 only in the long-wavelength channel in AMD are caused by metabolic changes at the fundus. The τ_3 histogram in ch 2 of AMD patients is completely contained in the more extended τ_3 histogram of healthy subjects. The difference of both histograms shows missing contributions of τ_3 around 2.5 ns in AMD. This lifetime corresponds well with the long component $\tau_2 = 2.4$ ns of melanin in biexponential fit. Thus, the depigmentation of RPE in AMD is also confirmed by τ_3 in ch 2, and the corresponding increased amplitude (α_3) can be attributed to an increased contribution of connective tissues.

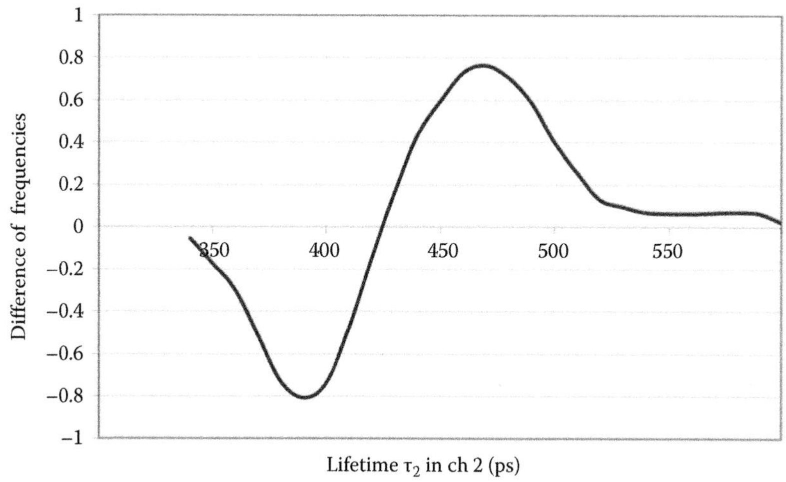

FIGURE 15.17 Difference of lifetime τ_2 histograms (AMD–healthy subjects) in channel 560–700 nm. Pixels with τ_2 of about 390 ps are missing in the selected ROI, and pixels with τ_2 of about 470 ps appear additionally.

TABLE 15.9 Intervals of Amplitudes α_1, α_2, and α_3 (ch 2, 560–700 nm) for Significant Discrimination of Late AMD from Healthy Subjects

	Significance (%)	Intervals of Amplitudes (ch 2) (%)	Sensitivity (%)	Specificity (%)
α_1	5	64, 65, 65.5, 64.5, 66.5, 76.5, 76, 66, 62.5, 63, 68.5, 68, 63.5, 60, 69, 61, 75.5, 76.5, 77, 77.5, 76, 75, 78, 69.5, 78.5, 79, 74.5	90.9	92.1
	1	64, 65, 65.5, 64.5, 66.5, 76.5, 76, 66, 62.5, 63, 68.5, 68, 63.5, 60, 69, 61, 75.5, 76.5, 77, 77.5, 76, 75, 78, 69.5, 78.5		
	0.5	64, 65, 65.5, 64.5, 66.5, 76.5, 76, 66, 62.5, 63, 68.5, 68, 63.5, 60, 69, 61, 75.5, 76.5, 77, 77.5, 76, 75, 78, 69.5		
	0.1	64, 65, 65.5, 64.5, 66.5, 76.5, 76, 66, 62.5, 63, 68.5, 68, 63.5, 60, 69, 61, 75.5, 76.5, 77, 77.5, 76		
α_2	5	28.5, 27.5, 29, 27, 29.5, 30, 30.5, 28, 26.5, 31, 21.5, 21, 20.5, 20, 26, 22, 19.5, 19, 22.5	90.9	94.7
	1	28.5, 27.5, 29, 27, 29.5, 30, 30.5, 28, 26.5, 31, 21.5, 21, 20.5, 20, 26, 22, 19.5		
	0.5	28.5, 27.5, 29, 27, 29.5, 30, 30.5, 28, 26.5, 31, 21.5, 21, 20.5, 20, 26		
	0.1	28.5, 27.5, 29, 27, 29.5, 30, 30.5, 28, 26.5, 31		
α_3	5	6, 6.5, 3, 5.5, 2.5, 5	90.9	81.6
	1	6, 6.5, 3, 5.5, 2.5, 5		
	0.5	6, 6.5, 3, 5.5, 2.5		
	0.1	6, 6.5, 3, 5.5		

15.6.3 FLIM in Glaucoma

The diagnostic value of FLIM was tested on 12 glaucoma patients in comparison with 38 healthy subjects. The ROI was 71 × 101 pixels located in the superior temporal quadrant and the macula was included in the ROI. Significant changes were found in the short-wavelength channel only for τ_3 and for α_3 as demonstrated in Table 15.10.

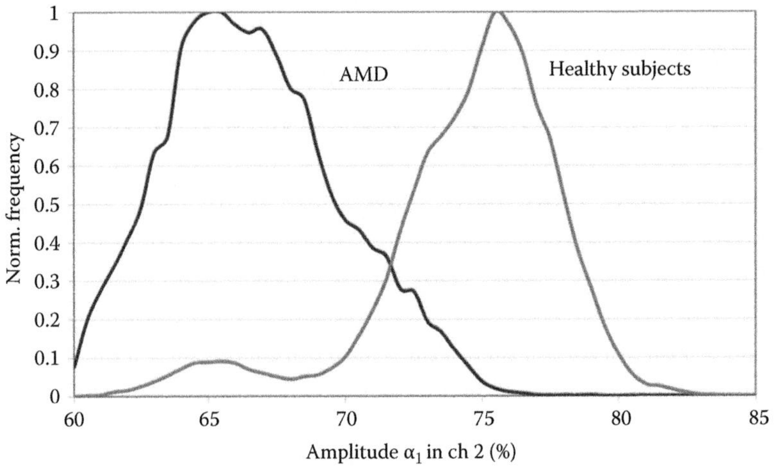

FIGURE 15.18 Histogram of amplitude α_1 in ch 2 in AMD patients and healthy subjects.

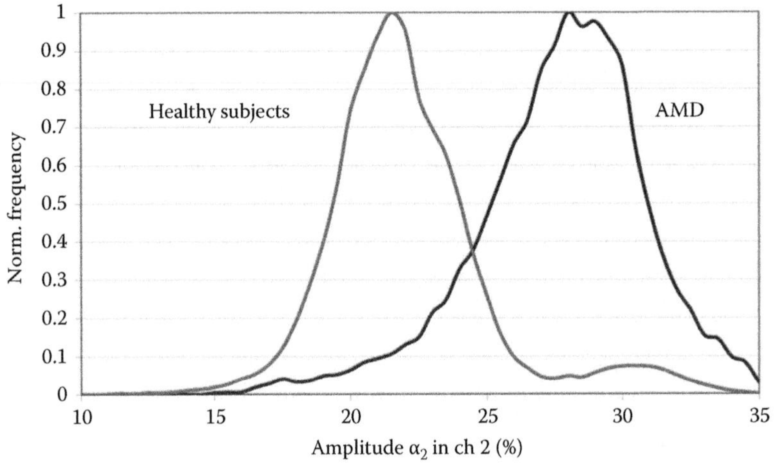

FIGURE 15.19 Histogram of amplitude α_2 in ch 2 in AMD patients and healthy subjects.

The histogram of τ_3 is given in Figure 15.21. The modal value of lifetime τ_3 is shifted from 3.30 ns in healthy subjects to 2.35 ns in glaucoma patients. The wide asymmetric distribution of the relative amplitude α_3 in ch 1 from 1.5% to 7% in healthy subjects is small and symmetric at 2% in glaucoma patients (Figure 15.22).

As the FLIM parameters of component 3 relate to the connective tissue and are influenced to a certain degree by the long lifetime of the crystalline lens, the shortening of τ_3 corresponds to a reduction of influences of components with long lifetime and the lower relative amplitude α_3 is interpretable as a thinning of the nerve fiber layer.

In the long-wavelength channel (560–700 nm), intervals with significant differences were detectable for τ_1 and especially for τ_3, as shown in Table 15.11. No difference in amplitudes was found between glaucoma patients and healthy subjects in ch 2.

The histograms of lifetime τ_1 in ch 2 for glaucoma patients and in healthy subjects are shown in Figure 15.23. The modal value of $\tau_1 = 70$ ps in healthy subjects is weakly shifted to $\tau_1 = 80$ ps in glaucoma patients.

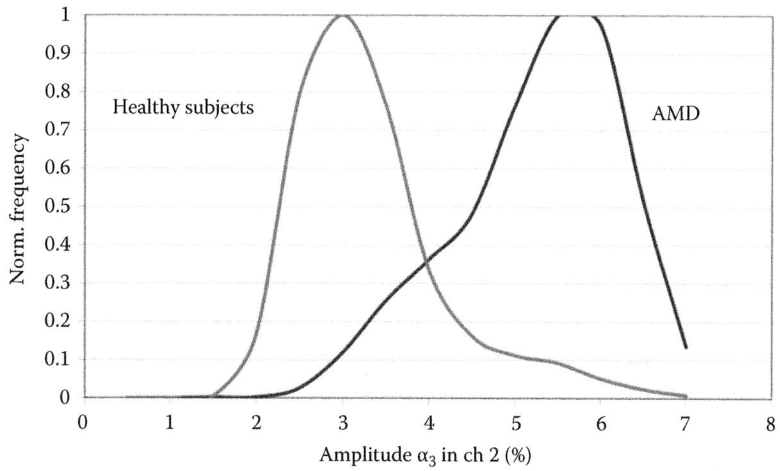

FIGURE 15.20 Histogram of amplitude α_3 in ch 2 in AMD patients and healthy subjects.

TABLE 15.10 Intervals of Significant Differences in Lifetime and Amplitude (ch 1, 490–560 nm) between Glaucoma Patients and Healthy Subjects

	Significance (%)	Intervals of Lifetime and Amplitude in Channel 1 (%)	Sensitivity (%)	Specificity (%)
τ_3	5	2450, 2400, 2500, 2550, 2600, 2650, 3700, 3650, 2700, 3750, 3600, 3800, 3500, 3550, 3900, 3450, 3850, 3400, 3950, 3350, 4000	83.3	94.7
	1	2450, 2400, 2500, 2550, 2600, 2650, 3700, 3650, 2700, 3750, 3600, 3800, 3500, 3550		
	0.5	2450, 2400, 2500, 2550, 2600, 2650, 3700, 3650, 2700		
	0.1	2450, 2400, 2500, 2550, 2600, 2650, 3700, 3650		
α_3	5	1, 1.5, 2	91.7	89.5
	1	1, 1.5		
	0.5	1, 1.5		
	0.1	1, 1.5		

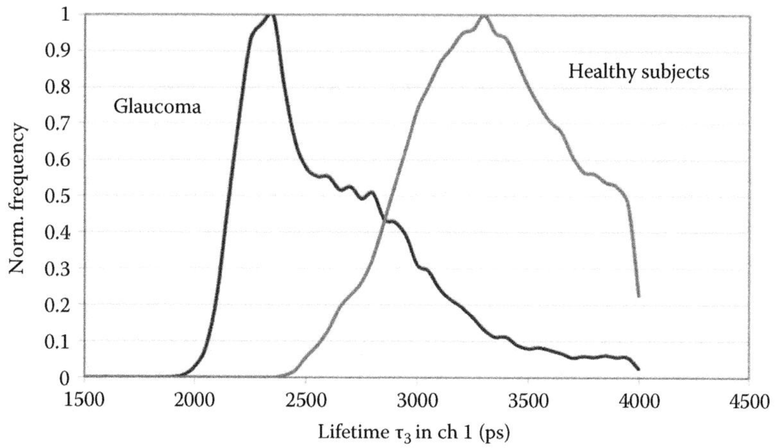

FIGURE 15.21 Histograms of lifetime τ_3 in glaucoma patients and healthy subjects (ch 1, 490–560 nm).

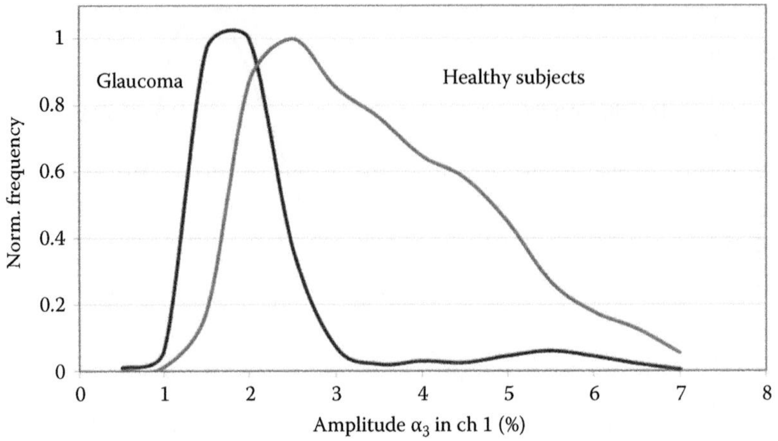

FIGURE 15.22 Comparison of histograms of α_3 for glaucoma patients and healthy subjects (ch 1, 490–560 nm).

TABLE 15.11 Intervals of Significant Differences in Lifetime (ch 2, 560–700 nm) between Glaucoma Patients and Healthy Subjects

	Significance (%)	Intervals of Lifetime and Amplitude in Channel 1 (%)	Sensitivity (%)	Specificity (%)
τ_1	5	60, 65	83.3	78.9
	1	60		
τ_3	5	1850, 1900, 1950, 2600, 2000, 2350, 2450, 2500, 2550, 2650	91.7	89.5
	1	1850, 1900, 1950		
	0.5	1850, 1900, 1950		
	0.1	1850, 1900, 1950		

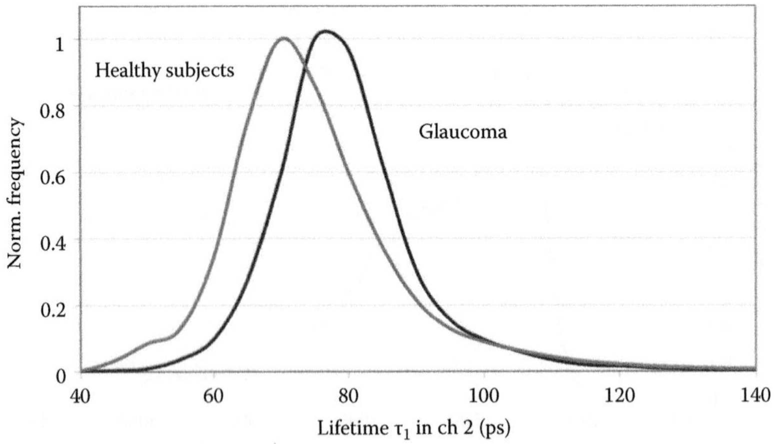

FIGURE 15.23 Histograms of lifetime τ_1 in glaucoma patients and healthy subjects (ch 2, 560–700 nm).

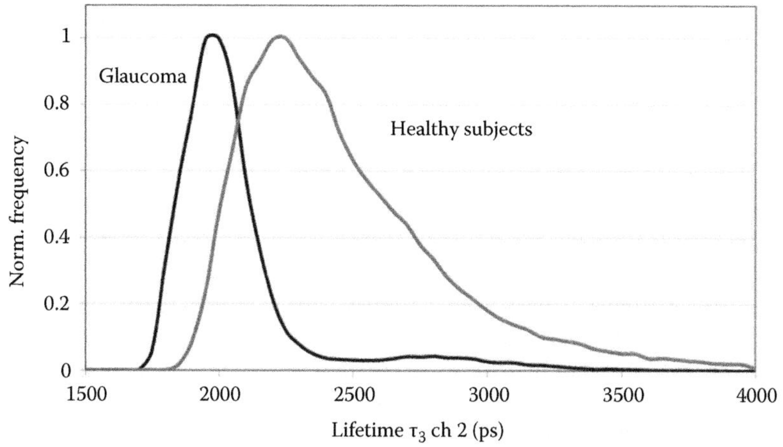

FIGURE 15.24 Histograms of lifetime τ_3 in glaucoma patients and healthy subjects (ch 2, 560–700 nm).

The modal value in histograms of τ_3 in ch 2 is shifted from 2.25 ns in healthy subjects to 2.00 ns in glaucoma patients (Figure 15.24). The change in relative amplitudes between both groups was negligible. Thus, the difference in both histograms of τ_3 is the result of unequal composition of fluorophores in healthy subjects and in glaucoma, probably in the nerve fiber layer.

15.7 Summary

Measurement of time-resolved autofluorescence is a new, promising method for detection of metabolic changes as first sign of pathological processes. However, the detectable signal is extremely weak under recommended laser illumination in order to avoid a permanent damage of the eye. As a result, the TCSPC and FLIM approach are ideal for diagnostic purposes of the eye. An online image registration system ensures the photon acquisition in the designated image pixel, spectral detection channel, and time channel. In the presence of several fluorophores, the interpretation of the fitting parameters (e.g., lifetimes and amplitudes) in terms of the tissue physiology becomes more challenging. However, comparing the local distribution of lifetimes and amplitudes with the anatomical structure of the fundus, a certain relation in triple-exponential fit was found between the shortest component (τ_1) and RPE. The component of the middle lifetime τ_2 corresponds to the neuronal retina (photoreceptors, ganglion cells). The component of the longest lifetime τ_3 corresponds to connective tissue in the nerve fiber layer. In the short-wavelength channel (490–560 nm), τ_3 is also influenced by the fluorescence of the crystalline lens. In addition, local pathological alterations can be detected using FLIM images in terms of the lifetime components τ, amplitudes α, or relative contribution Q.

Importantly, this FLIM-based diagnostic approach enables us to detect metabolic changes in early stages of diseases that are otherwise undetectable, especially in the absence of pathological signs. Here, the Holm–Bonferroni method, where the FLIM parameters are divided into intervals, is a powerful tool for evaluating differences in the shape of lifetime histograms in patients as compared with healthy subjects for effective diagnosis. The significance of differences of FLIM parameters of patients and healthy subjects is examined in each interval by the Wilcoxon rank-sum test.

This method was applied for the comparison of healthy subjects with diabetic patients, patients suffering from AMD, and glaucoma patients.

In diabetic patients, significant differences were found for the lifetimes τ_1, τ_2, and τ_3 in both spectral channels, when no signs of diabetic retinopathy are visible. The best values for sensitivity and specificity were found for τ_2 in the short-wavelength range, which corresponds to the neuronal retina (receptors, bipolar and ganglion cells). The elongated lifetime τ_2 in early stage of diabetic retinopathy is attributed to the shift of free NADH to the protein-bound form. No significant changes were found comparing the amplitudes.

In advanced AMD (AREDS stages ≥2), no changes were found for lifetimes in ch 1. But the amplitude α_1 is reduced pointing at a thinning of the RPE. The amplitude α_2 is raised in AMD and very good sensitivity (90.9%) and specificity (97.4%) are reached.

The best discrimination between AMD patients and healthy subject was reached in the long-wavelength channel, which overlaps with the lipofuscin emission. High significant differences were found in intervals of elongated lifetimes τ_1 and τ_2 in AMD by the accumulation of lipofuscin that has longer lifetime as melanin in RPE. Best sensitivity (81.3%) and specificity (97.4%) were reached for τ_2.

High significant discrimination was also possible using the amplitudes α_1, α_2, and α_3. The amplitude α_1 was reduced, but both α_2 and α_3 increased in AMD. These alterations are interpretable as complete change in the anatomical fundus structure of fluorescent substances in the long-wavelength range.

Where diabetes-specific changes occur predominantly in the short-wavelength channel for τ_2, which corresponds to the neuronal retina, the AMD-specific changes are detectable predominantly in the long-wavelength channel for τ_1 and τ_2, caused by depigmentation and accumulation of lipofuscin in RPE.

In contrast to these alterations, glaucoma-specific changes are detectable predominantly in the component with the longest decay time τ_3 in both spectral channels, which correspond to the nerve fiber layer. In both channels, the lifetime τ_3 is shortened, corresponding to a change in the composition of fluorophores in the nerve fiber layer. The decreased α_3 value corresponds to thinning in the nerve fiber layer in glaucoma.

Acknowledgments

Many thanks to all members of the Experimental Ophthalmology group Dr. Martin Hammer, Stefan Schenke, Susanne Jentsch, Sven Peters, Matthias Klemm, Silvio Quick, Lydia Deutsch, and Johannes Meyer for their active contribution in the development of fluorescence lifetime imaging as a new diagnostic method in ophthalmology.

References

ANSI. 2000. American National Standard for the safe use of laser ANSI Z 136.1-2000. Laser Institute of America, Suite 128, 13501 Ingenuite Drive, Orlando, FL, 2000.

AREDS Research Group. 2001. A randomized, placebocontrolled, clinical trial of high-dose supplementation with vitamins C and E, beta carotene, and zinc for age-related macular degeneration and vision loss: AREDS report no 8. *Archives of Ophthalmology* 119(10): 1417–1436.

Becker, W. 2005. *Advanced Time-Correlated Single Photon Counting Techniques*. Springer Series in Chemical Physics, Vol. 81. Berlin, Germany: Springer.

Berezin, M.Y. and S. Achilefu. 2010. Fluorescence lifetime measurements and biological imaging. *Chemical Reviews* 110(5): 2641–2684.

Bird, D.K., L. Yan, K.M. Vrotsos et al. 2005. Metabolic mapping of MCF10A human breast cells via multiphoton fluorescence lifetime imaging of the coenzyme NADH. *Cancer Research* 65(19): 8766–8773.

Borkman, R.F., A. Douhal, and K. Yoshihara. 1993. Picosecond fluorescence decay in photolyzed lens protein a-crystallin. *Biochemistry* 32(18): 4787–4792.

Colbert, A. and A.A. Heikal. 2005. Towards probing skin cancer using endogenous melanin fluorescence. *The Penn State McNair Journal*. p. 8. http://forms.gradsch.psu.edu/diversity/mcnair/mcnair_jrnl2004/files/08_colbert.pdf.

Cubeddu, R., F. Docchio, R. Ramponi, and M. Boulton. 1990. Time-resolved fluorescence spectroscopy of the retinal pigment epithelium: Age-related studies. *IEEE Journal of Quantum Electronics* 26(12): 2218–2225.

Cubeddu, R., P. Taroni, D.N. Hu, N. Sakai, K. Nakanishi, and J.E. Roberts. 1999. Photophysical studies of A2-E, putative precursor of lipofuscin, in human retinal pigment epithelial cells. *Photochemistry and Photobiology* 70(2): 172–175.

Deutsch, L. 2012. Evaluierung des Fluorescence Lifetime Imaging vom Augenhintergrund bei Patienten mit Diabetes mellitus. Dissertation, Department of Experimental Ophthalmology, University of Jena, Jena, Germany.

Dillon, J. and S.J. Atheron. 1998. Time resolved spectroscopic studies on the intact human lens. *Photochemistry and Photobiology* 51: 465–468.

Dimitrow, E., I. Riemann, A. Ehlers et al. 2009. Spectral fluorescence lifetime detection and selective melanin imaging by multiphoton laser tomography for melanoma diagnosis. *Experimental Dermatology* 18(6): 509–515.

Eldred, G.E. and M.L. Katz. 1988. Fluorophores of the human retinal pigment epithelium: Separation and spectral characterization. *Experimental Eye Research* 47(1): 71–86.

Geeraets, W.J. and E.R. Berry. 1968. Ocular spectral characteristics as related to hazards from lasers and other light sources. *American Journal of Ophthalmology* 66(1): 15–20.

Gudgin, E., R. Lopez-Delgado, and W.R. Ware. 1981. The tryptophan fluorescence lifetime puzzle. A study of decay times in aqueous solution as a function of pH and buffer composition. *Canadian Journal of Chemistry* 59(7): 1037–1044.

Haralampus-Grynaviski, N.M., L.E. Lamb, C.M. Clancy et al. 2003. Spectroscopic and morphological studies of human retinal lipofuscin granules. *Proceedings of the National Academy of Sciences of the United States of America* 100(6): 3179–3184.

Heikal, A.A. 2010. Intracellular coenzymes as natural biomarkers for metabolic activities and mitochondrial anomalies. *Biomarkers in Medicine* 4(2): 241–263.

Holm, S. 1979. A simple sequentially rejective multiple test procedure. *Scandinavian Journal of Statistics* 6(2): 65–70.

Katikaa, K.M., L. Pilona, K. Dippleb, K. Levinc, J. Blackwella, and H. Berberoglua. 2006. In-vivo time-resolved autofluorescence measurements on human skin. *Proceedings of SPIE* 6078: 83–93.

Klemm, M. 2012. Theoretische und experimentelle Untersuchungen zur Erfassung von Parametern des zellulären Stoffwechsels vom Augenhintergrund auf der Grundlage der zeitaufgelösten Autofluoreszenz. Manuscript dissertation. Department of Biomedical Technique and Informatics, Technical University Ilmenau, Ilmenau, Germany.

Koenig, K. 2008. Clinical multiphoton tomography. *Journal of Biophysics* 1(1): 13–23.

Koenig, K. and H. Schneckenburger. 1994. Laser-induced autofluorescence for medical diagnosis. *Journal of Fluorescence* 4(1): 17–40.

Meyer, J. 2012. Modellierung der integralen zeitaufgelösten Autofluoreszenz des Augenhintergrundes durch die spektralen Eigenschaften und Fluoreszenz-Lebensdauerparameter einzelner Fluorophore—Vergleich mit in vivo Messungen. Master of Engineering Thesis. Faculty SciTec, Study course Laser- und Optotechnologien, Ernst- Abbe- University of Applied Science Jena, Jena, Germany.

Niesner, R., P. Peker, P. Schlüsche, and K.H. Gericke. 2004. NAD(P)H fluorescence lifetime imaging of glucose stimulated MIN6 cells. Accessed January 15, 2014. http://www.pci.tu-bs.de/aggericke/Publikationen/MIN6-alt/Min6_paper.doc

Peters, S., M. Hammer, and D. Schweitzer. 2011. Two-photon excited fluorescence microscopy of ocular fundus for the interpretation of fundus autofluorescence analysis in vivo. *Proceedings of SPIE* 8086: 808604.

Quick, S. 2009. Untersuchungen zum klinischen Wert des Fluoreszenz Lifetime Imaging am menschlichen Augenhintergrund. Dissertation. Department of Experimental Ophthalmology, University of Jena, Jena, Germany.

Schneckenburger, H., M. Wagner, R. Weber, W.S. Strauss, and R. Sailer. 2004. Autofluorescence lifetime imaging of cultivated cells using a uv picosecond laser diode. *Journal of Fluorescence* 14(5): 649–654.

Schweitzer, D. 2009. Quantifying fundus autofluorescence. In *Fundus Autofluorescence*, N. Lois and J.V. Forrester (eds.), pp. 78–95. Philadelphia, PA: Lippincott Williams & Wilkins.

Schweitzer, D., M. Hammer, and F. Schweitzer. 2005. Grenzen der konfokalen Laser Scanning Technik bei Messungen der zeitaufgelösten Autofluoreszenz am Augenhintergrund *Biomedizinische Technik* 50(9): 263–267.

Schweitzer, D., S. Jentsch, S. Schenke, M. Hammer, C. Biscup, and E. Gaillard. 2007a. Spectral and time-resolved studies on ocular structures. *Proceedings of SPIE* 6628: 662807-1–662807-12.

Schweitzer, D., A. Kolb, M. Hammer, and E. Thamm. 2000. Tau-mapping of the autofluorescence of the human ocular fundus. *Proceedings of SPIE* 4164: 79–89.

Schweitzer, D., L. Leistritz, M. Hammer, M. Scibor, U. Bartsch, and J. Strobel. 1995. Calibration-free measurement of the oxygen saturation in retinal vessels of men. *Proceedings of SPIE* 2393: 210–218.

Schweitzer, D., S. Schenke, M. Hammer et al. 2007b. Towards metabolic mapping of the human retina. *Microscopy Research and Technique* 70: 410–419.

Skala, M.C. 2007. Multiphoton microscopy, fluorescence lifetime imaging and optical spectroscopy for the diagnosis of neoplasia. PhD dissertation, Duke University, Durham, NC.

Skala, M.C., K.M. Riching, D.K. Bird et al. 2007. In vivo multiphoton fluorescence lifetime imaging of protein-bound and free nicotinamide adenine dinucleotide in normal and precancerous epithelia. *Journal of Biomedical Optics* 12(2): 024014.

Takeuchi, S. and T. Tahara. 1997. Ultrafast fluorescence study on the excited singlet-state dynamics of all-trans-retinal. *Journal of Physical Chemistry A* 101(17): 3052–3060.

Tanaka, F., N. Tamai, and I. Yamazakil. 1989. Picosecond-resolved fluorescence spectra of d-amino-acid oxidase. A new fluorescent species of the coenzyme. *Biochemistry* 28(10): 4259–4262.

Thiagarajan, G., E. Shirao, K. Ando, A. Inoue, and D. Balasubramanian. 2002. Role of xanthurenic acid 8-O-b-D-glucoside, a novel fluorophore that accumulates in the brunescent human eye lens. *Photochemistry and Photobiology* 76(3): 368–372.

van de Kraats, J. and D. van Norren. 2007. Optical density of the aging human ocular media in the visible and the UV. *Journal of the Optical Society of America A* 24(7): 1842–1857.

Vishwasrao, H.D., A.A. Heikal, K.A. Kasischle, and W.W. Webb. 2005. Conformational dependence of intracellular NADH on metabolic state revealed by associated fluorescence anisotropy. *Journal of Biological Chemistry* 280(26): 25119–25126.

16

Pathogen Effects on Energy Metabolism in Host Cells

Márta Szaszák
University of Lübeck

Jan Rupp
University of Lübeck

16.1 Introduction

Pathogens reprogram host cell metabolism toward different metabolic pathways depending on the pathogen and host cell type. In addition, metabolic fuel is a prerequisite for the proliferating immune cells and their functional responses. Several bacteria also actively respond to the nutrient status of host cells by transforming to diverse types of altered metabolic forms such as dormant or persistent forms. These various metabolic states of bacteria contribute to antibiotic resistance and chronic infections, which are important clinical problems. The molecular pathways behind the essential metabolic changes in most pathogens and their host cells can be attributed to genetic modifications. However, in case of the obligate intracellular bacteria such as the *Chlamydia* spp., where genetic modification is limited, alternative techniques have to be used to characterize host and pathogen metabolic interaction. Two-photon fluorescence lifetime imaging microscopy (FLIM) of the metabolic coenzymes, NAD(P)H, allows for independent characterization of *Chlamydia* and host cell metabolic changes in living cells. The metabolic changes of *Chlamydia* during the developmental cycle can be monitored in real time by quantifying the fluorescence lifetime of protein-bound NAD(P)H and its relative amount in ratio to free NAD(P)H. Importantly, the technique provides a power tool for visualizing persistent infections in living cells and compartmentalized characterization of the metabolic changes in the host cell (e.g., cytosol, mitochondria, and nucleus).

16.2 Infection-Induced Metabolic Changes in Host Cells

16.2.1 Pathogen-Dependent Changes in Host Cell Metabolism

Several infectious diseases—including tuberculosis, malaria, or acquired immune deficiency syndrome (AIDS)—can cause wasting, an obvious sign of pathogen interference with host energy metabolism. Other infections such as periodontal and adenovirus infections are correlated with obesity (Dhurandhar et al. 1997; Saxlin et al. 2011). Additionally, both obesity and malnutrition in humans have been shown to increase susceptibility to infections (Falagas and Kompoti 2006; Schaible and Kaufmann 2007). The molecular mechanisms involve direct changes in host metabolism that are induced by the pathogen or the modification of immune responses and inflammatory pathways.

Several pathogens actively redirect host cell metabolism rather than passively rely on basal host cell metabolic activities. Viruses and other obligate intracellular pathogens rely on host cell metabolism to provide energy and macromolecular precursors for their survival and replication. The activated host cell metabolic networks may depend on the type of pathogens and their developmental stage, as well as the type of the host cell and its environmental conditions.

Metabolic effects of even those viruses that belong to the same family can be diverse as exemplified by human cytomegalovirus (HCMV) and herpes simplex virus type-1 (HSV-1), both of which belong to the family of herpesviruses. HCMV and HSV-1 infect the majority of the population and are capable of lifelong latency and reactivation. HSV-1 infection can cause diverse symptoms from cold sores to encephalitis, while HCMV infection causes severe disease in immunocompromised persons. Mass spectrometric analysis of the metabolites from infected cell extracts showed that HCMV increases glycolytic flux and fatty acid synthesis, while HSV-1 increases pentose phosphate intermediates for the synthesis of nucleotides. These metabolic changes were conserved across different host cell types (Vastag et al. 2011).

Many pathogens go through a unique developmental cycle inside their host cells. Different stages of pathogen development can be associated with various host cell metabolic changes. The obligate intracellular bacterium *Chlamydia trachomatis*, which causes genital tract and eye infections, alternates between elementary bodies and reticulate bodies. Microarray analysis of host cell transcriptional changes during early and middle phases of *C. trachomatis* infection showed that a gene involved in tryptophan biosynthesis was upregulated in early infection, while genes involved in glucose and sphingolipid metabolism had increased expression in middle-phase infection (Xia et al. 2003).

Cell type–dependent differences in host–pathogen metabolic interactions are demonstrated by the infection of T cells and macrophages with human immunodeficiency virus type 1 (HIV-1) that causes AIDS. HIV-1 infects activated CD4+ T cells and induces cell death. However, the virus also infects terminally differentiated macrophages and causes cell survival and prolonged low-level viral production. Metabolomic analysis of HIV-1-infected primary human CD4+ T cells and differentiated macrophages showed that infected T cells have a higher glucose uptake with increased levels of glycolytic intermediates. In contrast, infected macrophages exhibited decreased glucose uptake with decreased concentrations of glycolytic intermediates (Hollenbaugh et al. 2011).

Changes of host cell environment—including changes in temperature, oxygen availability, or cellular concentrations of growth factors, cytokines, or nutrients—can control pathogen survival in the intracellular environment and effect metabolism (Dietz et al. 2012). *Chlamydia pneumoniae* is an obligate intracellular bacterium that causes respiratory tract infections and has been linked to the development of atherosclerosis. In a low-oxygen environment, chlamydial replication is increased (Juul et al.

2007). Metabolic adaption of the host cell to hypoxia by enhanced glucose uptake may contribute to increased chlamydial growth (Rupp et al. 2007). However, a hypoxic environment can also affect intracellular pathogens the opposite way, inducing the formation of their dormant forms as in the case of the *Mycobacterium tuberculosis* (Shleeva et al. 2010).

The earlier described examples demonstrate that elucidating the diverse mechanisms underlying pathogen hijacking of host cell metabolism is important in finding new ways to diagnose and treat infectious diseases. Glucose and lipid metabolism are the most frequently studied pathways in infected cells, although metabolism of other macro- and micronutrients such as amino acids, nucleotides, vitamins, and trace elements also plays an important role in infections.

16.2.1.1 Glucose Metabolism

Several pathogens cover their metabolic needs by upregulating host cell glycolysis. However, in case of other pathogens, downregulation of cellular glycolysis has been observed.

After glucose enters a cell, it is metabolized to pyruvate through glycolysis in the cytoplasm, producing two ATP molecules per one molecule of glucose. Pyruvate is converted to acetyl coenzyme A in the mitochondria and begins the tricarboxylic acid (TCA) cycle by combining with oxaloacetic acid to form citrate. The NADH produced in the TCA cycle fuels oxidative phosphorylation where 36 ATP molecules are synthesized.

In infected cells, the upregulation of aerobic glycolysis—the Warburg effect—is very similar to the metabolic changes of cancer cells. In HCMV-infected cells, glucose is not completely broken down in the TCA cycle and by oxidative phosphorylation for full energy production, but is used biosynthetically, while a fraction of pyruvate is converted to lactate. Citrate from the TCA cycle is transported back to the cytoplasm where it is used for de novo fatty acid biosynthesis. This process requires increased glucose uptake, increased glycolysis, and active rerouting of glucose carbon in the form of citrate from the TCA cycle. The diversion of citrate from the TCA cycle (cataplerosis) requires induction of enzymes to promote glutaminolysis to maintain the TCA cycle (anaplerosis) and ATP production (Figure 16.1) (Yu et al. 2011a).

Enhanced glucose uptake can be mediated by several different mechanisms. The ubiquitous glucose transporter (GLUT1) is regulated at multiple levels of infections. Transcriptional upregulation and increased cell surface trafficking play key roles in the regulation of the number of GLUT1 transporter on the cell surface (Wang et al. 2008). In HCMV-infected cells, increased GLUT4 expression leads to the elimination and replacement of cell surface GLUT1, resulting in greater transport capacity. GLUT4 is localized in intracellular vesicles and activation of the PI3K/AKT pathway results in its trafficking to the cell surface—a vital process for virus survival. GLUT1 elimination is mediated by the HCMV major immediate-early protein, IE72 (Yu et al. 2011b).

Interestingly, GLUT1 is a cell surface receptor for the oncogenic deltaretroviruses HTLV-1 and HTLV-2, facilitating virus entry (Manel et al. 2003). Oncogenic viruses induce several signaling pathways that control crucial metabolic pathways such as PI3K, MAPK, p53, and hypoxia-inducible factor 1α (HIF1α). The increased glucose uptake enhances cell proliferation contributing to the tumorigenic effect of the viruses. In return, the tumor niche provides a highly favorable environment for oncogenic viral replication (Noch and Khalili 2012).

Regulation of glycolytic enzymes can also occur by different mechanisms such as controlling their transcription and mRNA stability or through allosteric activation. One of the key regulating enzymes of glycolysis is phosphofructokinase (PFK), which is regulated by several kinases that are upregulated during infection (Yu et al. 2011a).

Changes in glucose metabolism induce the consequent changes in glutamate metabolism. Replenishment of TCA cycle intermediates by glutaminolysis requires the activity of glutaminase (GLS) and glutamate dehydrogenase (GLUD).

In *Burkholderia pseudomallei* infection, glycolysis and TCA cycle are transcriptionally repressed (Chin et al. 2010). Hepatitis C virus (HCV) downregulates GLUT1 and GLUT2 cell surface expression

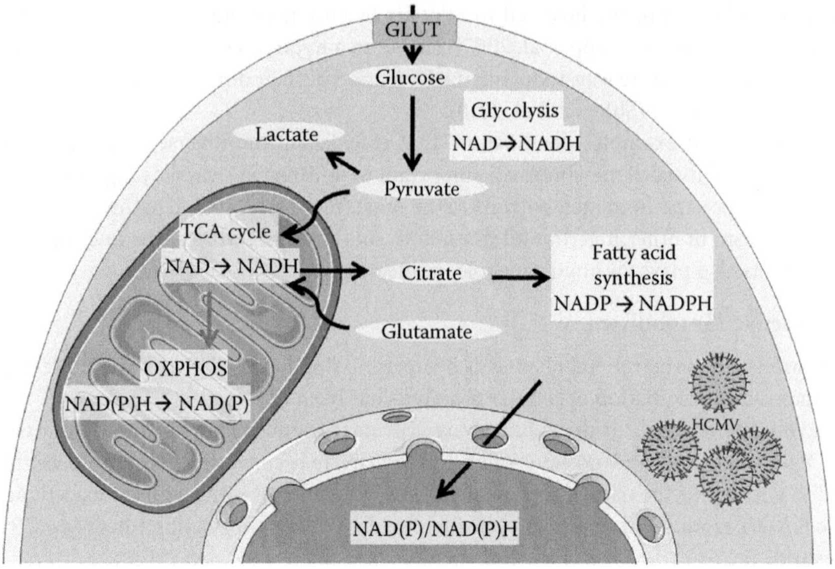

FIGURE 16.1 Regulation of host cell glucose and lipid metabolic pathways during HCMV infection. Glucose is transported by GLUT into the cytosol; it is normally broken down in the mitochondria and used for ATP synthesis (gray arrow). However, in infected cells, glucose is used for fatty acid synthesis by upregulation of glycolysis (black arrows). Glutamate replenishes the TCA cycle as citrate is directed for fatty acid synthesis (Yu et al. 2011a). The autofluorescent metabolic cofactors NADH and NADPH enable the visualization of cellular metabolic changes due to infection (Szaszák et al. 2011). The concentration of NAD(P)H is highest in the mitochondria. NAD(P)/H is freely diffusible between the nucleus and cytosol. *Abbreviations*: TCA, tricarboxylic acid; OXPHOS, oxidative phosphorylation. (Pictures of mitochondria, nucleus, cell, and virus were obtained from Servier Medical Art, Neuilly-sur-Seine, France, www.servier.com.)

(Kasai et al. 2009). In *Salmonella* Typhimurium infection, inflammasome-mediated caspase-1 activation leads to cleavage of enzymes in the glycolytic pathway (Shao et al. 2007).

16.2.1.2 Lipid Metabolism

In host–pathogen interactions, lipids play an important role not only as energy sources but also as membrane components and as signaling molecules. The dynamic membrane rearrangements during the uptake and trafficking of intracellular pathogens require the coordinated action of host cell lipid kinases and phosphatases (van der Meer-Janssen et al. 2010). Several pathogens remain in an intracellular vacuole, while others spread into the cytosol. The intracellular pathogen–containing vesicles typically include lipid species that are derived by the host cell.

One of the most well-studied viruses regarding host cell lipid metabolism is HCV, which causes acute and chronic liver disease with characteristic fatty deposits in the liver (Syed et al. 2010). HCV infection increases lipogenesis, reduces β-oxidation of lipids, and decreases lipoprotein secretion (Vescovo et al. 2012). HCV exploits lipid metabolism to accomplish several steps of its life cycle, while the HCV RNA replication strictly depends on intermediates of cholesterol synthesis (Kapadia and Chisari 2005). Microarray analysis, proteomic analysis, and complementary lipidomic analysis have shown the complex regulation of biosynthesis, degradation, and transport of host cell lipids in HCV infection (Heaton and Randall 2011).

Expressions of several genes involved in lipid metabolism are upregulated during HCV infection, while others are only posttranslationally regulated (Blackham et al. 2010; Diamond et al. 2010). Host cell-derived phosphatidylinositol-4-phosphate (PI4-P) synthesized by PI4-kinases has been shown to be a hallmark of virus-induced intracellular membranes (Alvisi et al. 2011).

Lipid acquisition by pathogens can also occur by capturing organelle-derived lipids (Elwell and Engel 2012). These organelles include the Golgi, multivesicular bodies, and lipid droplets (Herker and Ott 2012; Pierini et al. 2009; Saka and Valdivia 2010). *C. trachomatis* inclusions acquire sphingomyelin and cholesterol from Golgi-derived exocytic vesicles (Hackstadt et al. 1996). Lipid droplets are lipid storage compartments composed of a neutral lipid core surrounded by a phospholipid monolayer. During *C. trachomatis* infection, the number of lipid droplets increases, but no accumulation is observed: LDs are recruited and translocated into the inclusion from the host cytoplasm (Cocchiaro et al. 2008).

16.2.2 Transcriptional Regulation of Host Cell Metabolic Changes

Almost all transcription factor families regulate the transcription of metabolic enzymes. Several of them have already been shown to play a key role during infections as well.

The main transcriptional regulators of glucose metabolism are the forkhead box O (FOXO) proteins and their coactivators: CREB-binding protein (CBP) and peroxisome proliferator–activated receptor-γ coactivator 1α (PGC1A) (Desvergne et al. 2006). HIF1α is a crucial transcriptional regulator of glucose metabolism in hypoxia. Under hypoxic conditions, HIF1α is stabilized, while under normoxic conditions, hydroxylation of HIF1α by prolyl hydroxylases (PHDs) promotes its binding to the von Hippel–Lindau (VHL) tumor suppressor that targets HIF1α for ubiquitination and subsequent degradation (Ivan et al. 2001; Maxwell et al. 1999). It was suggested that the HIF1α positively mediates host cell metabolic changes in *C. pneumoniae*-infected cells under hypoxia when chlamydial growth is enhanced. Enhanced glucose uptake mediated by GLUT1 expression induced by HIF1α could benefit the metabolic needs of *C. pneumoniae* (Rupp et al. 2007). The protein deacetylase, sirtuin 1 (SIRT1), a major regulator of cellular lifespan, also contributes to the transcriptional regulation of glucose metabolism. SIRT1 promotes gluconeogenesis by deacetylating and activating PGC1A and FOXO1. In addition, it decreases glycolysis by deacetylating and repressing HIF1α, contributing to energy conservation (Houtkooper et al. 2012). SIRT1 activity is directly regulated by its substrate NAD (Michan and Sinclair 2007). SIRT1 transcriptional regulation occurs through the interaction with carboxy-terminal binding protein 1 (CTBP) and hypermethylated in cancer (HIC1) in response to nuclear NADH levels (Fjeld et al. 2003; Zhang et al. 2002, 2007). HIV-1 Tat protein directly interacts with the deacetylase domain of SIRT1 and blocks its deacetylase activity, which induces T-cell hyperactivation by activating the expression of NF-κB-responsive genes (Kwon et al. 2008). Several of the transcription factors that are involved in glucose metabolism also play a key role in lipid metabolism (Desvergne et al. 2006).

Transcriptional regulation of lipid metabolism is executed mainly by sterol regulatory element-binding proteins (SREBPs), peroxisome proliferator–activated receptors (PPARs), CCAAT/enhancer-binding proteins (C/EBPs), and the nuclear receptor family. SREBP-1c is the major transcription factor involved in fatty acid biosynthesis, while SREBP-2 is involved in cholesterol synthesis. PPARγ regulates fatty acid storage and adipogenesis. PPARα and PPARβ regulate fatty acid oxidation (Desvergne et al. 2006). HCV increases the activity of SREBPs and induces their proteolytic cleavage (Waris et al. 2007). Protease inhibitors that are used in the treatment of HIV infection, such as indinavir, decrease the level of nuclear SREBP-1 and PPARγ that may explain their lipodystrophic side effect (Caron et al. 2001).

16.2.3 Regulation of Autophagy in Response to Infection

Autophagy plays a crucial role in cellular metabolism and adaptation to infections. Autophagosomes are double-membrane-bound organelles of endomembranous origins that mature to autolysosomes

(Takahashi et al. 2009). The key regulators of autophagy are the serine/threonine kinase mammalian target of rapamycin (mTOR) and phosphatidylinositol-3-kinases (PI3Ks). The execution of autophagy is mediated by autophagy-related proteins (ATG proteins). Autophagy is initiated by the protein complex of ATG1 and PI3K with beclin-1 that leads to the activation of other ATG proteins such as MAP1LC3 (ATG8 or LC3) and ATG12 (Levine and Deretic 2007). During nutrient deprivation, autophagy provides metabolic substrates for ATP synthesis to preserve cellular bioenergetics. In addition, it also functions as a routine quality and quantity control by removing protein aggregates and damaged organelles to maintain cytoplasmic biomass (Dinkins et al. 2010).

In the case of infections, either autophagy is required for pathogen growth or it participates in pathogen elimination. In the latter case, it may function directly through eradication of the pathogen or indirectly by the breakdown of host cell factors required for pathogen replication, the inhibition of innate immune signaling, or the promotion of cell survival (Levine and Deretic 2007). Galectin 8 (LGALS8), a cytosolic lectin, plays a role in detecting and targeting bacteria-containing vacuoles for antibacterial autophagy. LGALS8 binds to host glycans on bacteria-containing vacuoles, and by recruiting NDP52, it activates autophagy (Thurston et al. 2012). Autophagy can also limit inflammation that accompanies infection (Shi et al. 2012).

Several pathogens have evolved strategies either to avoid elimination by autophagy or to actively hijack its components to promote their own growth. To use proteins and membranes from the autophagic pathway, numerous intracellular bacteria and viruses stimulate the accumulation of membranes that have autophagic markers and inhibit the maturation of autophagosomes. Replication of the pathogen in these compartments can provide protection against immune defense (Kirkegaard et al. 2004).

The nutritive function of autophagy can also favor pathogen replication by providing access to host cell metabolites (Wang et al. 2009). Autophagy contributes to the amount of cellular lipids by mobilizing cellular lipid stores, which allows cells to adapt to lipogenic stimuli (Rodriguez-Navarro and Cuervo 2010). Dengue virus has been shown to use autophagy to control cellular lipid metabolism through autophagy. The autophagy-dependent processing of lipid droplets and triglycerides during dengue virus infection releases free fatty acids, which increases β-oxidation to generate ATP (Heaton and Randall 2010). The interaction between autophagy and lipid metabolism is bidirectional since changes in the amount of cellular lipids also regulate autophagic activity (Rodriguez-Navarro and Cuervo 2010).

16.3 Immune Response and Metabolism

The immune system functions to recognize and eliminate pathogens by multiple different mechanisms. The inborn innate immune system provides the first line of defense as an immediate, antigen-nonspecific response. It also activates the adaptive immune system that provides an antigen-specific defense mechanism and grants long-lasting immunity. Metabolic fueling of proliferating immune cells is a prerequisite of immune responses. Autophagy influences both innate and adaptive immunity by immune regulation and antigen presentation.

16.3.1 Regulation of Innate Immune Responses by Metabolism

Functional cellular metabolism is indispensable for innate immune cell survival and function. Toll-like receptors—such as TLR4 that is the receptor for LPS—play a pivotal role in initiating innate immune responses and cytokine expression as well as mediating the interplay between metabolism and immunity (Shi et al. 2006). Polyunsaturated fatty acids are important modulators of macrophage and neutrophil responses and support proinflammatory functions (Calder 1998). In contrast, excess free fatty acids and triglycerides promote inflammatory responses by TLR4 activation (Afacan et al. 2012; Shi et al. 2006). Lipid metabolism in macrophages regulated by SREBP-1a is crucial for IL1β secretion and inflammasome activation (Im et al. 2011). Glutamine is a crucial nutrient for the proper phagocytic function and cytokine production of macrophages and neutrophils (Pithon-Curi et al. 2002; Wallace and Keast 1992).

16.3.2 Regulation of T-Cell Functions by Metabolism

Cells of the adaptive immune system, B cells and T cells, recognize specific antigens. T-cell metabolism changes dramatically to support specific functions. Resting T cells use the energy-efficient oxidative metabolism. Upon pathogen encounter, they change to glycolytic metabolism to stimulate their proliferation. This metabolic transition is governed by the PI3K/AKT pathway, GLUT1 expression, and AMP-activated protein kinase (AMPK) (Fox et al. 2005). After pathogen elimination, T cells return to oxidative metabolism to support T-cell memory. If metabolism does not succeed in matching the energetic demand, the functions of T cells are impaired or they undergo premature apoptosis resulting in suppressed adaptive immunity. On the other hand, metabolic surplus can prevent apoptosis, boost T-cell function, and lead to inflammatory diseases, autoimmunity, or neoplasia. Growth factors and cytokines play a crucial role in regulating T-cell metabolism (Michalek and Rathmell 2010).

The relatively constant number of naïve lymphocytes in the blood is maintained by the delicate balance between proliferation, differentiation, and cell death (Jameson 2002). This homeostasis is based on extrinsic signals that help to sustain a low rate of basal glucose consumption, which allows for maximal ATP synthesis utilizing oxidative metabolism. Chemokines, T-cell receptor stimulation, and IL7 are among the crucial extrinsic signals that maintain the quiescent state of naïve T cells (Michalek and Rathmell 2010). IL7 promotes glucose uptake and glycolysis through STAT5. In addition, IL7 causes a delayed, sustained activation of the PI3K/AKT pathway that contributes to GLUT1 trafficking (Wofford et al. 2008). Metabolic pathways are interconnected with survival in resting T cells as evidenced by the fact that IL7 is unable to promote survival under glucose-limiting conditions (Michalek and Rathmell 2010).

T-cell activation requires rapid cell growth, proliferation, and generation of effector T cells that is supported by the significant upregulation of aerobic glycolysis (Michalek and Rathmell 2010). Failing to increase glycolysis may result in a reduced T-cell proliferation, production of the effector cytokine IFNγ, and activation of proapoptotic pathways (Alves et al. 2006; Cham and Gajewski 2005). Transgenic overexpression of GLUT1 increases T-cell size, cytokine production, and proliferation upon activation in mice. In addition, it leads to readily activated memory-phenotype T cells and signs of autoimmunity. T-cell receptor stimulation alone does not necessarily lead to an increase in glucose metabolism; concurrent ligation of the costimulatory receptor CD28 or CD3 is necessary for upregulation of glucose uptake and glycolysis (Jacobs et al. 2008). After the initial upregulation of glycolysis, IL2 plays a role in sustaining the metabolism of activated T cells (Wofford et al. 2008). Notably, glycolysis also serves as a key metabolic regulator of the development of discrete effector populations of T cells. Blocking glycolysis inhibits the development of T_H17 cells but promotes the generation of Foxp3-expressing regulatory T cells (T_{reg} cells). HIF1α is selectively expressed in T_H17 cells and is a crucial regulator of the glycolytic changes (Shi et al. 2011).

After pathogen clearance, the majority of the antigen-specific T-cell clones undergo apoptosis. The surviving T cells differentiate into memory cells that subsist to protect against reinfection (D'Cruz et al. 2009). Loss of growth factor and cytokine signals after the immune response leads to a dramatic decrease in metabolism and results in cellular atrophy and apoptosis (Michalek and Rathmell 2010). Inactivation of the PI3K/AKT pathway leads to GLUT1 internalization and lysosomal degradation, as well as endocytosis of amino acid transporters (Wieman et al. 2007). Autophagy is activated to provide metabolic fuel by self-digestion of excess organelles and membranes (Lum et al. 2005). Interestingly, a subset of cells survives and becomes long-lived memory T cells. In this case, selective upregulation of IL7 receptor enhances nutrient uptake and maintains metabolism on a homeostatic level, generating a metabolic phenotype similar to that of naïve T cells with minimal biosynthetic demand and maximum ATP production (Kaech et al. 2002; Michalek and Rathmell 2010). The tumor necrosis factor (TNF) receptor-associated factor 6 (TRAF6) plays a crucial role in enhancing T-cell memory by promoting fatty acid metabolism (Pearce et al. 2009). Memory T cells rapidly grow and proliferate upon restimulation indicating that the metabolic capacity is different of naïve T cells, which is yet to be defined (Michalek and Rathmell 2010).

16.3.3 Inflammasomes and Metabolism

A key aspect of infections is the associated inflammation, and caspases are crucial regulatory enzymes of inflammation that are activated by inflammasomes, multimolecular complexes of the innate immune system. Inflammasomes are assembled around NOD-like receptors (NLRs) such as NLRP3 and contain the adaptor protein ASC and procaspase-1. NLRs detect pathogen-associated molecular patterns (PAMPs) or endogenous damage-associated molecular patterns (DAMPs) in intracellular compartments (Lamkanfi 2011). Inflammasome activation leads to caspase activation that converts inflammatory cytokines—including IL1β and IL18—into their active form.

Importantly, caspases control not only the secretion of cytokines but also cell death and energy metabolism through regulation of glycolysis and lipid metabolism (McIntire et al. 2009). Caspase-1 was shown to target multiple cellular pathways by processing more than 40 different proteins including several metabolic enzymes. For example, caspase-1 cleaves the glycolytic enzymes such as aldolase (ALDO), triosephosphate isomerase (TPI), glyceraldehyde-3-phosphate dehydrogenase (GAPDH), α-enolase (ENO), and pyruvate kinase (PK). The importance of this cleavage during infection was demonstrated for the *S.* Typhimurium infection, which impairs glycolysis in a caspase-1-dependent manner (Shao et al. 2007). Caspase-1 also promotes lipid biogenesis during infection by activating SREBPs (Gurcel et al. 2006). Inflammasome-induced lipid biogenesis through caspase-1 activation was shown to be crucial *C. trachomatis* replication (Abdul-Sater et al. 2009).

Inflammasomes are also linked to metabolism through the glucose-regulated transcription of IL1β, and the underlying mechanism for glucose effect on IL1β production remains unknown (Wen et al. 2012). Autophagy plays an important role in eliminating inflammasomes and consequently decreasing IL1β secretion, which in return limit inflammation. Assembled inflammasomes undergo ubiquitination that targets them for destruction through the interaction with the autophagic adaptor, p62, and blocking autophagy potentiates the activation of inflammasome (Shi et al. 2012).

16.4 Developmental Changes in Pathogen's Metabolism in an Intracellular Environment

Intracellular pathogens undergo developmental changes after entering the host cell. Pathogen growth, replication, and progeny formation require a changing gene expression profile during the developmental cycle. Expressions of different sets of immediate, early, and late genes is characteristic. Disruption of the developmental cycle causes pathogen death or its transformation to a metabolically altered from.

16.4.1 Altered Metabolic Forms of Bacteria and Viruses

Several extracellular bacteria are known to form spores to establish their long-term survival. Although the intracellular environment is strictly regulated, certain conditions can also induce metabolic changes in intracellular bacteria that induce bacterial transformation to a persistent form. These conditions include change in local nutrient and oxygen availability or the enhanced action of immune control mechanisms that are not sufficient to kill the bacteria. Such changes can enable the bacteria to survive inside the host organism for a prolonged time. However, this reversible state of the pathogen is characterized by decreased metabolic activity, altered morphology and gene expression, lack of replication, resistance to antibiotics, and prolonged survival. In the case of *M. tuberculosis*, this state is called dormancy and can cause latent tuberculosis, and *M. tuberculosis* replicates rapidly until the host develops a sufficient immune response (Russell et al. 2010), which is characterized by a granuloma formed around the infected macrophages in the tissues. The granuloma limits the dissemination of the infection. The center of the granuloma is low in oxygen and rich in lipids and proteins derived from dead cells and bacteria; *M. tuberculosis* can persist under these conditions for decades (Höner zu and Russell 2001). When exposed to a low-oxygen environment, *M. tuberculosis* upregulates genes that are controlled by the dosR regulon complex (Russell et al. 2010). Enzymes of anaerobic respiration and fermentation in *M. tuberculosis* play a key role in their survival under hypoxia. Throughout latency, there is a dynamic interplay between *M. tuberculosis* and the host immune system with a continuous recruitment of immune cells into the granuloma. When this process becomes dysregulated, reactivation occurs (Ehlers 2009).

Persistence is also of relevance in chlamydial infections. Different stimuli are able to induce persistence of *C. pneumoniae* and *C. trachomatis* in cell culture models including IFNγ, penicillin, heat shock, deprivation of amino acid, glucose, and iron, as well as compounds found in cigarette smoke, adenosine, or coinfection with HSV (Schoborg 2011). Antibiotics that are used in therapy can also induce persistence if they are applied in subinhibitory concentrations (Gieffers et al. 2004). The transcriptional profile of the bacteria depends on the method by which persistence is induced demonstrating the ability of bacteria to adapt to the environment for long-term survival (Klos et al. 2009). Persistently infected cells resist apoptosis and induce the release of the high-mobility group box 1 (HMGB1) protein, which is a proinflammatory molecule (Rödel et al. 2012).

Several viruses such as the herpes or Epstein–Barr are also capable of lifelong latency and reactivation. During latency, the viral metabolism is shut down and the viral protein expression is reduced. Host genetic background plays an important role in the development and reactivation of latent infections (Iwatsuki et al. 2004).

16.4.2 Clinical Relevance of Persistence

Diagnosis and treatment of persistent infections present a major challenge in clinical practice. As persistent and latent infections are often clinically asymptomatic, they can only be diagnosed by direct PCR or by demonstration of the immunological memory response since the pathogens are not able to be cultured (Ehlers 2009). Although these altered metabolic forms may not induce the typical symptoms of acute disease, they can induce chronic inflammation leading to chronic inflammatory diseases. For example, *C. pneumoniae* and HSV-1 infections were connected to coronary heart disease and Alzheimer's disease, while *M. tuberculosis* infection was linked to Crohn's disease and stroke. Since diagnosis occurs mainly through routine screening or when infection-induced chronic inflammation causes complications, it is difficult to estimate the role of acute, chronic persistent, reactivated, or a reinfection in the duration and the sequelae of the infection. Thus, the exact pathomechanisms of chronic inflammatory diseases associated with infections are still under debate (Shima et al. 2010; Wang and Kaltenboeck 2010).

Discrimination of chronic persistent infections from reactivating infections is also important to determine proper antimicrobial and immunomodulatory strategies (Ehlers 2009). Persistent and dormant forms of bacteria are antibiotic resistant; therefore, development of novel therapies is crucial. The effects of the environmental conditions must also be considered as they cannot only strongly affect bacterial metabolism but can also influence antibiotic treatment. Efficacy of first-choice antibiotics for the treatment of *C. trachomatis* infections—including azithromycin and doxycycline—is reduced under hypoxia. The upregulation of the multidrug transporter protein (MDR1) is presumed to play a role in the enhanced antibiotic efflux under low oxygen concentration, as MDR1 transcription is regulated by HIF1α (Shima et al. 2011). Concerning *M. tuberculosis* infections, it was shown that bacteria resist antimicrobial treatment in hypoxic areas of the granuloma in a guinea pig model (Lenaerts et al. 2007).

An in vitro multiple-stress dormancy model that implements low oxygen, high CO_2, low nutrient, and acidic pH conditions was suggested for screening of potential drugs against *M. tuberculosis* and other intracellular pathogens (Deb et al. 2009). Induction of aerobic glycolysis by the Kaposi's sarcoma herpesvirus (KSHV) is required to maintain latency in infected cells (Delgado et al. 2010). Consequently, targeting specific host metabolic pathways—such as glycolysis in case of KSHV infection—could be a novel treatment option.

16.5 Autofluorescence Imaging to Study Host–Pathogen Metabolic Interaction and Immune Cells

Imaging the autofluorescence of the metabolic coenzymes, NAD(P)H and flavin adenine dinucleotide (FAD), or amino acids and lipids by advanced microscopic techniques allows for studying host–pathogen metabolic interactions and the examination of metabolic changes in both pathogen and its host cell. Thus far, however, only a few pioneering studies have tapped into the potential of these techniques.

16.5.1 NAD(P)H

NAD(H) and its phosphorylated form NADP(H) are well known for their functions as metabolic coenzymes; however, NAD is also an important signaling molecule that participates in a wide range of signaling pathways including protein deacetylation, mono- and poly-ADP-ribosylation, and calcium mobilization (Berger et al. 2004). Hundreds of cellular proteins contain the characteristic NAD-binding domain, also known as the Rossmann fold (Rossmann and Argos 1978). The phosphorylated forms, NADP and NADPH, are formed by the NAD kinase (NADK) enzyme (Pollak et al. 2007b). The reduced forms, NAD(H) and NADP(H), are also essential metabolites for all pathogens that evolved diverse biosynthetic or transport systems to acquire NAD(P) (Gossmann et al. 2012). Importantly, only the reduced forms, NAD(P)H, are autofluorescent.

Two-photon FLIM of NAD(P)H has been used to study cellular metabolism in living cells by utilizing the autofluorescence properties of NAD(P)H (Bird et al. 2005; Chia et al. 2008; Skala et al. 2007a,b). At 730–750 nm excitation, the cell is dominated by NAD(P)H autofluorescence as measured by two-photon microscopy (Bird et al. 2005; Huang et al. 2002; Li et al. 2008; Skala et al. 2007b). The majority of cellular NAD(P)H autofluorescence originates from the mitochondria, but cytosolic and nuclear signal intensities are also detectable (Szaszák et al. 2011).

Although the autofluorescence spectra of NADH and NADPH are very similar, the two different coenzymes have distinct physiological functions. For example, NAD(H) is a cofactor for catabolic reactions such as glucose catabolism in the cytosol and the mitochondria, while the phosphorylated NADP(H) is a cofactor in anabolic reactions such as lipid biosynthesis and plays an important role in the cellular antioxidative defense system (Niesner et al. 2008; Pollak et al. 2007a). NADH is generated within the glycolytic pathway, while NADPH is produced by the enzymes of the pentose phosphate pathway in the cytosol of host cell. In the mitochondria, NAD(P)H cannot pass through the mitochondrial membrane. Only electrons from NAD(P)H are carried across the membrane for the TCA cycle and for oxidative phosphorylation, which oxidizes NADH to NAD. By contrast, free NAD(P)H from the cytosol can cross the nuclear envelope by diffusion through the nuclear pores; therefore, changes of cytosolic levels of free NAD(P)H are reflected in the nucleus (Figure 16.1). An important binding protein of NADH in the nucleus is the transcriptional corepressor, CTBP (Zhang et al. 2002). Consequently, nuclear NADH functions to control gene transcription by regulating CTBP (Zhang et al. 2007).

As the fluorescence lifetime of a molecule is not dependent on its total concentration but strongly depends on its respective protein binding within a given microenvironment, differences of the NAD(P)H fluorescence lifetimes are based on the different protein bindings of NAD(P)H and reflect its diverse physiological functions.

The ratio of free to protein-bound NAD(P)H is the established marker to sensor the metabolic activity of eukaryotic cells and has been used to monitor changes in host cell glycolysis or oxidative phosphorylation (Li et al. 2008, 2009; Skala et al. 2007a). As FLIM of NAD(P)H can determine compartmentalized redox changes in living cells, it is an ideal technique to study host and pathogen metabolic interaction (Szaszák et al. 2011).

16.5.2 FAD

FAD is a redox coenzyme involved in metabolic reactions. Unlike NAD(P)H, however, it is the oxidized form, FAD, that is autofluorescent. No single protein-binding domain was identified so far for FAD binding. In every FAD-binding protein family, a pyrophosphate-binding sequence moiety exists (Dym and Eisenberg 2001). FAD-binding proteins exist in various organisms (Macheroux et al. 2011).

Cellular autofluorescence imaged by two-photon microscopy at 890–910 nm is dominated by FAD fluorescence (Huang et al. 2002). Colocalization studies with the mitochondrial marker tetramethylrhodamine methyl ester (TMRM) showed that cellular FAD signal originates from the mitochondria without significant contribution from the cytosol or nucleus (Dumollard et al. 2004). Changes in autofluorescence of FAD indicate alterations of mitochondrial function (Heikal 2010). As several intracellular pathogens interfere with mitochondrial function, FAD fluorescence lifetime changes could be used as a marker of mitochondrial function in infected cells.

The mitochondrial NAD(P)H pool is in dynamic balance with the FAD pool (Huang et al. 2002). Imaging the autofluorescence intensities of the two coenzymes and calculating their ratio (redox ratio) can be used as a marker of the cellular redox status (Skala et al. 2007a).

16.5.3 Amino Acids

Cellular protein content and protein structure can be imaged through the autofluorescence of the amino acids tryptophan, tyrosine, and phenylalanine. Tryptophan can be excited at 560 nm by two-photon microscopy. Normalizing NAD(P)H fluorescence intensities to tryptophan fluorescence enables the quantification of cellular metabolism. In addition, the fluorescent lifetime changes of tryptophan are dependent on cellular tryptophan metabolism (Li et al. 2009).

16.5.4 Lipids

Oxidized lipid derivatives called lipofuscin also show autofluorescence and have broad emission spectra by two-photon excitation. Lipofuscin is highly abundant in the brain, the eye, and ageing tissues. The rate of lipofuscin accumulation in tissues correlates with oxidative stress. In the case of chronic degenerative diseases where reactive oxygen species formation plays a critical role such as Alzheimer's disease or age-related macular degeneration, lipofuscin imaging can be used to study the pathomechanism of the disease (Dowson et al. 1992; Von Rückmann et al. 1998). Autofluorescence of the eye's fundus is increased in cytomegalovirus (CMV) retinitis as a possible sign of enhanced lipofuscinogenesis (Yeh et al. 2010).

A further possibility to study lipids by their autofluorescence is the coherent anti-Stokes Raman spectroscopy (CARS), which enables the visualization of cellular lipid droplets (Robinson et al. 2010). Lipids are rich in C–H bonds, which have strong Raman-active vibrations at 2845 cm^{-1} creating a high-contrast image in CARS microscopy using Stokes laser at 892/896 nm (Nan et al. 2006; Robinson et al. 2010). Studies on lipid droplet dynamics during HCV infection when host cell lipid metabolism was inhibited demonstrated that lipid droplet aggregates are formed due to increased lipid storage. Coincidently, HCV RNA disperses in the cytosol and HCV replication is decreased (Lyn et al. 2009). Three different stages of CMV infection can be distinguished by visualizing distinct lipid droplet dynamics using CARS (Wong et al. 2011).

16.5.5 Autofluorescence Imaging of Immune Cells

The autofluorescence of immune cells can also be utilized to study cell dynamics (Heintzelman et al. 2000). Phenotypic characterization and in vivo dynamics of migrating immune cells have been described based on their autofluorescence (Gehlsen et al. 2010; Klinger et al. 2012; Steven et al. 2011). FLIM was used to enable optical fingerprinting of individual immune cells and their surrounding epithelial cells (Steven et al. 2009). Activated human primary T cells and malignant B lymphoma cells could be distinguished based on their autofluorescence intensity spectra (Pantanelli et al. 2009). Changes in FAD autofluorescence intensity were observed in neutrophils after toxic granulation was induced (Kim et al. 2009). However, functional studies to examine autofluorescence changes of NADP(H) or FAD in relation to the dramatic metabolic changes of immune cells following pathogen challenge have yet to be performed.

16.6 Application: Imaging *C. trachomatis* Metabolism and Its Interaction with Host Cells

C. trachomatis is an obligate intracellular bacterium that is the major causative agent of bacterial-based sexually transmitted diseases in the world. In addition, it is the leading cause of infectious blindness due to trachoma.

Although the genetic modification of *C. trachomatis* recently became possible, it is time and labor intensive (Kari et al. 2011). In addition, the obligate intracellular developmental cycle of the bacteria does not allow for independent cultivation and studies of the host cells. Thus, distinguishing host and *C. trachomatis* metabolic pathways is challenging. Current knowledge on chlamydial metabolism is restricted to microarray and RT-PCR analyses about the expression of metabolic genes during different intracellular developmental stages, the characterization of recombinant chlamydial metabolic enzymes, and the biochemical analysis of infected cells (Belland et al. 2003; Gérard et al. 2002; Harper et al. 2000a,b; Iliffe-Lee and McClarty 1999; Shaw et al. 2000). Raman microspectroscopy demonstrated amino acid uptake and protein synthesis in *C. trachomatis* after extracellular incubation (Haider et al. 2010). Increased glucose consumption and ATP levels were observed in cells infected with chlamydiae; however, these studies could not distinguish between host cell and bacterial metabolism (Ojcius et al. 1998; Yaraei et al. 2005).

NAD(P)H FLIM provides an ideal approach to distinguish and examine host and pathogen metabolism separately in living cells by analyzing subcellular compartments.

Figure 16.2a shows that the majority of cellular autofluorescence at 730 nm excitation originates from the mitochondria in noninfected cells. The cytosol and nuclei show reduced fluorescence, representing the compartmentalized distribution of NAD(P)H in the cells. The fluorescence lifetimes of protein-bound NAD(P)H are different in the mitochondria, cytosol, and nucleus as also described by Li et al. in similar measurements (Li et al. 2008) (Figure 16.2b, d, and e). The different values represent different environmental conditions and/or different protein bindings, thus different functions of NAD(P)H.

In *C. trachomatis*-infected cells, the chlamydial inclusion shows a strong autofluorescence signal (Figure 16.2c). The spectra of this fluorescence do not show any shift compared to the autofluorescence of the other cellular compartments, indicating that the fluorescence inside the chlamydial inclusion also originates primarily from NAD(P)H (Szaszák et al. 2011). Interestingly, even though *C. trachomatis* has several metabolic enzymes that require NAD(P) as coenzyme, no pathways for the biosynthesis of NAD and no NAD kinase for the synthesis of phosphorylated NAD have been found in the chlamydial genome so far. Although it seems obvious that a mechanism must exist in order to import NAD(P) from the host cell to that of environmental chlamydiae, no NAD(P) transporter has yet been identified in *C. trachomatis* according to sequence homology searches (Haferkamp et al. 2004). Thus, the source of NAD(P)H in the chlamydial inclusion is still not known.

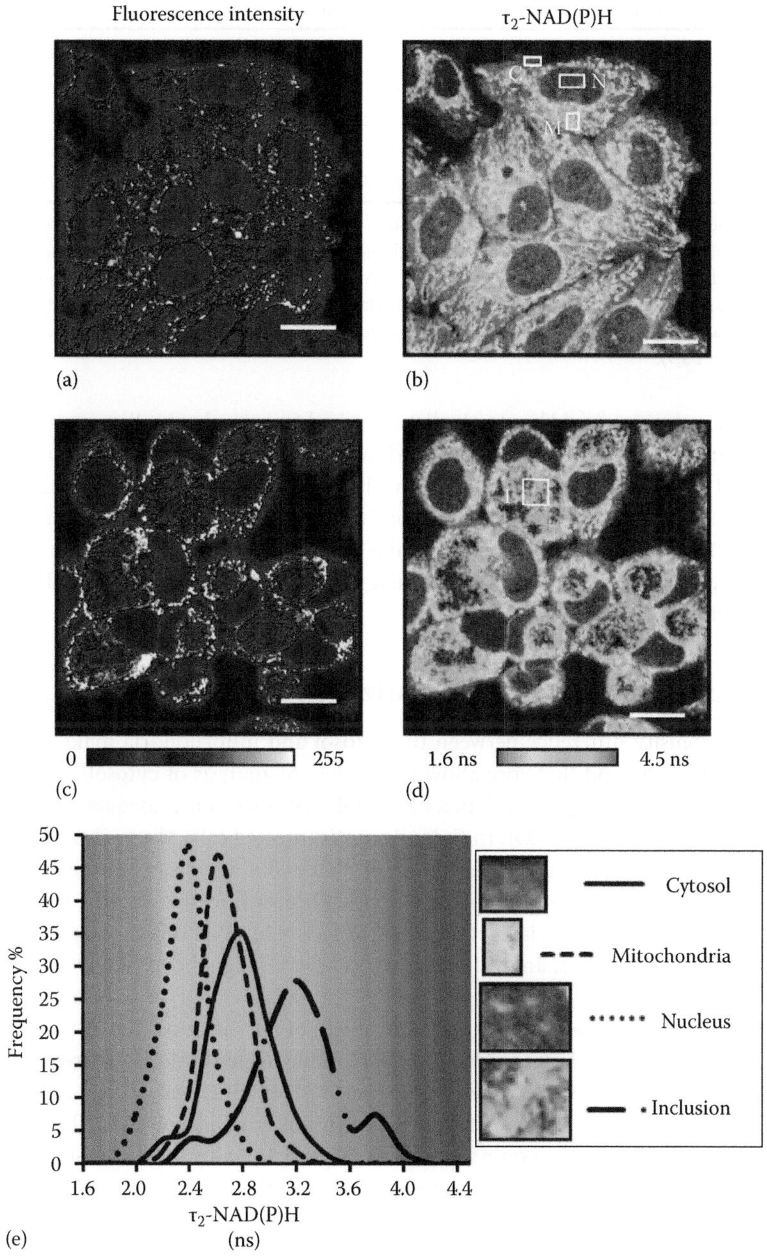

FIGURE 16.2 NAD(P)H fluorescence intensity signals and fluorescence lifetimes of protein-bound NAD(P)H in noninfected and *C. trachomatis*-infected HEp-2 cells. (a) NAD(P)H fluorescence signal intensities depicted in pseudo colors in noninfected HEp-2 cells. (b) Color-coded images of protein-bound NAD(P)H fluorescence lifetimes (τ_2) in noninfected HEp-2 cells. (c) NAD(P)H fluorescence signal intensities depicted in pseudo colors in *C. trachomatis*-infected HEp-2 cells. (d) Color-coded images of protein-bound NAD(P)H fluorescence lifetimes (τ_2) in *C. trachomatis*-infected HEp-2 cells. White squares show ROIs of cellular compartments used for quantitative analysis of τ_2-NAD(P)H (C, cytosol; M, mitochondria; N, nucleus; I, inclusion; scale bar = 20 μm). (e) Histogram of protein-bound NADP(H) fluorescence lifetime frequency distribution in the cytosolic, mitochondrial, and nuclear compartments of noninfected cells (n = 54) and in the chlamydial inclusion of infected cells (n = 54). Representative images show enlargement of ROIs used for analysis.

TABLE 16.1 Quantitative Analysis of NAD(P)H FLIM

	τ_1 (ns)	τ_2 (ns)	a_1 (%)	a_2 (%)	q_1 (%)	q_2 (%)
Cytosol	0.49 ± 0.11	2.69 ± 0.24	79.10 ± 1.62	20.90 ± 1.62	40.35 ± 3.53	59.65 ± 3.53
Mitochondria	0.38 ± 0.06	2.57 ± 0.17	81.86 ± 2.07	18.14 ± 2.07	40.09 ± 3.24	59.91 ± 3.24
Nucleus	0.43 ± 0.07	2.27 ± 0.17	83.63 ± 1.70	16.37 ± 1.70	49.65 ± 2.74	50.35 ± 2.74
Inclusion	0.46 ± 0.12	3.10 ± 0.34	80.46 ± 3.96	19.53 ± 3.96	38.47 ± 6.02	61.54 ± 6.02

Notes: Fluorescence lifetimes, relative amounts, and fluorescence quantum yields of free and protein-bound NAD(P)H in the cytosol, mitochondria, and nucleus of noninfected HEp-2 cells and in the *C. trachomatis* inclusion. τ_1, fluorescence lifetime of free NAD(P)H; τ_2, fluorescence lifetime of protein-bound NAD(P)H; a_1, relative amount of free NAD(P)H; a_2, relative amount of protein-bound NAD(P)H; q_1, fluorescence quantum yield of free NAD(P)H; q_2, fluorescence quantum yield of protein-bound NAD(P)H (n = 54; mean ± SD).

The chlamydial inclusions could be clearly distinguished from the host cell compartments due to the different fluorescence intensities of NAD(P)H inside and around the chlamydial inclusion. The selected region of interest (ROI) was then used to determine fluorescence lifetimes of NAD(P)H inside the chlamydial inclusion (Figure 16.2d). We observed a broad distribution of τ_2-NAD(P)H that indicates heterogeneity of NAD(P)H protein binding in each cellular compartment. τ_2-NAD(P)H was increased inside the chlamydial inclusion compared to other cellular compartments (Figure 16.2e). The mean fluorescence lifetime values are listed in Table 16.1.

16.6.1 Characterizing Compartmentalized Metabolic Changes in Host Cells

The fluorescence intensity difference between the cytosol and mitochondria makes it easy to distinguish these compartments and therefore allow for a separate analysis of cytosolic and mitochondrial NAD(P)H. As mitochondria size is small, pixel-by-pixel analysis is more adequate for analyzing mitochondrial signals than selecting an ROI. In *C. trachomatis*-infected cells, the analysis is challenging due to the altered morphology of the host cells. The large chlamydial inclusions fill most of the cytosol of the cells in the late stages of *C. trachomatis* infection (24 h post infection [hpi]), making it difficult to distinguish cytosol and mitochondria. Nevertheless, as free NAD(P)H is diffusible, cytosolic levels are reflected in the cell nucleus. Thus, analyzing nuclear NAD(P)H signal can provide information about metabolic changes in the cytosol.

To characterize host cell metabolism during later stages of the infection, the ratio of free to protein-bound NAD(P)H (a_1/a_2) and the values of τ_2-NAD(P)H in the nucleus were measured. The ratio of a_1/a_2 and the autofluorescence intensity were decreased, and τ_2-NAD(P)H was increased in the nucleus of *C. trachomatis*-infected cells at 24 hpi. The chemical inhibitor of glycolysis, 2-fluoro-deoxy-glucose (2FDG), causes similar changes in noninfected cells (Figure 16.3a through d). Glucose limitation by inhibition of glycolysis mimics cellular starvation. Consequently, decreased autofluorescence intensity and decreased relative amount of free NAD(P)H in the host cell nucleus indicates cellular starvation in *C. trachomatis*-infected cells.

16.6.2 Imaging the Developmental Cycle of *C. trachomatis*

C. trachomatis metabolism is strongly connected to its unique biphasic developmental cycle. *C. trachomatis* alternates between two metabolically different developmental forms that ensure its infectivity and replication. The infectious form, the elementary body, enters the host cell and differentiates into a metabolically active form, the reticulate body. *C. trachomatis* reticulate bodies grow and replicate within the host cell in an intracellular membrane-bound compartment called the chlamydial inclusion. Reticulate bodies redifferentiate to infectious elementary bodies in the late developmental cycle, and the elementary bodies induce cell lysis in order to be released.

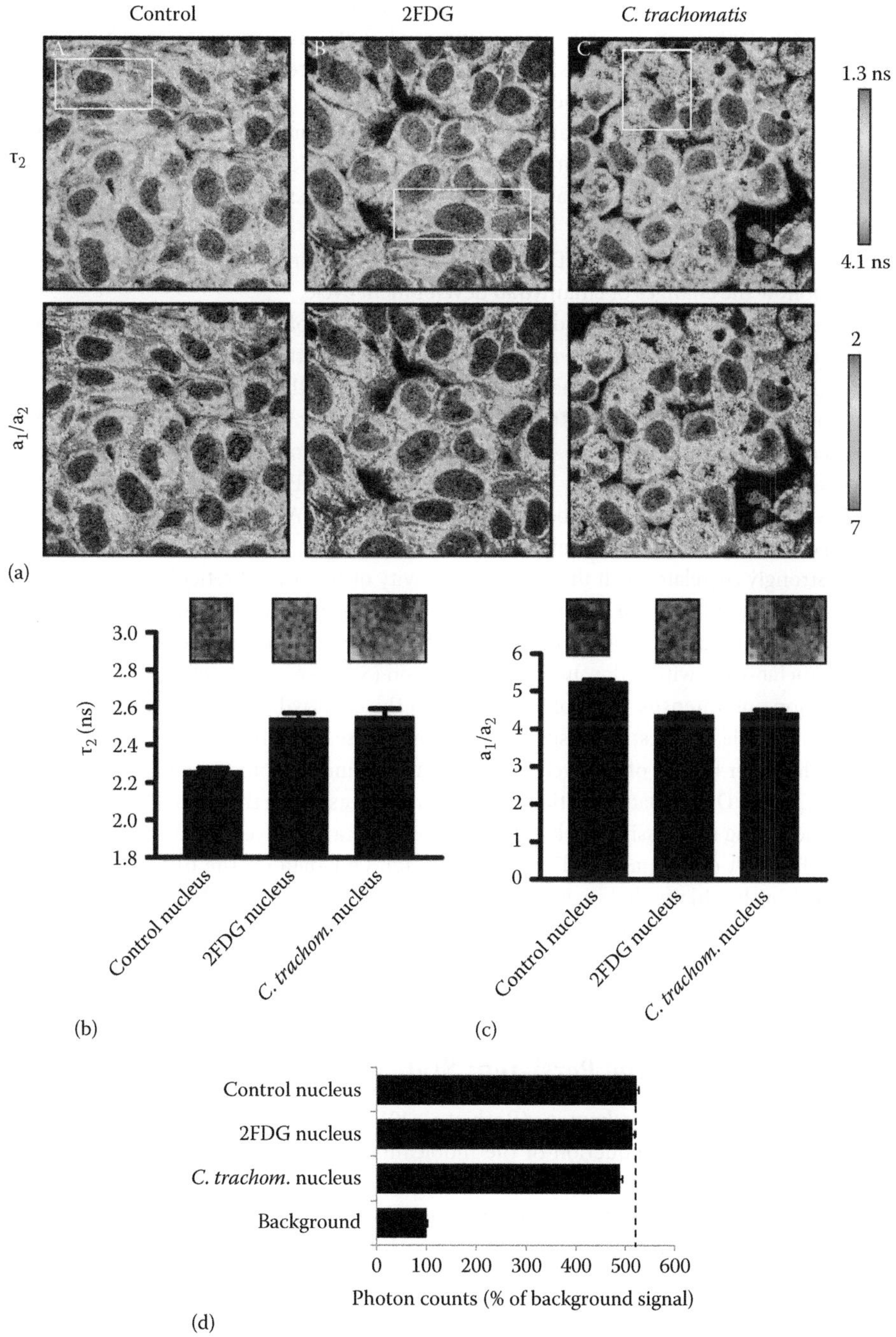

FIGURE 16.3 Cellular starvation in *C. trachomatis*-infected HEp-2 cells. (a) Color-coded images of protein-bound NAD(P)H fluorescence lifetime (τ_2) (upper panels) and relative amount of free to protein-bound NAD(P)H (a_1/a_2) (lower panels) in untreated HEp-2 cells (control) (A) or in HEp-2 cells treated with 5 mM 2FDG (B) or infected with *C. trachomatis* (C). (b) Quantitative analysis of protein-bound NAD(P)H fluorescence lifetime (τ_2) in the nucleus. (c) Quantitative analysis of the ratio of free to protein-bound NAD(P)H (a_1/a_2) in the nucleus. (d) Quantitative analysis of photon counts in the nucleus of the cells shown in ROIs of (A) through (C). Analysis was done using photon counts of 96–160 pixels per ROI.

C. trachomatis has several glucose-metabolizing enzymes and also an ADP/ATP transporter (Hatch et al. 1982; Stephens et al. 1998; Trentmann et al. 2007). However, the metabolic pathways of *C. trachomatis* are often truncated. Thus, *C. trachomatis* might directly import the substrates required to compensate for the incomplete metabolic pathways (Stephens et al. 1998). Using microarray technology, it was shown that the ADP/ATP translocase and the ATP requiring oligopeptide transporters are expressed as immediate-early genes. Furthermore, some metabolic enzymes such as the malate dehydrogenase (which requires NAD as a cofactor) are also expressed in the early phase of infection (Belland et al. 2003).

Fluorescence lifetime changes of NAD(P)H in the chlamydial inclusions during the developmental cycle strongly follow the changes of chlamydial development cycle.

Increased metabolism of reticulate bodies during the midphase of intracellular growth is directly correlated with an increase of τ_2-NAD(P)H in the early phase of infection, between 18 and 24 hpi. By contrast, τ_2-NAD(P)H does not change further during the late phase of the infection (48 hpi) when infectious elementary bodies are formed (Figure 16.4a through c). However, the distribution of τ_2-NAD(P)H is more heterogeneous at 48 hpi (Figure 16.4d). The ratio of free to protein-bound NAD(P)H (a_1/a_2) inside the chlamydial inclusion also decreases between 18 and 24 hpi (Figure 16.4e). The measurements of τ_2-NAD(P)H values during the developmental cycle indicate that NAD(P)H within *C. trachomatis* inclusions binds to different proteins as chlamydial metabolism changes. In addition, the results indicate that τ_2-NAD(P)H strongly correlates with the metabolic activity of chlamydial reticulate bodies and not with the inclusion size (Figure 16.4b and f). The increase in τ_2-NAD(P)H in the midphase of infection, accompanied by a decrease in the relative amount of free NAD(P)H, might indicate the presence of an oxidative type of energy metabolism within the chlamydial inclusion (Stephens et al. 1998).

The distribution and intensity of autofluorescence in the chlamydial inclusions also changes during the developmental cycle. Highest fluorescence in the midphase (18 hpi) of the chlamydial development is observed at the inner border of the inclusion. The fluorescence is homogeneously distributed within the inclusion at 24 hpi. During late infection (48 hpi), large areas inside the inclusion show no NAD(P)H fluorescence, indicating the possible replacement of metabolically active reticulate bodies by infectious but metabolically inert elementary bodies (Figure 16.4c). Our data on the distribution of NAD(P)H fluorescence within the chlamydial inclusion support the model that reticulate bodies first occupy the juxtamembrane space during the formation of the chlamydial inclusion and later move to the center for differentiation (Wilson et al. 2009). In the center of the inclusion, limited supply of essential metabolites may ultimately trigger their redifferentiation to infectious elementary bodies (Wilson et al. 2006).

16.6.3 Visualization of the Persistent State of *C. trachomatis*

IFNγ induces persistence of *C. trachomatis* (Roth et al. 2010). The molecular mechanism involves tryptophan depletion through the induction of the indoleamine 2,3-dioxygenase (IDO) by IFNγ (Taylor and Feng 1991). *C. trachomatis* persistence is defined as a viable but noncultivable developmental stage characterized by enlarged reticulate bodies with reduced metabolism in morphological aberrant inclusions (Beatty et al. 1993). The relative amount of protein-bound NAD(P)H (a_2-NAD(P)H) is largely reduced in the persistent state compared to inclusions of active infection 24 hpi (Figure 16.5). Therefore, a_2-NAD(P)H in the chlamydial inclusion is a sensitive marker of the persistence-induced chlamydial metabolic changes. In addition, the resolution of two-photon microscopy enables the visualization of the enlarged reticulate bodies inside the inclusions.

16.6.4 Measurement Setup and Methods

HEp-2 cells were grown on glass coverslips in 50 mm culture dishes and infected with 1 infectious unit (IFU) *C. trachomatis* per cell. Cover glasses were examined in a MiniCeM chamber for microscopy (JenLab, Jena, Germany) fitted to a heated stage, which enabled live cell imaging. The two-photon microscope

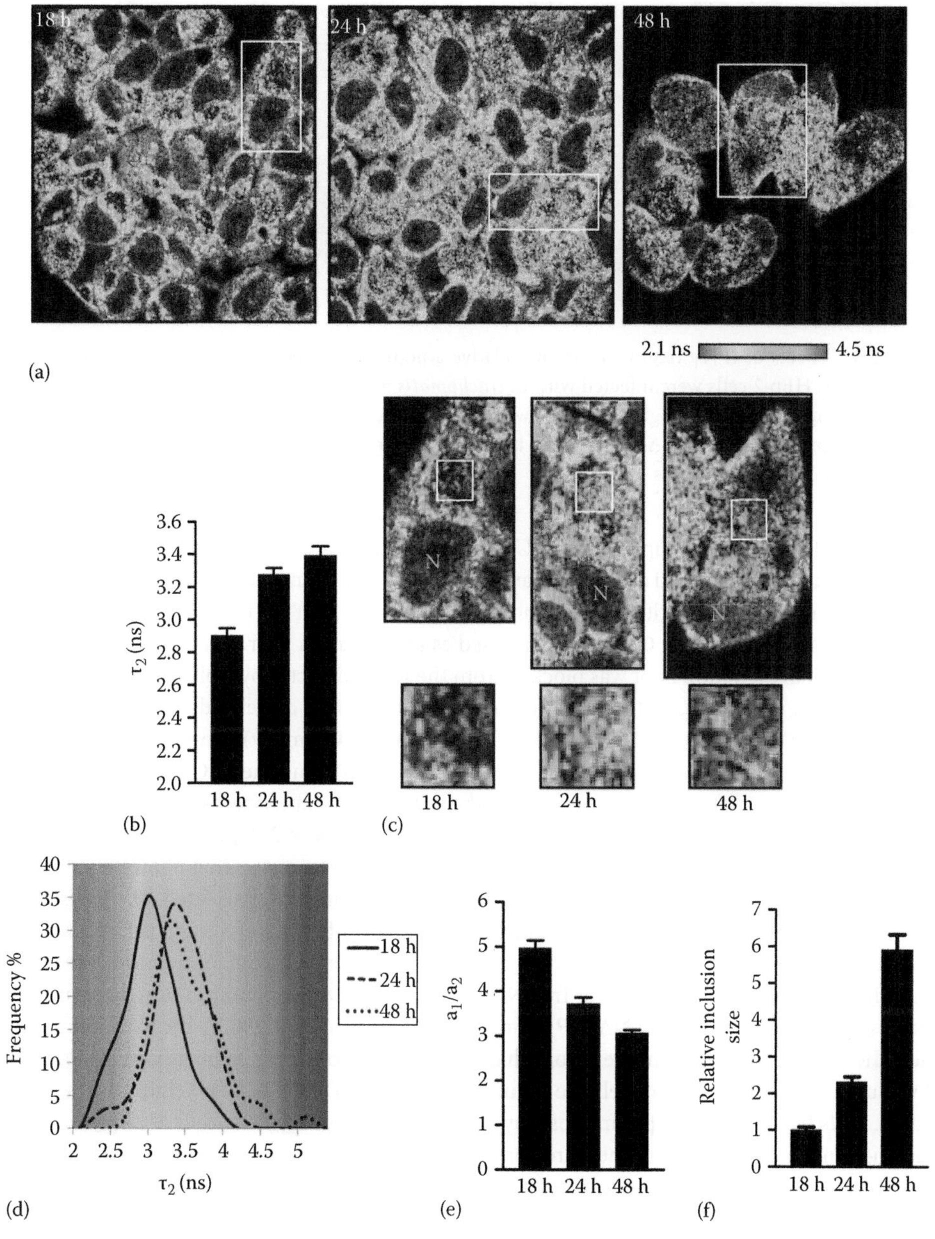

FIGURE 16.4 Changes of fluorescence lifetimes of protein-bound NAD(P)H in the inclusion during the development cycle of *C. trachomatis*. (a) Color-coded image of protein-bound NAD(P)H fluorescence lifetimes (τ_2) in *C. trachomatis*-infected HEp-2 cells after 18 h (left panel), 24 h (middle panel), and 48 h (right panel) infection. (b) Quantitative analysis of fluorescence lifetimes of protein-bound NAD(P)H (τ_2) inside the chlamydial inclusion after the indicated times of infection (n = 54; mean ± SEM). (c) Representative images of selected cells and ROIs inside the chlamydial inclusion that were used for analysis. Fluorescence lifetime of protein-bound NAD(P)H is color coded as in (a). (d) Histogram of protein-bound NADP(H) fluorescence lifetime frequency distribution inside the chlamydial inclusion of infected cells after the indicated time of infection (n = 54). (e) Quantitative analysis of the ratio of free to protein-bound NAD(P)H (a_1/a_2) inside the chlamydial inclusion after the indicated times of infection (n = 54; mean ± SEM). (f) Quantitative analysis of the sizes of chlamydial inclusions after the indicated times of infection.

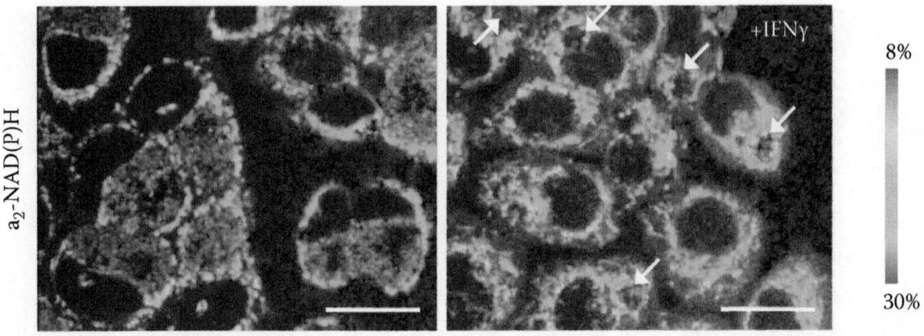

FIGURE 16.5 Effects of IFNγ treatment on the relative amount of protein-bound NAD(P)H inside the chlamydial inclusions. HEp-2 cells were infected with *C. trachomatis* for 24 h and treated with 10 units/mL IFNγ 24 h prior to and during the infection (right panels) or were left untreated (left panels). Color-coded images of the relative amount of protein-bound a_2-NAD(P)H (scale bar = 20 μm). Arrows point to persistent chlamydial inclusions harboring enlarged reticulate bodies.

(DermaInspect; JenLab) was equipped with a Chroma 640DCSPXR dichroic mirror (AHF analysentechnik AG, Tübingen, Germany) and a 40×/1.3 Plan-Apochromat oil-immersion objective (Zeiss, Göttingen, Germany). A tunable infrared titanium–sapphire femtosecond laser (710–920 nm tuning range; Mai Tai; Spectra-Physics, Darmstadt, Germany) was used as an excitation source at 730 nm for excitation of NAD(P)H. Residual excitation light was blocked from the FLIM detector by a blue emission filter (BG39, Schott AG, Mainz, Germany). FLIM data were collected by a time-correlated single-photon counting (TCSPC) system (PMH-100-0, SPC-830, Becker & Hickl, Berlin, Germany). Single-photon counting was done for 49.7 s per image. Fluorescence lifetimes were analyzed using the SPCImage software version: 2.9.5.2996 (Becker & Hickl). The average power for the cell imaging experiments was 12 mW, and the scan area for each image plane was 110 × 110 μm^2 corresponding to 256 × 256 pixels. Photon count rates at the beginning and at the end of image acquisition were monitored to ensure that photobleaching did not occur. For the image analysis, the ROI inside the chlamydial inclusion was selected. The ROIs contained 42–1147 pixels depending on the chlamydial inclusion size. The lifetimes of 5 × 5 pixels in the ROI were averaged earlier. The lifetime decay curves were fit to a double-exponential decay model, in which the fast-decaying component corresponds to free NAD(P)H [τ_1-NAD(P)H] and the slow-decaying component corresponds to protein-bound NAD(P)H [τ_2-NAD(P)H] (Bird et al. 2005). The instrument response function (IRF) was measured from the second-harmonic generation signal of beta-barium-borate crystal and it was used in the lifetime fit model. The mean values of τ_2-NAD(P)H of all pixels inside the ROIs were calculated. ROIs from three cells per microscopy field were analyzed from six different microscopy fields per chamber on three independent measurement days [n = 54 (3 × 6 × 3)]. To compare τ_2-NAD(P)H of different infection times and metabolic states, values were normalized to the average τ_2-NAD(P)H 24 hpi (100%) for each experimental day. For fluorescence intensity analysis of FLIM pictures, photon count values of selected ROIs inside the chlamydial inclusion were exported to Excel from SPCImage analysis software. Values of each pixel inside the ROIs from three different cells were used to create a histogram of fluorescence intensity distribution (Szaszák et al. 2011).

References

Abdul-Sater, A. A., E. Koo, G. Häcker, and D. M. Ojcius. 2009. Inflammasome-dependent caspase-1 activation in cervical epithelial cells stimulates growth of the intracellular pathogen *Chlamydia trachomatis*. *Journal of Biological Chemistry* 284(39): 26789–26796.

Afacan, N. J., C. D. Fjell, and R. E. W. Hancock. 2012. A systems biology approach to nutritional immunology—Focus on innate immunity. *Molecular Aspects of Medicine* 33(1): 14–25.

Alves, N. L., I. A. M. Derks, E. Berk, R. Spijker, R. A. W. van Lier, and E. Eldering. 2006. The Noxa/Mcl-1 axis regulates susceptibility to apoptosis under glucose limitation in dividing T cells. *Immunity* 24(6): 703–716.

Alvisi, G., V. Madan, and R. Bartenschlager. 2011. Hepatitis C virus and host cell lipids: An intimate connection. *RNA Biology* 8(2): 258–269.

Beatty, W. L., G. I. Byrne, and R. P. Morrison. 1993. Morphologic and antigenic characterization of interferon gamma-mediated persistent *Chlamydia trachomatis* infection in vitro. *Proceedings of the National Academy of Sciences of the United States of America* 90(9): 3998–4002.

Belland, R. J., G. Zhong, D. D. Crane et al. 2003. Genomic transcriptional profiling of the developmental cycle of *Chlamydia trachomatis*. *Proceedings of the National Academy of Sciences of the United States of America* 100(14): 8478–8483.

Berger, F., M. H. Ramırez-Hernandez, and M. Ziegler. 2004. The new life of a centenarian: Signalling functions of NAD(P). *Trends in Biochemical Sciences* 29(3): 111–118.

Bird, D. K., L. Yan, K. M. Vrotsos et al. 2005. Metabolic mapping of MCF10A human breast cells via multiphoton fluorescence lifetime imaging of the coenzyme NADH. *Cancer Research* 65(19): 8766–8773.

Blackham, S., A. Baillie, F. Al-Hababi et al. 2010. Gene expression profiling indicates the roles of host oxidative stress, apoptosis, lipid metabolism, and intracellular transport genes in the replication of hepatitis C virus. *Journal of Virology* 84(10): 5404–5414.

Calder, P. C. 1998. Dietary fatty acids and the immune system. *Nutrition Reviews* 56(1): S70–S83.

Caron, M., M. Auclair, C. Vigouroux, M. Glorian, C. Forest, and J. Capeau. 2001. The HIV protease inhibitor indinavir impairs sterol regulatory element-binding protein-1 intranuclear localization, inhibits preadipocyte differentiation, and induces insulin resistance. *Diabetes* 50(6): 1378–1388.

Cham, C. M., G. Driessens, J. P. O'Keefe, and T. F. Gajewski. 2008. Glucose deprivation inhibits multiple key gene expression events and effector functions in $CD8^+$ T cells. *European Journal of Immunology* 38(9): 2438–2450.

Cham, C. M. and T. F. Gajewski. 2005. Glucose availability regulates IFN-gamma production and p70S6 kinase activation in CD8+ effector T cells. *Journal of Immunology* 174(8): 4670–4677.

Chia, T. H., A. Williamson, D. D. Spencer, and M. J. Levene. 2008. Multiphoton fluorescence lifetime imaging of intrinsic fluorescence in human and rat brain tissue reveals spatially distinct NADH binding. *Optics Express* 16(6): 4237–4249.

Chin, C.-Y., D. Monack, and S. Nathan. 2010. Genome wide transcriptome profiling of a murine acute melioidosis model reveals new insights into how *Burkholderia pseudomallei* overcomes host innate immunity. *BMC Genomics* 11(1): 672.

Cocchiaro, J. L., Y. Kumar, E. R. Fischer, T. Hackstadt, and R. H. Valdivia. 2008. Cytoplasmic lipid droplets are translocated into the lumen of the *Chlamydia trachomatis* parasitophorous vacuole. *Proceedings of the National Academy of Sciences of the United States of America* 105(27): 9379–9384.

D'Cruz, L. M., M. P. Rubinstein, and A. W. Goldrath. 2009. Surviving the crash: Transitioning from effector to memory $CD8^+$ T cell. *Seminars in Immunology* 21(2): 92–98.

Deb, C., C.-M. Lee, V. S. Dubey et al. 2009. A novel in vitro multiple-stress dormancy model for *Mycobacterium tuberculosis* generates a lipid-loaded, drug-tolerant, dormant pathogen. *PLoS One* 4(6): e6077.

Delgado, T., P. A. Carroll, A. S. Punjabi, D. Margineantu, D. M. Hockenbery, and M. Lagunoff. 2010. Induction of the Warburg effect by Kaposi's sarcoma herpesvirus is required for the maintenance of latently infected endothelial cells. *Proceedings of the National Academy of Sciences of the United States of America* 107(23): 10696–10701.

Desvergne, B., L. Michalik, and W. Wahli. 2006. Transcriptional regulation of metabolism. *Physiological Reviews* 86(2): 465–514.

Dhurandhar, N. V., P. R. Kulkarni, S. M. Ajinkya, A. A. Sherikar, and R. L. Atkinson. 1997. Association of adenovirus infection with human obesity. *Obesity Research* 5(5): 464–469.

Diamond, D. L., A. J. Syder, J. M. Jacobs et al. 2010. Temporal proteome and lipidome profiles reveal hepatitis C virus-associated reprogramming of hepatocellular metabolism and bioenergetics. *PLoS Pathogens* 6(1): e1000719.

Dietz, I., S. Jerchel, M. Szaszak, K. Shima, and J. Rupp. 2012. When oxygen runs short: The microenvironment drives host–pathogen interactions. *Microbes and Infection* 14(4): 311–316. Available from: PM:22133978.

Dinkins, C., J. Arko-Mensah, and V. Deretic. 2010. Autophagy and HIV. *Seminars in Cell and Developmental Biology* 21(7): 712–718.

Dowson, J. H., C. Q. Mountjoy, M. R. Cairns, and H. Wilton-Cox. 1992. Changes in intraneuronal lipopigment in Alzheimer's disease. *Neurobiology of Aging* 13(4): 493–500.

Dumollard, R., P. Marangos, G. Fitzharris, K. Swann, M. Duchen, and J. Carroll. 2004. Sperm-triggered [Ca^{2+}] oscillations and Ca^{2+} homeostasis in the mouse egg have an absolute requirement for mitochondrial ATP production. *Development* 131(13): 3057–3067.

Dym, O. and D. Eisenberg. 2001. Sequence-structure analysis of FAD-containing proteins. *Protein Science* 10(9): 1712–1728.

Ehlers, S. 2009. Lazy, dynamic or minimally recrudescent? On the elusive nature and location of the mycobacterium responsible for latent tuberculosis. *Infection* 37(2): 87–95.

Elwell, C. A. and J. N. Engel. 2012. Lipid acquisition by intracellular Chlamydiae. *Cellular Microbiology* 14(7): 1010–1018.

Falagas, M. E. and M. Kompoti. 2006. Obesity and infection. *The Lancet Infectious Diseases* 6(7): 438–446.

Fjeld, C. C., W. T. Birdsong, and R. H. Goodman. 2003. Differential binding of NAD^+ and NADH allows the transcriptional corepressor carboxyl-terminal binding protein to serve as a metabolic sensor. *Proceedings of the National Academy of Sciences USA* 100(16): 9202–9207.

Fox, C. J., P. S. Hammerman, and C. B. Thompson. 2005. Fuel feeds function: Energy metabolism and the T-cell response. *Nature Reviews Immunology* 5(11): 844–852.

Gehlsen, U., G. Hüttmann, and P. Steven. 2010. Intravital multidimensional real-time imaging of the conjunctival immune system. In: *Research Projects in Dry Eye Syndrome*, H. Brewitt, ed., pp. 40–48. Hannover, Germany: Karger.

Gérard, H. C., J. Freise, Z. Wang et al. 2002. *Chlamydia trachomatis* genes whose products are related to energy metabolism are expressed differentially in active vs. persistent infection. *Microbes and Infection* 4(1): 13–22.

Gieffers, J., J. Rupp, A. Gebert, W. Solbach, and M. Klinger. 2004. First-choice antibiotics at subinhibitory concentrations induce persistence of *Chlamydia pneumoniae*. *Antimicrobial Agents and Chemotherapy* 48(4): 1402–1405.

Gossmann, T. I., M. Ziegler, P. Puntervoll, L. F. de Figueiredo, S. Schuster, and I. Heiland. 2012. NAD^+ biosynthesis and salvage—A phylogenetic perspective. *FEBS Journal* 279(18): 3355–3363.

Gurcel, L., L. Abrami, S. Girardin, J. Tschopp, and F. G. van der Goot. 2006. Caspase-1 activation of lipid metabolic pathways in response to bacterial pore-forming toxins promotes cell survival. *Cell* 126(6): 1135–1145.

Hackstadt, T., D. D. Rockey, R. A. Heinzen, and M. A. Scidmore. 1996. *Chlamydia trachomatis* interrupts an exocytic pathway to acquire endogenously synthesized sphingomyelin in transit from the Golgi apparatus to the plasma membrane. *The EMBO Journal* 15(5): 964.

Haferkamp, I., S. Schmitz-Esser, N. Linka et al. 2004. A candidate NAD^+ transporter in an intracellular bacterial symbiont related to Chlamydiae. *Nature* 432(7017): 622–625.

Haider, S., M. Wagner, M. C. Schmid et al. 2010. Raman microspectroscopy reveals long-term extracellular activity of chlamydiae. *Molecular Microbiology* 77(3): 687–700.

Harper, A., C. I. Pogson, M. L. Jones, and J. H. Pearce. 2000a. Chlamydial development is adversely affected by minor changes in amino acid supply, blood plasma amino acid levels, and glucose deprivation. *Infection and Immunity* 68(3): 1457–1464.

Harper, A., C. I. Pogson, and J. H. Pearce. 2000b. Amino acid transport into cultured McCoy cells infected with *Chlamydia trachomatis*. *Infection and Immunity* 68(9): 5439–5442.

Hatch, T. P., E. Al-Hossainy, and J. A. Silverman. 1982. Adenine nucleotide and lysine transport in *Chlamydia psittaci*. *Journal of Bacteriology* 150(2): 662–670.

Heaton, N. S. and G. Randall. 2010. Dengue virus-induced autophagy regulates lipid metabolism. *Cell Host and Microbe* 8(5): 422–432.

Heaton, N. S. and G. Randall. 2011. Multifaceted roles for lipids in viral infection. *Trends in Microbiology* 19(7): 368–375.

Heikal, A. A. 2010. Intracellular coenzymes as natural biomarkers for metabolic activities and mitochondrial anomalies. *Biomarkers in Medicine* 4(2): 241–263.

Heintzelman, D. L., R. Lotan, and R. R. Richards-Kortum. 2000. Characterization of the autofluorescence of polymorphonuclear leukocytes, mononuclear leukocytes and cervical epithelial cancer cells for improved spectroscopic discrimination of inflammation from dysplasia. *Photochemistry and Photobiology* 71(3): 327–332.

Herker, E. and M. Ott. 2012. Emerging role of lipid droplets in host/pathogen interactions. *Journal of Biological Chemistry* 287(4): 2280–2287.

Hollenbaugh, J. A., J. Munger, and B. Kim. 2011. Metabolite profiles of human immunodeficiency virus infected CD4+ T cells and macrophages using LC-MS/MS analysis. *Virology* 415(2): 153–159. Available from: PM:21565377.

Höner zu, B. K. and D. G. Russell. 2001. Mycobacterial persistence: Adaptation to a changing environment. *Trends in Microbiology* 9(12): 597–605.

Houtkooper, R. H., E. Pirinen, and J. Auwerx. 2012. Sirtuins as regulators of metabolism and healthspan. *Nature Reviews Molecular Cell Biology* 13(4): 225–238.

Huang, S., A. A. Heikal, and W. W. Webb. 2002. Two-photon fluorescence spectroscopy and microscopy of NAD(P)H and flavoprotein. *Biophysical Journal* 82(5): 2811–2825.

Iliffe-Lee, E. R. and G. McClarty. 1999. Glucose metabolism in *Chlamydia trachomatis*: The 'energy parasite' hypothesis revisited. *Molecular Microbiology* 33(1): 177–187.

Im, S.-S., L. Yousef, C. Blaschitz et al. 2011. Linking lipid metabolism to the innate immune response in macrophages through sterol regulatory element binding protein-1a. *Cell Metabolism* 13(5): 540–549.

Ivan, M., K. Kondo, H. Yang et al. 2001. HIFα targeted for VHL-mediated destruction by proline hydroxylation: Implications for O_2 sensing. *Science* 292(5516): 464–468.

Iwatsuki, K., T. Yamamoto, K. Tsuji et al. 2004. A spectrum of clinical manifestations caused by host immune responses against Epstein-Barr virus infections. *Acta Medica Okayama* 58: 169–180.

Jacobs, S. R., C. E. Herman, N. J. MacIver et al. 2008. Glucose uptake is limiting in T cell activation and requires CD28-mediated Akt-dependent and independent pathways. *The Journal of Immunology* 180(7): 4476–4486.

Jameson, S. C. 2002. Maintaining the norm: T-cell homeostasis. *Nature Reviews Immunology* 2(8): 547–556.

Juul, N., H. Jensen, M. Hvid, G. Christiansen, and S. Birkelund. 2007. Characterization of in vitro chlamydial cultures in low-oxygen atmospheres. *Journal of Bacteriology* 189(18): 6723–6726.

Kaech, S. M., S. Hemby, E. Kersh, and R. Ahmed. 2002. Molecular and functional profiling of memory CD8 T cell differentiation. *Cell* 111(6): 837–851.

Kapadia, S. B. and F. V. Chisari. 2005. Hepatitis C virus RNA replication is regulated by host geranylgeranylation and fatty acids. *Proceedings of the National Academy of Sciences of the United States of America* 102(7): 2561–2566.

Kari, L., M. M. Goheen, L. B. Randall et al. 2011. Generation of targeted *Chlamydia trachomatis* null mutants. *Proceedings of the National Academy of Sciences of the United States of America* 108(17): 7189–7193.

Kasai, D., T. Adachi, L. Deng et al. 2009. HCV replication suppresses cellular glucose uptake through down-regulation of cell surface expression of glucose transporters. *Journal of Hepatology* 50(5): 883–894.

Kim, Y. S., H. H. Park, H.-W. Rhee, J.-I. Hong, and K. Han. 2009. Neutrophils with toxic granulation show high fluorescence with bis (Zn^{2+}-dipicolylamine) complex. *Annals of Clinical and Laboratory Science* 39(2): 114–119.

Kirkegaard, K., M. P. Taylor, and W. T. Jackson. 2004. Cellular autophagy: Surrender, avoidance and subversion by microorganisms. *Nature Reviews Microbiology* 2(4): 301–314.

Klinger, A., R. Orzekowsky-Schroeder, D. von Smolinski et al. 2012. Complex morphology and functional dynamics of vital murine intestinal mucosa revealed by autofluorescence 2-photon microscopy. *Histochemistry and Cell Biology* 137(3): 269–278.

Klos, A., J. Thalmann, J. Peters, H. C. Gerard, and A. P. Hudson. 2009. The transcript profile of persistent Chlamydophila (*Chlamydia*) *pneumoniae* in vitro depends on the means by which persistence is induced. *FEMS Microbiology Letters* 291(1): 120–126.

Kwon, H.-S., M. M. Brent, R. Getachew et al. 2008. Human immunodeficiency virus type 1 Tat protein inhibits the SIRT1 deacetylase and induces T cell hyperactivation. *Cell Host and Microbe* 3(3): 158–167.

Lamkanfi, M. 2011. Emerging inflammasome effector mechanisms. *Nature Reviews Immunology* 11(3): 213–220.

Lenaerts, A. J., D. Hoff, S. Aly et al. 2007. Location of persisting mycobacteria in a Guinea pig model of tuberculosis revealed by r207910. *Antimicrobial Agents and Chemotherapy* 51(9): 3338–3345.

Levine, B. and V. Deretic. 2007. Unveiling the roles of autophagy in innate and adaptive immunity. *Nature Reviews Immunology* 7(10): 767–777.

Li, D., W. Zheng, and J. Y. Qu. 2008. Time-resolved spectroscopic imaging reveals the fundamentals of cellular NADH fluorescence. *Optics Letters* 33(20): 2365–2367.

Li, D., W. Zheng, and J. Y. Qu. 2009. Two-photon autofluorescence microscopy of multicolor excitation. *Optics Letters* 34(2): 202–204.

Lum, J. J., D. E. Bauer, M. Kong et al. 2005. Growth factor regulation of autophagy and cell survival in the absence of apoptosis. *Cell* 120(2): 237–248.

Lyn, R. K., D. C. Kennedy, S. M. Sagan et al. 2009. Direct imaging of the disruption of hepatitis C virus replication complexes by inhibitors of lipid metabolism. *Virology* 394(1): 130–142.

Macheroux, P., B. Kappes, and S. E. Ealick. 2011. Flavogenomics—A genomic and structural view of flavin-dependent proteins. *FEBS Journal* 278(15): 2625–2634.

Manel, N., F. J. Kim, S. Kinet, N. Taylor, M. Sitbon, and J.-L. Battini. 2003. The ubiquitous glucose transporter GLUT-1 is a receptor for HTLV. *Cell* 115(4): 449–459.

Maxwell, P. H., M. S. Wiesener, G.-W. Chang et al. 1999. The tumour suppressor protein VHL targets hypoxia-inducible factors for oxygen-dependent proteolysis. *Nature* 399(6733): 271–275.

McIntire, C. R., G. Yeretssian, and M. Saleh. 2009. Inflammasomes in infection and inflammation. *Apoptosis* 14(4): 522–535.

Michalek, R. D. and J. C. Rathmell. 2010. The metabolic life and times of a T-cell. *Immunological Reviews* 236(1): 190–202.

Michan, S. and D. Sinclair. 2007. Sirtuins in mammals: Insights into their biological function. *Biochemical Journal* 404: 1–13.

Nan, X., A. M. Tonary, A. Stolow, X. S. Xie, and J. P. Pezacki. 2006. Intracellular imaging of HCV RNA and cellular lipids by using simultaneous two-photon fluorescence and coherent anti-stokes Raman scattering microscopies. *ChemBioChem* 7(12): 1895–1897.

Niesner, R., P. Narang, H. Spiecker, V. Andresen, K-H. Gericke, and M. Gunzer. 2008. Selective detection of NADPH oxidase in polymorphonuclear cells by means of NAD(P)H-based fluorescence lifetime imaging. *Journal of Biophysics* 2008: 602639.

Noch, E. and K. Khalili. 2012. Oncogenic viruses and tumor glucose metabolism: Like kids in a candy store. *Molecular Cancer Therapeutics* 11(1): 14–23.

Ojcius, D. M., H. Degani, J. Mispelter, and A. Dautry-Varsat. 1998. Enhancement of ATP levels and glucose metabolism during an infection by *Chlamydia* NMR studies of living cells. *Journal of Biological Chemistry* 273(12): 7052–7058.

Pantanelli, S. M., Z. Li, R. Fariss, S. P. Mahesh, B. Liu, and R. B. Nussenblatt. 2009. Differentiation of malignant B-lymphoma cells from normal and activated T-cell populations by their intrinsic autofluorescence. *Cancer Research* 69(11): 4911–4917.

Pearce, E. L., M. C. Walsh, P. J. Cejas et al. 2009. Enhancing CD8 T-cell memory by modulating fatty acid metabolism. *Nature* 460(7251): 103–107.

Pierini, R., E. Cottam, R. Roberts, and T. Wileman.2009. Modulation of membrane traffic between endoplasmic reticulum, ERGIC and Golgi to generate compartments for the replication of bacteria and viruses. *Seminars in Cell and Developmental Biology* 20(7): 828–833.

Pithon-Curi, T. C., A. G. Trezena, W. Tavares-Lima, and R. Curi. 2002. Evidence that glutamine is involved in neutrophil function. *Cell Biochemistry and Function* 20(2): 81–86.

Pollak, N., C. Dolle, and M. Ziegler. 2007a. The power to reduce: Pyridine nucleotides-small molecules with a multitude of functions. *Biochemical Journal* 402: 205–218.

Pollak, N., M. Niere, and M. Ziegler. 2007b. NAD kinase levels control the NADPH concentration in human cells. *Journal of Biological Chemistry* 282(46): 33562–33571.

Robinson, I., M. A. Ochsenkühn, C. J. Campbell et al. 2010. Intracellular imaging of host-pathogen interactions using combined CARS and two-photon fluorescence microscopies. *Journal of Biophotonics* 3(3): 138–146.

Rödel, J., C. Große, H. Yu et al. 2012. Persistent *Chlamydia trachomatis* infection of HeLa cells mediates apoptosis resistance through a *Chlamydia* protease-like activity factor-independent mechanism and induces high mobility group box 1 release. *Infection and Immunity* 80(1): 195–205.

Rodriguez-Navarro, J. A. and A. M. Cuervo. 2010. Autophagy and lipids: Tightening the knot. *Seminars in Immunopathology* 32(4): 343–353.

Rossmann, M. G. and P. Argos. 1978. The taxonomy of binding sites in proteins. *Molecular and Cellular Biochemistry* 21(3): 161–182.

Roth, A., P. König, G. van Zandbergen et al. 2010. Hypoxia abrogates antichlamydial properties of IFN-γ in human fallopian tube cells in vitro and ex vivo. *Proceedings of the National Academy of Sciences of the United States of America* 107(45): 19502–19507.

Rupp, J., J. Gieffers, M. Klinger et al. 2007. *Chlamydia pneumoniae* directly interferes with HIF-1α stabilization in human host cells. *Cellular Microbiology* 9(9): 2181–2191.

Russell, D. G., B. C. VanderVen, W. Lee et al. 2010. *Mycobacterium tuberculosis* wears what it eats. *Cell Host and Microbe* 8(1): 68–76.

Saka, H. A. and R. H. Valdivia. 2010. Acquisition of nutrients by *Chlamydiae*: Unique challenges of living in an intracellular compartment. *Current Opinion in Microbiology* 13(1): 4–10.

Saxlin, T., P. Ylöstalo, L. Suominen-Taipale, S. Männistö, and M. Knuuttila. 2011. Association between periodontal infection and obesity: Results of the Health 2000 Survey. *Journal of Clinical Periodontology* 38(3): 236–242.

Schaible, U. E. and S. H. Kaufmann. 2007. Malnutrition and infection: Complex mechanisms and global impacts. *PLoS Medicine* May; 4(5): e115.

Schoborg, R. V. 2011. *Chlamydia* persistence—A tool to dissect *chlamydia*-host interactions. *Microbes and Infection* 13(7): 649–662.

Shao, W., G. Yeretssian, K. Doiron, S. N. Hussain, and M. Saleh. 2007. The caspase-1 digestome identifies the glycolysis pathway as a target during infection and septic shock. *Journal of Biological Chemistry* 282(50): 36321–36329.

Shaw, E. I., C. A. Dooley, E. R. Fischer, M. A. Scidmore, K. A. Fields, and T. Hackstadt. 2000. Three temporal classes of gene expression during the *Chlamydia trachomatis* developmental cycle. *Molecular Microbiology* 37(4): 913–925.

Shi, C.-S., K. Shenderov, N.-N. Huang et al. 2012. Activation of autophagy by inflammatory signals limits IL-1 [beta] production by targeting ubiquitinated inflammasomes for destruction. *Nature Immunology* 13(3): 255–263.

Shi, H., M. V. Kokoeva, K. Inouye, I. Tzameli, H. Yin, and J. S. Flier. 2006. TLR4 links innate immunity and fatty acid–induced insulin resistance. *Journal of Clinical Investigation* 116(11): 3015–3025.

Shi, L. Z., R. Wang, G. Huang et al. 2011. HIF1α–dependent glycolytic pathway orchestrates a metabolic checkpoint for the differentiation of TH17 and Treg cells. *The Journal of Experimental Medicine* 208(7): 1367–1376.

Shima, K., G. Kuhlenbäumer, and J. Rupp. 2010. *Chlamydia pneumoniae* infection and Alzheimer's disease: A connection to remember? *Medical Microbiology and Immunology* 199(4): 283–289.

Shima, K., M. Szaszák, W. Solbach, J. Gieffers, and J. Rupp. 2011. Impact of a low-oxygen environment on the efficacy of antimicrobials against intracellular *Chlamydia trachomatis*. *Antimicrobial Agents and Chemotherapy* 55(5): 2319–2324.

Shleeva, M. O., E. G. Salina, and A. S. Kaprelyants. 2010. Dormant forms of mycobacteria. *Microbiology* 79(1): 1–12.

Skala, M. C., K. M. Riching, D. K. Bird et al. 2007a. In vivo multiphoton fluorescence lifetime imaging of protein-bound and free nicotinamide adenine dinucleotide in normal and precancerous epithelia. *Journal of Biomedical Optics* 12(2): 024014.

Skala, M. C., K. M. Riching, A. Gendron-Fitzpatrick et al. 2007b. In vivo multiphoton microscopy of NADH and FAD redox states, fluorescence lifetimes, and cellular morphology in precancerous epithelia. *Proceedings of the National Academy of Sciences of the United States of America* 104(49): 19494–19499.

Stephens, R. S., S. Kalman, C. Lammel et al. 1998. Genome sequence of an obligate intracellular pathogen of humans: *Chlamydia trachomatis*. *Science* 282(5389): 754–759.

Steven, P., F. Bock, G. Hüttmann, and C. Cursiefen. 2011. Intravital two-photon microscopy of immune cell dynamics in corneal lymphatic vessels. *PLoS One* 6(10): e26253.

Steven, P., M. Müller, N. Koop, C. Rose, and G. Hüttmann. 2009. Comparison of Cornea Module and DermaInspect for noninvasive imaging of ocular surface pathologies. *Journal of Biomedical Optics* 14(6): 064040.

Syed, G. H., Y. Amako, and A. Siddiqui. 2010. Hepatitis C virus hijacks host lipid metabolism. *Trends in Endocrinology and Metabolism* 21(1): 33–40.

Szaszák, M., P. Steven, K. Shima et al. 2011. Fluorescence lifetime imaging unravels *C. trachomatis* metabolism and its crosstalk with the host cell. *PLoS Pathogens* 7(7): e1002108.

Takahashi, Y., C. L. Meyerkord, and H.-G. Wang. 2009. Bif-1/endophilin B1: A candidate for crescent driving force in autophagy. *Cell Death and Differentiation* 16(7): 947–955.

Taylor, M. W. and G. S. Feng. 1991. Relationship between interferon-gamma, indoleamine 2,3-dioxygenase, and tryptophan catabolism. *FASEB Journal* 5(11): 2516–2522.

Thurston, T. L. M., M. P. Wandel, N. von Muhlinen, Á. Foeglein, and F. Randow. 2012. Galectin 8 targets damaged vesicles for autophagy to defend cells against bacterial invasion. *Nature* 482(7385): 414–418.

Trentmann, O., M. Horn, A. C. T. van Scheltinga, H. E. Neuhaus, and I. Haferkamp. 2007. Enlightening energy parasitism by analysis of an ATP/ADP transporter from chlamydiae. *PLoS Biology* 5(9): e231.

van der Meer-Janssen, Y. P. M., J. van Galen, J. J. Batenburg, and J. B. Helms. 2010. Lipids in host–pathogen interactions: Pathogens exploit the complexity of the host cell lipidome. *Progress in Lipid Research* 49(1): 1–26.

Vastag, L., E. Koyuncu, S. L. Grady, T. E. Shenk, and J. D. Rabinowitz. 2011. Divergent effects of human cytomegalovirus and herpes simplex virus-1 on cellular metabolism. *PLoS Pathogens* 7(7): e1002124. Available from: PM:21779165.

Vescovo, T., A. Romagnoli, A. B. Perdomo et al. 2012. Autophagy protects cells from HCV-induced defects in lipid metabolism. *Gastroenterology* 142(3): 644–653.

Von Rückmann, A., K. G. Schmidt, F. W. Fitzke, A. C. Bird, and K. W. Jacobi. 1998. Dynamics of accumulation and degradation of lipofuscin in retinal pigment epithelium in senile macular degeneration. *Klinische Monatsblatter fur Augenheilkunde* 213(1): 32–37.

Wallace, C. and D. Keast. 1992. Glutamine and macrophage function. *Metabolism* 41(9): 1016–1020.

Wang, C.-M. and B. Kaltenboeck. 2010. Exacerbation of chronic inflammatory diseases by infectious agents: Fact or fiction? *World Journal of Diabetes* 1(2): 27.

Wang, Y., L. M. Weiss, and A. Orlofsky. 2009. Host cell autophagy is induced by *Toxoplasma gondii* and contributes to parasite growth. *Journal of Biological Chemistry* 284(3): 1694–1701.

Wang, Z. Q., W. T. Cefalu, X. H. Zhang et al. 2008. Human adenovirus type 36 enhances glucose uptake in diabetic and nondiabetic human skeletal muscle cells independent of insulin signaling. *Diabetes* 57(7): 1805–1813.

Waris, G., D. J. Felmlee, F. Negro, and A. Siddiqui. 2007. Hepatitis C virus induces proteolytic cleavage of sterol regulatory element binding proteins and stimulates their phosphorylation via oxidative stress. *Journal of Virology* 81(15): 8122–8130.

Wen, H., J. P. Y. Ting, and L. A. J. O'Neill. 2012. A role for the NLRP3 inflammasome in metabolic diseases—Did Warburg miss inflammation? *Nature Immunology* 13(4): 352–357.

Wieman, H. L., J. A. Wofford, and J. C. Rathmell. 2007. Cytokine stimulation promotes glucose uptake via phosphatidylinositol-3 kinase/Akt regulation of Glut1 activity and trafficking. *Molecular Biology of the Cell* 18(4): 1437–1446.

Wilson, D. P., P. Timms, D. L. S. McElwain, and P. M. Bavoil. 2006. Type III secretion, contact-dependent model for the intracellular development of *Chlamydia*. *Bulletin of Mathematical Biology* 68(1): 161–178.

Wilson, D. P., J. A. Whittum-Hudson, P. Timms, and P. M. Bavoil. 2009. Kinematics of intracellular Chlamydiae provide evidence for contact-dependent development. *Journal of Bacteriology* 191(18): 5734–5742.

Wofford, J. A., H. L. Wieman, S. R. Jacobs, Y. Zhao, and J. C. Rathmell. 2008. IL-7 promotes Glut1 trafficking and glucose uptake via STAT5-mediated activation of Akt to support T-cell survival. *Blood* 111(4): 2101–2111.

Wong, C. S. Y., I. Robinson, M. A. Ochsenkühn, J. Arlt, W. J. Hossack, and J. Crain. 2011. Changes to lipid droplet configuration in mCMV-infected fibroblasts: Live cell imaging with simultaneous CARS and two-photon fluorescence microscopy. *Biomedical Optics Express* 2(9): 2504–2516.

Xia, M., R. E. Bumgarner, M. F. Lampe, and W. E. Stamm. 2003. *Chlamydia trachomatis* infection alters host cell transcription in diverse cellular pathways. *Journal of Infectious Diseases* 187(3): 424–434.

Yaraei, K., L. A. Campbell, X. Zhu, W. Conrad Liles, C. Kuo, and M. E. Rosenfeld. 2005. Effect of *Chlamydia pneumoniae* on cellular ATP content in mouse macrophages: Role of Toll-like receptor 2. *Infection and Immunity* 73(7): 4323–4326.

Yeh, S., F. Forooghian, L. J. Faia et al. 2010. Fundus autofluorescence changes in cytomegalovirus retinitis. *Retina* 30(1): 42.

Yu, Y., A. J. Clippinger, and J. C. Alwine. 2011a. Viral effects on metabolism: Changes in glucose and glutamine utilization during human cytomegalovirus infection. *Trends in Microbiology* 19(7): 360–367.

Yu, Y., T. G. Maguire, and J. C. Alwine. 2011b. Human cytomegalovirus activates glucose transporter 4 expression to increase glucose uptake during infection. *Journal of Virology* 85(4): 1573–1580.

Zhang, Q., D. W. Piston, and R. H. Goodman. 2002. Regulation of corepressor function by nuclear NADH. *Science* 295(5561): 1895–1897.

Zhang, Q., S.-Y. Wang, C. Fleuriel et al. 2007. Metabolic regulation of SIRT1 transcription via a HIC1: CtBP corepressor complex. *Proceedings of the National Academy of Sciences of the United States of America* 104(3): 829–833.

Index

A

C

D

E

F

O

P

R

S

T

W

Printed and bound by CPI Group (UK) Ltd, Croydon, CR0 4YY
21/10/2024
01777046-0008